Edited by
Mario Leclerc and Jean-François Morin

Design and Synthesis of Conjugated Polymers

Related Titles

Martín, N./Giacalone, F. (eds.)

Fullerene Polymers

Synthesis, Properties and Applications

2009

ISBN: 978-3-527-32282-4

Elias, Hans-Georg

Macromolecules

Series: Macromolecules (Volume 1–4)

2009

ISBN: 978-3-527-31171-2

Xanthos, M. (ed.)

Functional Fillers for Plastics

second, updated and enlarged edition

2010

ISBN: 978-3-527-32361-6

Pascault, J-P./Williams, R. J. J. (eds.)

Epoxy Polymers

New Materials and Innovations

2010

ISBN: 978-3-527-32480-4

Mathers, R. T./Meier, M. A. R. (eds.)

Green Polymerization Methods

Renewable Starting Materials, Catalysis and Waste Reduction

2010

ISBN: 978-3-527-32625-9

Gujrati, P. D./Leonov, A. I. (eds.)

Modeling and Simulation in Polymers

2010

ISBN: 978-3-527-32415-6

Janssen, L./Moscicki, L. (eds.)

Thermoplastic Starch

A Green Material for Various Industries

2009

ISBN: 978-3-527-32528-3

Severn, J. R./Chadwick, J. C. (eds.)

Tailor-Made Polymers

Via Immobilization of Alpha-Olefin Polymerization Catalysts

2008

ISBN: 978-3-527-31782-0

Matyjaszewski, Krzysztof/Müller, Axel H. E. (eds.)

Controlled and Living Polymerizations

From Mechanisms to Applications

2009

ISBN: 978-3-527-32492-7

Edited by Mario Leclerc and Jean-François Morin

Design and Synthesis of Conjugated Polymers

WILEY-VCH Verlag GmbH & Co. KGaA

The Editors

Prof. Mario Leclerc
CERMA
Département de Chimie
Université Laval
Quebec City
Quebec, G1V 0A6
Canada

Prof. Jean-François Morin
CERMA
Département de Chimie
Université Laval
Quebec City
Quebec, G1V 0A6
Canada

Library of Congress Card No.: applied for

British Library Cataloguing-in-Publication Data
A catalogue record for this book is available from the British Library.

Bibliographic information published by the Deutsche Nationalbibliothek
The Deutsche Nationalbibliothek lists this publication in the Deutsche Nationalbibliografie; detailed bibliographic data are available on the Internet at http://dnb.d-nb.de.

Cover Design Grafik-Design Schulz, Fußgönnheim
Typesetting Laserwords Private Limited, Chennai
Printing and Binding betz-druck GmbH, Darmstadt

Printed in the Federal Republic of Germany
Printed on acid-free paper

ISBN: 978-3-527-32474-3

Contents

Design and Synthesis of Conjugated Polymers. Edited by Mario Leclerc and Jean-François Morin

ISBN: 978-3-527-32474-3

Preface

As often happens in science, the discovery of conducting polymers was somehow accidental. A student of the Shirakawa's laboratory was working on Ziegler-Natta polymerization of acetylene. By accident, this student prepared a many molar concentration of the catalyst instead of the usual millimolar concentration and obtained a polyacetylene thin film which looked like a metallic foil instead of the usual dark, powdery material. During a visit to the University of Tokyo, Prof. MacDiarmid (University of Pennsylvania) met Prof. Shirakawa and invited him to come to Philadelphia to investigate in more detail this new form of polyacetylene. In collaboration with Prof. A.J. Heeger, this team reported in 1977 that upon partial oxidation or reduction (so-called doping reaction), the conductivity of polyacetylene increases more than a billionfold (up to $100\,S\ cm^{-1}$). They eventually received the 2000 Nobel Prize in chemistry for this discovery. Unfortunately, this first conducting polymer is difficult to process and is unstable in the presence of oxygen or water.

Rapidly, many chemists investigated more stable conjugated polymers. For instance, in the beginning of the 1980s, a lot of studies were devoted to electropolymerized polypyrroles, polythiophenes, and polyanilines. This method has the advantage of being able to prepare thin films of these infusible and insoluble conjugated polymers in one step. However, the ultimate goal was and still remains the development of polymeric materials that combine the electrical and optical properties of metals or semiconductors with the processing advantages and mechanical properties of traditional polymers. For this purpose, substituted conjugated polymers were prepared since the unsubstituted parent polymers are not melt or solution-processable owing to strong interchain interactions and chain stiffness. First experiments indicated that the presence of bulky substituents may induce a twisting of the backbone which gave processable but poorly conjugated materials with reduced electrical properties. A breakthrough occurred in 1985–1986 with the syntheses of highly conjugated, conducting, and processable poly(3-alkylthiophene)s. Following these first studies on poly(3-alkylthiophene)s, it became quite clear that the synthesis of well-defined head-to-tail coupled (which should give the lowest steric hindrance from the side chains and, possibly, a more efficient three-dimensional packing) poly(3-alkylthiophene)s would lead to a significant improvement in the performance of these polymeric materials. Therefore,

Design and Synthesis of Conjugated Polymers. Edited by Mario Leclerc and Jean-François Morin

ISBN: 978-3-527-32474-3

in attempts to bring more reliable synthetic procedures to the field of electronic materials, a variety of synthetic tools were utilized and allowed significant advances in this research field. For instance, these investigations have led to the first preparations of well-defined regioregular poly(3-alkylthiophene)s in 1992. The synthesis of air-stable highly conducting poly(3,4-ethylenedioxythiophene) is another example of a rational design of a conjugated polymer with optimized structural and physical properties.

In the meantime, through collaborations with physicists and engineers, the focus shifted from the synthesis of metallic-like conductors to the design of stable semiconducting polymers. This new driving force is based on the aim to set up the so-called plastic electronics where microelectronic devices can be printed on different substrates. Such applications include the fabrication of light-emitting diodes, field-effect transistors, photovoltaic cells, and the list continues to grow. For instance, the relatively young field of chemical and biological sensors based on semiconducting conjugated polymers is expanding rapidly.

Along these lines, it is the aim of this book to review advances in the controlled and well-designed synthesis of important classes of processable conjugated polymers. This book includes chapters on the synthesis of well-defined and versatile polyacetylenes (by Liu, Lam, and Tang), polyarylenes (Sakamoto, Rehahn, and Schlüter), polythiophenes (Osaka and McCullough), poly(*p*-phenylenevinylene)s (Pang), and ladder conjugated polymers (Grimsdale and Müllen). It also covers recent progress made in the synthesis of the bright poly(aryleneethynylene)s (VanVeller and Swager) and polyfluorenes (Zhao, Liu, and Huang) together with chapters on newcomers such as silole-containing conjugated polymers (Chen and Cao) and poly(2,7-carbazole) derivatives (Boudreault, Morin, and Leclerc). This book highlights recent state-of-the-art synthetic contributions but since this research field is based on close interactions between chemists, physicists, and engineers, we do hope that scientists with different backgrounds will enjoy reading these new advances in the synthesis of conjugated polymers. They will certainly find in the numerous reported chemical structures some inspirations for their own research activities.

Finally, we cannot end this preface without expressing our gratitude to all those who have contributed to this book. This obviously includes all authors for their hard, timely, and excellent work but also people at Wiley-VCH who gave us a well appreciated support.

Quebec City, August 25, 2009

Mario Leclerc
Jean-François Morin

List of Contributors

Pierre-Luc T. Boudreault
CERMA
Département de Chimie
Université Laval
Quebec City
Quebec, G1V 0A6
Canada

Yong Cao
South China University of
Technology Institute of Polymer
Optoelectronic Materials &
Devices
Key Lab of Specially Functional
Materials of Ministry of Education
381 Wushan Rd.
Guangzhou 510640
China

Junwu Chen
South China University of
Technology Institute of Polymer
Optoelectronic Materials &
Devices
Key Lab of Specially Functional
Materials of Ministry of Education
381 Wushan Rd.
Guangzhou 510640
China

Andrew C. Grimsdale
Nanyang Technological
University
School of Materials
Science and Engineering
50 Nanyang Avenue
Singapore 639798

Wei Huang
Nanjing University of Posts &
Telecommunications (NUPT)
Jiangsu Key Laboratory for
Organic Electronics &
Information Displays and
Institute of Advanced
Materials (IAM)
9 Wenyuan Road
Nanjing 210046
China

Jacky W. Y. Lam
The Hong Kong University of
Science & Technology
Department of Chemistry
Clear Water Bay
Kowloon
Hong Kong
China

Design and Synthesis of Conjugated Polymers. Edited by Mario Leclerc and Jean-François Morin

ISBN: 978-3-527-32474-3

Mario Leclerc
CERMA
Département de Chimie
Université Laval
Quebec City
Quebec, G1V 0A6
Canada

Jianzhao Liu
The Hong Kong University of Science & Technology
Department of Chemistry
Clear Water Bay
Kowloon
Hong Kong
China

Shujuan Liu
Nanjing University of Posts & Telecommunications (NUPT)
Jiangsu Key Laboratory for Organic Electronics & Information Displays and Institute of Advanced Materials (IAM)
9 Wenyuan Road
Nanjing 210046
China

Richard D. McCullough
Carnegie Mellon University
Department of Chemistry
4400 Fifth Ave.
Pittsburgh
PA 15213
USA

Jean-François Morin
CERMA
Département de Chimie
Université Laval
Quebec City
Quebec, G1V 0A6
Canada

Klaus Müllen
Max Planck Institute for Polymer Research
Ackermannweg 10
Mainz 55128
Germany

Itaru Osaka
Carnegie Mellon University
Department of Chemistry
4400 Fifth Ave.
Pittsburgh
PA 15213
USA

and

Hiroshima University
Department of Applied Chemistry
1-4-1 Kagamiyama
Higashi-hiroshima
Hiroshima 739-8527
Japan

Yi Pang
The University of Akron
Department of Chemistry
Maurice Morton Institute of Polymer Science
Akron
190 E. Buchtel Commous
OH 44325
USA

Matthias Rehahn
Darmstadt University of Technology
Ernst-Berl-Institute for Chemical Engineering and Macromolecular Science
Petersenstraße 22
64287 Darmstadt
Germany

Junji Sakamoto
ETH Zürich
Department of Materials
Laboratory of Polymer Chemistry
HCi G 523
Wolfgang Pauli Strasse 10
8093 Zürich
Switzerland

A. Dieter Schlüter
ETH Zürich
Department of Materials
Laboratory of Polymer Chemistry
HCI J541
Wolfgang Pauli Strasse 10
8093 Zürich
Switzerland

Timothy M. Swager
Massachusetts Institute of Technology
77 Massachusetts Avenue
Cambridge
MA 02139
USA

Ben Zhong Tang
The Hong Kong University of Science & Technology
Department of Chemistry
Clear Water Bay
Kowloon
Hong Kong
China

and

Zhejiang University
Department of Polymer Science & Engineering
Zhe Da Road
Hangzhou 310027
China

Brett VanVeller
Massachusetts Institute of Technology
77 Massachusetts Avenue
Cambridge
MA 02139
USA

Qiang Zhao
Nanjing University of Posts & Telecommunications (NUPT)
Jiangsu Key Laboratory for Organic Electronics & Information Displays and Institute of Advanced Materials (IAM)
9 Wenyuan Road
Nanjing 210046
China

1
Synthesis and Functionality of Substituted Polyacetylenes

Jianzhao Liu, Jacky W. Y. Lam, and Ben Zhong Tang

1.1
Introduction

Polyacetylene (PA) is the archetypal conjugated polymer. The seminal discovery of the metallic conductivity of its doped form has triggered a huge surge of interest in conductive polymers and has spawned an exciting area of research on synthetic metals.[1)] As a result of rapid advances in the area, we are now on the threshold of a "plastic-electronics era" that previously could only be imagined in science fiction.

Structurally, PA is a linear polyene chain $[-(HC{=}CH)_n-]$. The existence of two hydrogen atoms in its repeat unit offers ample opportunity to decorate the backbone with pendants: replacement of hydrogen in each repeat unit by one or two substituents yields mono- (**1**) and disubstituted PAs (**2**), respectively (Scheme 1.1). The pendant and backbone can interact with each other: for example, the former perturbs the electronic conjugation of the latter, while the latter influences the molecular alignment of the former. Proper structural design may tune the backbone–pendant interplay into harmony and synergy, generating new substituted PAs with novel functionalities. While PA is electrically conductive, introduction of such pendant as mesogen, chromophore, photosensitive double bond, or naturally occurring building block may endow it with such new functional properties as electro-optic activity, photonic responsiveness, and biologic compatibility.

Considerable efforts have been devoted to the synthesis of the substituted PAs and study of their properties [1–9]. Attachment of polar groups to the polyene backbone enables the integration of functional pendants into the PA chain through various functional-group transformations, which was, however, a difficult task at the early stage of PA chemistry. Difficulties were ascribed to the poisoning of the metathesis catalysts by the functional groups [10, 11]. The later discovery of the functionality-tolerant organorhodium catalysts was a thrilling advance in the

1) Heeger, MacDiarmid, and Shirakawa won the Nobel Prize in Chemistry in 2000 for their pioneer work in the area.

Design and Synthesis of Conjugated Polymers. Edited by Mario Leclerc and Jean-François Morin

ISBN: 978-3-527-32474-3

$$\text{H–C≡C–R} \xrightarrow{\text{Transition-metal catalyst}} \text{–(C(H)=C(R))}_n\text{–} \quad \textbf{(1)} \quad \text{Monosubstituted polyacetylene}$$

$$\text{R–C≡C–R}' \xrightarrow{\text{Transition-metal catalyst}} \text{–(C(R)=C(R'))}_n\text{–} \quad \textbf{(2)} \quad \text{Disubstituted polyacetylene}$$

Scheme 1.1 Polymerizations of mono- and disubstituted acetylenes.

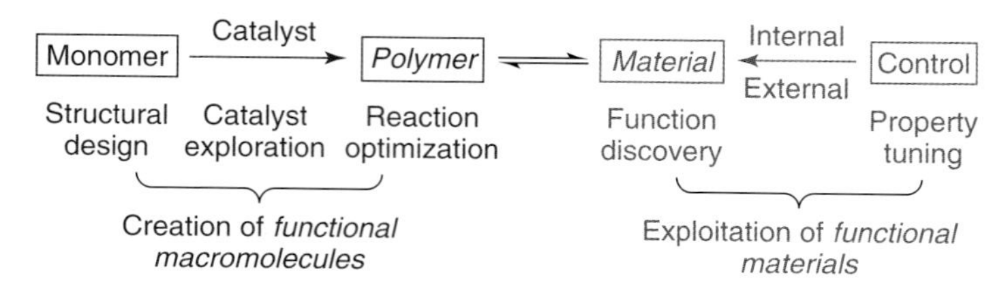

Chart 1.1 Development of polyacetylene-based functional macromolecules and materials.

area [12, 13], but the catalysts work for only limited types of monomers (e.g., phenylacetylene derivatives and monosubstituted 1-alkynes) through an insertion mechanism [14]. The syntheses of functional PAs via the metathesis route remained difficult, with the polymerizations of disubstituted functional acetylenes being particularly intractable.

Our research group has worked on PA functionalization through a systematic approach. Our efforts have covered design of monomer structures, exploration of new catalysts, optimization of reaction conditions, discovery of functions, and manipulation of properties by internal and external means (Chart 1.1). Our strategy has worked well and we have generated a wide variety of new PA-based functional macromolecules and materials. In this chapter, we briefly summarize our efforts and results on the development of the new functional PAs.

1.2 Polymer Syntheses

1.2.1 Catalysts

In general, in the presence of Ziegler–Natta catalysts, the pure PA was obtained as an intractable black powder that is very difficult to make into samples of a shape suitable for measurement of spectra and physical properties because of its insolubility, infusibility, and instability. To address the processability and stability problem encountered by unsubstituted PA and explore new functionalities, polymer scientists tried to synthesize functionalized substituted PAs. This task is critically dependent

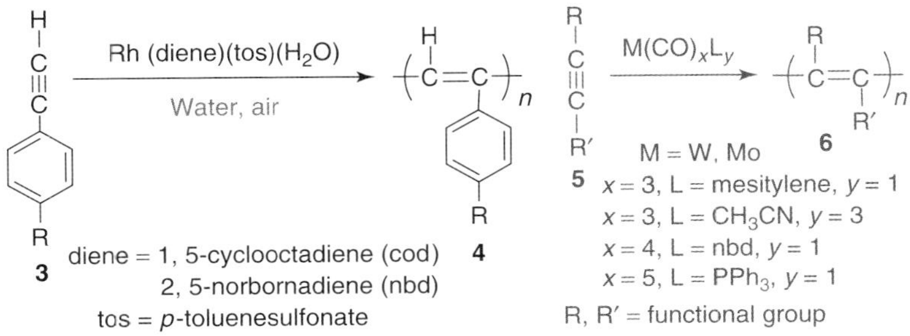

Rh (diene)(tos)(H_2O)
Water, air
3
4
diene = 1, 5-cyclooctadiene (cod)
2, 5-norbornadiene (nbd)
tos = *p*-toluenesulfonate

$M(CO)_xL_y$
5
6
M = W, Mo
$x = 3$, L = mesitylene, $y = 1$
$x = 3$, L = CH_3CN, $y = 3$
$x = 4$, L = nbd, $y = 1$
$x = 5$, L = PPh_3, $y = 1$
R, R′ = functional group

Scheme 1.2 Polymerizations of acetylenes by water-soluble Rh and air-stable Mo, and W-based carbonyl complexes.

on the exploration of effective catalysts. At an early stage, polymerization of substituted acetylenes was studied by using conventional Ziegler–Natta catalysts aiming at synthesizing processable and functional PAs [4]. However, many attempts led to the conclusion that only sterically unhindered monosubstituted acetylenes can be polymerized to give insoluble polymers and/or soluble oligomers. Traditional ionic and radical initiators were known to be inefficient to polymerize substituted acetylenes. Later, Mo, W, Ta, and Nb-based metathesis catalysts were successfully developed [1, 3, 7, 11]. The main difficulty with these systems was the "incompatibility" of the polar groups with these metathesis catalysts. The functionality tolerance of the Rh-based insertion catalysts thus attracted our attention. To expand the variety of the Rh catalysts, we prepared a series of Rh complexes Rh(diene)L (L = ligand). Importantly, we have developed a catalyst system that can initiate acetylene polymerizations in water and air, giving stereoregular polymers (**4**) in high yields (Scheme 1.2) [15].

Like their older versions, the new Rh catalysts are also only effective for the polymerizations of specific types of monomers. To develop more general systems, we turned our attention back to metathesis catalysts. $MoCl_5$ and WCl_6 are classical metathesis catalysts, but the early attempts to use them to polymerize functional acetylenes often resulted in the formation of insoluble or oligomeric products in low yields. We envisioned that the metal halides would be functionality tolerant, for they could cyclopolymerize functional diynes into cyclic polyenes [5]. Our prediction proved correct: through catalyst–substrate matching and reaction-condition optimization, we accomplished our goal of using the metal halides to catalyze the linear polymerizations of various functional acetylenes.

The metal halides are, however, air- and moisture-sensitive. In 1989, Tang found that several stable metal carbonyl complexes catalyzed the polymerizations of nonfunctional 1-alkynes. We extended his effort in the area and prepared a series of $M(CO)_xL_y$ complexes. Delightfully, many functional acetylenes (**5**) could be polymerized by these metal carbonyl complexes (Scheme 1.2) [16].

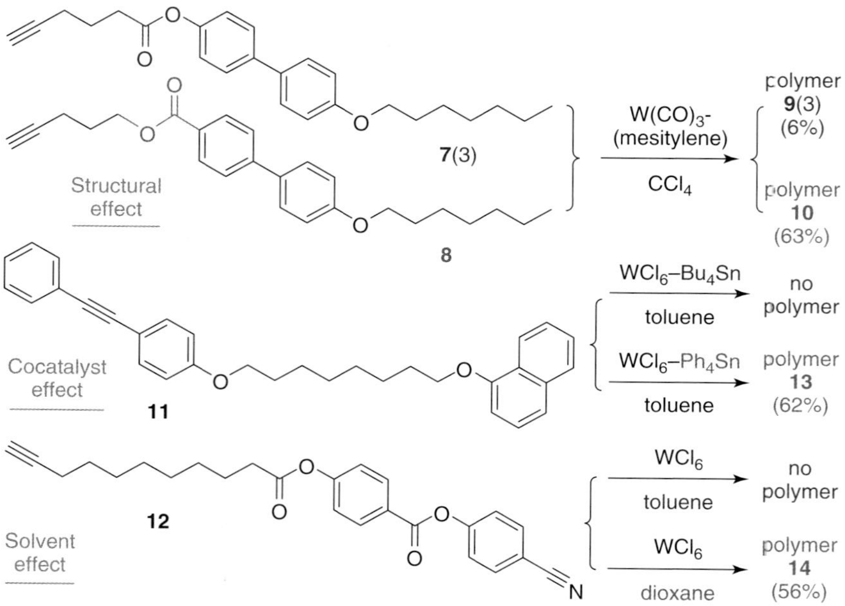

Scheme 1.3 Effects of substrate, cocatalyst, and solvent on acetylene polymerizations.

1.2.2
Polymerization Behaviors

In order to match the catalysts with particular substrates, we prepared hundreds of functional acetylenes. The substrate–catalyst matching is structurally sensitive: a seemingly subtle variation in the functional group can greatly influence the behavior of polymerization of a monomer by a given catalyst. For example, **7**(3) and **8** differ only in the orientation of the ester unit but show distinct polymerizability by $W(CO)_3$(mesitylene): **8** can be effectively polymerized but **7**(3) cannot (Scheme 1.3) [16]. Polymerization conditions (cocatalyst, solvent, etc.) can affect the fate of a specific substrate–catalyst pair. For instance, **11**–WCl_6 may be a bad or good combination, depending on whether tetra(*n*-butyl)tin or tetraphenyltin is used as a cocatalyst [17]. Polymerization of **12** does not occur at all in toluene but proceeds well in dioxane.

In general, rhodium catalysts can efficiently polymerize various monosubstituted acetylenes (e.g., phenylacetylene and 1-alkyne derivatives with polar groups) to give polymers with high molecular weights (MWs). However, the polymerization of highly polar monomers, particularly those with acidic protons, is still a difficult proposition. For example, polymerization of acetylenic monomers containing carboxy groups has been a daunting job. The direct polymerization of 4-ethynylbenzoic acid was deemed very difficult, if not impossible. Our group has succeeded in direct polymerizations of a series of highly polar phenylacetylene derivatives. The

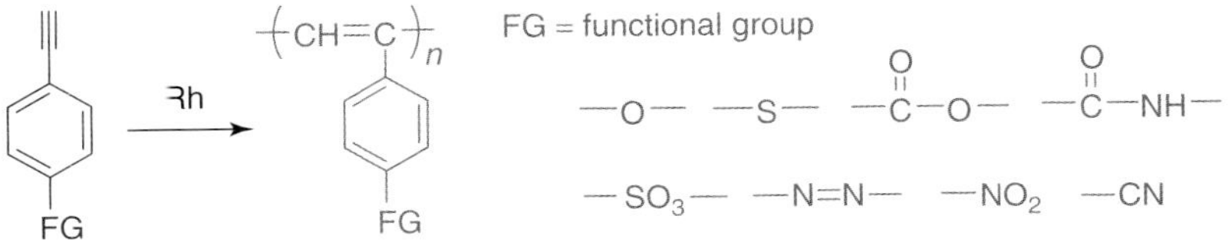

$$\text{15 (FG-substituted phenylacetylene)} \xrightarrow{\text{Rh}} \text{16: } \mathrm{+CH{=}C+_n}\text{ (with FG-substituted phenyl pendant)}$$

FG = functional group

—O— —S— —C(=O)—O— —C(=O)—NH—

$—SO_3—$ —N=N— $—NO_2$ —CN

$—NH_2$ —OH —SH —COOH

Scheme 1.4 Polymerization of functionalized acetylenes.

polymerizations of phenylacetylene derivatives **15** (Scheme 1.4) with various polar groups (e.g., oxy, carboxyl, hydroxyl, azo, cyano, thiol, amino, and nitro) catalyzed by organorhodium complexes, $[Rh(cod)Cl]_2$ (cod = 1,5-cyclooctadiene), $[Rh(nbd)Cl]_2$, and $Rh^+(nbd)[C_6H_5B^-(C_6H_5)_3]$ (nbd = norbornadiene), afforded corresponding polymers **16** with high MWs (M_w up to 488 500) and low polydispersity indexes (PDIs = M_w/M_n, down to 1.03, where M_n and M_w stand for number and weight-average MWs) [1, 2, 18].

As a result of optimization of the polymerization reactions, many functional monomers are transformed into polymers of high MWs (up to several million) in high yields (up to 100%). Substituted PAs containing various functional groups (e.g., crown ether, sulfonate, amide, dipeptide, saccharide, nucleoside, sterol, cyano, thiophene, vinyl, phthalimide, indole, siloxane, and silole) have been obtained [1, 2]. Examples are given in Charts 1.2 and 1.3.

On the basis of our results, a substrate–catalyst matching map is drawn as shown in Chart 1.4. Functionalized 1-arylacetylenes are polymerizable only by the Rh-based catalysts, although their nonfunctional cousins can be polymerized by other catalysts. Polymerizations of functionalized 1-alkynes are effected by the W-based catalysts, with the Rh ones giving moderate results. Propiolates can be effectively polymerized by the Mo and Rh catalysts. The Rh complexes cannot polymerize disubstituted acetylenes. Polymerizations of alkyl 2-alkynoates and alkyl 3-arylpropiolates are effected by the Mo catalysts, while those of aryl 3-arylpropiolates, 1-aryl-1-alkynes, and diarylacetylenes are initiated by the W catalysts. While polymerization of monosubstituted acetylenes (Chart 1.4, nos. 1–3) normally proceeds at room temperature in polar solvents (e.g., dioxane), disubstituted acetylenes (nos. 4–8) generally polymerize at higher temperatures (e.g., ~80 °C) in nonpolar solvents (e.g., toluene).

1.2.3
Polymer Reactions

Polymer reactions can be used as a tool to further functionalize the preformed acetylenic polymers or create functional polymers inaccessible by direct polymerizations. For example, the polymer **45** comprises hydrophobic poly(phenylacetylene) (PPA) skeleton and hydrophilic amine pendant. Because of this structural feature, the polymers are not soluble in common organic solvents as well as aqueous media. Ionization of **45** by hydrochloric acid furnishes polycation **46** (Scheme 1.5)

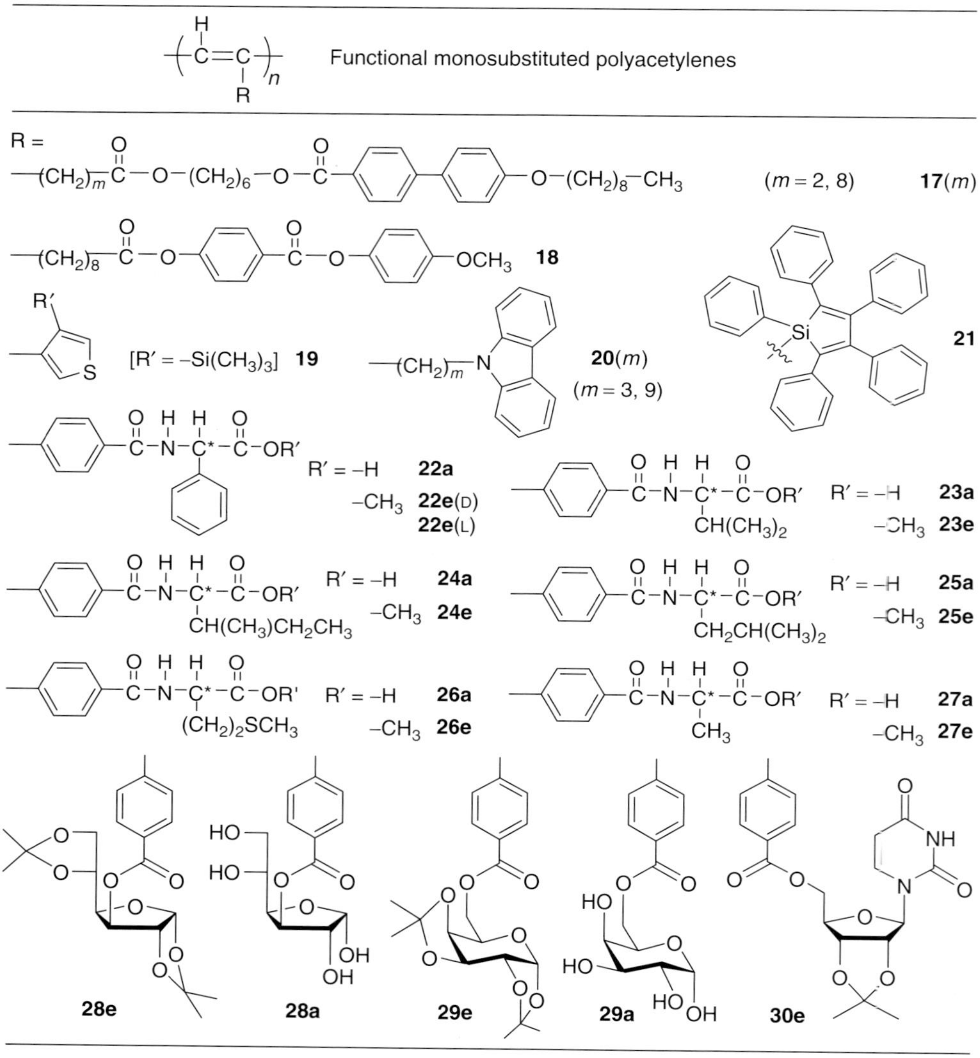

Functional monosubstituted polyacetylenes

Chart 1.2 Examples of monosubstituted PAs.

[18], which is soluble in water. While polymer reactions normally do not proceed to 100% completion, surprisingly, cleaving the ether groups of **29e** by hydrolysis can give **29a** with cytocompatabile sugar pedants which do not contain any ether residue.

It should be noted that it is not possible to gain access to functionalized disubstituted PAs by the direct polymerizations, owing to the inherent intolerance of the metathesis catalysts to polar groups. However, polymer-reaction approach can address this problem. By click chemistry, many highly polar functional groups

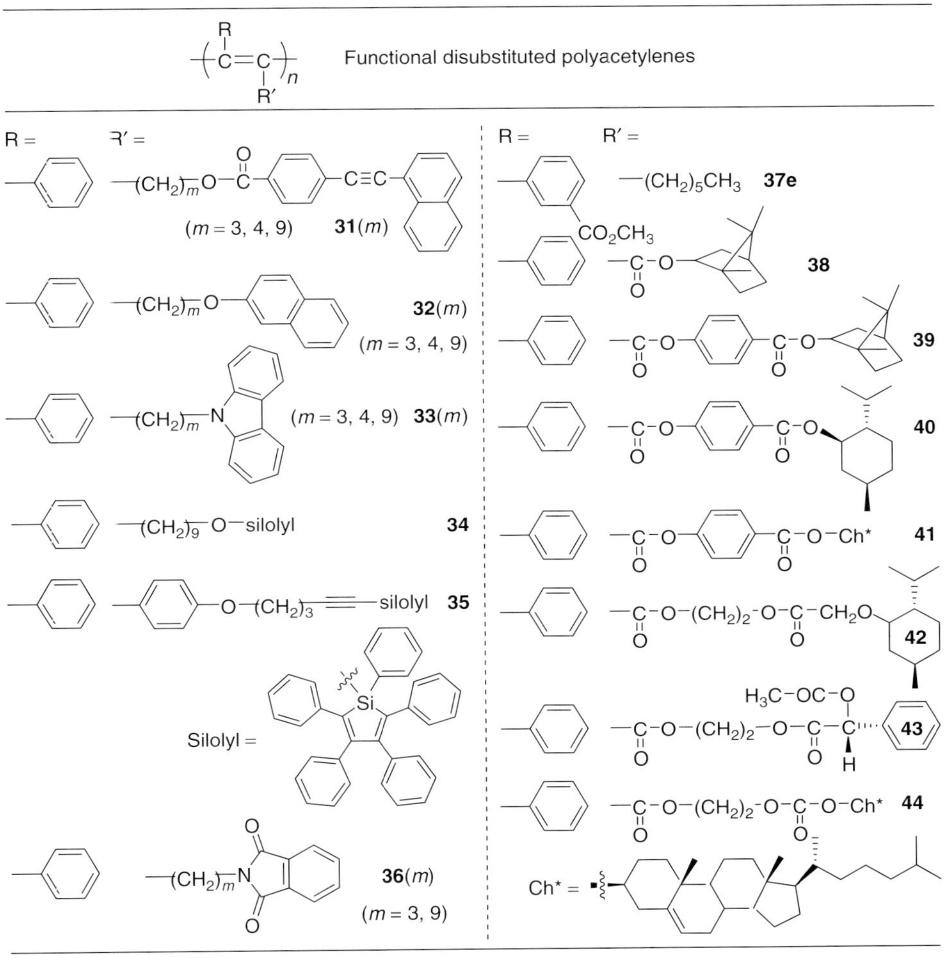

Functional disubstituted polyacetylenes

Chart 1.3 Examples of disubstituted PAs.

can be conveniently attached to disubstituted PA **48** to furnish **49** [19]. **47** and **37e** can undergo nucleophilic substitution and hydrolysis reactions, respectively, to provide imidazole- and carboxy-functionalized disubstituted PAs **50** and **37a** [20, 21]. Deprotection of imide **36**(9) yields amine **51**, which can further be ionized by hydrobromic acid to give polycation **52** [22, 23].

1.3
Functional Properties

The substituted PAs comprising conjugated backbones and functional pendants are expected to show unique properties. Studies of these properties will help

No.	Substrate: Acetylene monomer carrying functional group (FG)		Catalyst: Mo	W	Rh
1	H—≡—Ar—FG	1-arylacetylene	○	○	●
2	H—≡—$(CH_2)_m$—FG	1-alkyne	○	●	◎
3	H—≡—CO_2—FG	Propiolate	●	◎	●
4	$CH_3(CH_2)_m$—≡—CO_2—$(CH_2)_m$—FG	Alkly 2-alkynoate	●	○	○
5	Ar—≡—CO_2—$(CH_2)_m$—FG	Alkyl 3-arylpropiolate	●	○	○
6	Ar—≡—CO_2—Ar—FG	Aryl 3-arylpropiolate	○	●	○
7	FG—Ar—≡—$(CH_2)_m$—FG	1-aryl-1-alkyne	○	●	○
8	Ar—≡—Ar—FG	Diarylacetylene	○	●	○

Degree of substrate–catalyst matching: ● = excellent, ◎ = good, ○ = bad

Chart 1.4 Catalyst–substrate matching map.

understand how the polymers behave and aid our efforts to develop the new macromolecules for technologically useful materials. PA is notoriously intractable and unstable. In contrast, most of our functionalized PAs are soluble, thanks to the suppression of backbone interactions by the pendant groups. All the disubstituted PAs are very stable: for example, **32**(4) does not decompose when heated in air at high temperatures (e.g., 200 °C) for a prolonged time (e.g., 12 hours). The pendants shroud the polyene backbone, protecting it from thermolytic attack [24].

The functionalized poly(1-alkyne)s and polypropiolates are also stable. For example, **9**(3) does not degrade when it is heated in air at temperatures up to 200 °C (Figure 1.1a). Similarly, M_w of **9**(9) does not change when it is exposed to UV irradiation or stored in air (Figure 1.1b). Functionalized poly(1-arylacetylene)s, however, are only moderately stable and degrade when heated to >100 °C or stored under ambient conditions [17]. This makes them unsuitable for long-term uses but potentially attractive as materials that must decompose during and/or after use (e.g., photo- and biodegradable polymers).

1.3.1 Electrical Conductivity and Photoconductivity

The introduction of substituents onto the PA backbone has been proved to be an excellent solution to the problem of processability and stability. Nevertheless, the electrical conductivity of their doped form is usually much lower than that of doped PA, and still in the semiconductor region, which may be attributed to the twisted backbones forced by substituents and, hence, decreased overlap of π-orbitals.

Although the substituted PAs have the low absolute values of electrical conductivity, their relative conductivities are still higher than those of the conventional single-bonded polymers and may be further enhanced by external stimuli. Of

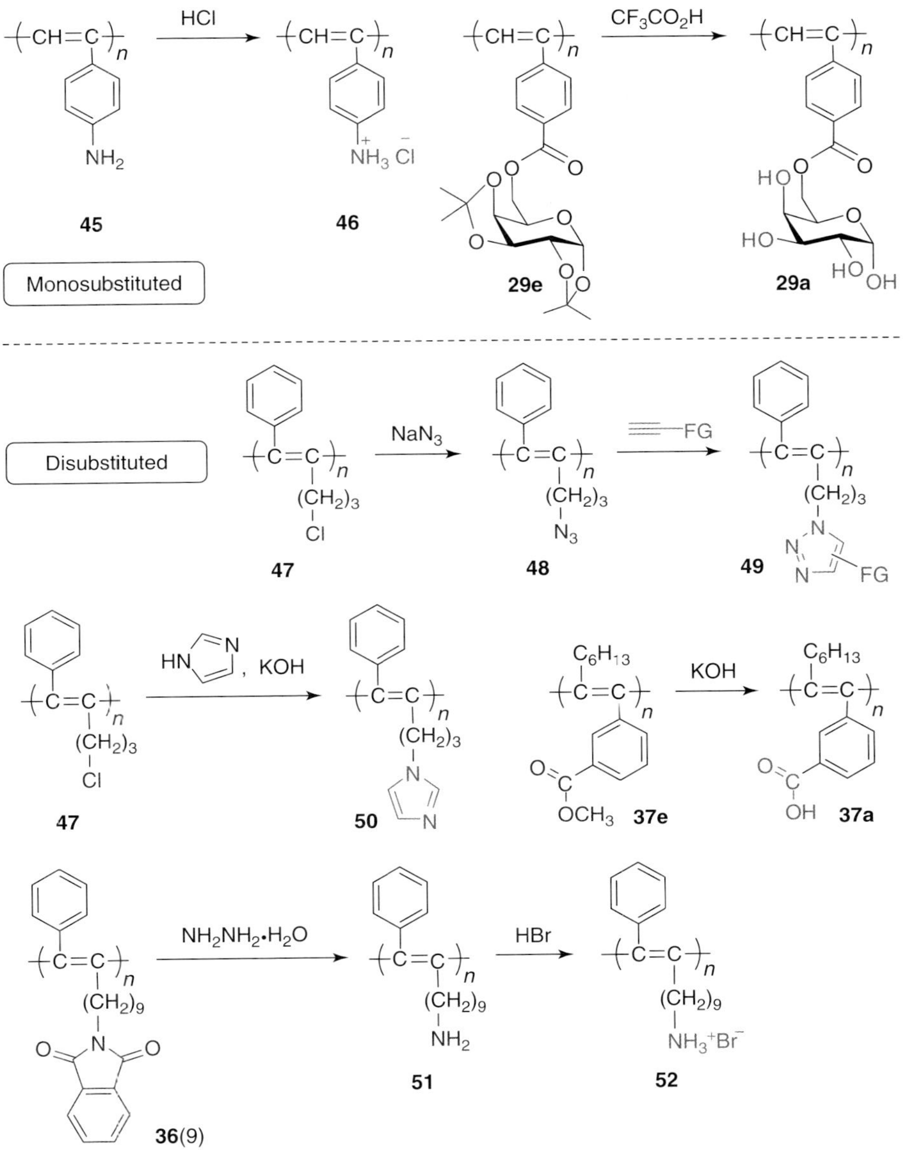

Scheme 1.5 Examples of polymer reactions.

particular interest and technological implication is the enhancement in electrical conductivity by photoirradiation, commonly known as *photoconductivity* (PC). Much work has been done on the photoconduction in substituted PAs [25]. It has been found that (i) the PAs containing electron-donating substituents exhibit higher PC than those with electron-accepting ones; (ii) the PC is further improved

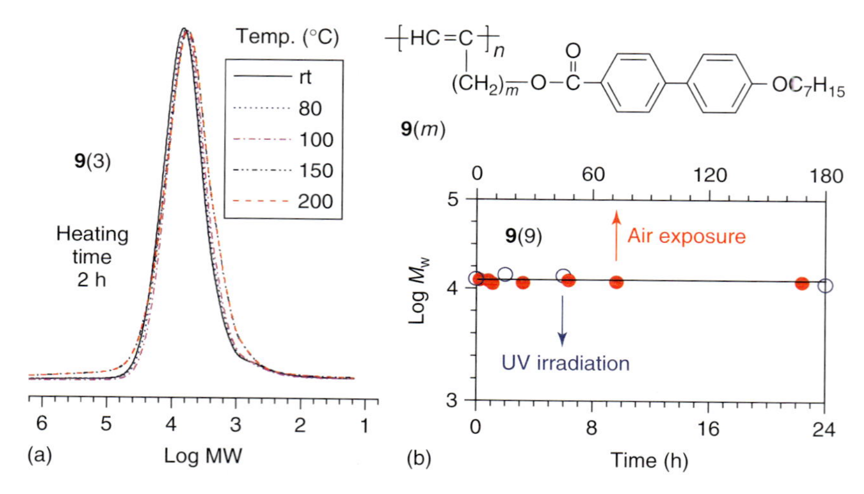

Figure 1.1 Resistances of **9**(*m*) against (a) thermolysis and (b) oxidation and photolysis in air.

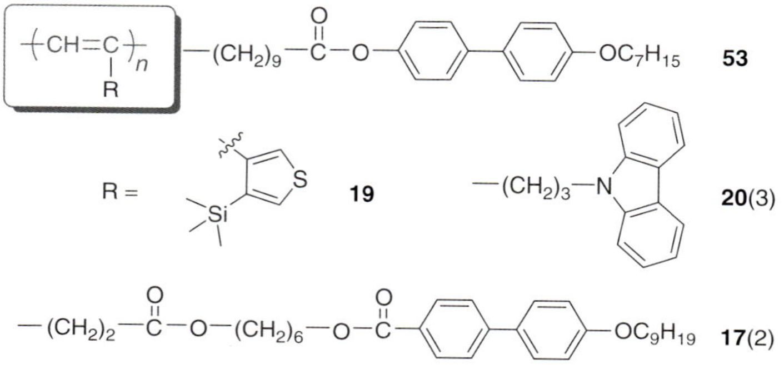

Chart 1.5 Examples of photoconductive PAs.

when the donor substituents are simultaneously good hole-transporters; and (iii) the photoconduction becomes more efficient when the donor substituents are mesogenic and can be packed in an ordered fashion.

Some examples of photoconductive PAs are shown in Chart 1.5. **19** and **20**(3) with donor substituents have high conductivity, with the value for latter one being higher because **20**(3) is simultaneously a good hole-transporter (Figure 1.2) [25]. A liquid crystalline polyacetylene (LCPA) is expected to have a high PC because of the high charge mobility in its closely packed mesogen phase. Interfaces between amorphous and mesogen phases also favor photoconductions, for excitons dissociate efficiently at, and carriers move rapidly along, the interfaces. Indeed, LCPA **53** shows higher PC than PPA, a non-LC photoconductor [26]. LCPA **17**(2) shows even higher PC, because of its better-ordered LC aggregates formed in the photoreceptor fabrication process.

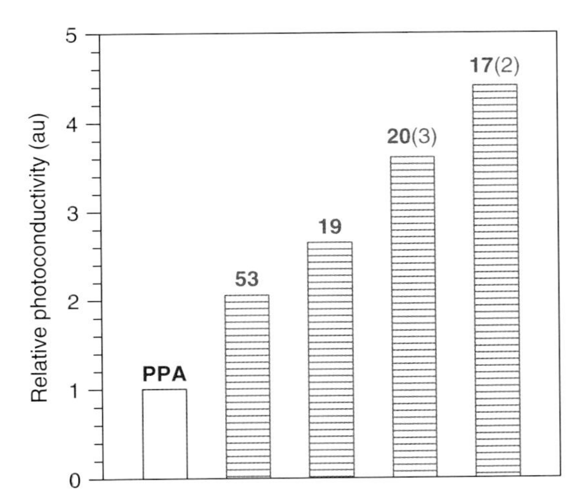

Figure 1.2 Photoconductivity of some selected PAs.

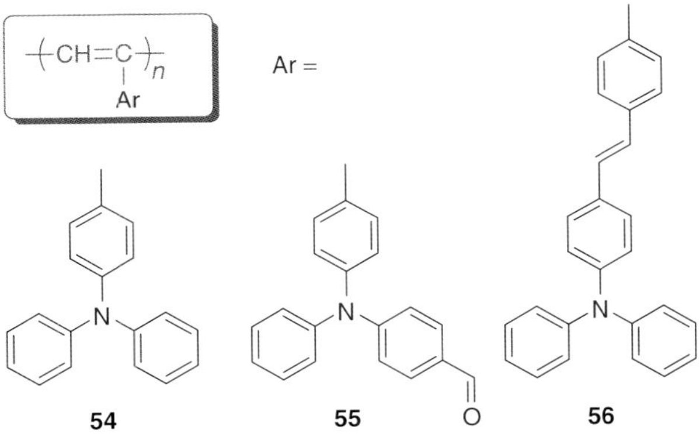

Chart 1.6 Triphenylamine-functionalized PAs.

Triphenylamine (TPA) is a typical donor and a famous building block for hole-transport materials. With an anticipation to obtain PAs with good photoconduction performances, we incorporated TPA into PAs (Chart 1.6), which were used as chare-generation materials (CGMs) [27]. For example, the half-discharge time ($T_{1/2}$) of **54** is 0.09 seconds and the photosensitivity (S) is 1010.1 $mm^2\ mW^{-1}\ s^{-1}$, which are quite high among substituted PAs. The photoreceptor with **55** as CGM shows a lower S value than the device based on **54**. This is easy to understand because the formyl group attached to the TPA moiety is electron withdrawing, which decreases the hole-transport efficiency. The photoreceptor with **56** as CGM gives the lowest S value among the three TPA-containing PAs. This result can be explained in terms of charge-generation efficiency. **56** has the highest fluorescence quantum efficiency among the polymers. This indicates that the photogenerated geminate pairs have a high probability to recombine. Consequently, the number of

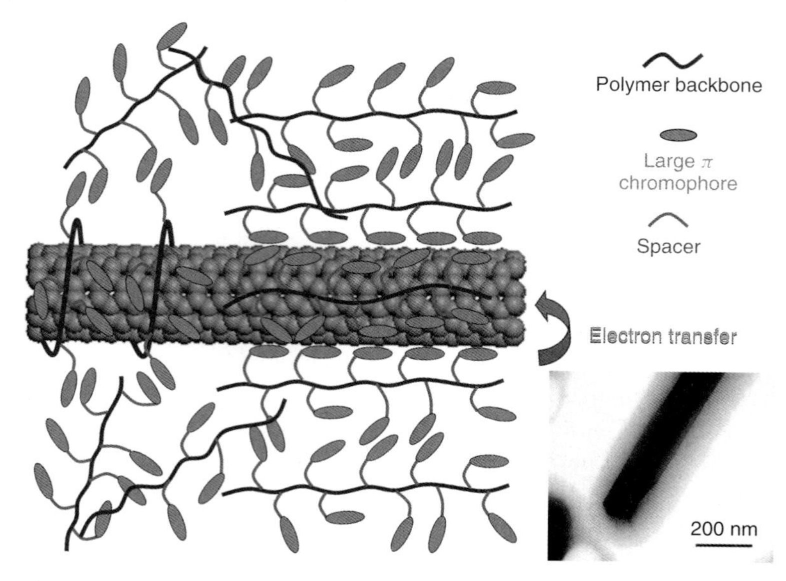

Figure 1.3 Schematic illustration of hybridization of CNT with PAs. Inset: TEM image of the hybrid of **54** and MWCNT.

the photogenerated free charges in the **56**-based photoreceptor becomes smaller, hence its observed low PC.

Like most conjugated polymers, substituted PAs can be considered as p-type semiconductors, and their hybridizations with n-type carbon nanotube (CNT) and inorganic semiconductors may offer nanohybrids with enhanced PC performances. Our group has found that substituted PAs bearing aromatic pendants can efficiently wrap CNTs and thus enhance the solubility. On the basis of our systematic study, the solubilization mechanism of CNTs may be summarized as sketched in Figure 1.3 [28–30]. The polymer chains may thread onto and/or wrap around the CNT shells due to strong π–π stacking and donor–acceptor interaction of their aromatic pendants and conjugated backbones with the CNT wall surfaces. The polymer coating is thickened by the deposition of polymer chains via pendant–pendant electronic interactions and physical chain entanglements. The affinity of the polymer chains with the solvent molecules brings the polymer-coated or -wrapped CNTs into the organic solvent media, leading to the dissolution of CNTs.

Employing the p–n junction strategy and the CNT-solubilizing method mentioned above, we fabricated photoreceptor devices using PA/CNT nanohybrids as CGMs. As expected, the photoreceptor devices displayed improved PCs, in comparison to those of the devices based on the parent polymers. Amazingly, **55**/multiwalled carbon nanotube (MWCNT)-based photoreceptor shows a $T_{1/2}$ value as low as 0.01 seconds [27]. Its corresponding S value is as high as ~9091 $mm^2\ mW^{-1}\ s^{-1}$, about 40 times that of the **55**-based device. This dramatically enhanced PC is due to the MWCNT component in the photoreceptor. As discussed above, the MWCNTs behave as an electron acceptor. The existence of the MWCNTs in the charge-generation layer improves charge-generation efficiency through a mechanism of photoinduced charge transfer from **55** to MWCNTs (Figure 1.3). Meanwhile, the MWCNTs offer one-dimensional electron-transport channels. Their conductive networks quickly transport the photogenerated electrons to the surface of the photoreceptor to neutralize the surface charge, thus resulting in steeply decreased surface potential and transiently short $T_{1/2}$ value. The device with **54**/MWCNT hybrid as CGM also shows enhanced PC, but the enhancement is less than that for the **55**/MWCNT-based device.

Nanohybridization of inorganic semiconductors with organic conjugated PAs offers the possibility to create new hybrids with combined advantages of the two components, that is, the high charge mobility of the inorganics and the ready processability of the organics, which may serve as active materials in the fabrication of photoreceptors. With great efforts, we developed a facile strategy for the hybridization of CdS nanorods with PPA chains. Utilizing an approach of "preassembly plus copolymerization," [31] CdS–PPA nanohybrid **59** has successfully been prepared (Figure 1.4). This hybrid is completely soluble in common organic solvents and can form homogeneous films when its solutions are cast on solid substrates. Because of the high solubility and film-forming capability of the hybrid, a series of **59**-based photoreceptors were prepared by a solution-casting process and their PCs were evaluated. One of the devices shows a $T_{1/2}$ value of 2.63 seconds and an S value of 34.3 $mm^2\ mW^{-1}\ s^{-1}$, which are better than those of all the devices using the parent form of PPA ($T_{1/2}$ = 3.67 seconds and S = 24.5 $mm^2\ mW^{-1}\ s^{-1}$) and CdS ($T_{1/2}$ = 4.74 seconds and S = 19.0 $mm^2\ mW^{-1}\ s^{-1}$) as well as their physical blend PPA/CdS ($T_{1/2}$ = 4.30 seconds and S = 21.0 $mm^2\ mW^{-1}\ s^{-1}$) as CGMs.

Similarly, functional perovskite hybrid **60** generated between PA ammonium **52** and $PbBr_2$ shows a better photoconduction result than its parent polymer **52** (Scheme 1.6) [22]. The PC performance of the photoreceptor of **52** is poor, as evidenced by its high $T_{1/2}$ (10 seconds) and low S (9.1 $mm^2\ mW^{-1}\ s^{-1}$) values. In the solid film, the ammonium cations and bromide anions of the polysalt form a network of randomly distributed electrostatic centers. The photogenerated charges are readily attracted to, and detained by, the network of the ions, which reduces their likelihood of contributing to photoconduction. When **52** is hybridized with lead bromide, its once-free ammonium cations now become confined in an ordered fashion through their coordination with the sheets of the corner-sharing metal halide octahedrons. The interactions between the fixed ions anchored to the inorganic frameworks and the mobile charges photogenerated on the conjugated PA backbones are weaker than those in the polysalt film. Some of the charges may find their paths to the electrode to discharge the surface potential. As a result, the hybrid **60** displays an improved PC performance ($T_{1/2}$ = 3.96 seconds and S = 23.0 $mm^2\ mW^{-1}\ s^{-1}$), in comparison to its polysalt precursor.

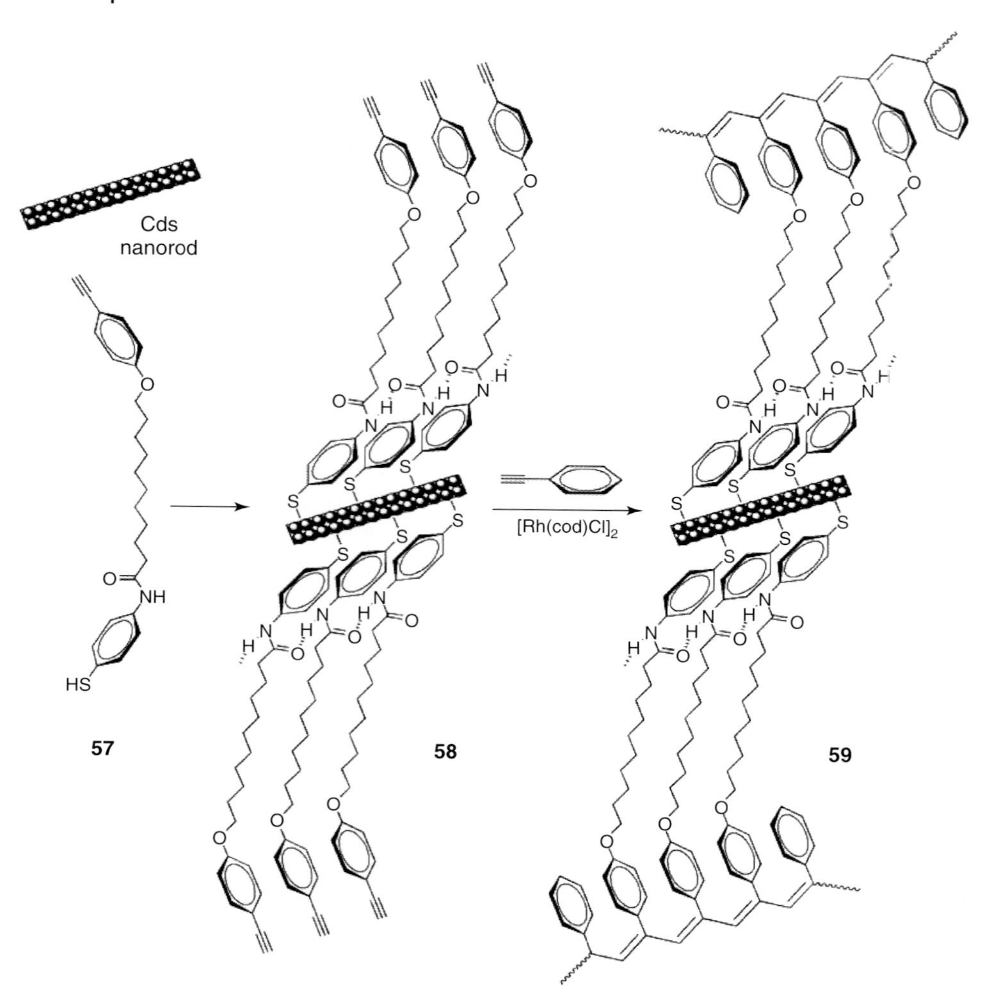

Figure 1.4 Assembly of the molecules of monomer **57** on the surface of CdS nanorod gives composite **58**, whose copolymerization with phenylacetylene yields CdS–PPA hybrid **59**.

1.3.2
Liquid Crystallinity

Liquid crystals and PAs are optically and electronically active, respectively; melding the two at molecular level may spawn LCPA hybrids with both optical and electronic activities. The LCPAs may also be third- and second-order nonlinear optically (NLO) susceptible because conjugated polymer chains and polarized mesogenic groups, are respectively, $\chi^{(3)}$- and $\chi^{(2)}$-active. The common recipe for the structural design

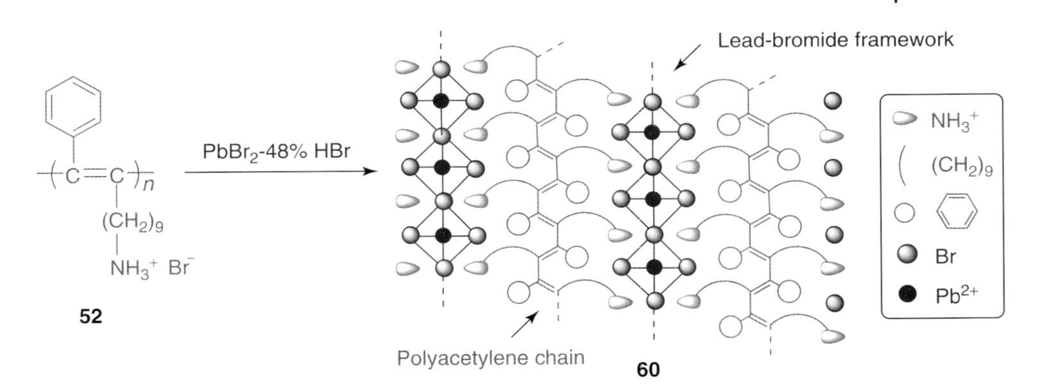

Scheme 1.6 Preparation of PA–perovskite hybrid.

of side chain LC polymers has been "mesogenic pendant + flexible backbone." Rigid polymer backbones have generally been believed to be detrimental to the packing arrangements of the mesogens in the side chains. As a result of this general belief, LC polymers with rigid backbones have seldom been developed. The study of LCPAs with rigid polyene backbones is of interest, and successful development of such polymers may open a new avenue for the exploration of new types of LC polymers and help deepen our understanding on liquid crystals. Since stiff-chain polymers can be oriented by external forces, the LCPAs may show anisotropic mechanical properties, which may contribute to the search for high-performance engineering materials.

Inspired by these attractive prospects, many research groups have worked on the development of LCPAs. Through careful design of flexible spacers, functional bridges, mesogenic cores, and functional tails, we have synthesized a series of highly soluble and thermally stable mono- and disubstituted thermotropic LCPAs and systematically studied their LC behaviors [1, 2, 32–38]. Typical examples are shown in Chart 1.7. We have systematically studied their LC behaviors, especially the structure–property relationships in the LCPA systems. The packing arrangements in the mesophases of the LCPAs and their mesomorphic transitions have been found to vary with the molecular structures of their mesogenic pendants. Some examples of the structure–property relationships are given in Table 1.1.

Polymer **64** is a poly(1-phenyl-1-alkyne) derivative containing a mesogenic unit with a biphenyl (BP) core. It displays an S_A mesophase in the temperature range of 172–158 °C when cooled from its isotropic melt. Its cousin with a phenylcyclohexyl core (**65**) exhibits a nematic (N) mesophase at much lower temperatures (108–90 °C), although it differs from **64** by one ring structure in the mesogenic core (cyclohexyl in **65** versus phenyl in **64**). Polymers **62** and **63** have almost identical molecular structures, expect for that the former and latter have ester and ether bridges, respectively. This seemingly subtle structural difference affects the mesogenic packing arrangements, with the former ordered in a bilayer structure in an interdigitated fashion, namely, interdigitated smectic A phase (S_{Ad}), while the latter packs in a monolayer structure (S_A; Table 1.1, nos. 3 and 4).

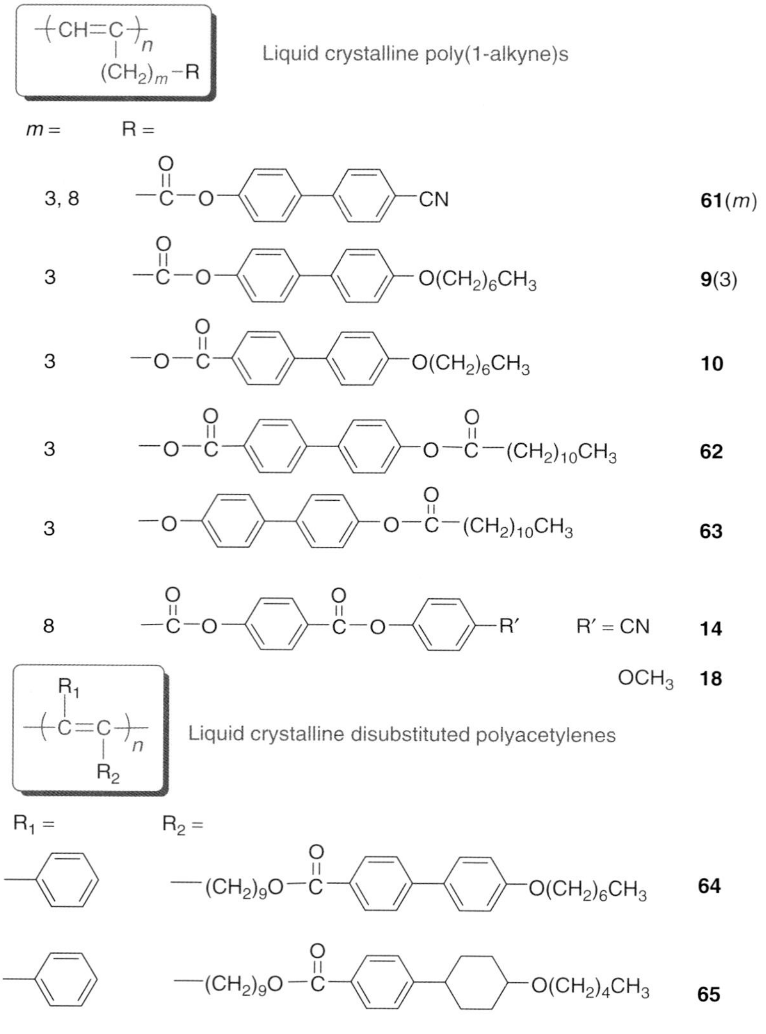

Liquid crystalline poly(1-alkyne)s

m = R =

3, 8 — **61**(m)

3 — **9**(3)

3 — **10**

3 — **62**

3 — **63**

8 — R′ = CN **14**; OCH_3 **18**

Liquid crystalline disubstituted polyacetylenes

R_1 = R_2 =

64

65

Chart 1.7 Examples of liquid-crystalline PAs.

The elemental composition or molecular formulae for **9**(3) and **10** are exactly the same; the two polymers differ only in the orientation of their ester bridge. This small difference changes the mesogenic packing: the mesostructure of **10** involves mixed monolayer and bilayer packing arrangements, whereas the mesophase of **9**(3) is associated with a pure bilayer mesogenic alignment. An increase in the length of the flexible spacer encourages better mesogenic order and induces a transition from nematicity to smecticity in the poly(1-alkyne)-based system **61**(m) (Table 1.1, nos. 7 and 8). The increase in the spacer length also widens the LC temperature range (ΔT); in the case of **61**(m), ΔT for **61**(8) is approximately three

Table 1.1 Effects of molecular structures on the mesogenic packing of liquid-crystalline PAs[a].

No.		Structural variation		Mesophase	$T_c - T_m$(°C)	ΔT(°C)
1	Mesogenic core	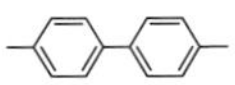	**64**	S_A	172–158	14
2		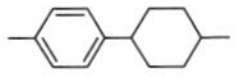	**65**	N	108–90	18
3	Functional bridge	—O—C(=O)—	**62**	S_{Ad}	199–162	37
4		—O—	**63**	S_A	180–122	58
5	Bridge orientation	—O—C(=O)—	**10**	$S_A + S_{Ad}$[b]	174–100	74
6		—C(=O)—O—	**9**(3)	S_{Ad}	206–127	79
7	Flexible spacer[c]	$—(CH_2)_3—$	**61**(3)	N	189–152	37
8		$—(CH_2)_8—$	**61**(8)	S_{Ad}	190–81	109
9	Functional tail	—CN	**14**	S_A	195–80	115
10		$—OCH_3$	**18**	S_A–N[d]	141–114–96	45

[a] Data taken from the DSC thermograms recorded under nitrogen in the first cooling scan.
[b] Coexistent mono- (S_A) and bilayer (S_{Ad}) mesophases.
[c] For mesogenic poly(1-alkyne)s.
[d] Consecutive S_A and N mesophase transitions.
Abbreviations: S_A, Smectic A phase; N, nematic phase; S_{Ad}, interdigitated smectic A phase; T_c, clearing point; T_m, melting point; ΔT, $T_c - T_m$ (temperature range of the liquid-crystalline phase).

times higher than that for **61**(3). The polymer with polar cyano tails (**14**) undergoes an enantiotropic S_A transition over a wide temperature range (115 °C), but its counterpart with less polar methoxy tails (**18**) goes through enantiotropic S_A and N transitions in a narrow temperature range (Table 1.1, nos. 9 and 10).

The most distinct structural feature of the LCPAs is the rigidity of their double-bond backbones, in comparison to the flexibility of the single-bond backbones of nonconjugated polymers. According to the researchers of LC polymers, a stiff backbone is generally regarded as a structural defect that distorts the packing arrangements of mesogens, which is why the flexible backbones of polysiloxane and polyacrylate, for example, have often been used in the design and synthesis of "conventional" LC polymers. Is a rigid backbone only destructive or can it play any constructive role? It is envisioned that an oriented rigid main chain might induce unique alignments of its mesogenic side chains. The stiff backbone of an LCPA possesses a long relaxation time, and a mechanically oriented LCPA system

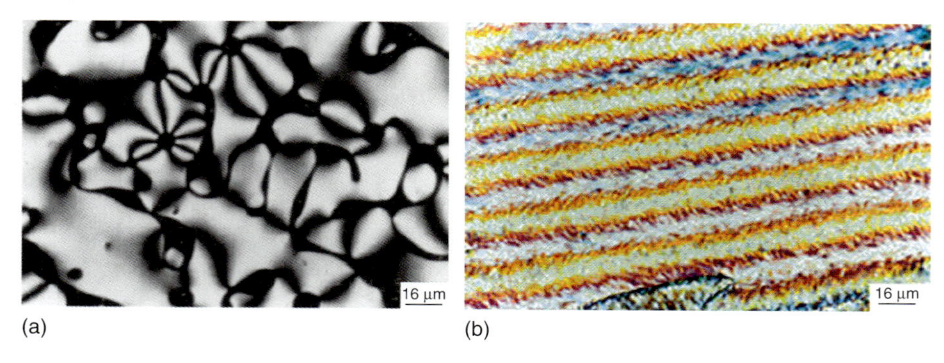

Figure 1.5 (a) Schlieren textures with disclination strengths of 3/2 and 2 observed after a rotationally agitated **18** has been annealed at 136 °C for 5 minutes. (b) Induced alignments of **18** by mechanical shear.

may not quickly return to the preperturbed state, thus offering the opportunity to generate macroscopic anisotropy by the application of a mechanical force.

To examine its response to mechanical perturbation, a rotational force was applied to **18** when it was cooled from the isotropic state to the N phase. Interestingly, unusual high-strength disclination strengths of 3/2 and 2 are observed (Figure 1.5a) [37], which have been found in a few systems of main-chain LC polymers with rigid backbones but never in any systems of side chain LC polymers with flexible backbones. The high disclination values are demonstrative of the backbone rigidity of the LCPA, because only a stiff backbone can survive the severe director field distortions involved in the formation of the high-strength disclinations. When a shear force is applied to the N phase of **18** at ~140 °C, inversion walls are generated. Natural cooling of the system to room temperature results in the formation of well-ordered parallel microbands (Figure 1.5b) [37]. In general, a main-chain LC polymer with a rigid backbone can be easily aligned by a shear force, while a side chain LC polymer with a flexible backbone can be readily processed by solution processes. Evidently, **18** possesses the combined advantages of main- and side chain LC polymers: it can be readily aligned by a shear force and can in the meanwhile be easily processed because of its excellent solubility in common organic solvents.

1.3.3 Luminescence

The study of luminescence properties of π-conjugated polymers is a hot topic of great current interest. Whereas PA is an archetypal π-conjugated polymer, little work had been done on the development of light-emitting polyacetylenes (LEPAs) in the early time, because PA itself is a very poor luminophore. Modifications of molecular structure of PA, such as attachments of the pendant groups with different electronic and steric effects, have been used to tune the conjugation length along

Table 1.2 Photoluminescence of PAs bearing biphenyl (BP) pendant.

No.	Polymer	Skeleton	Emitting center	Emission color	Quantum yield
1	PA	$-(CH=CH)_n-$	Backbone	IR	Low
2	PPAs	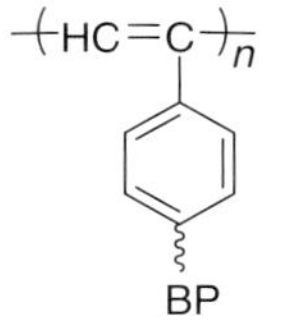 $-(HC=C)_n-$ with phenylene ~ BP	Pendant	UV	Low
3	Poly(1-alkyne)s	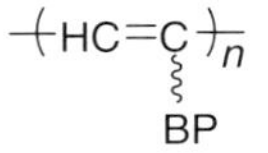 $-(HC=C)_n-$ ~ BP	Pendant	UV	Medium
4	Poly(propiolate)s	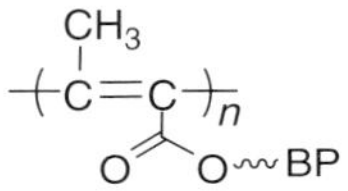 $-(C(CH_3)=C)_n-$, C(=O)O ~ BP	Pendant	UV	High
5	Poly(1-phenyl-1-alkyne)s	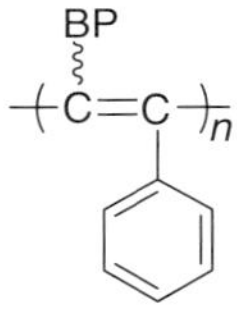 BP ~ $-(C=C)_n-$ with phenyl	Skeleton	Blue	High
6	Poly(diphenylacetylene)s	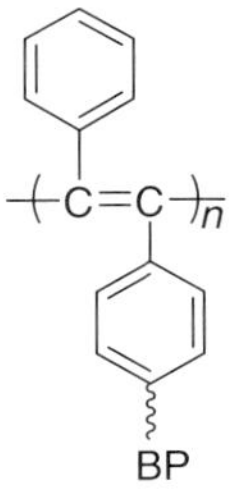 phenyl $-(C=C)_n-$ phenylene ~ BP	Skeleton	Green	High

the polyene backbone and the electronic interaction between the polymer chains. These approaches have worked well and led to the generation of a large variety of LEPAs with high photoluminescence (PL) and good electroluminescence (EL) efficiencies. In the LEPAs, the emitting center can be pendant or backbone, the energy transfer can be from pendant to backbone or from backbone to pendant, and the emission color can be violet, blue, green, or red – all tunable by changing the backbone and pendant structures.

A typical example is the study of the PL performances of PAs bearing a BP pendant, as shown in Table 1.2 [1, 2, 23, 33, 34, 39–42]. The data for the parent form of (unsubstituted) PA are also given in the table for the purpose of comparison; it emits very weakly in the infrared spectral region due to the existence of exciton traps in the polyene backbone. Similarly, PPA itself is a poor light emitter owing to the photogenerated radical defects in the conjugated skeleton. The PPA derivative with BP pendant emits UV light with a low Φ_F value, because the light emitted from the BP pendant is partially quenched by the defects in the PPA skeleton (Table 1.2, no. 2). When the PPA skeleton is changed to a poly(1-alkyne) one, the Φ_F value is increased due to the alleviated quenching by the poly(1-alkyne) skeleton. Disubstituted poly(propiolate)s, however, luminesce efficiently (Table 1.2, no. 4), because of their photochemically stable, defect-free skeleton structure.

From the result of disubstituted poly(propiolate)s, it is envisioned that if the hydrogen atom in the repeat unit of a poly(1-alkyne) is replaced by a bulky aromatic substituent, the backbone would become twisted and more stable. The steric effect would shorten the effective conjugation length of the resultant disubstituted PAs, hence widening their bandgap and shifting their skeleton emission from IR to visible. Indeed, poly(1-phenyl-1-alkyne)s emit strongly in the blue region (Table 1.2, no. 5). Meanwhile, the polymer skeleton absorbs the UV light emitted from the BP pendant, which helps enhance the light emission efficiency. Following this strategy, BP-containing poly(diphenylacetylene) derivatives have been synthesized, whose emission efficiencies are comparably high but their emission colors are redshifted to green.

The molecular engineering endeavors change not only the location of the emitting center, as discussed above (Table 1.2), but also the direction of the energy transfer, as exemplified in Chart 1.8 [41, 42]. In polymer **21**, an efficient fluorescence energy transfer (FRET) process occurs from the green light-emitting silole pendant to the red light-emitting polyene backbone, because of the direct electronic communication between the pendant and the backbone. The PA backbone of **21** emits faintly in the long-wavelength region, in agreement with the early observation that pure PA is a weak IR emitter. In polymer **34**, however, the FRET is from the poly(1-phenyl-1-alkyne) skeleton to the silole pendant. The blue light emitted from the skeleton excites the pendant, resulting in the emission of green light. Polymer **35** emits green light efficiently, because the excited states of the silole pendant and the poly(diphenylacetylene) skeleton both undergo radiative transitions in a similar spectral region.

Aggregation of conjugated polymer chains in the solid state often results in the formation of less-emissive or nonemissive species, such as excimers and exciplexes, which partially or completely quench the luminescence of the polymers. This aggregation-caused quenching effect has been a thorny problem in the development of efficient polymer light-emitting diodes (PLEDs), because the conjugated polymers are commonly used as thin solid films in the EL devices

During our search for efficient light-emitting materials, we found a group of highly luminescent organometallic molecules called *siloles in the solid state.* They exhibit an invaluable effect of aggregation-induced emission (AIE) [43]: they are practically nonluminescent when molecularly dissolved in solution but become strongly emissive when aggregated in poor solvents or fabricated into solid films. 1,1,2,3,4,5-Hexaphenylsilole (HPS) is a typical paradigm of AIE luminogen

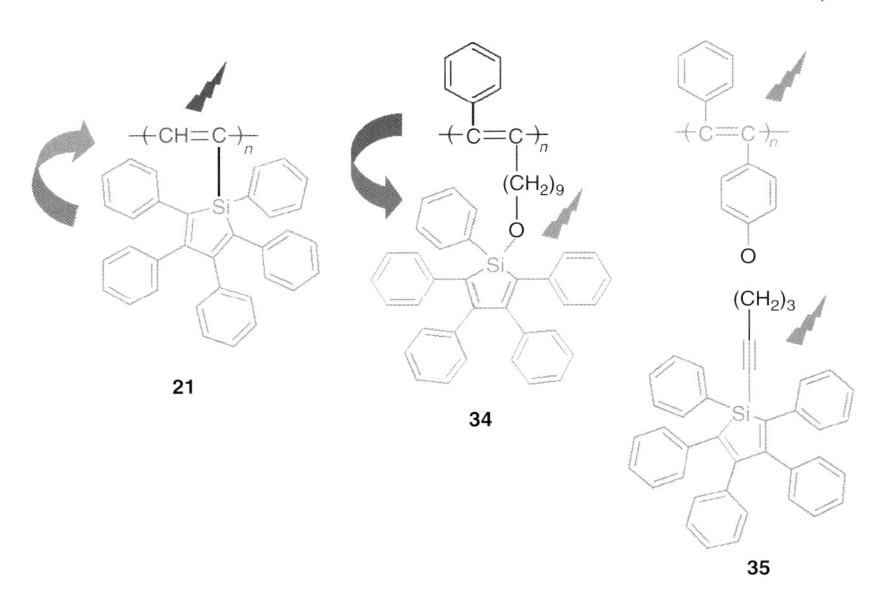

Chart 1.8 Energy transfers in luminescent PAs.

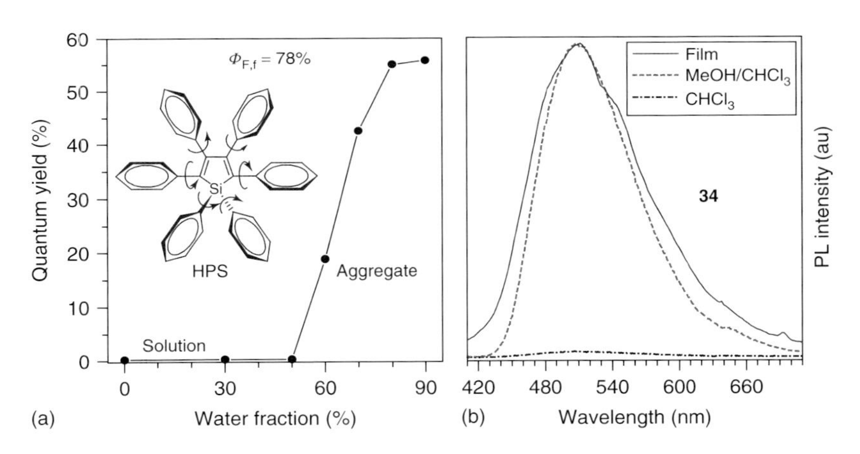

Figure 1.6 (a) Quantum yield (Φ_F) of HPS versus composition of acetonitrile/water mixture. The quantum yield of its film ($\Phi_{F,f}$) is shown for comparison. (b) Photoluminescence spectra of **34** in chloroform (molecularly dissolved solution), a methanol/chloroform mixture (9 : 1 by volume; nanoaggregate suspension), and solid state (thin film).

(Figure 1.6a). Its Φ_F value in pure acetonitrile is as low as 0.2%. The Φ_F almost remains unchanged when up to 50% of water is added into the acetonitrile solution but starts to swiftly increase afterward due to the formation of nanoaggregates. When the water content is 90 vol%, the Φ_F rises to ~56%, which is ~280 times higher than that in the acetonitrile solution [44]. In the film state, the Φ_F is even higher and goes to ~78% [45]. Through experimental and theoretical studies [46, 47], the AIE mechanism can be understood as follows. In the dilute solution at room temperature, the active intramolecular rotations of the peripheral phenyl rings of HPS around the axes of the single bonds linked to the central silole core nonradiatively annihilate the excitons, thereby making the silole molecules nonemissive. In the aggregates, the propeller shape of the silole molecules prevents them from forming excimeric species but the physical constraint in the solid state restricts their intramolecular rotations, which blocks the nonradiative relaxation channels and populates the radiative decay, thus making the silole molecules luminescent.

Siloles are thus a group of excellent molecules for LED applications [44, 48]. Low MW compounds, however, have to be fabricated into thin films by relatively expensive techniques such as vacuum sublimation and vapor deposition, which are not well suited to the manufacture of large-area devices. One way to overcome this processing disadvantage is to make high-MW polymers, which can be readily processed from their solutions into thin solid films over large areas by simple spin coating or doctor's blade techniques. Motivated by these considerations, we incorporated the silole moiety into the poly(1-phenyl-1-alkyne) as a pendant, with the aim of generating AIE-active polymers. HPS-containing poly(1-phenyl-1-alkyne) **34** is an example of such polymers. The solution of **34** is virtually nonluminescent (Figure 1.6b) [41], because in the FRET process of this polymer, the PL from the emissive poly(1-phenyl-1-alkyne) skeleton is quenched by the nonemissive HPS pendant. Thanks to its AIE attribute, the Φ_F value of its nanoaggregates in the methanol/chloroform mixture with 90% methanol is ~46-fold higher than that in the chloroform solution.

Retrostructural analysis of another AIE-active 1,1,2,2-tetraphenylethene (TPE) developed by us reveals that there is one common structural feature between TPE, and disubstituted poly(1-phenyl-1-octyne) **66** and poly(diphenylacetylene) **67**: one or more aromatic peripheries are connected to an olefinic core [49–53]. In a dilute solution of TPE, intramolecular rotations of the aromatic rings against the olefinic core effectively and nonradiatively deactivate the excited states, thus making the molecules nonemissive ("off"). In the aggregate state, such rotations are restricted. The blockage of the nonradiative decay channels turns the light emissions "on." Therefore, it is expected that **66** and **67** shown in Chart 1.9 might show an aggregation-induced emission enhancement (AIEE) effect, since the polymers themselves are already emissive in solution. As anticipated, **66** and **67** are AIEE active, in which the emission intensities are progressively increased with the increasing water fraction in the THF/water mixture (Figure 1.7) [54, 55]. Spectral profile of **66** remains unchanged [54], while the emission spectrum of **67** gradually redshifts [55].

Further studies of the light-emitting behaviors of these disubstituted PAs reveal that their PL spectra are commonly very broad and lack of fine structures, characteristic of excimer emissions. According to the "$n = 3$ rule" shown in Chart 1.10, molecules with phenyl rings spatially separated by three carbon atoms (e.g., 1,3-diphenylpropane and PS) [56, 57] can form intramolecular excimers

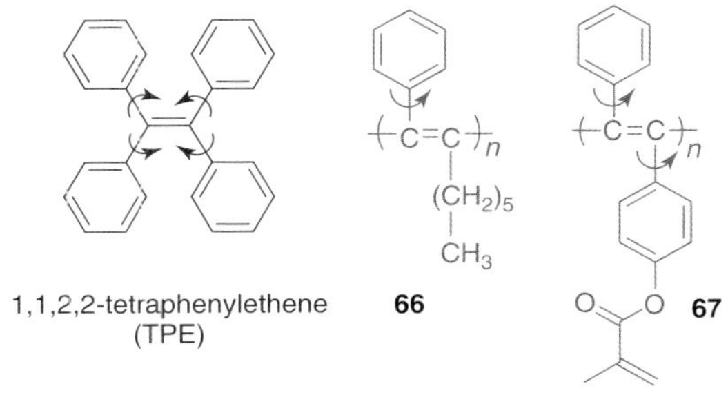

1,1,2,2-tetraphenylethene (TPE) **66** **67**

Chart 1.9 Intramolecular rotations in TPE and disubstituted PAs.

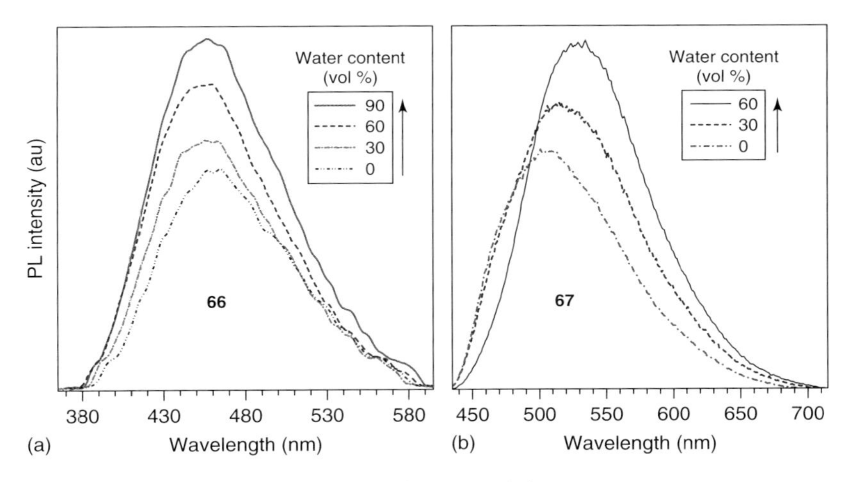

Figure 1.7 Photoluminescence spectra of (a) **66** and (b) **67** in THF/water mixtures.

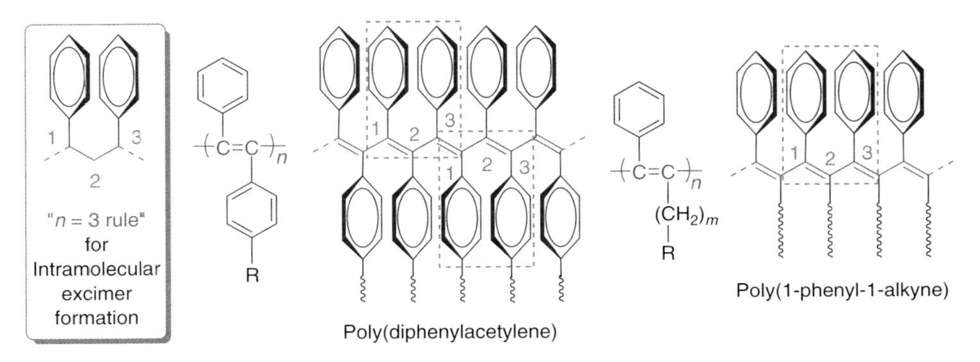

Chart 1.10 Formation of intramolecular excimers in disubstituted PAs.

that emit in the redder spectral regions, in comparison to their "monomer" emissions [54, 55]. Both the poly(diphenylacetylene) and poly(1-phenyl-1-alkyne) derivatives follow this "$n = 3$ rule." In general, this rule has been used to explain why light emissions of organic dyes are quenched in the aggregate state, while we have, however, found that the poly(diphenylacetylene) and poly(1-phenyl-1-alkyne) derivatives show an "abnormal" AIEE phenomenon: their light emissions are enhanced by aggregate formation.

Through the systematic investigations assisted by theoretical calculations, we have gained some insights into the relationship between the AIEE effect and the intramolecular excimer formation. In the poly(diphenylacetylene) and poly(1-phenyl-1-alkyne) skeletons, some "$n = 3$" intramolecular or intrachain excimer pairs have already formed in the solution state, as indicated by the broad and structureless spectra of the polymer solutions, although because of the polyene backbone twisting, the phenyl ring pairing is disrupted at some places, leaving some phenyl rings unpaired and standing alone as monomers. Upon formation by aggregration, the volume shrinkage of the poly(diphenylacetylene) chains in the solvent mixture with poor solvating power puts the phenyl rings in closer vicinities. This enhances the $\pi-\pi$ stacking interactions of the phenyl rings and restricts their intramolecular rotations: the former effect populates the excimer species, hence redshifting the emission spectrum, while the latter effect boosts its emission intensity, hence the observed AIEE phenomenon. In the poly(1-phenyl-1-alkyne) system, however, the population of the excimers does not change too much with aggregate formation, hence no apparent redshift in the PL spectrum. A unique feature to be noted in this system is that the twisted polyene backbone and long side chains weaken the interactions between the polymer strands, which helps prevent the light emission of polymer from being quenched by the formation of interchain excimers.

The LEPAs enjoy good spectral stability: for example, **32**(4) emits a blue light, whose PL profile does not vary after it has been heated to 200 °C in air or exposed to UV irradiation under nitrogen or under vacuum (Figure 1.8a) [58]. The efficient PL encouraged us to check EL of the LEPAs. The PLED of **32**(4) emits a blue light of 460 nm, whose EL peak is symmetrically shaped with no sideband. Its external quantum efficiency (η_{EL}) reaches 0.85%, comparable to those of some of the best blue light-emitting polymers. Although the device configurations of our PLEDs have not been optimized, many of them already display an η_{EL} of 0.5–0.86%, much higher than that (0.01%) of nonfunctional poly(1-phenyl-1-octyne) (**66**). Polyfluorenes are the best-known blue light-emitting polymers, but their spectral stabilities are low. A broad peak centered at ~520 nm appears when a current density as low as 1.8 mA cm^{-2} passes through the poly(9,9-dioctylfluorene) (PDOF) device (Figure 1.8b). The EL spectrum of **32**(4) does not change when a current density as high as 127 mA cm^{-2} passes through its device. This demonstrates that disubstituted PAs are thus spectrally very stable.

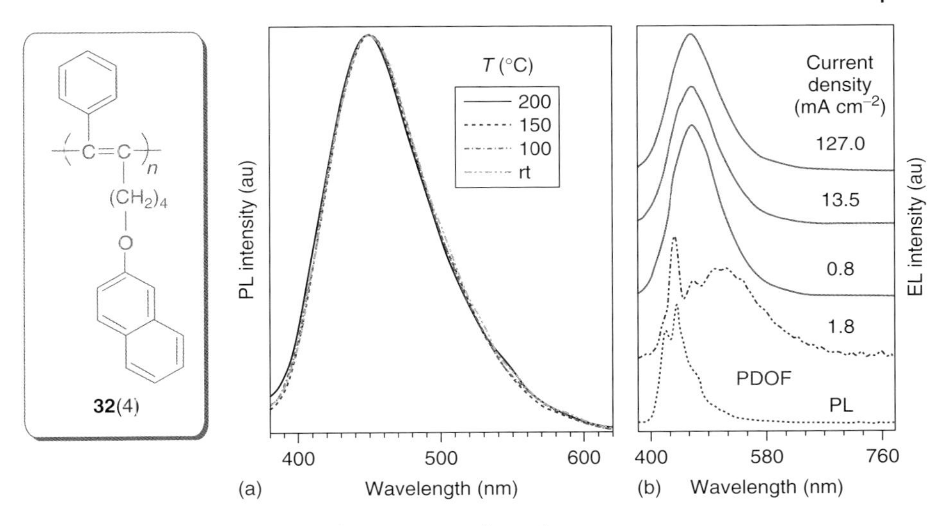

Figure 1.8 (a) Spectral stability of **32**(4) against thermolysis. (b) Comparison of its EL spectral stability with that of poly(9,9-dioctylfluorene) (PDOF).

1.3.4
Fluorescence Sensing

Fluorescent conjugated acetylenic polymers have been actively used for fluorescent sensors by virtue of their specific PL responses to interactions with analytes. PL of some LEPAs can make different responses to different metal ions. For example, when one equivalent of Cu^{2+} ion is added to a solution of **37a**, its PL drops to half the original value(Figure 1.9) [21]. The effect of Fe^{3+} is more striking: it completely quenches the emission. In contrast, the PL becomes stronger in the presence of Al^{3+}. These distinct PL responses associated with the specific polymer–metal interactions enable the LEPA to act as a chemosensor.

We developed a sequential chemosensor based on imidazole-containing LEPA **68** (Scheme 1.7). Among the different kinds of metal ions, only Cu^{2+} ion can completely and efficiently quench or turn off the strong fluorescence of **68**, with a detection limit as low as 1.48 ppm [20]. The associated Stern–Volmer quenching constant (K_{sv}) is as high as 3.7×10^5 M^{-1}, because of the high affinity of Cu^{2+} ion toward imidazole group, which enables the polymer to differentiate Cu^{2+} from other metallic ions. The Cu^{2+}-quenched light emission of **68** can be turned on by the addition of CN^- ion, thus allowing the polymer to function as a unique dual-response sequential ionosensor for cyanide detection.

1.3.5
Patterning and Imaging

If a processable polymer enjoys both photosensitive and light-emitting properties, it will be an excellent photoresist candidate for the fabrication of a luminescent

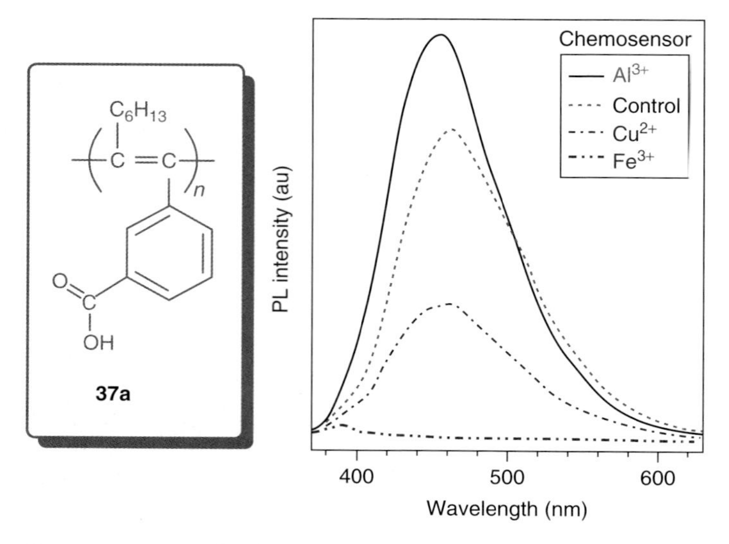

Figure 1.9 Tuning photoluminescence of THF solutions of **37a** by ionic species.

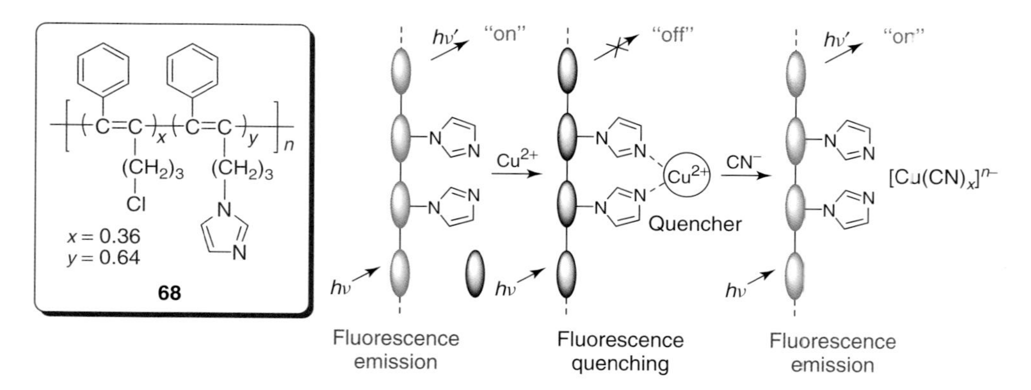

Scheme 1.7 Sensing Cu^{2+} and CN^- ions by an imidazole-functionalized PA.

pattern by the photolithography technique. Given this reason, many LEPAs fulfill the requirements and are thus promising photoresist materials for applications in photonic and electronic devices, and biological sensing and probing chips (e.g., LEDs, liquid crystal display, and medicinal diagnostic biochip).

UV irradiation of **69** readily initiates photo-polymerization of its acrylic pendants (Figure 1.10a) [55]. Development of the exposed films gives well-defined photoresist patterns (Figure 1.10c,d). The patterned lines glow under UV illumination, because its poly(diphenylacetylene) skeleton is highly emissive in a green color. Disubstituted PAs are resistant to thermolysis. Some of them, however, are sensitive to photo-oxidation, which quenches their PL. For example, UV irradiation of a film of **20**(4) in air through a mask quenches the luminescence of the exposed region

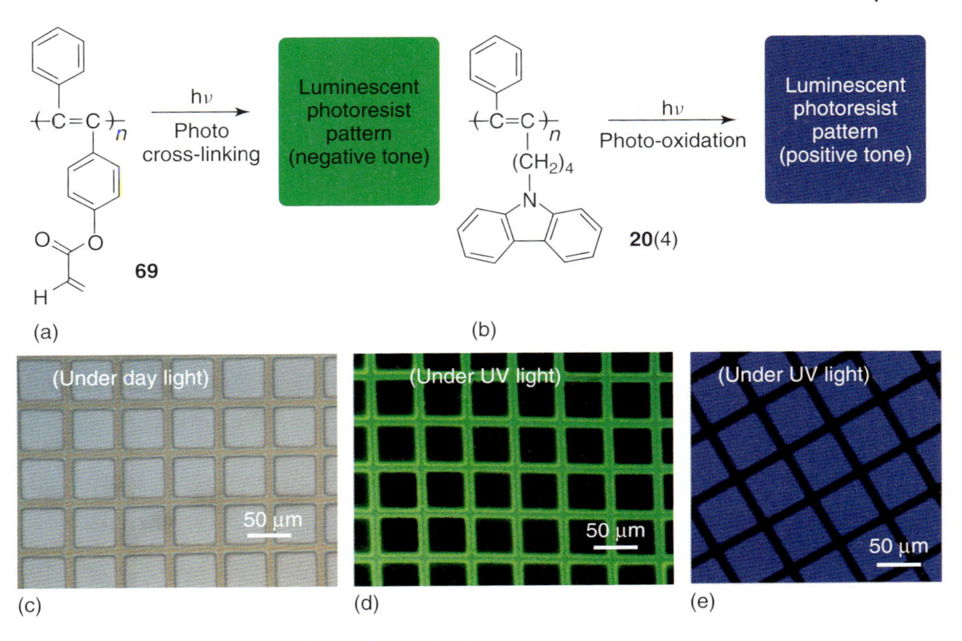

Figure 1.10 (a) Photo-cross-linking of **69** and (b) photo-oxidation of **20**(4) lead to the formation of (c and d) negative and (e) positive photoresist images.

(black lines) (Figure 1.10e), while the unexposed area remains emissive in a blue color [59]. A PL image is thus directly drawn without a developing process.

1.3.6 Chromism

The polyene backbone of a segment in a PA chain can adopt *E-s-E*, *E-s-Z*, *Z-s-E*, or *Z-s-Z* conformation [4, 10, 34]. Because the energy needed for the conformational conversion is small, the conformation of a PA segment is anticipated to be tunable by changing its surrounding environment. In toluene and chloroform, **28e** absorbs weakly and strongly, respectively (Figure 1.11a). The optical transition between these two states can be manipulated continuously and reversibly by varying the ratio of the solvents in the polymer solution. Most segments of **28e** may adopt a more planar conformation in chloroform while the opposite is true in toluene, thereby accounting for the solvatochromism.

The light transmission spectrum of **19** resembles that of an optical cutoff filter, which redshifts with an increase in its concentration (c) (Figure 1.11b). Its cutoff wavelength (λ_c) changes semilogarithmically with c over a wide spectral range [60], which is possibly due to nanocluster formation. This offers an easy way to make a plastic optical filter, whose λ_c can be readily tuned by varying its PA content.

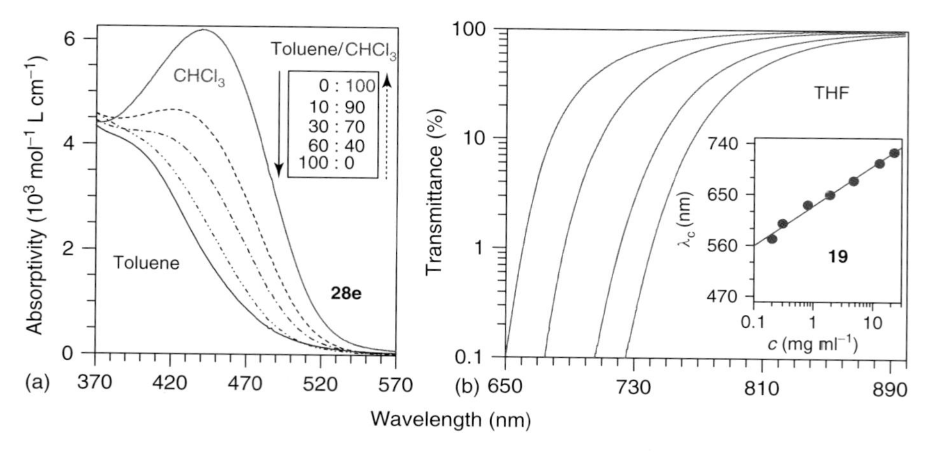

Figure 1.11 (a) Solvato- and (b) concentratochromisms of **28e** and **19**.

1.3.7 Optical Activity

Tang synthesized one of the few optically active polyacetylenes (OAPAs) in the late 1980s [61]. He continued his pursuit of OAPAs after he had moved to Hong Kong in the mid-1990s. While the early OAPAs contained no polar groups owing to the difficulty of synthesizing them, the recent advancements in the polymer syntheses allow us to design a diverse palette of new OAPAs by incorporating various chiral units into the PA chains.

We have concentrated our efforts on the development of OAPAs with naturally occurring building blocks such as amino acids, saccharides, nucleosides, and sterols [62–73]. Figure 1.12a shows circular dichroism spectra of a pair of polymers containing D- [**22e**(D)] and L-α-phenylglycine units [**22e**(L)]. The positive and negative Cotton effects in the long-wavelength region reveal that the chain segments of the PAs carrying the pendants of opposite chirality form spirals of opposite helicity. Obviously, the backbone helicity (M/P) is determined by the pendant chirality (D/L) under the same environmental conditions. When solvent of the polymer solution is changed from chloroform to THF, the Cotton effect is greatly decreased, indicating that the helical preference of the chain segment can be tuned by an environmental variation. The helical chain segments are believed to be mainly stabilized by the intra- and interstrand hydrogen bonds between the chiral pendants (Figure 1.12b,c). The noncovalent stabilization can be broken by external perturbation and the system will reach a new equilibrium. This dynamic process enables the OAPAs to cope with the variations in their surrounding environments, as do helical biopolymers such as proteins.

Further examples of the helicity tuning by external stimuli are given in Figure 1.13 [68]. The ellipticities ([θ]) of **23e** are positive and negative in THF and $CHCl_3$, respectively; that is, the majority of its chain segments form coils of opposite helicity

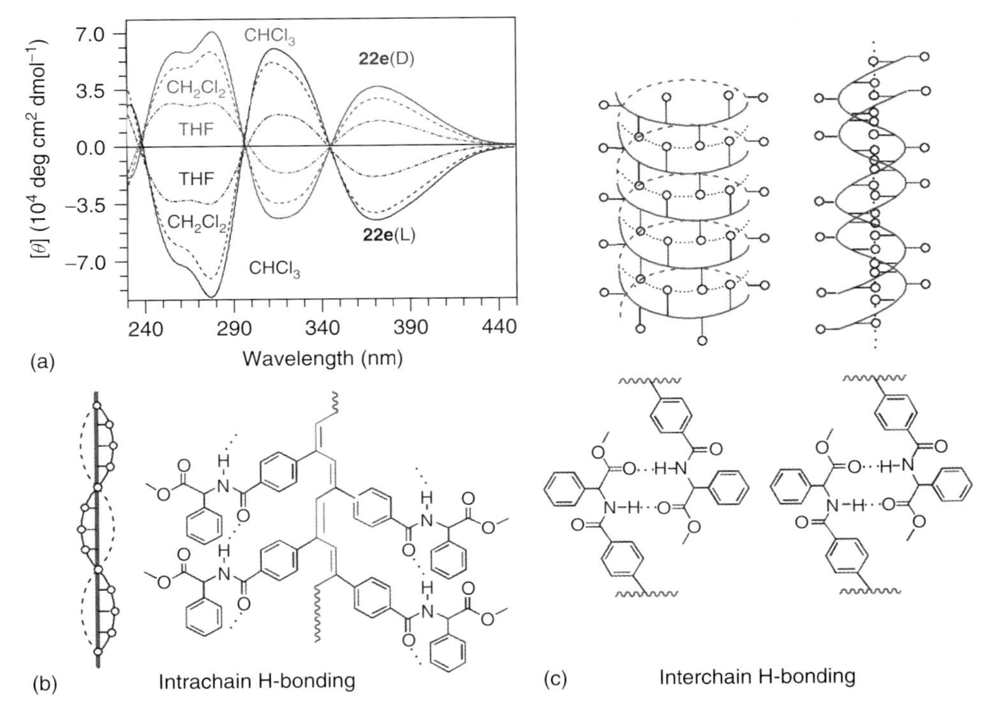

Figure 1.12 (a) Chain helicity determined by pendant chirality and manipulated by solvent variation. (b) Single and (c) double helical strands of **22e** stabilized by intra- and interchain hydrogen bonds.

in these two solvents. In other words, the helical preference can be completely reversed by simply changing the solvent. The helicity change is continuous and reversible, as revealed by the $[\theta]$–solvent plot in Figure 1.13a. Ellipticity changes rapidly when the amount of chloroform is increased from ~65 to ~80%. This is indicative of a cooperative process: once some segments are associated via noncovalent interactions (e.g., hydrogen bond), their neighboring segments will be zipped up quickly along the preferred direction. The ellipticity of **23e** monotonically decreases with increasing temperature because the thermally activated chain randomization induces the segments to unwind.

Addition of KOH to a solution of **23a** progressively weakens its $[\theta]$, because the ionic interaction of K^+ with CO_2^- breaks the intra- and interchain hydrogen bonds (Figure 1.13c) [64]. The original $[\theta]$ can be regained when the KOH solution is neutralized by HCl; this illustrates the reversibility of helicity tuning with variation in the pH. Intriguingly, $[\theta]$ can also be tuned by achiral additives. With continuous addition of glycine (an achiral essential amino acid) to the solution of **23a**, its $[\theta]$ also continuously increases. The glycine molecules may bind to the L-valine pendants, causing an increase in the pendant bulkiness and thereby inducing further twists in the chain segments.

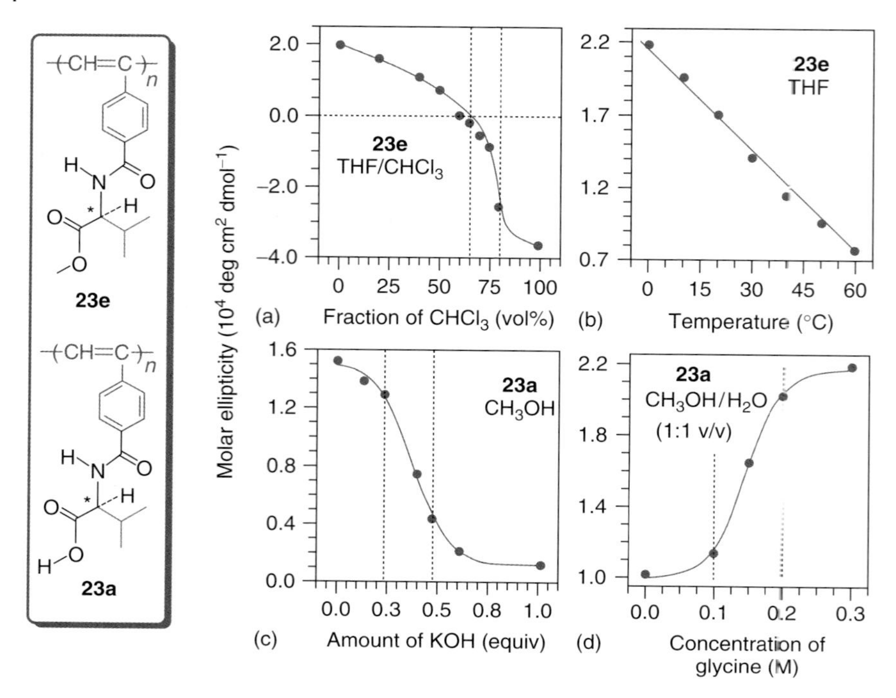

Figure 1.13 Tuning ellipticity of **23e** (a,b) and **23a** (c,d) at ~375 nm by (a) solvent, (b) temperature, (c) pH, and (d) additive.

1.3.8
Supramolecular Assembly

After studying the chain helicity as associated with the secondary structure of the OAPAs, we investigated their higher-order structures. Noticing the unique amphiphilicity of the OAPAs originating from their hydrophobic backbones and hydrophilic pendants, we explored the possibility of utilizing them as "Legos" to construct biomimetic hierarchical structures.

The OAPAs can self-organize into a variety of morphologies reminiscent of natural structural motifs such as vesicles, tubules, helixes, and honeycombs. Thus, evaporation of a solution of **30e** (an OAPA with uridine pendants) yields vesicles, whilst that of **24a** (L-isoleucine pendants) gives tubules with coexisting vesicles (Figure 1.14) [62]. In the polar solvents, the OAPAs cluster into bi- or multilayered spherical aggregates or vesicles with hydrophobic cores and hydrophilic coronas. Under appropriate conditions, the vesicles further associate into the tubular structures.

The tunability of chain helicity of the OAPAs through internal and external perturbations suggests that their self-assembling morphologies can be manipulated by changing their molecular structures and environmental conditions, because the

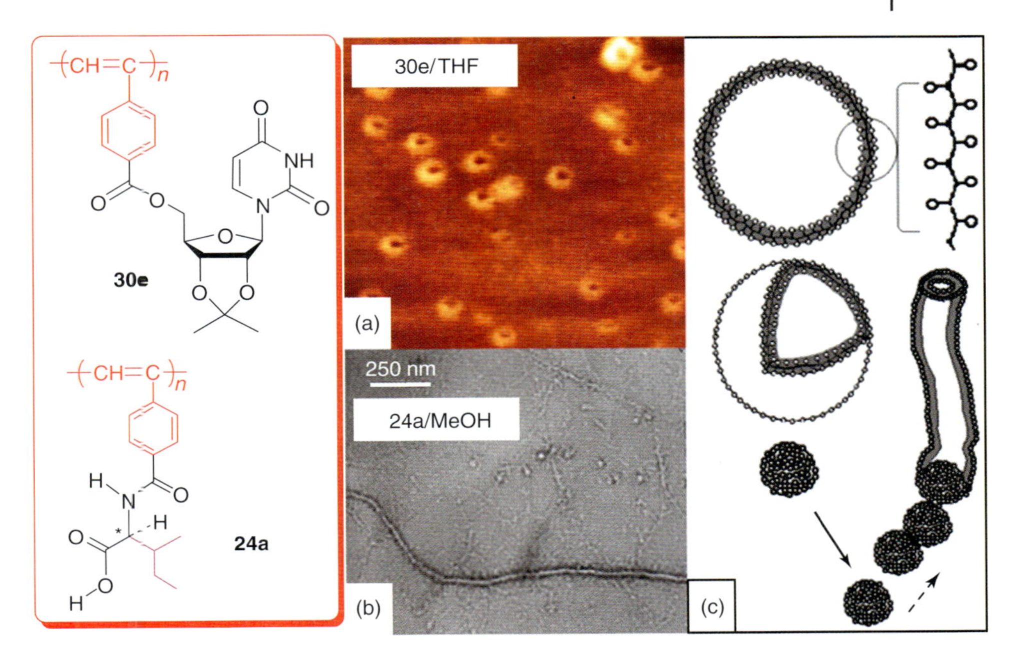

Figure 1.14 (a) Vesicles and (b) nanotubes formed by natural evaporation of PA solutions (10–19 μM), and (c) proposed processes for the formation of the nanostructures.

variations in the secondary structure should affect the higher-order structures. This proves to be the case. Evaporation of a methanol solution of **23e** gives a string of pearls (Figure 1.15a). Helical ropes are obtained when its structure (internal) and solvent (external) are changed to **23a** and THF, respectively (Figure 1.15b,c). Addition of KOH to the methanol solution of **23a** changes the helical cable to random thread (Figure 1.15d). The association of K^+ ions with CO_2- groups breaks the hydrogen bonds. The charged polyelectrolyte chains repulse each other and are difficult to associate into multistrand helical cables, hence the observed thin random coils [64].

Evaporation of more concentrated solutions gives structures of bigger sizes: for example, a large honeycomb pattern is obtained from a thick solution of **25a** (Figure 1.16a) [74]. Precipitation of a concentrated solution of **26e** into its poor solvents yields micron-size helical fibers with both left- and right-handed twists (Figure 1.16b) [75].

1.3.9
Optical Nonlinearity

Conjugated organic materials exhibiting strong NLO properties and fast response time have attracted considerable interest because of their potential high-tech

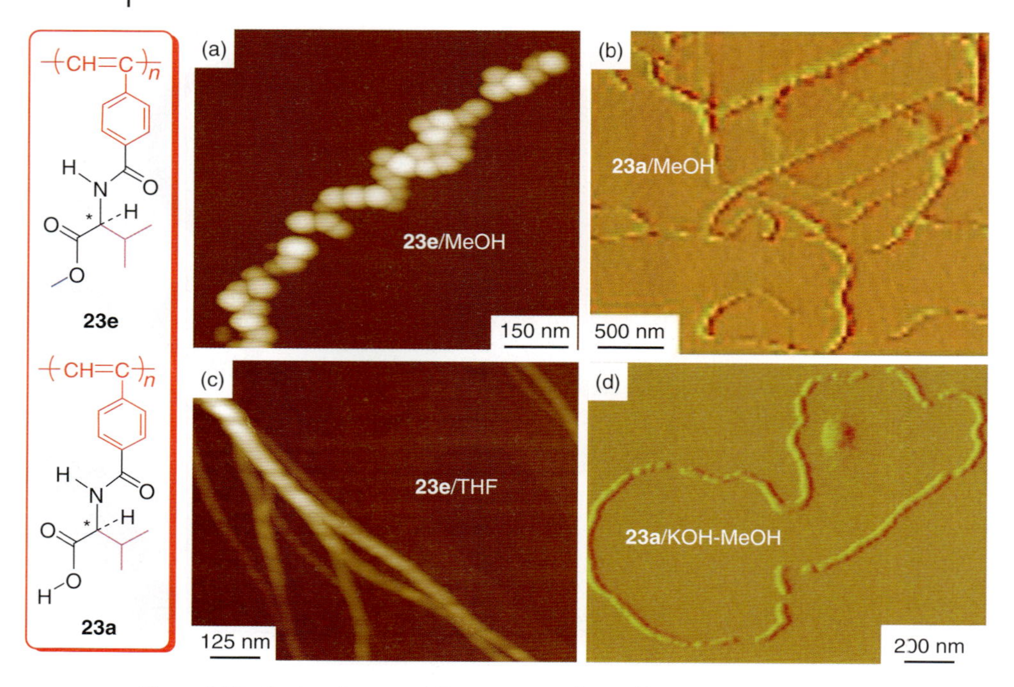

Figure 1.15 Tuning the assembling structures formed by natural evaporation of the solutions of **23(a,c)** and **23(b,d)** (~ 11.6–12.2 μM) through internal and external perturbations.

applications in various optical and photonic devices. Of particular interest is the optical limiter, which is a novel optical material that transmits light of normal intensity but attenuates light of high power. In other words, it allows mild light to pass through but prevents harsh light from being transmitted. The rapid advancements in the laser-based technologies and the growing enthusiasm in the Space exploration have motivated intensive research efforts in the development of new optical limiting materials with novel structures and improved performances.

Substituted PAs with various functional groups have been found to enjoy excellent optical limiting and NLO properties. Many research groups have studied their third-order NLO properties derived from their π-conjugated nature. Azobenzene is a well-known second-order NLO-active chromophore. Incorporating such a chromophore into PAs might offer materials with novel optical properties. We synthesized a series of PAs with NLO properties [76–80]. Figure 1.17 shows the optical limiting properties of the polymers **70** and **71** with azobenzene as the pendant at the same linear transmittance ($T = 75\%$) [76]. All the polymers are good optical limiters, especially for those with short spacer lengths. For example, the limiting threshold and limiting amplitude of **70**(3) are 0.245 and 0.349 J cm^{-2}, respectively, which are 1.3 and 1.2 times lower than those of **70**(8). The optical limiting properties are also affected by the ring substituent: the polymer

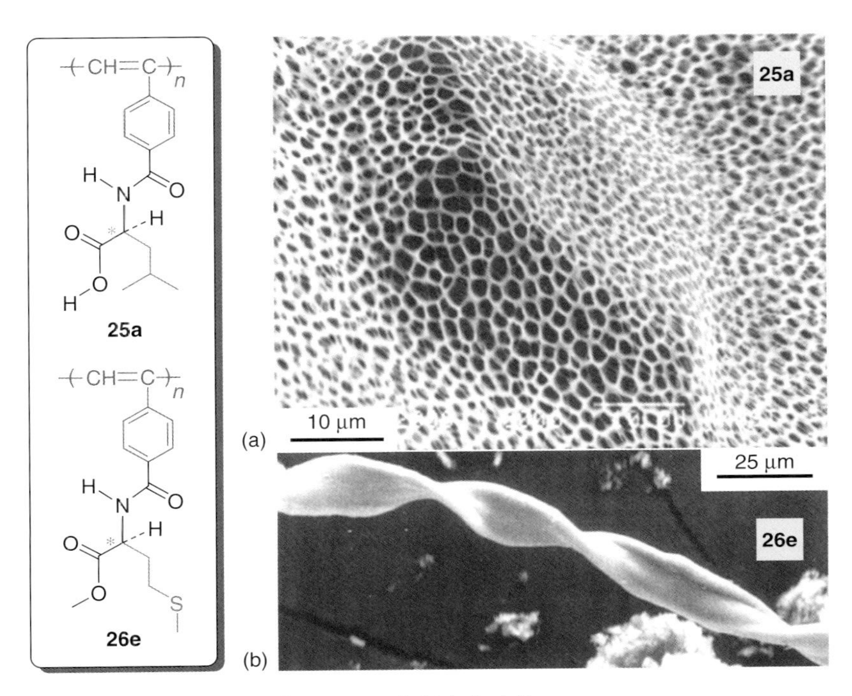

Figure 1.16 (a) Honeycomb pattern and (b) helical fiber formed by (a) evaporation of **25a**/MeOH solution (29 μM) on glass slide and (b) precipitation of **26e**/DMF solution (100 mM) into acetone/ether mixture.

containing bromo substituent shows inferior performance, as understood from the comparison of the limiting threshold (0.295 J cm^{-2}) and saturation fluence (0.393 J cm^{-2}) of **71**(3) with those of **70**(3).

Fullerenes are known to limit intense optical pulses via a reverse saturable absorption mechanism. The fullerene tips and the graphene sheets of the CNTs also undergo NLO absorption processes, while their cylindrical bodies, although with high aspect ratios, are light-scattering centers. Both the NLO absorption and the light scattering make the CNTs promising for optical limiting applications. C_{60} and CNTs, however, have notoriously poor solubility and processability. Hybridization of C_{60} and CNTs with processable π-conjugated polymers is a promising way to enhance the solubility of the carbon allotropes, with the possibility of generating new materials with novel properties. Our group has done a pioneering work in wrapping CNTs with conjugated polymers. In 1999, our group successfully prepared PPA/CNT nanohybrid by *in situ* polymerization technique [81], and thereafter established a platform to prepare a series of PA/CNT and PA/C_{60} nanohybrids by physical and chemical methods as shown in Scheme 1.8 [82, 83].

The optical limiting performances of PPA/CNT and **72**/C_{60} nanohybrids are shown in Figure 1.18 as examples. The data for the parent forms, that is, PPA and C_{60}, are given in the same figure for comparison [81, 83]. When a THF

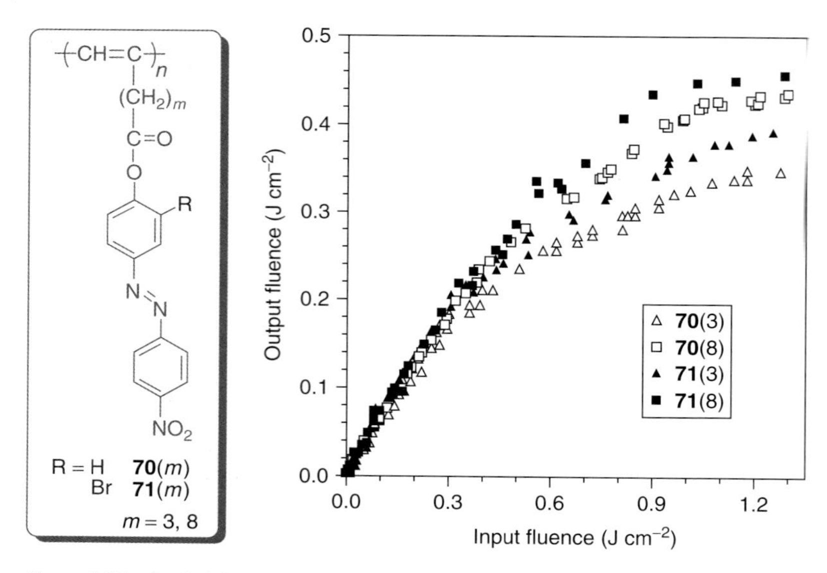

Figure 1.17 Optical limiting responses to 8-ns, 532-nm laser light, of THF solutions of **70**(*m*) and **71**(*m*) with a linear transmission of 75%.

solution of PPA is shot by the 8-ns, 532-nm laser light, the transmitted fluence linearly increases in the region of low incident fluence (linear transmittance 75%). The output starts and continues to deviate from the linear-transmission line from the input by about 1.7 J cm^{-2}, implying that the intense illumination gradually bleaches the PPA to transparency, probably by the laser-induced photolysis of the main chains. The PPA/CNT solutions, however, respond to the optical pulses in a strikingly different way. The linear transmittance of a dilute PPA/CNT solution (0.4 mg ml^{-1}) is only 57% (Figure 1.18a), although its concentration is only one-tenth of that (4 mg ml^{-1}) of PPA, probably because of the optical losses caused by the nanotube absorption and scattering. As the incident fluence increases, the PPA/CNT solution becomes opaque, instead of transparent, with its transmitted fluence eventually leveling off or saturating at 1.85 J cm^{-2} (saturation fluence). Clearly, the CNTs have endowed the PPA/CNT with optical limiting power. While PPA is liable to photolysis, the PPA/CNT is stable at very high incident fluence. The energy-sinking and radical-trapping functions of aromatic rings often protect polymers from photodegradation, and the extensively conjugated graphitic aromatic system of the CNTs may have enhanced the resistance of the main chains of PPA against the harsh laser irradiation. As the concentration of the PPA/CNT solution increases, its saturation fluence decreases. Increasing the concentration to 0.6 mg ml^{-1} readily decreases the saturation fluence to as low as 0.45 J cm^{-2}. Similarly, the **72**/C_{60} also effectively limits the strong laser pulses. Compared to the solution of the parent C_{60}, the **72**/C_{60} solutions show higher or same linear transmittance but lower saturation fluence [83].

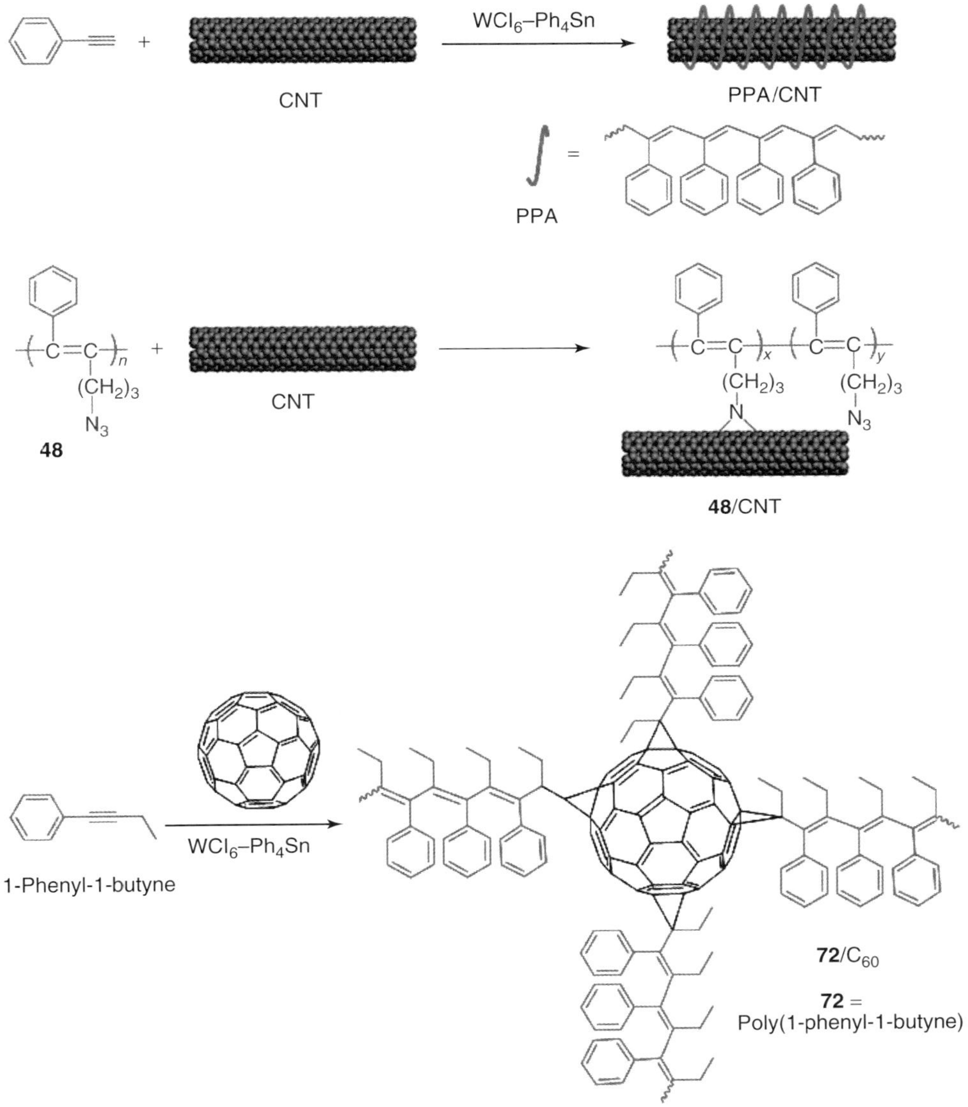

Scheme 1.8 Nanohybridizations of CNT and C_{60} with polyacetylenes.

Much effort has been devoted to the development of second-order NLO materials, particularly to the efficient translation of large molecular first hyperpolarizability to high second harmonic generation (SHG) coefficient (d_{33}). The greatest obstacle in the area has been the chromophoric aggregation in the thin solid films. The NLO dyes are usually highly polarized by the D–A push–pull interactions. During the film formation, the chromophores with large dipole moments tend to compactly

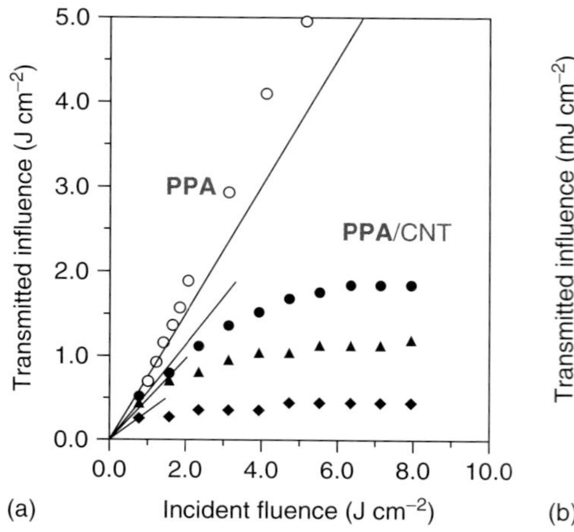

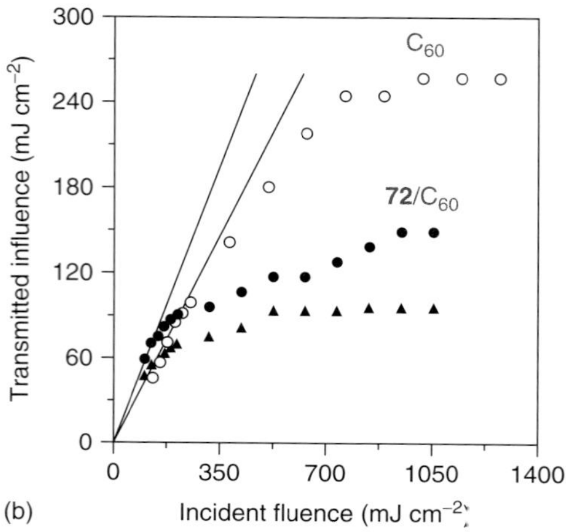

Figure 1.18 (a) Optical limiting responses to 8-ns, 532-nm laser light, of THF solutions of the PPA/CNT hybrid shown in Scheme 1.8. Concentration (c; mg ml^{-1})/linear transmittance (T; %): 0.4/57 (●), 0.5/48 (▲), 0.6/34 (♦). The optical limiting responses of a THF solution of PPA are shown for comparison (c (mg ml^{-1})/T (%): 4.0/75 (○)). (b) Optical limiting responses of a toluene solution of C_{60} (○) and THF solutions of **72**/C_{60} (● and ▲); c (mg ml^{-1})/T (%): 0.16/43 (○), 1.50/56 (●), 3.00/43 (▲).

pack owing to the strong intermolecular electrostatic interactions, leading to the diminishment or cancellation of the NLO effects in the solid state. Researchers have been working on the development of different strategies to overcome this thorny aggregation problem.

Attachment of azobenzene pedants to part of the chain segments of a disubstituted PA may help alleviate the problem because, in addition to the twisted polyene backbone, the chain segments without the azobenzene pendants may serve as isolation buffers to hamper the aggregate formation of the NLO chromophores. It is, however, difficult to synthesize disubstituted PAs with polar azobenzene pendants because of the lack of efficient catalyst systems. We have taken a polymer-reaction approach to prepare the polymers inaccessible by the direct polymerizations. For example, utilizing click reaction, we have successfully synthesized disubstituted PAs with varying contents of functional azobenzene groups, examples of which are shown in Scheme 1.9. As expected, the resultant PAs (**75**) exhibited high d_{33} values (up to ~130.5 pm V^{-1}) [19].

1.3.10
Biological Compatibility

Decorating the PA backbones with the pendants of naturally occurring building blocks may impart biocompatibility to the π-conjugated polymers. The wrapping of the unnatural polymer chains by the natural molecular coats may result in the "naturalization" of the synthetic polymers, thereby generating cytophilic molecular

		d_{33} (pm V^{-1})
75a	R′ = NO_2, R″ = H	72.4
75b	R′ = NO_2, R″ = Ph	130.5

Scheme 1.9 NLO-active polyacetylenes.

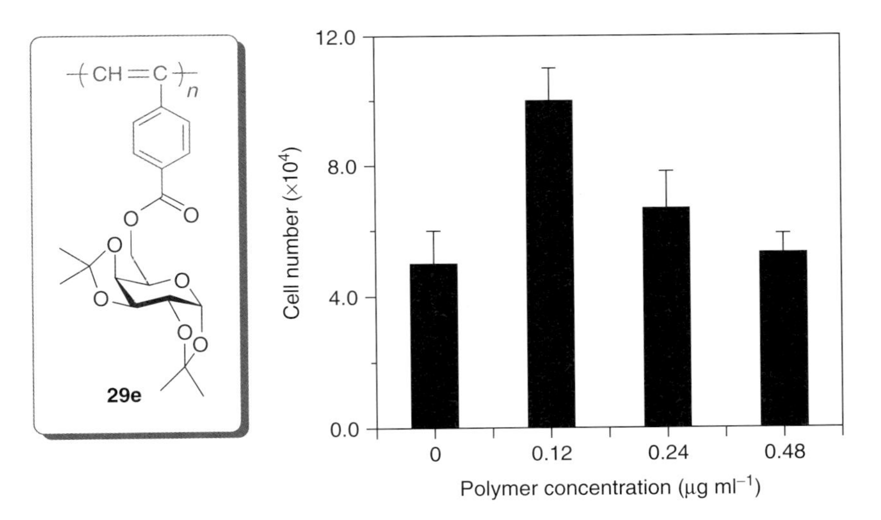

Figure 1.19 Growth of living HeLa cells in the incubation media containing **29e** with different concentrations.

wires. We have systematically investigated the cytotoxicity of a series of PPA derivatives bearing pendant groups of natural origin (amino acids, sugars, nucleosides, etc.) and found them all biocompatible. The bioactivity data for a sugar-containing PPA derivative (**29e**) are plotted in Figure 1.19 [84]. The HeLa cells were subcultured onto the microtiter plates, into which a THF solution of **29e** was added a few hours after the cells had been seeded. The cells were stained with trypan blue and counted with a hemocytometer. As can be seen from Figure 1.19, in the presence of a

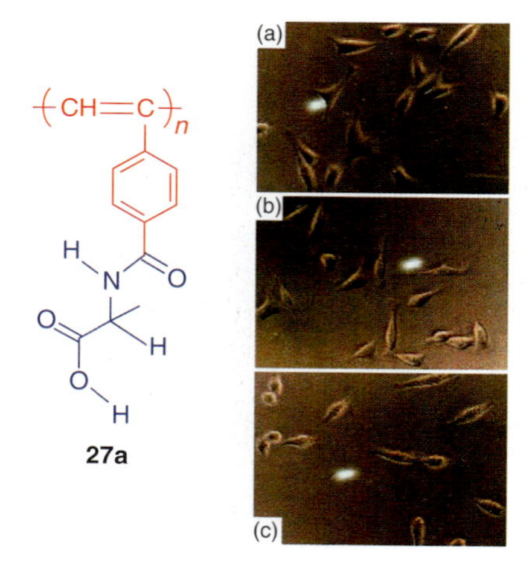

Figure 1.20 HeLa cell adhesion to the microtiter plates precoated with different amounts of **27a** (μg cm^{-2}): (a) 0 (control), (b) 15.8, and (c) 22.2. Incubation time: one day.

small amount of polymer **29e**, the cells grow faster. At a polymer concentration of 0.12 μg ml^{-1}, the cell population is 20-fold higher than that of the control, in which the polymer concentration is zero, although the cell growth returns to normal when the polymer concentration is increased. This indicates that the polymer is cytophilic and can stimulate the growth of the living cells at a low feed.

Figure 1.20 shows the biocompatibility data of a PPA derivative decorated by L-alanine pendants (**27a**) [69]. The HeLa cells were subcultured onto the microtiter plates precoated with polymer **27a**, and the adhesion and growth of the living cells were observed with an optical microscope. After incubation for one day, the cells were found to adhere to, and grow on, the plates as they did in the control experiment, in which the plates without the polymer coatings were used. The polymer exhibited no toxicity to the cells; in other words, it was biocompatible. Even when the polymer coating density was increased to as high as ~22 mg cm^{-2}, no dead cells were found throughout the experiment, demonstrative of excellent cytocompatibility of the polymer. Similar results were obtained for the PPA derivatives carrying other amino acid pendants, such as **25a** and **76** (Figure 1.21). This proves that the incorporation of naturally occurring building blocks into the PA structure is a versatile strategy for conferring biocompatibility on the conjugated polymers.

1.4
Conclusions and Prospects

In this work, a large variety of functional moieties are integrated into the molecular structure of PA by the direct polymerizations of functional monomers and the

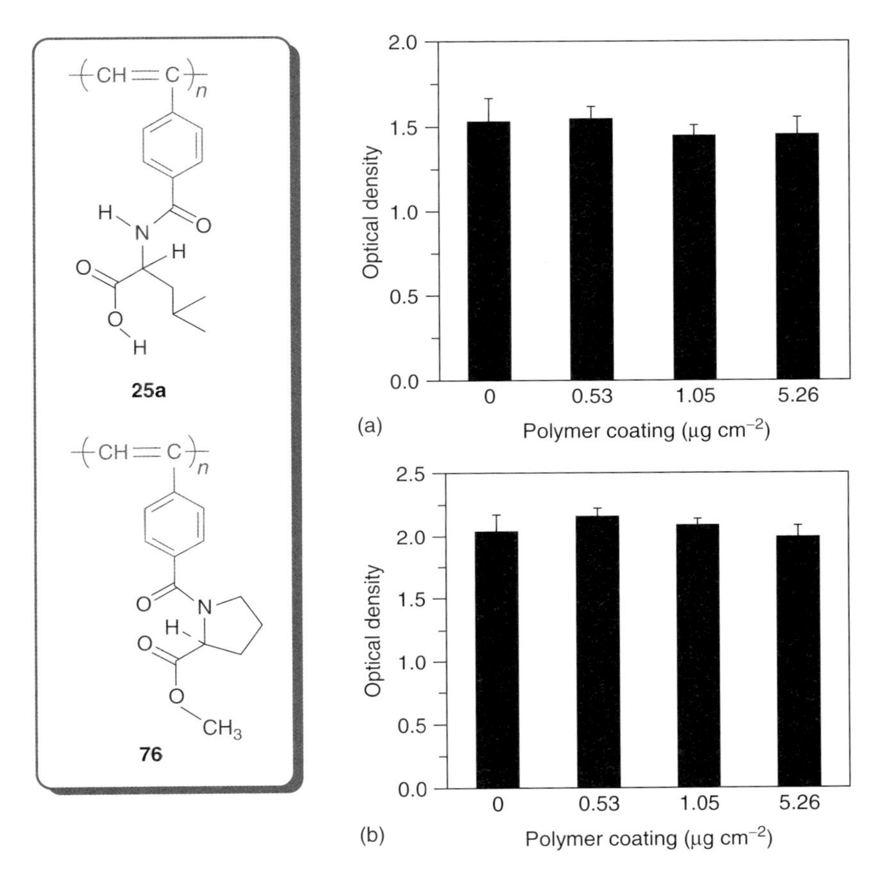

Figure 1.21 Growth of HeLa cells on the microtiter plates precoated with (a) **25a** and (b) **76** after three-day incubation.

polymer reactions of preformed polymers. The functionalization routes explored and the structural insights gained in this study offer versatile synthetic tools and valuable designer guidelines for further developments in the area. Our success in the polymer synthesis proves that the conjugated polyene backbone can be used as a new structural scaffold for the construction of novel functional macromolecules.

The conjugated backbones and functional pendants as well as their synergistic interplays endow PA with an array of unique functionalities. The mesogenic pendants confer liquid crystallinity on PA, while the polyene backbones allow the LCPA chains to be aligned by simple shearing. The energy transfers between the chromophoric pendants and the conjugated skeletons enable the PL colors and efficiencies of the LEPAs to be tuned by molecular engineering. Some LEPAs enjoy an invaluable AIEE nature and are thus promising candidates for PLED applications. The bulky pendants sheath the conjugated skeletons, which weakens chain interactions and boosts spectral stabilities. The chiral pendants induce the polyene backbones to spiral, with the chain helicity being determined by the pendant chirality. The conformational sensitivity of polyene backbone to environmental surrounding, on the other hand, permits the preferred handedness of their chain segments to be modulated by external perturbations continuously and reversibly. The OAPA chains self-assemble into biomimetic hierarchical structures, thanks to the amphiphilicity stemming from their hydrophobic backbones and hydrophilic pendants. Some substituted PAs with various polarized groups and nanohybrids of PAs with CNT and C_{60} are excellent optical limiting materials. Incorporation of naturally occurring building blocks into the PA structure is a versatile strategy for conferring biocompatibility on the conjugated polymers.

Many of the functionalized PAs are robust and processable, in addition to their simple syntheses and novel functional properties. They may find a wide range of applications as specialty materials in, for example, information storage, photovoltaic cell, chemical sensing, optical display, photoresist pattern, luminescent imaging, optical filtration, nonlinear optics, chiral separation, light polarization, biomimetic morphosynthesis, controlled drug delivery, and tissue engineering. Efforts to exploit the technological applications of the functionalized PAs are currently underway in our laboratories.

Acknowledgments

This work was partly supported by the Research Grants Council of Hong Kong (603008, 602707, 602706, 601608), the National Basic Research Program of the Ministry of Science & Technology of China (2002CB613401), and the National Science Foundation of China (20634020). B.Z.T. thanks the support from the Cao Guangbiao Foundation of Zhejiang University.

References

1. Lam, J.W.Y. and Tang, B.Z. (2005) *Acc. Chem. Res.*, **38**, 745.
2. Lam, J.W.Y. and Tang, B.Z. (2003) *J. Polym. Sci., Part A: Polym. Chem.*, **41**, 2607.
3. Masuda, T. (2007) *J. Polym. Sci., Part A: Polym. Chem.*, **45**, 165.
4. Simionescu, C.I. and Percec, V. (1982) *Prog. Polym. Sci.*, **8**, 133.
5. Choi, S.K., Gal, Y.S., Jin, S.H., and Kim, H.K. (2000) *Chem. Rev.*, **100**, 1645.
6. Mayershofer, M.G. and Nuyken, O. (2005) *J. Polym. Sci., Part A: Polym. Chem.*, **43**, 5723.
7. Schrock, R.R., Luo, S.F., Lee, J.C., Zanetti, N.C., and Davis, W.M. (1996) *J. Am. Chem. Soc.*, **118**, 3883.
8. Moore, J.S., Gorman, C.B., and Grubbs, R.H. (1991) *J. Am. Chem. Soc.*, **113**, 1704.
9. Yashima, E., Maeda, K., and Furusho, Y. (2008) *Acc. Chem. Res.*, **41**, 1166.
10. Simionescu, C.I. and Percec, V. (1980) *J. Polym. Sci., Part C: Polym. Symp.*, **67**, 43.
11. Masuda, T. and Higashimura, T. (1984) *Acc. Chem. Res.*, **17**, 51.
12. Furlani, A., Licoccia, S., and Russo, M.V. (1986) *J. Polym. Sci., Part A: Polym. Chem.*, **24**, 991.
13. Yang, W., Tabata, M., Yokota, K., and Shimizu, A. (1991) *Polym. J.*, **23**, 1135.
14. Kishimoto, Y., Eckerle, P., Miyatake, T., Kainosho, M., Ono, A., Ikariya, T., and

Noyori, R. (1999) *J. Am. Chem. Soc.*, **121**, 12035.
15. Tang, B.Z., Poon, W.H., Leung, S.M., Leung, W.H., and Peng, H. (1997) *Macromolecules*, **30**, 2209.
16. Xu, K.T., Peng, H., Lam, J.W.Y., Poon, T.W.H., Dong, Y.P., Xu, H.Y., Sun, Q.H., Cheuk, K.K.L., Salhi, F., Lee, P.P.S., and Tang, B.Z. (2000) *Macromolecules*, **33**, 6918.
17. Law, C.W., Lam, J.W.Y., Dong, Y.P., and Tang, B.Z. (2004) *Polym. Prepr.*, **45**, 839.
18. Yuan, W.Z., Tang, L., Zhao, H., Jin, J.K., Sun, J.Z., Qin, A.J., Xu, H.P., Liu, J.H., Yang, F., Zheng, Q., Chen, E.Q., and Tang, B.Z. (2009) *Macromolecules*, **42**, 52.
19. Zeng, Q., Li, Z., Li, Z., Ye, C., Qin, J., and Tang, B.Z. (2007) *Macromolecules*, **40**, 5634.
20. Zeng, Q., Cai, P., Li, Z., Qin, J.G., and Tang, B.Z. (2008) *Chem. Commun.*, 1094.
21. Tong, H., Lam, J.W.Y., Haussler, M., and Tang, B.Z. (2004) *Polym. Prepr.*, **45**, 835.
22. Xu, H.P., Sun, J.Z., Qin, A.J., Hua, J.L., Li, Z., Dong, Y.Q., Xu, H., Yuan, W.Z., Ma, Y.G., Wang, M., and Tang, B.Z. (2006) *J. Phys. Chem. B*, **110**, 21701.
23. Hua, J.L., Li, Z., Lam, J.W.Y., Xu, H.P., Sun, J.Z., Dong, Y.P., Dong, Y.Q., Qin, A.J., Yuan, W.Z., Chen, H.Z., Wang, M., and Tang, B.Z. (2005) *Macromolecules*, **38**, 8127.
24. Dong, Y.P., Lam, J.W.Y., Peng, H., Cheuk, K.K.L., Kwok, H.S., and Tang, B.Z. (2004) *Macromolecules*, **37**, 6408.
25. Tang, B.Z., Chen, H.Z., Xu, R.S., Lam, J.W.Y., Cheuk, K.K.L., Wong, H.N.C., and Wang, M. (2000) *Chem. Mater.*, **12**, 213.
26. Kang, E.T., Ehrlich, P., Bhatt, A., and Anderson, W. (1984) *Macromolecules*, **17**, 1020.
27. Zhao, H., Yuan, W.Z., Tang, L., Sun, J.Z., Xu, H.P., Qin, A.J., Mao, Y., Jin, J.K., and Tang, B.Z. (2008) *Macromolecules*, **41**, 8566.
28. Yuan, W.Z., Mao, Y., Zhao, H., Sun, J.Z., Xu, H.P., Jin, J.K., Zheng, Q., and Tang, B.Z. (2008) *Macromolecules*, **41**, 701.
29. Yuan, W.Z., Sun, J.Z., Liu, J.Z., Dong, Y.Q., Li, Z., Xu, H.P., Qin, A.J., Haeussler, M., Jin, J.K., Zheng, Q., and Tang, B.Z. (2008) *J. Phys. Chem. B*, **112**, 8896.
30. Yuan, W.Z., Sun, J.Z., Dong, Y.Q., Haussler, M., Yang, F., Xu, H.P., Qin, A.J., Lam, J.W.Y., Zheng, Q., and Tang, B.Z. (2006) *Macromolecules*, **39**, 8011.
31. Xu, H.P., Xie, B.Y., Yuan, W.Z., Sun, J.Z., Yang, F., Dong, Y.Q., Qin, A., Zhang, S., Wang, M., and Tang, B.Z. (2007) *Chem. Commun.*, 1322.
32. Ye, C., Xu, G.Q., Yu, Z.Q., Lam, J.W.Y., Jang, J.H., Peng, H.L., Tu, Y.F., Liu, Z.F., Jeong, K.U., Cheng, S.Z.D., Chen, E.Q., and Tang, B.Z. (2005) *J. Am. Chem. Soc.*, **127**, 7668.
33. Lam, J.W.Y., Dong, Y., Cheuk, K.K.L., Luo, J., Xie, Z., Kwok, H.S., Mo, Z., and Tang, B.Z. (2002) *Macromolecules*, **35**, 1229.
34. Lam, J.W.Y., Luo, J., Dong, Y., Cheuk, K.K.L., and Tang, B.Z. (2002) *Macromolecules*, **35**, 8288.
35. Lam, J.W.Y., Kong, X., Dong, Y., Cheuk, K.K.L., Xu, K., and Tang, B.Z. (2000) *Macromolecules*, **33**, 5027.
36. Kong, X., Lam, J.W.Y., and Tang, B.Z. (1999) *Macromolecules*, **32**, 1722.
37. Kong, X. and Tang, B.Z. (1998) *Chem. Mater.*, **10**, 3352.
38. Tang, B.Z., Kong, X., Wan, X., Peng, H., Lam, W.Y., Feng, X.-D., and Kwok, H.S. (1998) *Macromolecules*, **31**, 2419.
39. Huang, Y.M., Lam, J.W.Y., Cheuk, K.K.L., Ge, W.K., and Tang, B.Z. (1999) *Macromolecules*, **32**, 5976.
40. Lam, J.W.Y., Dong, Y., Kwok, H.S., and Tang, B.Z. (2006) *Macromolecules*, **39**, 6997.
41. Chen, J., Xie, Z., Lam, J.W.Y., Law, C.C.W., and Tang, B.Z. (2003) *Macromolecules*, **36**, 1108.
42. Chen, J., Kwok, H.S., and Tang, B.Z. (2006) *J. Polym. Sci., Part A: Polym. Chem.*, **44**, 2487.
43. Luo, J.D., Xie, Z.L., Lam, J.W.Y., Cheng, L., Chen, H.Y., Qiu, C.F., Kwok, H.S., Zhan, X.W., Liu, Y.Q., Zhu, D.B., and Tang, B.Z. (2001) *Chem. Commun.*, 1740.
44. Chen, J., Law, C.C.W., Lam, J.W.Y., Dong, Y., Lo, S.M.F., Williams, I.D.,

Zhu, D., and Tang, B.Z. (2003) *Chem. Mater.*, **15**, 1535.

45. Yu, G., Yin, S.W., Liu, Y.Q., Chen, J.S., Xu, X.J., Sun, X.B., Ma, D.G., Zhan, X.W., Peng, Q., Shuai, Z.G., Tang, B.Z., Zhu, D.B., Fang, W.H., and Luo, Y. (2005) *J. Am. Chem. Soc.*, **127**, 6335.
46. Peng, Q., Yi, Y., Shuai, Z., and Shao, J. (2007) *J. Am. Chem. Soc.*, **129**, 9333.
47. Peng, Q., Yi, Y., Shuai, Z., and Shao, J. (2007) *J. Chem. Phys.*, **126**, 114302.
48. Chen, H.Y., Lam, W.Y., Luo, J.D., Ho, Y.L., Tang, B.Z., Zhu, D.B., Wong, M., and Kwok, H.S. (2002) *Appl. Phys. Lett.*, **81**, 574.
49. Tong, H., Hong, Y.N., Dong, Y.Q., Haussler, M., Lam, J.W.Y., Li, Z., Guo, Z.F., Guo, Z.H., and Tang, B.Z. (2006) *Chem. Commun.*, 3705.
50. Hong, Y.N., Haussler, M., Lam, J.W.Y., Li, Z., Sin, K.K., Dong, Y.Q., Tong, H., Liu, J.Z., Qin, A.J., Renneberg, R., and Tang, B.Z. (2008) *Chem. Eur. J.*, **14**, 6428.
51. Cheng, K.H., Zhong, Y.C., Xie, B.Y., Dong, Y.Q., Hong, Y.N., Sun, J.Z., Tang, B.Z., and Wong, K.S. (2008) *J. Phys. Chem. C*, **112**, 17507.
52. Tong, H., Hong, Y.N., Dong, Y.Q., Haussler, M., Li, Z., Lam, J.W.Y., Dong, Y.P., Sung, H.H.Y., Williams, I.D., and Tang, B.Z. (2007) *J. Phys. Chem. B*, **111**, 11817.
53. Dong, Y.Q., Lam, J.W.Y., Qin, A.J., Liu, J.Z., Li, Z., and Tang, B.Z. (2007) *Appl. Phys. Lett.*, **91**, 011111.
54. Qin, A.J., Jim, C.K.W., Tang, Y.H., Lam, J.W.Y., Liu, J.Z., Mahtab, F., Gao, P., and Tang, B.Z. (2008) *J. Phys. Chem. B*, **112**, 9281.
55. Yuan, W.Z., Qin, A.J., Lam, J.W.Y., Sun, J.Z., Dong, Y.Q., Haussler, M., Liu, J.Z., Xu, H.P., Zheng, Q., and Tang, B.Z. (2007) *Macromolecules*, **40**, 3159.
56. Hirayama, F. (1965) *J. Chem. Phys.*, **42**, 3163.
57. Yanari, S.S., Bovey, F.A., and Lumry, R. (1963) *Nature*, **200**, 242.
58. Lam, J.W.Y., Qin, A., Dong, Y., Hong, Y., Jim, C.K.W., Liu, J., Dong, Y., Kwok, H.S., and Tang, B.Z. (2008) *J. Phys. Chem. B*, **112**, 11227.
59. Lam, J.W.Y., Peng, H., Haussler, M., Zheng, R., and Tang, B.Z. (2004) *Mol. Cryst. Liq. Cryst.*, **415**, 43.
60. Tang, B.Z., Poon, W.H., Peng, H., Wong, H.N.C., Ye, X.S., and Monde, T. (1999) *Chin. J. Polym. Sci.*, **17**, 81.
61. Tang, B.Z. and Kotera, N. (1989) *Macromolecules*, **22**, 4388.
62. Li, B.S., Chen, J., Zhu, C.F., Leung, K.K.L., Wan, L., Bai, C., and Tang, B.Z. (2004) *Langmuir*, **20**, 2515.
63. Li, B.S., Cheuk, K.K.L., Ling, L., Chen, J., Xiao, X., Bai, C., and Tang, B.Z. (2003) *Macromolecules*, **36**, 77.
64. Li, B.S., Cheuk, K.K.L., Salhi, F., Lam, J.W.Y., Cha, J.A.K., Xiao, X., Bai, C., and Tang, B.Z. (2001) *Nano Lett.*, **1** 323.
65. Li, B.S., Cheuk, K.K.L., Yang, D., Lam, J.W.Y., Wan, L.J., Bai, C., and Tang, B.Z. (2003) *Macromolecules*, **36**, 5447.
66. Li, B.S., Kang, S.Z., Cheuk, K.K.L., Wan, L., Ling, L., Bai, C., and Tang, B.Z. (2004) *Langmuir*, **20**, 7598.
67. Cheuk, K.K.L., Lam, J.W.Y., Chen, J., Lai, L.M., and Tang, B.Z. (2003) *Macromolecules*, **36**, 5947.
68. Cheuk, K.K.L., Lam, J.W.Y., Lai, L.M., Dong, Y., and Tang, B.Z. (2003) *Macromolecules*, **36**, 9752.
69. Cheuk, K.K.L., Li, B.S., Lam, J.W.Y., Xie, Y., and Tang, B.Z. (2008) *Macromolecules*, **41**, 5997.
70. Lai, L.M., Lam, J.W.Y., Cheuk K.K.L., Sung, H.H.Y., Williams, I.D., and Tang, B.Z. (2005) *J. Polym. Sci, Part A: Polym. Chem.*, **43**, 3701.
71. Lai, L.M., Lam, J.W.Y., Qin, A., Dong, Y., and Tang, B.Z. (2006) *J. Phys. Chem. B*, **110**, 11128.
72. Lai, L.M., Lam, J.W.Y., and Tang, B.Z. (2006) *J. Polym. Sci., Part A: Polym. Chem.*, **44**, 6190.
73. Lai, L.M., Lam, J.W.Y., and Tang, B.Z. (2006) *J. Polym. Sci., Part A: Polym. Chem.*, **44**, 2117.
74. Salhi, F., Cheuk, K.K.L., Sun, Q., Lam, J.W.Y., Cha, J.A.K., Li, G., Li, B., Luo, J., Chen, J., and Tang, B.Z. (2001) *J. Nanosci. Nanotechnol.*, **1**, 137.
75. Cheuk, K.K.L., Li, B.S., and Tang, B.Z. (2002) *Curr. Trends Polym. Sci.*, **7**, 41.
76. Yin, S., Xu, H., Su, X., Gao, Y., Song, Y., Lam, J.W.Y., Tang, B.Z., and Shi, W. (2005) *Polymer*, **46**, 10592.

77. Yin, S.C., Xu, H.Y., Su, X.Y., Li, G., Song, Y.L., Lam, J., and Tang, B.Z. (2006) *J. Polym. Sci., Part A: Polym. Chem.*, **44**, 2346.
78. Yin, S.C., Xu, H.Y., Shi, W.F., Gao, Y.C., Song, Y.L., and Tang, B.Z. (2006) *Dyes Pigm.*, **71**, 138.
79. Yin, S.C., Xu, H.Y., Shi, W.F., Gao, Y.C., Song, Y.L., Wing, J., Lam, Y., and Tang, B.Z. (2005) *Polymer*, **46**, 7670.
80. Wang, X., Wu, J., Xu, H., Wang, P., and Tang, B.Z. (2008) *J. Polym. Sci., Part A: Polym. Chem.*, **46**, 2072.
81. Tang, B.Z. and Xu, H. (1999) *Macromolecules*, **32**, 2569.
82. Li, Z., Dong, Y., Haussler, M., Lam, J.W.Y., Dong, Y., Wu, L., Wong, K.S., and Tang, B.Z. (2006) *J. Phys. Chem. B*, **110**, 2302.
83. Tang, B.Z., Xu, H., Lam, J.W.Y., Lee, P.P.S., Xu, K., Sun, Q., and Cheuk, K.K.L. (2000) *Chem. Mater.*, **12**, 1446.
84. Li, B., Zhou, J., Xie, Y., and Tang, B.Z. (2001) *Polym. Prepr.*, **42**, 543.

2
Suzuki Polycondensation: A Powerful Tool for Polyarylene Synthesis

Junji Sakamoto, Matthias Rehahn, and A. Dieter Schlüter

2.1
Introduction

The key feature of polyarylenes is the direct connection of their aromatic repeat units by CC bonds. In numerous other classes of conformationally rigid polymers such as aromatic polyesters, polyamides, and polyimides, these units are coupled through functional groups like esters, amides, and imides, respectively. This seemingly small structural difference poses a fundamental difference in terms of synthesis (and properties)! Since the discipline's early days, organic chemistry offers how to produce esters, amides, and imides by CO and CN bond formation reactions. Only recently however, it came up with solutions on how to connect two sp^2-hybridized C atoms to give CC bonds. Thus, polyesters and alike had already been developed, over decades, to considerable maturity when the first reports on CC bond formation relevant for polyarylene synthesis appeared in organic chemistry literature. Of the several CC bond formation reactions through cross-coupling [1], it was the one by Kumada, Tamao, and Corriu [2] which was applied to the synthesis of polyphenylenes by Yamamoto *et al.* in 1978 [3]. Another important cross-coupling protocol was subsequently invented by Suzuki, Miyaura, and coworkers in 1979 [4]. It allowed to connect two specific sp^2-hybridized C atoms more efficiently and under milder conditions. These results together with the increasing demand to develop a robust synthetic protocol for polyarylenes triggered the discovery [5] and subsequent worldwide development of what is now called *Suzuki polycondensation* (SPC). For a comprehensive picture of SPC including a historical background, the reader is referred to a recent pertinent review article [6]. The present chapter focuses on the SPC-based synthesis aspects which include (i) issues relevant to those considering to use SPC, (ii) an overview of recent developments of which future impact on SPC is expected, as well as (iii) representative polymer classes that were and are being synthesized by SPC. Regarding the latter aspect, the class of polyfluorenes (PFs) is only briefly touched because it is the subject of a separate

Design and Synthesis of Conjugated Polymers. Edited by Mario Leclerc and Jean-François Morin

ISBN: 978-3-527-32474-3

chapter of this book. Finally, a brief outlook attempts to foresee the next steps to follow.

2.2
General Remarks

SPC is generally believed to be of the step growth type and follows the main chemical equations in Scheme 2.1. They refer to the so-called AA/BB and AB approaches, a differentiation with important implications. In the AA/BB case, two aromatic monomers are combined, resulting in polyarylene backbones, which contain the two aromatic residues in an alternating fashion. This allows for considerable structural variability. In the AB case, polyarylenes are obtained, which have repeat units from one kind of aromatic unit only. This limits structural variability somewhat as will become clear in a few lines. A second implication is that in the former approach two different monomers are required, each of which carrying either two boronic acids (or esters) or two leaving groups such as halogen or triflate, whereas in the latter, the monomer carries both functional groups at

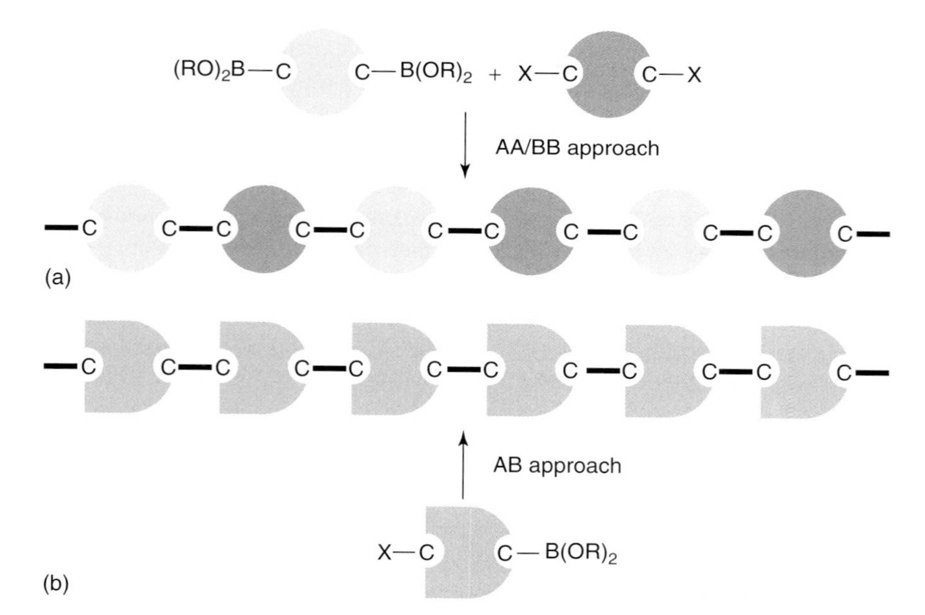

Scheme 2.1 The two different approaches to SPC based on AA and BB-type (a) and AB-type monomers (b). The boronic acid functionality can be employed as such or as a cyclic ester (e.g., pinacol), which is much more often done. The leaving group X is in most cases Br but can also be I or OTf, and since very recently even Cl. The black and gray symbols represent aromatic units, prominent examples being benzene and fluorene, although others have also been used.

the same time. Although this may be considered to be of minor importance, it has fundamental consequences, for example, with regard to ease of monomer synthesis. Analyzing the monomers that have so far been employed into SPC, it becomes evident that the synthesis of AA and BB monomers tends to be easier than that of their AB-type congeners. The reason for this is the often symmetric structure of AA and BB and the fact that the commonly used bifunctional boronic acid monomer **14a** (Figure 2.5, Section 2.4.1) is commercially available and can either be used as such or be modified in a rather simple procedure to the corresponding esters such as **14b** and **14c**.

AB monomers, in contrast, are intrinsically nonsymmetric and often require multistep procedures. On the other hand, the AB approach is the only one that provides access to polymers with chain directionality; the monomers are incorporated in *head/tail* fashion, whereas in AA/BB, each of the two monomers is incorporated in both possible orientations. The AA/BB approach has the additional attraction that entire families of closely related polyarylenes can be obtained by typically keeping the boronic acid component constant and varying the dihalogenide (or ditriflate). These advantages are the main reason why more than 90% of all SPCs were and are still being performed according to AA/BB. The advantages are so strong that even the approach's most serious disadvantage is accepted as the price to be paid. This disadvantage is the difficult to realize stoichiometric balance between the two monomers. Apart from AB monomers with their built-in stoichiometric balance between the functional groups required for cross-coupling, AA and BB monomers need to be combined in an exact mole ratio of 1 : 1 in order to achieve high molar mass product. Carothers equation describes the dramatic loss in molar mass caused by a stoichiometry mismatch. Although one may think that just some simple measures have to be taken to sufficiently account for this, these measures may turn into a real challenge when it comes to realization of SPC. As with any other step growth polymerization, greatest care has to be taken regarding monomer purity, storage, weighing, transfer to reactor, stirring, and so on. In many publications using SPC, such aspects that are so critical to success, are not described. Even worse, it seems that sometimes, they are not sufficiently considered in the experimental realization. This is why the reader will be briefly guided through this area in Section 2.3.

Irrespective of the approach chosen, one would expect oligomers to form at the beginning of growth, which condense to progressively larger units until high molar mass chains are reached. In a recent mass spectrometric study aiming at identification of end groups, it could be shown that this is in fact the case. Figure 2.1a displays the chemical equation [7] and Figure 2.1b, the corresponding MALDI-TOF mass spectrum of a sample with an intentionally low molar mass, that is, a chloroform soluble part. As can be seen well, all different oligomers resulting from the various possible combinations occur and they all carry the required functional groups at both their termini (B(pin) and OTf, respectively) enabling them to grow further [7b].

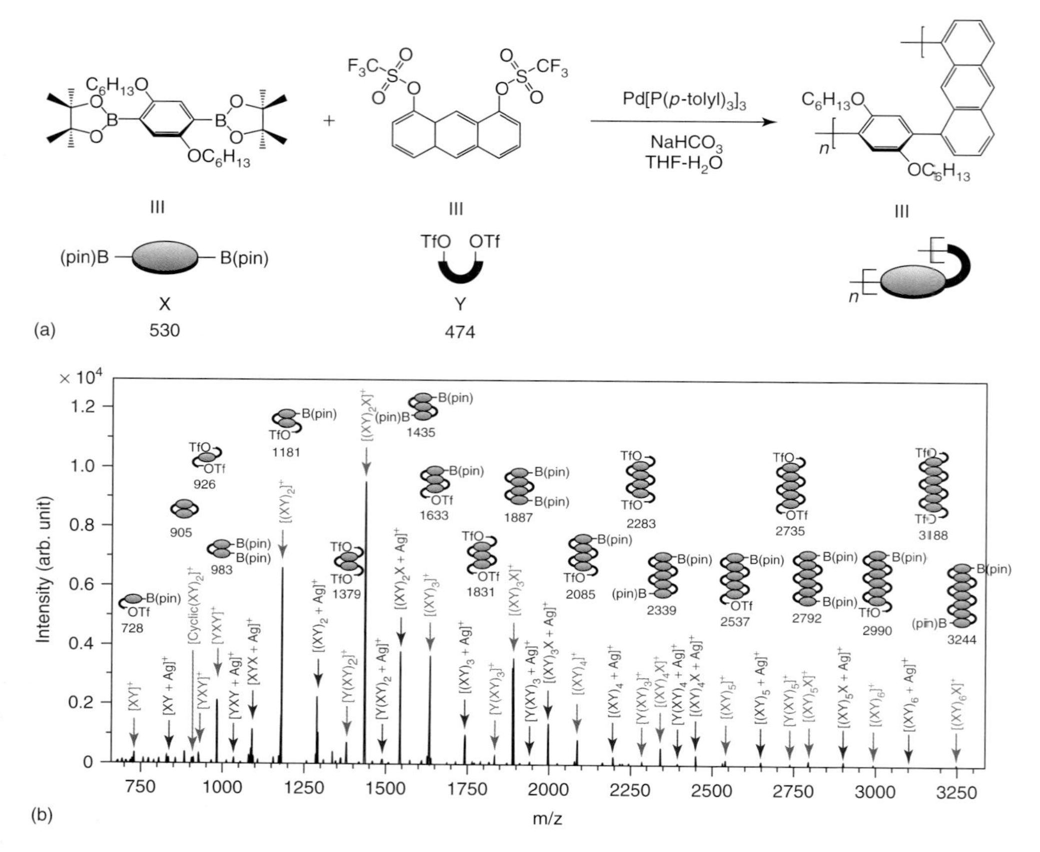

Figure 2.1 An example of SPC using AA/BB monomers (a phenylene pinacol diboronate and anthracene-1,8-ditriflate) (a) and the MALDI-TOF mass spectrum of the chloroform soluble product as a text book example for how step growth polymerization proceeds (b). The full set of oligomers up to $n = 6$ are assigned. The mass spectrum was recorded with a DCTB matrix and in the presence of AgOTf (DCTB = *trans*-2-[3-(4-*tert*-butylphenyl)-2-methyl-2-propenylidene]malononitrile).

2.3
How to Do an SPC and Aspects of Characterization

2.3.1
Monomer Purity, Stoichiometry, and Solvents

It is not the author's intention to go much into synthetic details in this chapter. A few comments regarding the most critical issues are nevertheless appropriate. As with any other step growth polymerization, monomer purity is a key for SPC,

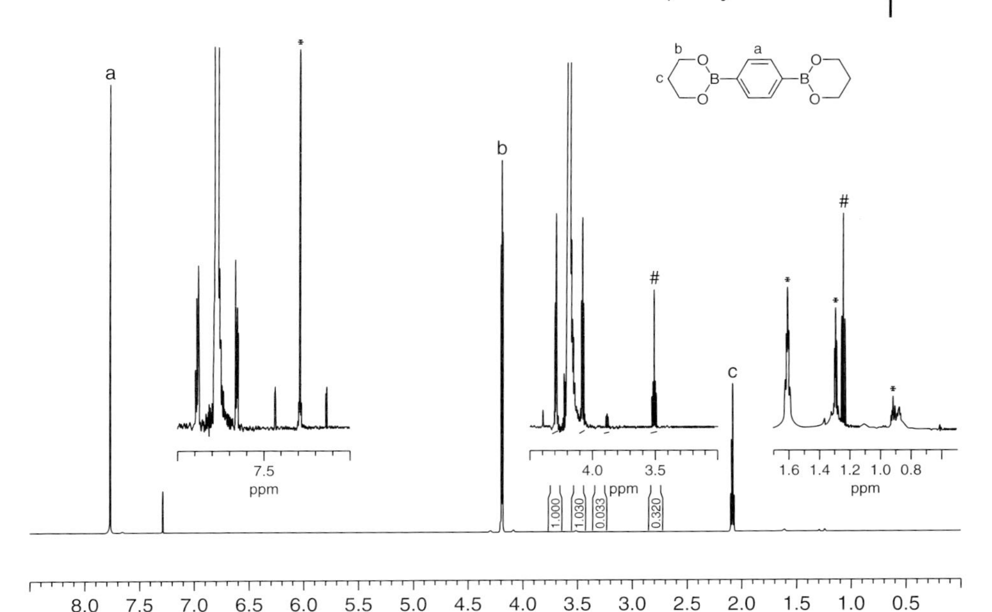

Figure 2.2 A 700-MHz ^{1}H NMR spectrum of a benzene-1,4-diboronic ester taken in the laboratory to illustrate the level of purity considered sufficient. The signal of the NMR solvent ($CDCl_3$) and impurities contained in it (!) are marked (*). The signals of residual solvent (diethylether) the monomer was recrystallized from are also marked (#). The insets show strongly amplified regions to check for trace impurities, which are practically absent (except for some unknown compound causing the signal at $\delta = 3.9$ ppm). The ^{13}C satellites, each of which has 0.5% of the corresponding main signal's intensity, are used to determine the content of ether by integration to be 0.3 mol% in the present case.

specifically as far as the AA/BB approach is concerned.[1] The degree of purity should be quantified by high-resolution NMR spectroscopy (Figure 2.2), GC, or HPLC analysis and be much greater than 99%. If the presence of certain impurities cannot be avoided, for example, for noncrystallizable monomers, at least their nature must be known and their mole fraction be considered when it comes to matching stoichiometry.

In cases where the amount and/or the nature of the impurities are unclear, a series of test polycondensations need to be performed in which the proportions of both monomers are gradually changed. From the respective molar masses, the proportion that best mirrored the exact 1 : 1 stoichiometry can then be

1) For AB it just needs to be ensured that the impurity does not interfere with the polycondensation reaction, by acting as terminating agent, like monofunctional compounds would do.

back-concluded. Bo *et al.* reported such a series in which the weight average degree of polymerization (P_w) differed by a factor of more than 10 by just changing the apparent monomer ratio from 1.000 : 1.000 to 1.000 : 1.005 [8]. It should be mentioned here that boronic acids tend to form partially and fully dehydrated condensation products [9], which can render realization of a certain stoichiometry a nightmare.[2] Instead, the corresponding cyclic esters are used in a widespread manner, whereby specifically, the pinacol esters have the important advantage to withstand silica gel column chromatography purification.[3][4] It is not absolutely certain whether the esters hydrolyze to the acids prior to coupling or are directly involved in the catalytic cycle [10]. In any case, free boronic acids tend to be more reactive than their ester analogs, which shows in the strengths of bases needed in both cases, with the stronger ones for the latter. There are, however, limits to this because protodeboronation [11], a detrimental side reaction, is also facilitated by stronger bases [12]. It should be stated that the aspect of relative reactivity of boronic acid versus ester is also influenced by the rather complicated solubility issue. Whereas the esters tend to be highly soluble in practically all organic solvents used in SPC, acids may only be sparingly soluble in both the aqueous and the organic phases. As a result, the higher reactivity of the acids can be counteracted by their lower solubility.

Naturally, the easier an exactly matching stoichiometry is realized, the larger the amount of monomer. For a research laboratory also investigating complex monomers requiring multistep syntheses, 500 mg of monomer is often a reasonable compromise between synthetic effort and the obtaining of meaningful results. We find that SPC on the 5 g scale and above causes the least problems and those on the 50 mg scale produce basically meaningless results; even if the most serious measures for precision experiments are taken in all steps, including those to avoid spilling of the stirred polymerization mixture on the walls of the apparatus and alike, high molar masses are mostly never reached. If the same experiments are done on the different gram scales, however, all of a sudden respectable molar masses are obtained. This shows a delicacy in the evaluation of the intrinsic potential of SPC for a given system. For a given set of monomers and conditions, one should refrain from concluding whether high molar masses can be achieved or not, if this is only based on small-scale experiments. Apropos intrinsic potential of SPC, often respectable molar masses are discussed referring to polymers that were isolated in yields of far below than 100%, say 60–70%. According to Carothers equation, a high molar mass can only be reached for extremely high conversions,

2) Trimeric cyclic condensation products of boronic acids, the boroxines, can be prepared and isolated. Their use, instead of free boronic acids, can help the stoichiometry problem. This argument does not obviously apply to BB monomers.

3) Other cyclic esters of boronic acids are used in those cases where they are crystalline and purification by column chromatography is obsolete.

4) Depending on the silica gel activity, partial hydrolysis can be observed for pinacol esters to varying degrees. The degree of hydrolysis is easier to control with this ester than with other esters commonly used.

which normally should go hand in hand with high yields. Mass losses during work-up of 5% can be considered normal, and in very special cases, perhaps even 10%.[5] Losses of 30–40%, however, are an indication that substantial amounts of the lower molecular weight chains were removed by fractionation. For an adequate description of actual and representative molar masses, any numbers referring to yields of 60–70% can neither be reasonably used nor compared with one another.[6]

Having discussed so much about the need for matching stoichiometry, it may sound like a contradiction that several laboratories doing SPC use the boron-based monomer in a slight excess. This is a measure of precaution taken in small-scale experiments in order to account for traces of invading oxygen or humidity contained in the boron monomer (see Section 2.3.3). It is the authors' belief however, that this is not normally required for larger scale experiments performed under optimal conditions. Slow monomer addition, when readily adjusted to the conversion, is always an option when performing step growth polymerizations. Whether it helps in SPC to reduce any eventual loss of boron functionality is not clarified.

Most SPCs are carried out in biphasic mixtures of organic solvents like toluene, xylene, THF, or dioxane and an aqueous medium containing the base, while other solvent systems, in particular, the homogenous ones, have been less explored. From the very early days, it was an issue inasmuch as whether the heterogeneity of this reaction mixture would affect the progress of the polycondensation. Conversion-dependent partitions of the involved components between the two phases together with interfacial effects were expected to render the polymerization process complicated and difficult to follow, for example, by kinetic studies. No report on a kinetic study has appeared yet, pointing toward possible future research.

Finally, it should be pointed out that it can be important to remove the bromo and boron functionality present at the chain ends. This is commonly done by an end-capping procedure in which a corresponding monofunctional reagent and new catalyst are added to the reaction mixture prior to work-up. For bromo end groups, phenylboronic acid is used and for boronic acid end groups bromobenzene or *p*-fluorobromobenzene is used. Depending on the molar mass, the efficiency of these processes can be determined by NMR spectroscopy. The removal of boron functionality ensures complete and time-independent solubility of the product. Boronic acids can easily condense and thus hamper molar mass determination.

5) Some losses may be due to low molar mass cyclic products, which do not precipitate during polymer recovery.

6) According to Carothers equation, polymers with, for example, P_n's of 50 and 100 require conversions of 98 and 99%, respectively. Claiming a $P_n = 50$ and stating a yield of 65% for example, therefore, does not make sense.

2.3.2
Brief Note on Optimization

By far most of the synthesis optimization has been done in industry. Given its considerable technological importance, PF was used as the prototype polymer for these studies. One of the earliest studies on improving the molar mass of PF used phase-transfer catalysts [13]. It was assumed that these catalysts facilitate the transfer of the anionic boron species from the aqueous into the organic layer, where the coupling reaction between the boronate and the halide–palladium complex takes place. Best results were obtained using tetra-*n*-butylammonium halides, benzyltriethylammonium halides, and tricaprylammonium chlorides (Aliquat® 336). The reaction leading to polymer **3** was studied most intensively (Scheme 2.2).

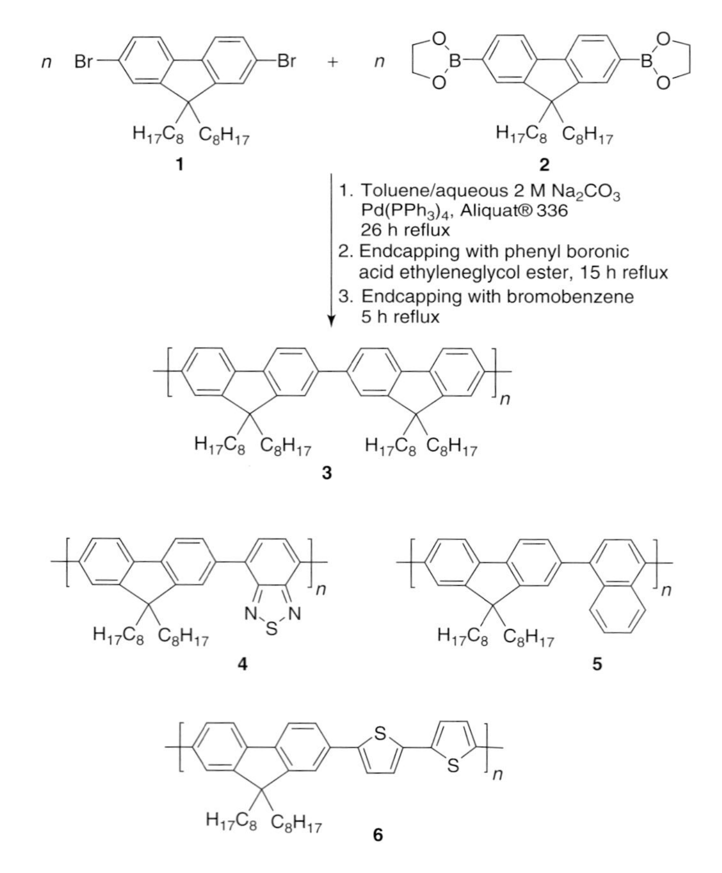

Scheme 2.2 Synthesis of PF **3** from monomers **1** and **2** and structures of PFs **4–6**. These PFs are among those whose synthesis was optimized in industry.

The gel permeation chromatography (GPC) molar masses of the fibrous polymer **3** (100% yield) was $M_w \approx 148\,000\,g\,mol^{-1}$, $M_n \approx 48\,000\,g\,mol^{-1}$, and its inherent viscosity amounted to $[\eta] = 1.50\,dl\,g^{-1}$. Without Aliquat 336 polymer **3** had $M_w \approx 13\,000\,g\,mol^{-1}$, $M_n \approx 9000\,g\,mol^{-1}$, and $[\eta] = 0.22\,dl\,g^{-1}$. Polymers **4–6** were synthesized similarly. Their inherent viscosities amounted to 0.94, 0.39, and 1.84 dl g^{-1}, respectively, indicative of significant molar masses. Later Towns and O'Dell described disadvantages of this procedure [14]. They include (i) slowing of the reaction when toluene is used, (ii) discoloration of the product, (iii) decomposition of the catalyst, (v) poor reproducibility of the polymerization, and (vi) foaming. These authors claimed in the same patent to having overcome these problems by the use of "organic bases"[7] instead of the commonly used inorganic ones. When studying the same reaction to polymer **3** (Scheme 2.2) in the presence of "organic bases" after only 2 hours reaction time, peak GPC molar masses of 204 000–370 000 were observed. Even higher molar mass can be found in another patent [15]. Monomers **7** and **8** were polymerized in a reaction mixture containing dioxane, toluene, water, K_3PO_4, and 0.025 mol% of a tris-*o*-tolylphosphine palladium catalyst (Scheme 2.3). By adding a small additional quantity of monomer **7** after 6 hours, performing an end-capping procedure after 7 hours using bromobenzene, treating the reaction mixture with aqueous NaCN followed by twofold precipitation of the raw polymer resulted in an 88% yield of polymer **9** as a colorless solid. GPC analysis gave $M_w = 814\,000\,g\,mol^{-1}$ and $M_n = 267\,000\,g\,mol^{-1}$.

When monomers **7** and **8** of an "extraordinary" purity were treated under the given conditions in the presence of 0.0125 mol% of palladium catalyst, polymer **9** was isolated in 93% yield and with $M_w = 1\,400\,000\,g\,mol^{-1}$ and $M_n = 410\,000\,g\,mol^{-1}$! The attention of the reader is also drawn to similar work by Towns [16] and a recent publication by Liu *et al.* in the open literature [17]. It seems that major improvements can be achieved when the factors that were delineated in Section 2.3.1 are sufficiently considered.

Finally, a recent excellent study by Goodson *et al.* should be mentioned, who, for the monomers shown in Scheme 2.4, reported one of the very few serious optimization studies in the open literature including solvent, base, and catalyst system [18, 19]. Although this study does not deal with polyarylenes, it is nevertheless an outstanding reading containing numerous helpful hints for those who are planning to do SPC. Because of the comprehensive character of this publication, the present section does not exhaustively cover solvent and base effects [20]. Although the optimized conditions reported naturally apply only to the set of monomers used, Goodson's study provides valuable guidelines on how to approach other cases. The molar masses (better: the average degrees of polymerization!) achieved here are

7) This includes alkylammonium hydroxides, alkylammonium carbonates, alkylammonium biscarbonates, alkylammonium borates, 1,4-diazabicyclo[2.2.2]octane (DABCO), dimethylaminopyridine, pyridine, trialkylamines, and alkylammonium fluorides.

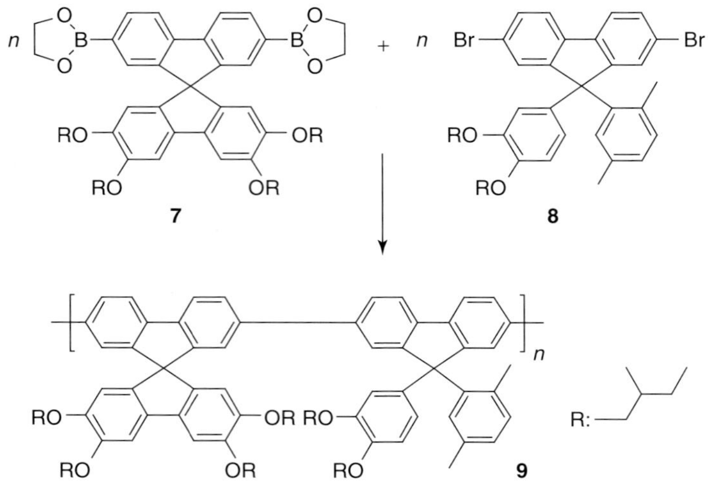

Scheme 2.3 An especially high molar mass case: The polymerization of monomers **7** and **8** to give spiro-type PF **9** with GPC molar masses M_w = 1 400 000 g mol^{-1} and M_n = 410 000 g mol^{-1}.

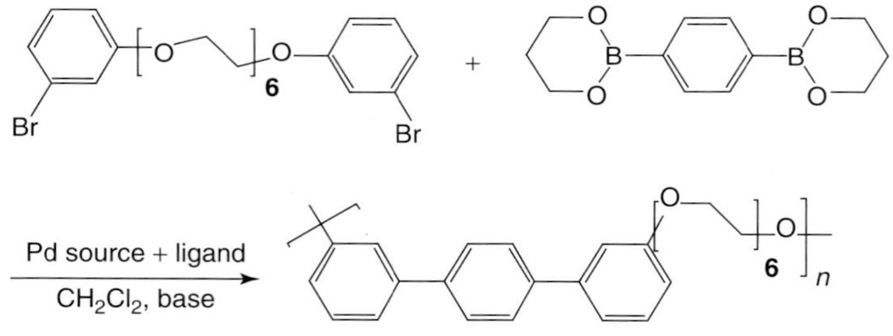

Scheme 2.4 A fully optimized SPC, which furnishes the polymer with molar masses of up to M_n = 64 000 and M_w = 180 000.

in a similarly high range as reported in some other cases [21] and in the patent literature aforementioned using different conditions, and it would be interesting to see whether the Goodson conditions can lead to even further improvements there. Since this note deals with optimization, the reader is also referred to Section 2.4.6, which briefly touches upon microwave techniques in SPC, which is another option for improvement.

2.3.3
Reduced Catalyst Amount and Product Purification

Almost all SPCs published to date in open literature use 1–3 mol% of catalyst, mostly $Pd(PPh_3)_4$, $Pd[P(p\text{-tolyl})_3]_3$, or $Pd[P(o\text{-tolyl})_3]_2$. A few reports use Pd(II) sources. In the light of avoiding eventual traces of leftover catalyst in the polymeric product, irrespective of whether this refers to P or Pd (see below), much smaller ratios are desired [22]. Buchwald *et al.* and others have shown that catalyst loadings of 0.001 mol% can still give very good results for Suzuki–Miyaura cross-coupling (SMC) [23]. Another even more extreme case was recently reported by Leadbeater *et al.*, who showed that trace amounts of palladium contaminants down to a level of 50 ppb in commercially available sodium carbonate, can catalyze SMC affording decent yields of biaryl products under microwave conditions in water at 150 °C [24]. In a recent example in the open literature, reduced catalyst amounts were applied to polymer synthesis. For a poly(*meta-para*-phenylene) [21] (see also Section 2.5.3) high molar masses could be reproducibly obtained if only 0.03 mol% of $Pd[P(p\text{-tolyl})_3]_3$ was used. Compared to an experiment where 0.3 mol% of the same catalyst was employed, the molar masses decreased only slightly from $P_w = 390$ to $P_w = 376$ and are, thus, still very high (Kandre and Schlüter, unpublished results). Residual traces of Pd are of great concern in the organic electronics industry because they interfere with the electronically excited states of the polymers. If present in the active polymer layer of an organic light-emitting diode (OLED), for example, Pd traces enhance the rate of nonradiative quenching and intersystem crossing from singlet excitons to long-living triplet states with dramatic consequences on the intensity and the spectral properties of the emitted light. It is reasonable to assume that the loadings used in industry are generally much below the ones typically used in academia. It seems that some companies have reached loadings as low as 0.01 mol%. Another issue of serious concern is oxygen. Phosphine-based Pd catalysts, which are still the ones used most often, need to be prepared fresh and can only be stored for a few days in a high-quality glove box before the first detrimental effects on molar masses are observed. It is essential to do the entire polymerization, which can last for a couple of days, under rigorous exclusion of air. For significantly shorter polymerization durations, see Section 2.4.6. This exclusion is not only about the sensitivity of phosphine toward oxidation, which may result in colloidal precipitation of Pd metal, but there is also evidence that oxygen can give rise to homocouplings of boronic acid and boronic ester monomers [25]. Homocouplings [26] would lead to a mismatch of the number of functional groups and, thus, result in a lower than possible molar mass and, of course, sequence defects in the backbone, which by no means can be healed. Addition of excess boronate monomer may be indicated in such a case. Another oxidative removal of boronic groups was reported during which hydroxyl groups are established [4, 27].

Dehalogenation of organic halides (ArX) in the presence of Pd catalyst and alcohol is also known to be a possible side reaction. It can proceed via β-hydride elimination of $Ar\text{-}Pd\text{-}OCH_2R$ giving rise to ArH and RCHO [28]. It is obvious that a careful optimization of reaction conditions is necessary for SPC, so as to minimize all these side reactions that can terminate propagation, and thus have a direct impact on molar masses as well as end functional group pattern (see below) [29].

Once the polymerization is completed, the polymers are typically recovered by drying the crude product, taking the soluble material up in an organic solvent, and precipitate the dissolved polymer into a nonsolvent (often MeOH or acetone). After filtration this procedure is repeated until the product has reached the desired quality with regards to color (in many cases colorless) or trace elements contained. Sometimes products are fractionated in order to obtain fractions with improved molar mass, narrower molar mass distribution, and/or lesser impurities [30]. However, a successful practical realization requires some training. The precise procedures for fractionations carried out in the industry are unfortunately intellectual property and not easily available.

For applications in OLEDs and other devices, it is important that the polyarylenes used, which are mostly of the PF type, have a negligible Pd content. Such polymers synthesized by academia in many cases contain undetermined but large quantities of Pd and other undesired elements. If synthesized by industry, Pd contents below approximately 1 ppm have been reached. These contents were determined by laser ablation inductively coupled plasma mass spectrometry (ICP-MS), a powerful technique for quantitative trace element analysis. For the above, poly(*meta-para*-phenylene) synthesized with 0.03 mol% of $Pd[P(p\text{-tolyl})_3]_3$; the lowest Pd content by ICP-MS was 2.8 ppm (Günther, Kandre, and Schlüter, unpublished results) after only two rounds of careful precipitation. If an even lower Pd content is needed, treatment with aqueous NaCN [18] or Pd scavengers similar to the one advertised by Krebs [31] can be considered. The latter were developed in order to remove metallic nanoparticles/catalysts specifically for polymer applications. This is achieved by dissolving the nanoparticles or catalyst residues, presumably atom by atom, into strongly colored complexes, which cannot only be detected and quantified by UV spectroscopy, but also be conveniently removed in the presence of a polymer because of these two components' respective solubility behavior. The azothioformamide **10** complexes Pd(0) to give **11**, which has a rather

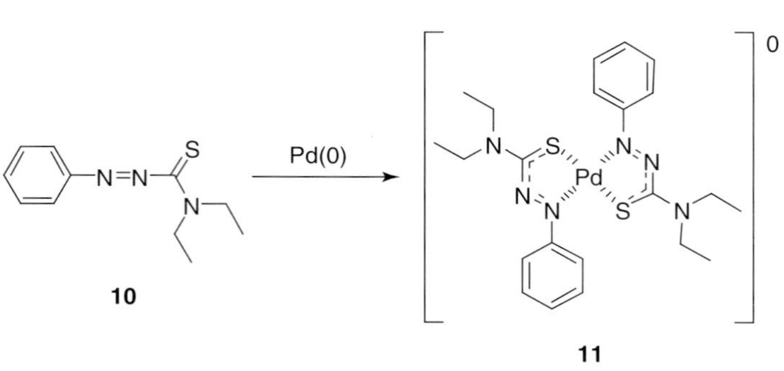

Scheme 2.5 Solubilization of Pd nanoparticles/catalyst by the light orange scavenger **10**, leading to the dark brown–green complex **11**.

characteristic UV absorption at the longest wavelength ($\lambda_{max} = 800$ nm) with a rather high extinction coefficient (Scheme 2.5). It should be mentioned, however, that any Pd chemically bonded to the polymer escapes this scavenging.

The P content has also been quantified by laser ablation ICP-MS and can reach values below 35 ppm although this will not be the end of the story. The P content is to be seen in context with ligand scramblings through which aryl groups on catalyst ligands are exchanged with the terminal aryls of growing polyarylene chains (Scheme 2.6). For more details on this matter, the reader is referred to [6b].

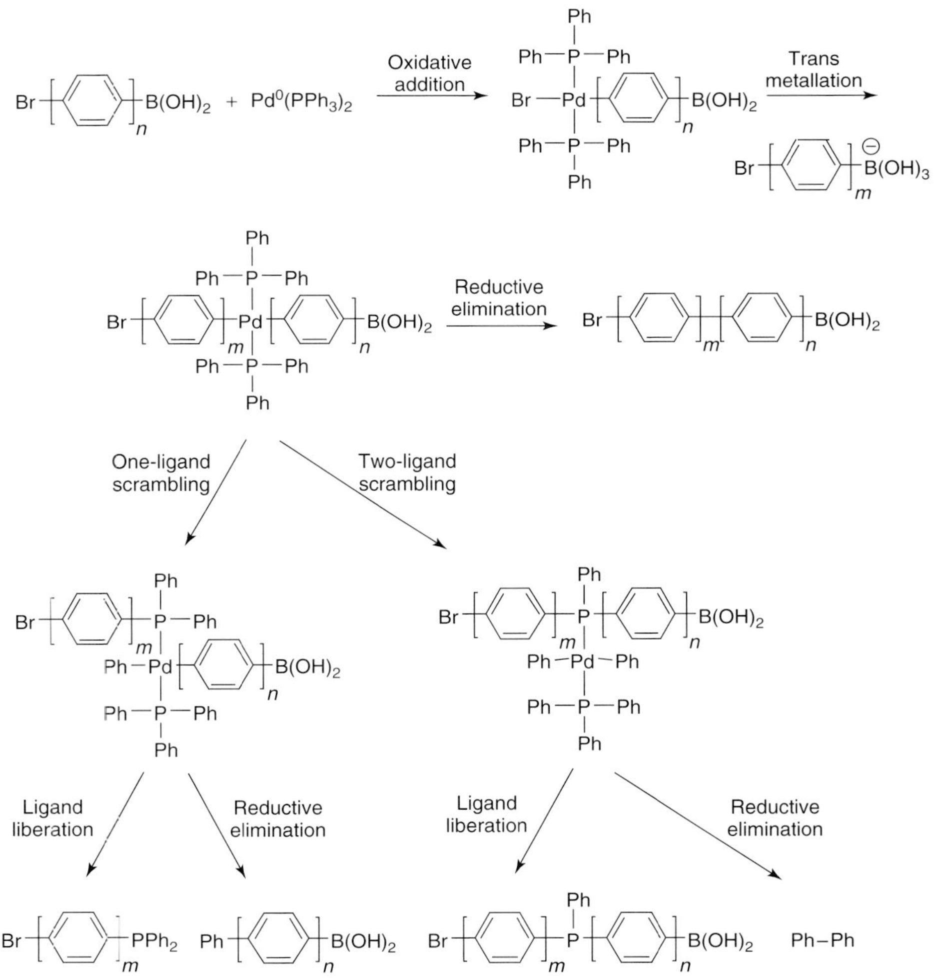

Oxidative addition
Trans metallation
Reductive elimination
One-ligand scrambling
Two-ligand scrambling
Ligand liberation
Reductive elimination
Ligand liberation
Reductive elimination

Scheme 2.6 Rationalization scheme explaining the incorporation of ligand-derived phosphorus into the polymer backbone during SPC. Not all options are shown.

2.3.4 Molar Mass Determination

Toward the end of Section 2.3, a word on molar mass determination is appropriate. Polyarylenes and other conjugated polymers like poly(arylene ethynylene)s are considered as rigid rods. Such macromolecules have a larger hydrodynamic volume than flexible ones of the same contour length. For practical reasons the molar masses and distributions of polymers are commonly determined by GPC, a relative method that separates according to hydrodynamic volume. Absolute methods such as light- and small-angle neutron scattering are applied only in exceptional cases. Narrowly distributed samples of different molar mass standard polymers (e.g., polystyrene (PS) or polymethylmethacrylate (PMMA)) are used for calibration. A GPC molar mass of a polyarylene derivative of 50 000 g mol^{-1}, means that this sample has the same hydrodynamic volume as a sample of the standard polymer used with an actual molar mass of 50 000. With the larger hydrodynamic volume of rodlike polymers, the actual molar mass of the polyarylene under consideration must obviously be lower; the number 50 000 represents just an upper limit. With this in mind, all publicized GPC molecular weights of polyarylene derivatives should be considered with great care. Further complications may arise caused by the rodlike nature of polyarylenes. This increases their propensity to form aggregates and it needs to be confirmed that the elution curves actually refer to molecularly dissolved polymer.

Fortunately, there is a thorough study available in which the actual molar mass of the poly(*para*-phenylene) (PPP) **12** in Figure 2.3 (R^1 = methyl, R^2 = dodecyl, and R^3 = 3,5-di-*tert*-butylphenyl) was determined by light scattering, osmometry, and GPC using universal calibration and compared with the masses obtained from PS calibration [32]. The universal calibration was done on the basis of the Mark–Houwink–Sakurada equation using nine fractions of **12** with molar masses ranging between $27 < M_w < 189$ kg mol^{-1}. A calibration based on the wormlike chain model gave very similar results. As seen from Figure 2.4, both these calibrations lead to almost superimposable elution curves at a much lower molar mass than PS. A quantification of this difference reveals that PS overestimates the real molar mass of PPP **12** by a factor of almost 2. Although substituents certainly have an effect on the stiffness and thereby the hydrodynamic volume of a polymer chain, it is reasonable to assume that this effect will normally be small. As a rule of thumb, GPC molar masses of PPP derivatives should, therefore, generally be corrected by this factor of 2 to lower values in order to have a more realistic estimate of the actual molar mass. For very large substituents such as in the dendronized polyarylenes discussed in Section 2.5.4 an additional aspect comes into play. If a given polymer's mass per unit length is much larger than the standard's, it can happen that GPC calibrated with PS underestimates the real molar mass even for rodlike polymers. It should be noted, however, that this underestimation by GPC is a clear exemption from the rule.

A more recent study by Harre and Wegner highlights the problem of aggregation for molar mass determination even for dilute solutions of PPPs [33]. Freshly

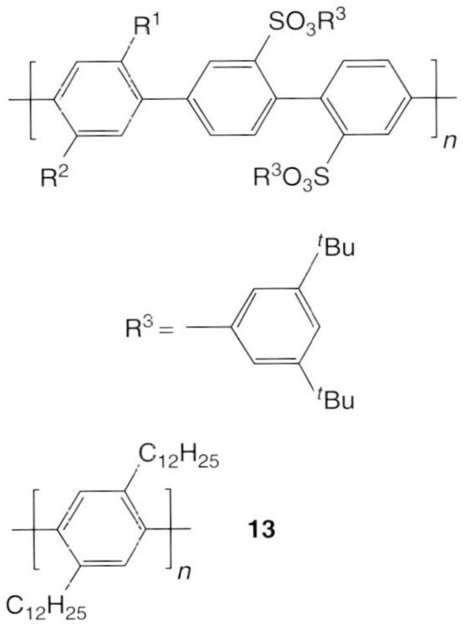

Figure 2.3 Chemical structures of PPPs **12** and **13**.

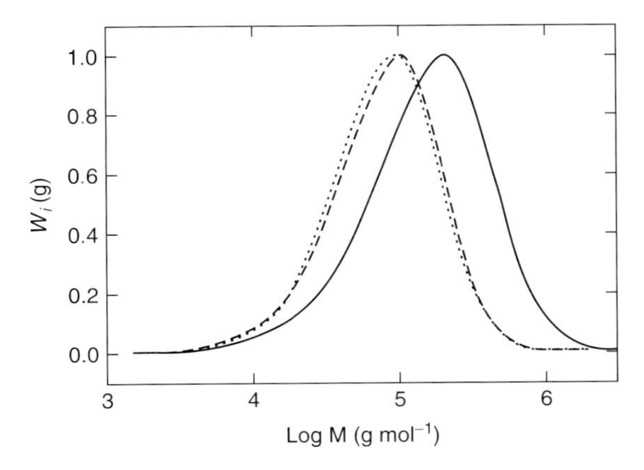

Figure 2.4 GPC elugram of polymer **12** (M_w = 113 kg mol^{-1}) in THF : PS calibration (solid line), universal calibration based on Mark–Houwink–Sakurada equation (dashed line), and universal calibration based on the wormlike chain model (dotted line). Reproduced from [32] with permission by the publisher.

prepared solutions of PPP **13** in toluene are metastable at room temperature with regards to the formation of aggregates composed of up to 100 individual macromolecules. This aggregation process has an induction period of more than 10 hours at room temperature. The kinetics of aggregation was investigated by making use of a fast capillary membrane osmometer. Aggregation follows an Avrami–Evans type formalism and suggests that clusters of lyotropic liquid crystalline phases are formed. The long induction period of aggregate formation in dilute toluene solution allows applying conventional techniques of molar mass determination like membrane osmometry and GPC. A relationship $[\eta] = 1.94 \times 10^{-3} M^{0.94}$ was found for **13** in toluene at 20 °C and a persistence length of 15.6 nm was derived applying the Bohdanecky formalism. This gives evidence of the wormlike nature of the nonaggregated **13** in dilute solution.

This result also explains why difficulties are frequently encountered if one tries to process solutions of polyarylenes by spin coating, ink jet printing, roll-to-roll printing and so on. Unless care is taken of inevitably ongoing aggregation processes in the solutions irreproducible results will be obtained with regard to the homogeneity or – more generally – morphology of the resulting films and patterns. This, in turn, has severe consequences for the reproducibility and control of the photonic and electronic properties of the processed materials.

2.4 Methodological Developments

2.4.1 Boronic Acid/Boronate Monomers

Already, shortly after the initial successful cases of SPC, which were done with AB monomers, the first AA/BB approaches were also reported [34, 35] and many others followed. Often they were and are being done using the commercially available free boronic acid **14a** or its esters such as **14b** and **14c** (Figure 2.5). Other boron-based monomers still require synthesis. The corresponding procedures, like the one for the often used monomer **15**, typically involve halogen metal exchange followed by quenching of the resulting bismetallated intermediates with trialkylborates, whereby compound **16** is the perhaps most often used borate for this purpose. The most recent application of this borate led to monomer **19**, which was used for the synthesis of low molar mass poly(*para*-phenylene-ethynylene)s [36]. Because of the aggressive nature of the lithiated or Grignard-type intermediates, quite a few substitution patterns of monomers are excluded by this protocol from the very beginning. It is therefore of continued importance that other routes of access to boronic esters are being developed. In this context, the work by Miyaura [37] and later, Masuda [38] should be highlighted, who reported mild Pd-catalyzed incorporations of boronic pinacol ester groups to aromatic units by reacting the corresponding aromatic halides with pinacoldiboron **17** and the more atom efficient pinacolborane **18**, respectively. Miyaura additionally and interestingly showed that monomer **14c** can be easily prepared in attractive yields from either 1,4-dibromobenzene or 4-iodobromobenzene. An additional advantage is that both compounds **17** and **18** are commercially available – although not really cheap. Another strategy that bears potential is the direct borylation of alkanes, alkenes, and – most importantly – arenes by transition metal–catalyzed CH activation [39], although this has not yet found widespread application in monomer synthesis (Scheme 2.7) [40]. Finally and perhaps more as a curiosity, it should be mentioned

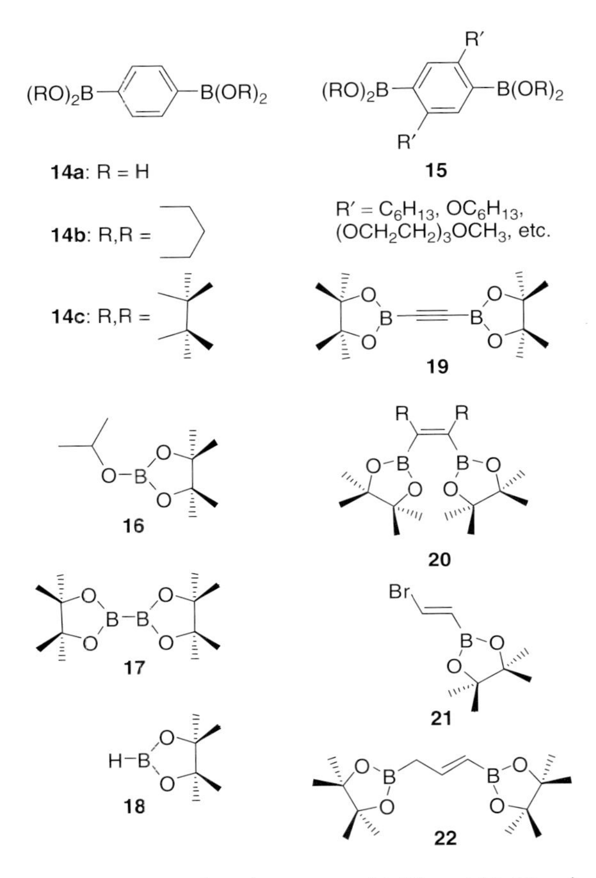

Figure 2.5 Boron-based monomers **14**, **15**, and **19–22** and reagents **16–18** for the incorporation of the pinacolboron group into monomers (and other compounds).

that compound **17** had already been used in the early 1990s to convert mono- and disubstituted acetylenes into compound **20**, a potential SPC monomer [41], by a Pt-catalyzed route. Its AB-type *trans* congener **21** was reported even earlier and obtained by the addition of BBr_3 to acetylene [42]. A γ-borylallylboron compound **22**, which was prepared by metathesizing pinacol allyl- and vinylboronate, might also be an interesting candidate for SPC [12].

So far we have dealt with rather simple boron-based monomers. There is a recent interesting discovery on masking groups for aryl boronic acids by Suginome [43] and shortly thereafter by Burke [44], which has the potential to facilitate synthesis of more complex monomers. The commercially available 1,8-diaminonaphthalene[8] and *N*-methyliminiodiacetic acid (MIDA) [45], respectively, are used in these cases.

8) This protection was chosen because it survives purification by column chromatography completely and can also be completely removed.

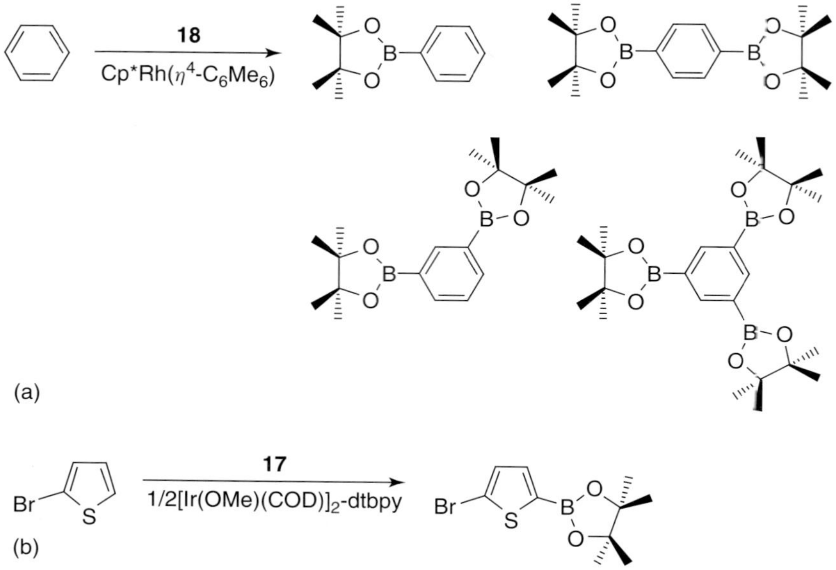

Scheme 2.7 Direct borylation via transition metal–catalyzed C – H activation of arenes to furnish potential SPC monomers. (a) Reaction between benzene and **18** in the presence of a catalyst precursor Cp*Rh(η^4 – C_6Me_6) leading to a mixture of mono-, *m*- and *p*-di-, and 1,3,5-tri-borylated compounds. The relative yield varies depending on the stoichiometric ratio between benzene and **18** as well as the reaction time. (b) Selective 5-borylation of 2-bromothiophene with **17** in the presence of a catalyst precursor 1/2[Ir(OMe)(COD)]$_2$ with a ligand 4,4'-di-*tert*-butyl-2,2'-bipyridine (dtbpy).

The protected boron is not attacked itself under cross-coupling conditions. For the sake of completeness it should be mentioned that placeholder/masking strategies for the other coupling functionality, the halogen or triflate, have already been known for a long time [46].

2.4.2
Boron-Based Ate Complexes

Tetracoordinated ate complexes of boron have often been described in the literature [47]. It has been known for more than 20 years that they can be employed into cross-coupling reactions without the simultaneous presence of a base, which otherwise is a characteristic feature of SMC and, of course, also SPC. Although ate complexes have the attraction of being isolable in the crystalline form (which is of advantage when it comes to realization of a certain stoichiometry; see Section 2.3) and can even be both water and air stable, their coupling efficiencies were so far not in a range making them promising candidates for polymer applications. Conversions of beyond 99% per bond formation step are required if high molar mass polymer should be formed and such values were never reported. Very recently, Yamamoto, Miyaura *et al.* disclosed an interesting finding which may

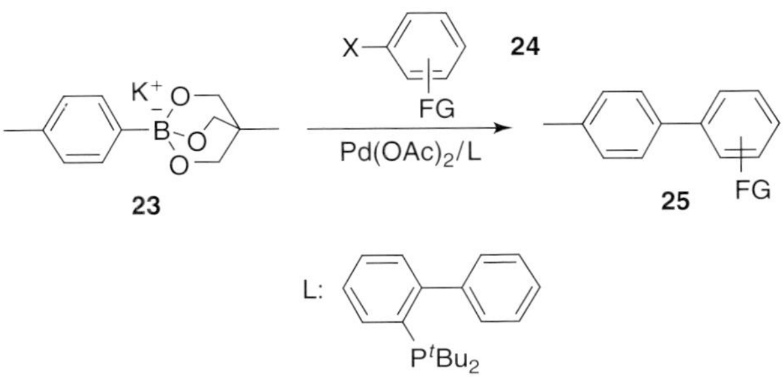

Scheme 2.8 SMC of the triolborate **23** with the haloaromatic **24** to give biphenyl **25**.

actually find its way into polymer synthesis. Their laboratory reported on the novel stable triolborate **23** that is sufficiently soluble in organic solvents and shows extremely high conversions in model cross-couplings of the sort shown in Scheme 2.8 [48].

Both for donor and acceptor substituted aryl bromides **24** quantitative conversions could be reached, and even for **24** with X = Cl and X = OTf, conversions of 90% were reported. For X = Cl, the presence of the phosphine ligand L was necessary. In addition, the couplings could be carried out in homogenous DMF/water solution at room temperature and were finished within a few hours. As for the initial SMC protocol published [4], the boron component was used in a small excess (1.1 equiv.), presumably to account for inadvertent deboronation. For polymer synthesis, where the AA and BB monomer have to be present in the exact 1 : 1 stoichiometry, net excess of a component is not normally acceptable.[9] Although this use of a slight excess of the boronate poses a certain risk in the applicability of this new protocol,[10] attempts to utilize it in polymer synthesis are nevertheless worthwhile.

2.4.3 Halo and Related Monomers

The majority of monomers used in SPC carry two bromides. There are a few examples with aryliodides. Some of them seem to furnish better results in terms of molar masses, [49] but there is also a report in which such an improvement was not

9) In a polycondensation it is essential that at high conversion both functional groups are present to the exact amount. Otherwise, growth is self-finished at a stage where this is not intrinsically necessary. If during growth, an initially present, slight excess of a component is reduced by destruction of the functional group (e.g., deboronation), it leads to chain termination which renders the achievement of high molar mass product less likely. If the loss of boronate functionality is caused by homocouplings, excess of bromide chain ends will be the result. They can be used to increase molar mass by adding diboronate monomers toward the end of conversion and let the polymerization continue for a while.

10) It is not entirely clear whether there are other reasons except deboronation that suggest this excess. The use of Pd(II) in SPC requires its reduction to the catalytically active Pd(0). This reduction could be the result of boronic acid homocouplings, which directly impact stoichiometry.

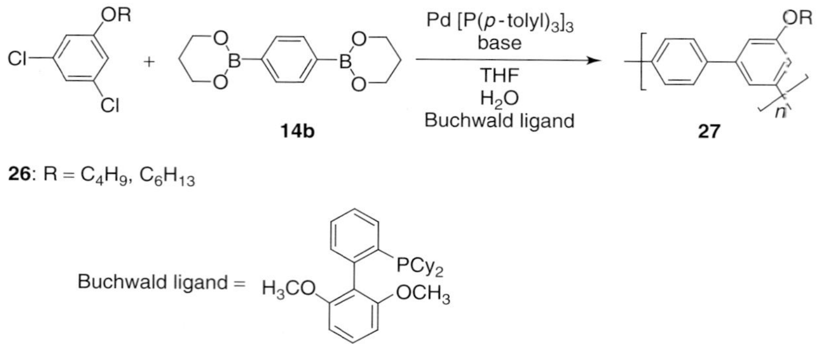

Scheme 2.9 The first example of an SPC using dichloroaromatic monomers. This success was possible through ligands specifically designed to overcome the relative inertness of the aromatic C – Cl bond toward Pd-mediated cross-coupling reactions.

observed [18]. There is too little information available specifically regarding direct comparisons of dibromo versus diiodo monomers of the exact structure carried out in the same laboratory to draw general conclusions. It is therefore presently advisable to conduct studies on a case to case basis. From the industry point of view, chloro-functionalized monomers have advantages. They include commercial availability and low cost. Unfortunately, aryl chlorides are still a challenge for all transition-metal-mediated cross-coupling reactions. This is due to the considerable stability of the C – Cl bond, although attractive developments have been reported during the last decade (see Section 2.4.4). Using a Buchwald ligand it has been very recently shown that the dichloromonomers **26** (Scheme 2.9) polymerize with boronic ester monomer **14b** to poly(*para-meta*-phenylene) **27** in satisfyingly high average molar masses of M_w = 24 000 g mol^{-1} and in yields of 95% referring to purified and freeze-dried material [50]. This molar mass corresponds to an attractive P_w = 94 and will stimulate broader research in this promising direction.

Besides halo aromatics, sulfonyloxy aromatics have also been tried [7, 51]. This is principally an interesting variation because it allows use of the amply available hydroquinones, resorcines, and related aromatic diols as precursors for the corresponding monomers. The authors are not aware of a case, however, in which SPC involving sulfonyloxy monomers gave a high molar mass product. In the open literature, trifluoromethyl, perfluoroalkyl, and imidazolyl sulfonated aromatics have been subjected mostly to SMC [52] and only scarcely to SPC [53].

2.4.4 Catalysts

Recent years have witnessed an enormous development in catalysts for SMC aiming at higher turnover numbers, milder conditions, and higher efficiencies and also allowing for the use of the much less reactive chloroaromatics as well as for sterically demanding cross-couplings. Examples of new ligands specifically addressing this matter [54] include Buchwald's biaryl-based phosphines [55], Beller's diadamantyl phosphines [56], Fu's tri-*tert*-butylphosphine [57], and Hartwig's pentaphenylated ferrocenyl phosphines [58]. The first application of a Buchwald ligand to SPC of a dichloro monomer was mentioned in Section 2.4.3. The authors are not aware of a case where any of the new ligands has led to improved molar masses in an SPC with dibromo monomers. This aspect has also been dealt with in the above mentioned work by Goodson (Section 2.3.2) [18] and suggests that the catalysts designed and optimized for SMC are not necessarily the best for SPC also. We feel that more exploratory work in this direction could be beneficial to the field. In addition, phosphine ligands are generally subject to ligand scrambling and oxidation by inadvertently present traces of air (see also Section 2.3.3). Whereas the former side reaction leads to incorporation of P atoms into the polymer chains [6b], the latter can result in colloidal precipitation of Pd metal. It has therefore always been a concern as to how to avoid P-based ligands in SPC. Use of supported catalysts [59] for SPC may be interesting to be considered. These catalysts can not only be reused in large-scale preparations or flow reactor applications, but may also be used to facilitate purification of SPC products. Liu *et al.* describe efficient water-soluble phosphine ligands for the catalytically active Pd species. They may be beneficial for SPC in the heterogenous reaction medium of THF and water [60]. The corresponding Pd complexes not only give improved molar masses when compared with $Pd(PPh_3)_4$ but also lower residual Pd in the final products after extraction with scavengers. Last but not least, it should be noted that recent progress in this field is so remarkable that catalysts are now available for highly sterically demanding SMC [61], enantioselective SMC [62], and SMC via C – O bond activation [63]. These developments could well have an impact on SPC and further its scope in polymer synthesis.

2.4.5
Chain Growth SPC

In Section 2.2, SPC was introduced as being of the step growth kind. Many experimental facts support this view. Rather convincing evidence is the fact that no initiation is needed and a series of oligomers are seen at the early stage of polymerization, whose molar masses are continuously shifted to higher values as reaction time passes following step growth statistics. Also, if a chain mechanism would be operative it had to be explained as to how the Pd, after each CC bond formation event, should pass toward the chain terminus in order to continue its action on the very same chain. In the AB case, the chain end is always the same (halogenide). Since there is evidence for Pd to be able to rest on π systems, one could possibly imagine that a Pd(0) fragment, stabilized by a ligand, moves from its last reaction site across the newly added terminal phenylene unit and then adds oxidatively into the existing CBr bond. This would result in continued growth of the same chain. For the AA/BB approach, however, such a picture cannot hold

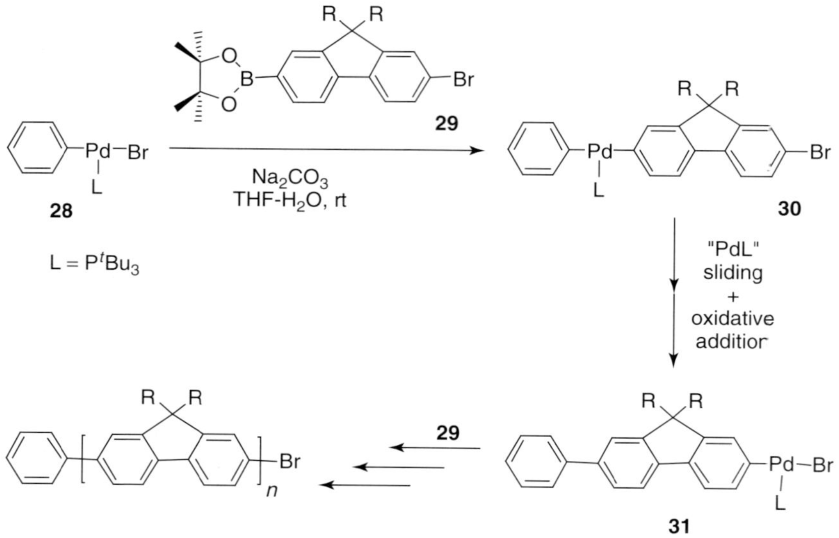

Scheme 2.10 Mechanism proposed for a catalyst transfer polycondensation using the Pd complex **28** as initiator and the SPC monomer **29**. After the CC bond formation in **30** the Pd residue shifts over the π part of the fluorene unit to oxidatively add into the terminal CBr bond, thus creating species **31** prone to undergo transmetallation with the next monomer to collide with the chain end.

because the nature of the terminal functional group changes after each addition from halogenide to boronic acid or ester.

Initiated by the work of Yokozawa [64] there has been quite some activity with regard to this so-called catalyst transfer polycondensation [65], which uses AB-type monomers and basically follows the above picture of sliding metals over π systems [66] and by this switch the polymerization mechanism from step to chain growth. Impressive results were obtained mostly for thiophene-based monomers and Kumada-type protocols. For certain cases and low molar masses it appears that the typical fingerprints for a chain growth can be proven.

Recently, Yokozawa reported the first two cases in which SPC monomers were polymerized initiated by the Pd complex **28** (Scheme 2.10) [67]. Also, here the molar mass linearly depended on conversion, the polydispersity index was low, and the phenyl group of the initiator was actually found as an end group in each chain of the product. In addition, a monofunctional arylbromide did not get involved in the polymerization [68]. Thus, it seems that for AB monomers one has to consider chain mechanisms as an option [69]. At present it is too early, however, to draw any general conclusions. One may well question why molar masses are rather low in all catalyst transfer polycondensations. In the above case, number of average molar masses of $M_n = 18\ 000\,\mathrm{g\ mol^{-1}}$ are reported corresponding to degrees of polymerization of only $P_n \sim 20$. It remains an important open question as to why the polymerizations were not investigated for higher molar masses. Insufficient solubility cannot be the reason in all cases, although it should be emphasized that the polymerizations are carried out at relatively low temperatures in order not to sacrifice narrow distributions and regioregularity (were applicative). The key characteristic of a chain growth polymer similar to the products of anionic, ring-opening metathesis, and controlled radical polymerizations is and remains high molar mass, often ranging up to millions. It is appropriate at this point to mention that in the case of regioregular poly(3-alkylthiophene), high molar masses have been achieved by Merck chemists through the Kumada-type, Ni(0)-catalyzed polycondensation [70]. GPC weight average molar masses of $M_w \sim 400\ 000\,g\ mol^{-1}$ in chlorobenzene were seen, which suggests a chain growth nature of the polymerization [71]. Hiorns and Bettignies were the first to report high molar masses for such polymers using a similar approach in the open literature [72]. In the end it may well be that different mechanisms are operating more or less simultaneously. Together with the heterogeneity of the reaction this could help explain why quite a few high molar mass SPC products show rather broad molar mass distributions. PDI values of 4–5 are not uncommon. Further clarification regarding this delicate issue is obviously required. This should include an *in situ* spectroscopic or mass spectrometric proof of the active chain end [65f].

2.4.6
Microwave and Technical Scale Microreactor Applications

There are two more engineering-type methodological developments to be mentioned in connection with SPC, the application of microwaves and of the microreactor technique. Whereas the first is still in its infancy, the latter has reached technical scale maturity. The microwave technique has led to astounding improvements for numerous chemical reactions. This includes SMC for which countless publications have appeared [73]. It has recently also been applied to SPC, so far although met with limited success only. Scherf *et al.* reported the first microwave SPCs on a rather complex monomer system (Scheme 2.11a). They applied two different microwave conditions and compared the respective outcomes with the results obtained when the same SPC was carried out under conventional conditions. The microwave conditions were (i) aqueous K_2CO_3/THF (150 W, 12 minutes) and (ii) solid KOH/dry THF (300 W, 10 minutes). The polymers were obtained in yields of 60 and 72% and had $M_n = 14\ 200$ and 29 900 respectively [74]. Under conventional, presumably not fully optimized conditions, the yields were about 80% and $M_n = 4000–11\ 000$. Thus, microwaves seem to have had a positive effect on the molar masses although an unequivocal conclusion is difficult to draw; all products are still more or less in the oligomeric regime. The drastically shortened overall reaction times (minutes instead of days), which is commonly observed in microwave applications is perhaps the main advantage also in this case. Shortened reaction times were observed in a more recent work too [75], which again only deals with oligomeric material.

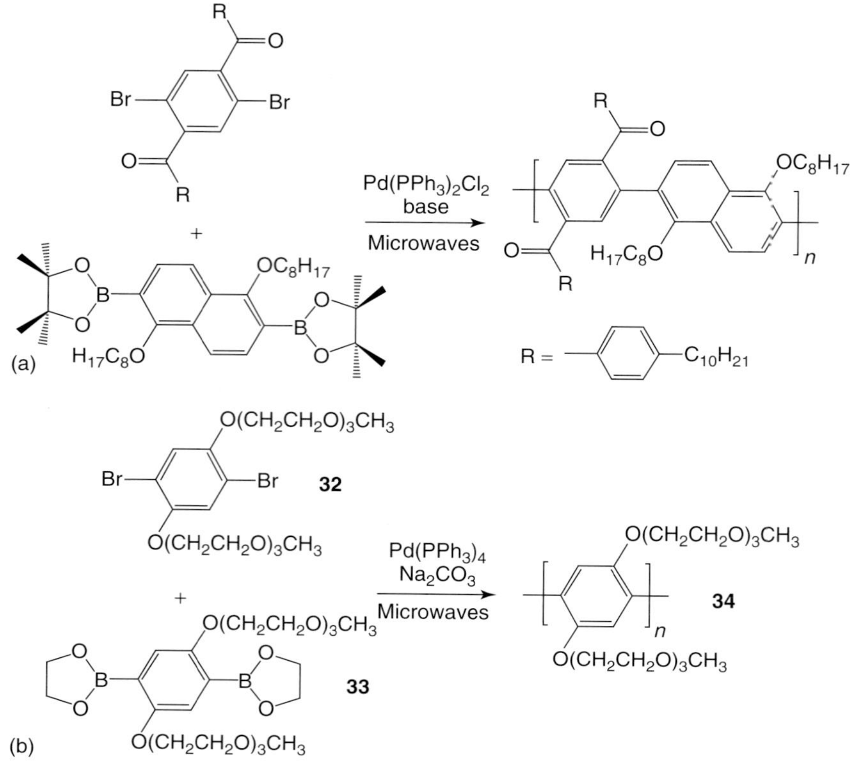

Scheme 2.11 Microwave assisted SPC: the first report in the open literature by Scherf *et al.* (a) and a recent preliminary study by one of the authors (M. R.) using monomers **32** and **33** to give polymer **34** (b).

A recent preliminary study by one of the authors (M.R.) shows that molar masses can be significantly improved by microwaves (Scheme 2.11b) (Weiss, Wittmeyer, and Rehahn, unpublished results). If the components diboronic ester **33**, dibromide **32**, Na_2CO_3, and $Pd(PPh_3)_4$ were mixed in the ratio 1 : 1 : 5 : 0.01 and treated in THF/water for 1–10 minutes with microwaves (150 W) at 9 bar, polymer **34** was obtained in yields exceeding 82% with M_n = 21 500–24 000. These masses correspond to P_n = 54–60, which represents an increase compared to conventional conditions by roughly a factor of 3. It should be noted at this point that this concrete improvement rested upon the homogeneity of the reaction medium and the use of a preformed catalyst, like the one mentioned. Obviously, microwaves have some potential even regarding molar masses but much more work is needed before the full scope of this technique in SPC is understood. It would be interesting to see how microwaves affect the outcome in one of the optimized cases, where already under conventional conditions molar masses in the range of 100 000 are reached.

Can the mass be tripled and is homogeneity a critical issue also here? As long as the molar mass cannot be increased in those cases, there is always some ambiguity that the masses in the above cases increased only because of an accidentally better match of stoichiometry and not because of microwaves.

In recent years, microreactors were specifically designed and adjusted to optimally match the commonly biphasic SPC conditions. This aimed at providing a technology that allows a continuous small but nevertheless technical scale production of polyarylenes. Some of these developments took and are still taking place at the Institut für Mikrotechnik, Mainz, Germany (IMM). Sufficient mass transfer in biphasic reactions is an obvious issue and two concepts were developed at IMM to ensure a sufficient interfacial area during SPC. The first is based on the repeated breakage of droplets with the help of a so-called redispersion microreactor consisting of exchangeable redispersion units separated by multichannels and an inspection window. This window allows continuous monitoring and investigating of flow patterns. The second concept uses a metal foam filled tube reactor as a kind of static mixer. The microstructure of the foam is chosen with a high porosity such that a pressure drop as low as possible is achieved. All details are of course proprietary information, but it appears that higher conversions in SPC are reached with the second concept. The reader can only be referred to some conference proceedings [76]. There also seems to be a successful development of a fixed-bed column reactor in which SPC can be done in an astoundingly short residence time of 20 minutes leading to relatively narrowly distributed, high molar mass products.

The authors do not claim to be insiders into what industry does in terms of SPC. There are however a number of indications [77] that certain polyarylenes, specifically of the PF type (Section 2.5.2), are being synthesized on a technical scale. This is said to be in the range of a few hundred kilograms up to tons per year for low molar mass products, and 50–100 kg per year for high molar mass ones [78], scales which match perfectly well with the potential of microreactors. Several companies are active in the field of organic electronics, some of which have polyarylenes in their portfolio. It is reasonable to assume that many of them are being synthesized by SPC [77]. The Yamamoto protocol [79] has the main disadvantage that it is more difficult and costly to remove residual traces of transition metal, which for this method is Ni rather than Pd. Ni is used in stoichiometric amounts and even if catalyzed versions are run, a reducing agent is required in stoichiometric amounts. Additionally there are limitations with regard to achievable sequences.

2.5 Selected Classes of Polyarylenes and Related Polymers

2.5.1 Poly(*para*-phenylene)s

This section covers only recently published articles on PPPs synthesized by SPC. Examples reported before 2001 are contained in the earlier review articles [6b].

In the years since then, the main topics focusing on these rigid rod polymers included (i) the aggregation behavior of amphiphilic representatives in solution, (ii) the use of charged versions as models to study polyelectrolyte properties, and (iii) molecular reinforcement. Also, some cases were reported for which synthetic or other aspects were in the center of interest. Moreover, conformationally rigid and rodlike polyelectrolytes are ideal models for polyelectrolytes in general, because the solution properties are controlled by intermolecular Coulombic interactions. The shape (chain conformation) of these charged macromolecules does not change with the ionic strength of the surrounding medium due to their inherent rigidity. Excluded volume effects have to be taken into consideration at higher concentration only. On the basis of earlier work by Rulkens [80] and Bockstaller [81] on PPP **12** (Figure 2.3) the aggregation behavior of the corresponding hydrolyzed version with free acid sulfonate groups, **35** (Figure 2.6) was investigated. In highly dilute aqueous solutions these polymers self-organize into anisotropic cylindrical micelles. The cross-section of these micelles contains ~15 individual macromolecules and their length amounts to approximately five times the contour length of the individual constituents. They can arrange into even more ordered motifs like closed loops and lasso-like structures, a phenomenon most likely driven by entropically motivated avoidance of fringes of dangling chains [82]. Other studies on related matters use nonchargeable low molar mass amphiphilic PPP **37** [83]. Additional potentially lengthwise segregating PPP derivates were reported by Valiyaveettil *et al.* [84].

The substituents of the above polymers were designed such that they can segregate lengthwise into hydrophilic and hydrophobic parts. Conformation **35a** visualizes this: both sulfonic acid groups simultaneously point toward the opposite side as opposed to dodecyl chains. If this segregation along the backbone is rendered impossible by introducing two short lateral substituents with terminal charges like in **36**, formation of cylindrical aggregates is suppressed. Polymer **36** and a variety of related systems could therefore be investigated as molecularly dispersed rigid rod polyelectrolytes and the results be compared with theory [85]. In the same context, PPP **38** is also an interesting development. It has the additional advantage that it is readily soluble in water without any ionic group and that the charge parameter can be systematically increased by increasing protonation or quaternization of the amino functionalities [86].

Other rather complex substitution patterns were realized in PPPs **39**–**41** [87], although high degrees of polymerization could not be achieved in all cases.

Finally, the aspect of molecular reinforcement [88] should be touched upon. Eisenbach *et al.* showed that by mixing of charged PPP-based rods, for which PPP **42** can serve as an example, with countercharged coiled macromolecules the mechanical properties of these ionomeric nanocomposites depend strongly on the molecular parameters of the rod component, that is, the degree of polymerization and the rod volume fraction [89].

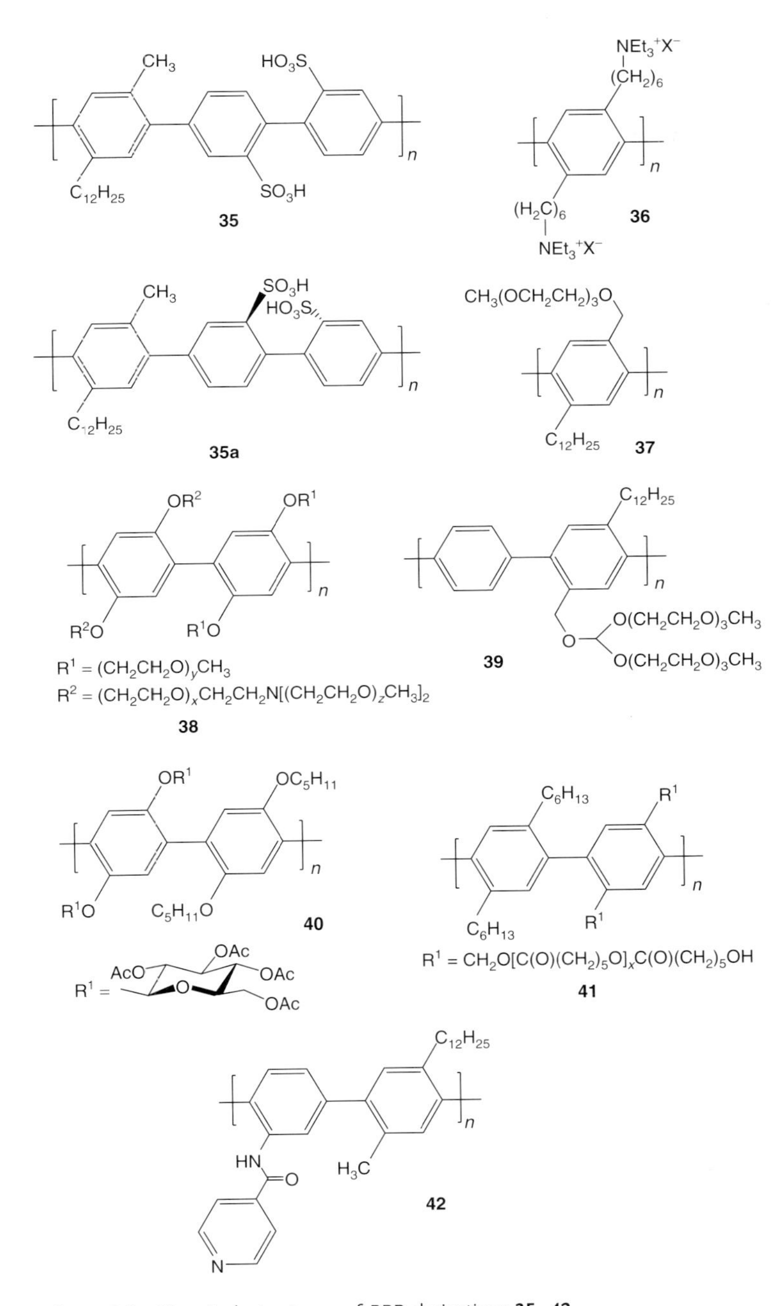

Figure 2.6 Chemical structures of PPP derivatives **35–42** with lateral substituents, that is, designed so as to induce specific aggregation behavior, test polyelectrolyte theories and the concept of molecular reinforcement.

2.5.2 Polyfluorenes

PFs represent a subclass of PPPs. Their basic chemical structure is shown in Figure 2.7. The *para*-phenylene repeating units are bridged pair-wise by methylene groups (the C-9). This reduces the tilt angle between the bridged phenylene units to almost 0°. Exceptional electrical, optical, and optoelectronic properties are the result: PFs can act simultaneously as semiconductors and efficient electroluminescent emitters. The bridge also forces the backbone out of linearity. Each repeat unit's director is kinked by an angle 2α resulting in a reduced persistence as compared with PPP. Another important feature of the PFs is that lateral substituents can be introduced at position C-9. The electronic properties of PF are largely dictated by the dihedral angle between consecutive repeat units, which themselves are nearly planar. Substituents in the ortho positions to the connecting CC bond force the dihedral angle invariably into high values, too high to allow for sufficient remaining cross-conjugation. On the other hand, in order to get blue instead of green emission the dihedral angle should not be too small either. It seems that the steric repulsion between the ortho hydrogens leads to an optimum dihedral angle. Any solubilizing or shielding substituents should not interfere with this delicate interplay. At C-9 their direct influence is the least. This is why in practically all cases this position is the one used for attaching substituents to PF.

Soon after the discovery of the first polymer-based OLEDs by Holmes *et al.* [90], PF derivatives were tested in such devices [91]. After considerable optimization [92], PF derivatives have gained commercial relevance [93]. PFs are typically synthesized by either the Yamamoto protocol [79] or SPC. On the basis of press releases [77–79], SPC is in fact an industrially applied method and seems to have gained the higher importance of the two. Because of the stiff competition in the field of OLEDs, an enormous variety of derivatives and modifications as well as copolymers and blends were published in both the open and patent literature, which cannot possibly be covered here. For more details, the reader may consult a pertinent recent review [6c, 93d] or consult Chapter 9 of the present book.

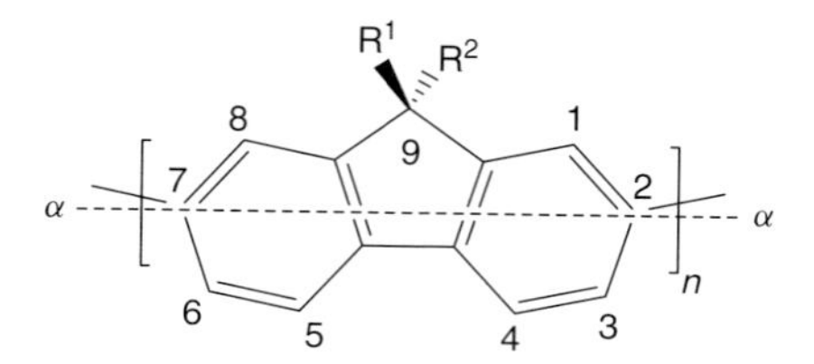

Figure 2.7 Chemical structure of a PF substituted at its 9-position with two groups R^1 and R^2 that serve the purpose to increase the parent polymer's poor solubility and processability. The bridging of the two phenylene units by the C-9 carbon causes the PF repeat unit to deviate from linearity by the angle α.

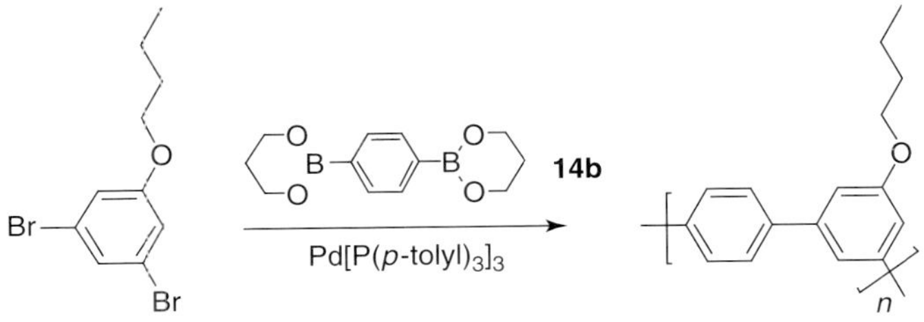

43 → 44 (14b, Pd[P(p-tolyl)3]3)

Scheme 2.12 Synthesis of poly(*para-meta*-phenylene) **44** from monomers **43** and **14b**. The high molar mass part has interesting mechanical properties rivaling those of the tough, amorphous polycarbonate.

2.5.3
Poly(*para-meta*-phenylene)s

Recently, it was shown that SPC can provide access to polymers with attractive mechanical properties [21]. Scheme 2.12 shows the synthetic route to the polyarylene **44** with alternatingly *para-* and *meta*-connected phenylene units. Short butyloxy chains at every other phenylene unit mediated solubility and processability. The raw polymer had $M_w = 83\,000\,g\ mol^{-1}$ ($P_w = 370$) with a relatively broad molar mass distribution and was obtained on the 3.5 g scale. Fractionating precipitation afforded a 1 g fraction with $M_w = 255\,000\,g\ mol^{-1}$ ($P_w = 1140$), whose NMR spectra are shown in Figure 2.8. Despite the respectable molar mass, narrow signals were obtained, which could be unequivocally assigned to the proposed structure. Melt-compression molded films (thickness ~150 μm) were flexible and tough and could be deformed well beyond the yield point. For comparison, stress–strain curves were also obtained, of comparably thick melt-compression molded films of common atactic PS, PMMA, and commercial polycarbonate (PC, MacrolonLQ-2847) (Figure 2.9). Although the films of polyarylene **44** were of a somewhat lower stiffness, they featured a nominal stress at a break point similar to the above polymers, but, remarkably, exhibited a macroscopic elongation at a break point that surpasses those of the brittle bulk polymers a-PS and PMMA, and approaches that of PC, which is reputed for its toughness.

During melt-compression molding the films of **44** turned brownish. This was believed to be a result of residual traces of catalyst in the sample. The amount of catalyst [Pd[P(*p*-tolyl)$_3$]$_3$] was therefore systematically reduced to 0.03%, which resulted in a colorless, fully transparent yet high molar mass material. Such a film cast from chloroform and then annealed in a press at 230 °C for 5 minutes (thus, well above $T_g = 166\,°C$) could be stretched at 180 °C easily by 700–800% (Figure 2.10) (Feldman *et al.* in preparation). This example shows well that SPC is a powerful polymerization protocol, which even allows entry into the domain of high performance polymers, a field of exploration that had not been seriously considered

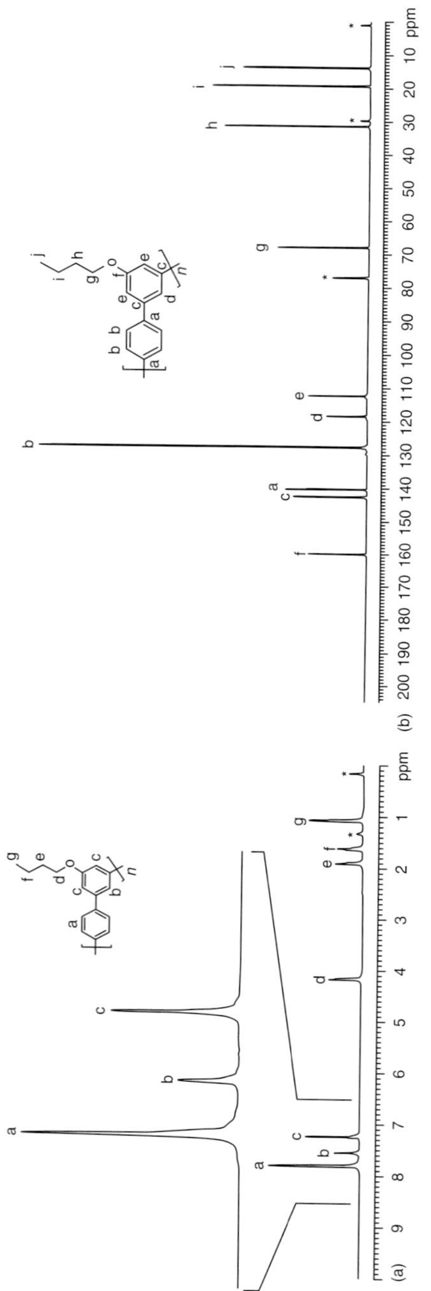

Figure 2.8 (a) 700 MHz ^1H and 1(b) 176 MHz ^{13}C NMR spectra with signal assignment of the fraction of polymer **44** with $M_w = 255\,000$ ($P_w = 1140$) to show the level of structure perfection achieved.

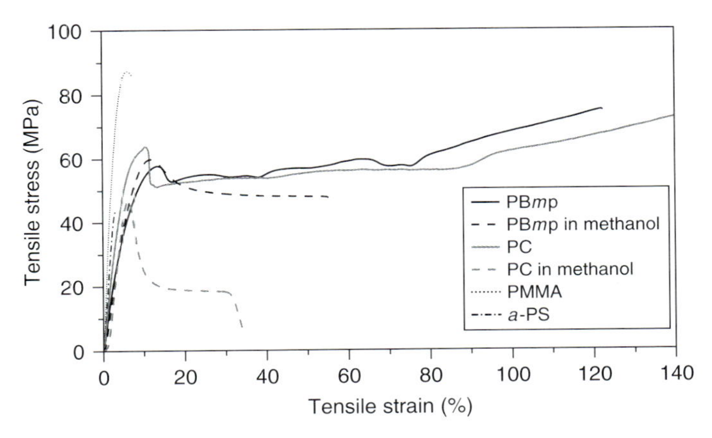

Figure 2.9 Stress–strain curves, recorded at room temperature of melt-compression molded films of poly(butyloxy-*para*-*meta*-phenylene) (PB*m*P) (**44**) (high molar mass part). For reference purposes, corresponding curves of a-PS, PMMA, and PC are also shown, illustrating the excellent mechanical properties of the new polyarylene. In addition, plotted are the stress–strain curves of PB*m*P and PC samples recorded while immersed in methanol, showing the superior stress-cracking resistance of the former polymer toward this polar organic liquid.

Figure 2.10 Polymer **44** after stretching at 180 °C by 800%.

in the initial years after the method's emergence. Last but not least, properly substituted poly(*para*-*meta*-phenylene)s can be considered polymeric versions of foldamers [94]. For example, if hydrophilic substituents are introduced to the hydrophobic main chain of **44** instead of the butyloxy groups, solvophobically driven helix formation is expected in a polar solvent [95].

2.5.4
Shielded Polyarylenes

Introducing sterically demanding substituents like dendrons of varying generations to each repeat unit of polymers has been considered a concept of insulating backbones from the surrounding environment and, thus, shielding it (Figure 2.11).

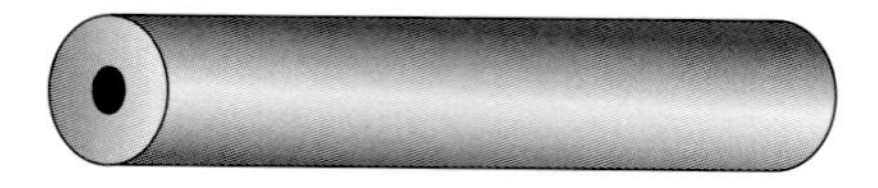

Figure 2.11 Cartoon of an active backbone (in black) surrounded by a shielding dendritic layer (gray).

Quite a few dendronized [96] PPPs were synthesized in the 1990s using SPC [97]. The first experiments proving the existence of shielding effects [98] were reported by Aida *et al.*, who showed that the collisional quenching of the photoluminescence of poly(*para*-phenylene-ethynylene)s can increasingly be suppressed by attaching dendrons of progressively higher generations (up to four) [99].

This concept was then also applied to poly(*para*-phenylene-vinylene)s [100] poly(triacetylene)s [101], poly(acetylene)s [102], PFs (see Section 2.5.2), polythiophenes [103], and other polyarylenes [104]. Except for the PF cases, SPC was not used much as a synthesis method. Only in recent years, it was employed again, more often. Besides dendronization, perhaps the most commonly used shielding concept is to protect the backbones by threading cyclic guests like cyclodextrines onto them. Anderson presented a few such elegant cases, mostly but not exclusively based on SPC [105]. Scheme 2.13 shows the concept and a concrete realization.

Shielding not only prevents aggregation but also possibly reduces collisional quenching of fluorescent backbones. It also has an impact on the polymers' solvatochromic behavior and conformational effects [105b]. Percec used the case of a dendronized polyacetylene to point out that the steric load around a backbone created by a dendritic shell can also efficiently stabilize it against intramolecular *cis-trans* isomerization and electrocyclic reactions [102b]. Along the same lines, intermolecular chemical reactions would be hindered.

2.5.5 Miscellaneous

Given the technological importance of regioregular polythiophenes, it is not surprising that this class of semiconducting polymers was not only approached by Kumada-type polymerizations (see brief mentioning in Section 2.4.5) but also by SPC. In 1998, Guillerez *et al.* [106] reported on the synthesis of poly(3-octylthiophene) **46** starting from AB monomer **45** in a yield of 55% and with $M_w \sim 27\,000$. The regioregularity of this fraction was estimated to be 96–97% head to tail (Scheme 2.14). Shortly thereafter, Janssen *et al.* widened the repertoire of thiophene-containing SPC polymers by copolymers of the kind **47** [29] and **48** [107] and also reported on another route to **46** starting from the AB monomer **49** with iodo and pinacolboronate functional groups [108]. Although the exact numbers are difficult to compare, the molar masses of the Janssen protocol are not only in the same range as the one by Guillerez but are also comparable to what has been published in the open literature by McCullough using Kumada chemistry [109]. The study by Janssen is an interesting reading because it carefully addresses the

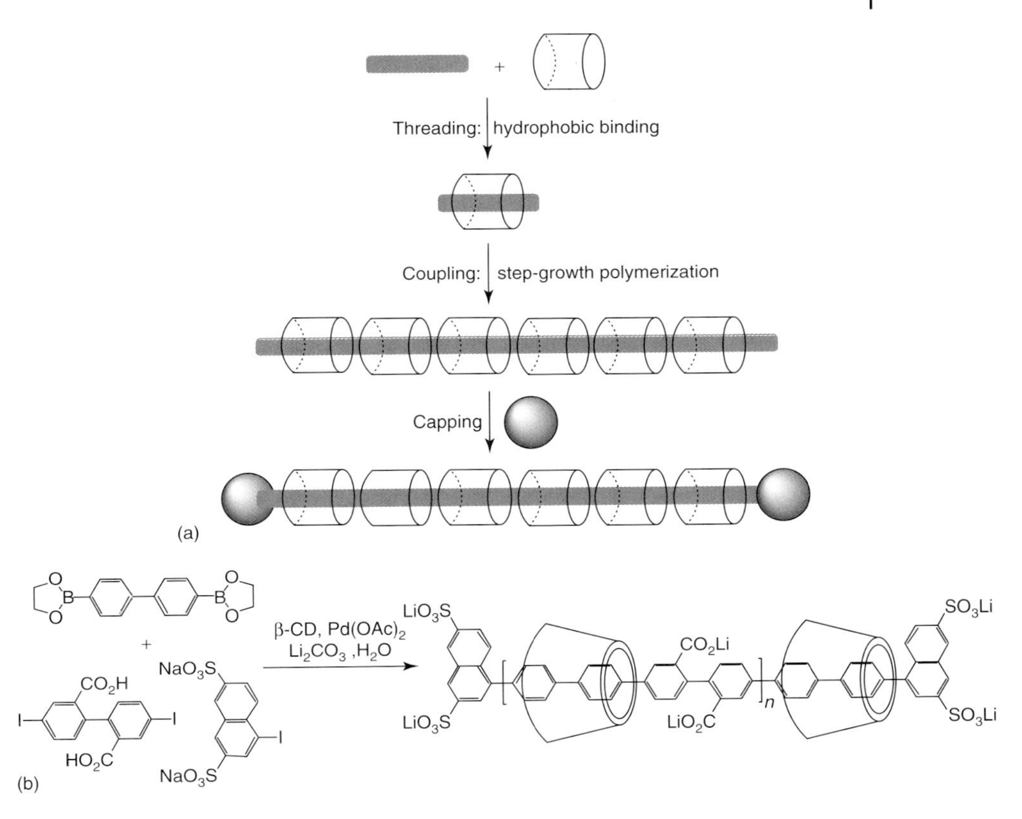

Scheme 2.13 A concept to shield, for example, a poly(*para*-phenylene) backbone (a) by doing an aqueous SPC in the presence of β-cyclodextrines (β-CD) (b). Threading efficiencies of 1.1 β-CD per repeat unit were achieved.

issue of the impact of catalysts on end group patterns. The reader is also referred to a recent review by McCullough [110]. In 2006, the poly(3-alkylthiophene) story was continued by a contribution from Higgins and the former Merck chemist McCulloch, who facilitated access to monomer **51** by treating the starting compound **50** with an Ir catalyst and (pin)B-B(pin) in the presence of a bipyridine ligand to obtain **51** in a yield of 97% (see also Section 2.4.1). This work describes an important step because it avoids the use of organolithium reagents, which are otherwise required to introduce boronic acids or boronate esters. McCulloch also described a new generation of bulky, electron-rich Pd(0)-phosphane catalysts being highly effective for the SPC preparation of regioregular polyalkylthiophenes [40]. Finally, in 2008, Bo's lab described a new Pd(0) complex stabilized by three tri(2-thienyl)phosphine ligands and showed it to be superior to $Pd(PPh_3)_4$ in SPC of AB thiophene-based SPC monomers [111]. Analyzing all these works somewhat critically one has to conclude that a real breakthrough of a high molar mass combined with a high degree of regioregularity is still awaiting to be achieved.

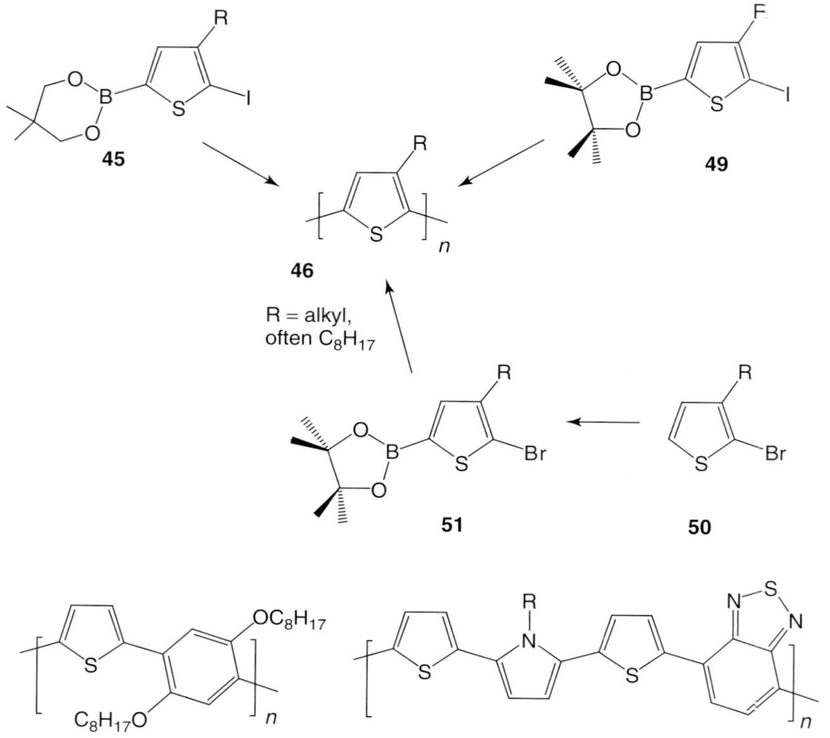

Scheme 2.14 Routes to regioregular polythiophene **46** and structures of thiophene-containing copolymers **47** and **48**.

While dealing with electron-rich aromatics like thiophenes it should be mentioned that there are also a number of polymers with ferrocenes as part of the backbone [112].

Hyperbranched polymers have already been synthesized early in SPC history [113]. Recent examples include the hyperbranched poly(phenylene-ethynylene)s **53**, which were obtained from the SPC AB_2 monomers **52** (Figure 2.12) [114]. The same laboratory has also described triphenylamine-based hyperbranched polymers using the AB_2 monomers **54** and **55**. Branching unit **54** was polymerized without an additional component and also starting from a short oligofluorene core; AB_2 monomer **55** was polymerized without an additional component. The polymers obtained were tested for their hole transporting capability in thin layer electrooptical devices.

Tieke *et al.* have developed a whole class of conjugated polymers containing the 1,4-dioxo-3,6-diphenylpyrrolo[3,4-*c*]pyrrole (DPP) unit. DPP and its derivatives are brilliant red technical pigments. Upon incorporation into polymers **56–58** (Figure 2.12) highly luminescent materials were obtained with photoemission

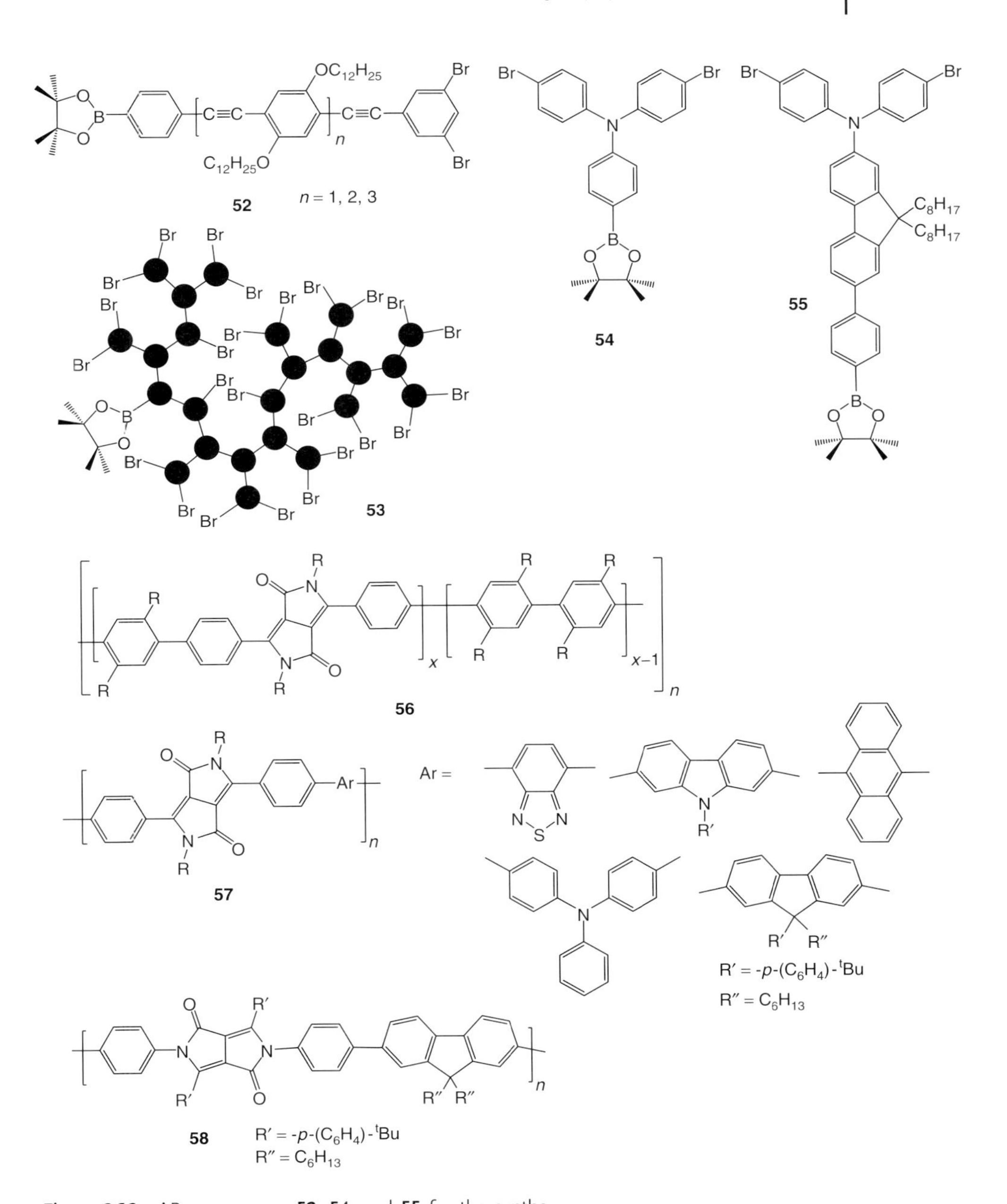

Figure 2.12 AB_2 monomers **52**, **54**, and **55** for the synthesis of hyperbranched polymers like **53** (from **52**) and structures of the highly luminescent polymers **56–58**. The black dots in **53** represent the entire carbon skeleton of monomer **52**.

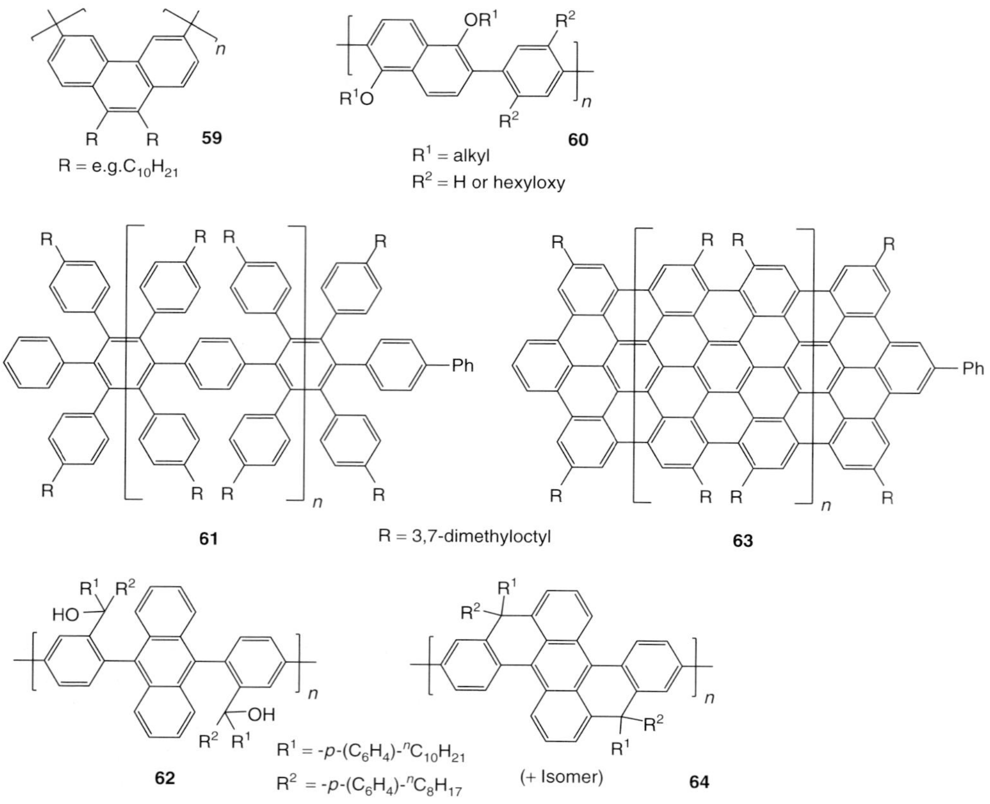

Figure 2.13 Structures of poly(phenanthrenylene)s **59**, poly(naphthylenephenylene) **60**, and the precursor polymers **61** and **62** for the laterally extended conjugated targets **63** and **64**, respectively.

maxima between $\lambda = 525-600$ nm and photoluminescence quantum yields up to 86% [115]. Müllen and Kijima reported on polyphenanthrylenes **59** [116] and poly(naphthylene phenylene)s of the kind **60** (Figure 2.13), respectively [117], and Rehahn and Klemm on phenanthroline [112, 118] and bipyridine-containing polyarylenes, respectively [119]. Finally, a few cases should be addressed in which unsaturated units like allenes, acetylenes, and olefins were incorporated between aromatic units thus furnishing poly(arylene allenylene)s [120], poly(arylene ethynylene)s [121], and poly(arylene vinylene)s [122]. Although the classical methods for synthesizing simple representatives of the latter two kinds of polymers are in most cases superior, it is interesting to note that such polymers can now also been accessed by SPC. This widens the available representatives, although considerable optimization work is still required in several of the mentioned cases to evaluate the full potential of SPC. The poly(phenylene-ethynylene) by Therien [36] is of special interest in that its synthesis uses monomer **19** (Figure 2.5), which is the simplest possible acetylene monomer in SPC. Unfortunately, the NMR spectra of the polymer cannot yet be completely assigned to the molecular structure.

Having now highlighted some developments leading to single-stranded polymers, which are supposed to remain as such, this short section will be wrapped up by two examples from Müllen's laboratory that provide access to more spatially extended conjugated polymers. Also, here, single-stranded structures, namely **61** and **62** are the initial goal. However, these polymers are designed such that they can be converted by some follow-up chemistry in the laterally more extended congeners **63** and **64**, respectively [123]. The broader context of the former work is to be seen in an attempt to create model structures for graphene [124]. A two-step approach was also applied to an *m*-phenylene precursor to obtain a helicene-type structure by Nishide *et al.* [125].

2.6 Conclusions and Outlook

Since its first report in 1989 [5], SPC has been developed into a powerful synthetic tool, which provides access to numerous polyarylenes and related unsaturated polymers of high molar mass. Some of these polymers have gained technical scale application, which underscores the importance and potential of the method. The key to this success is the SMC reaction, which, for conversion, is one of the absolutely best chemical reactions known so far. In SPC, degrees of polymerization of 1000 can be achieved; this translates into unbeatable conversions per bond formation step of 99.99%. With such a performance, SPC and, thus, SMC easily compete or may be even superior to catalyzed Huisgen dipolar cycloaddition of azides with acetylenes, which is sometimes referred to as *click* reaction [126].

To play the full scope of SPC for the synthesis of high molar mass polymers, reaction conditions have to be optimized. This chapter provides a guideline as to how to do this successfully. There are some cases in the literature where not even the simplest optimization was done and there is no doubt therefore, that the corresponding results are suboptimal. Although there may in fact be intrinsic reasons why in certain cases only poor results are obtained, for example, when electron-rich monomers are used, better results could have been achieved if the method had been used properly. Every hour that is invested into monomer purification, for example, pays off 10-fold when it comes to polymerization.

While some directions of SPC have reached maturity, others are still in their infancy. The use of chloroaromatics as monomers is one of the latter cases [50]. Also, regarding the interesting approach to regioregular polythiophene, the authors are not sure whether the optimum has already been reached. It seems that newly developed catalysts considering, *e.g.* the retarded oxidative addition could help improve here. Furthermore, the recent progress in asymmetric SMC that is mentioned [62], sterically demanding SMC [61], as well as orthogonal SMC [127] will also impact SPC and broaden its scope for polymer synthesis. Other aspects where SPC could have considerable impact have practically not been explored at all. This comprises the use of cheap metals like Fe instead of Pd [128]; the synthesis of high molar mass poly(*meta*-arylene)s with the potential to fold-up into helices with defined pores [94]; the synthesis of tactic polyarylenes [129], block copolymers by catalyst transfer polycondensation [68]; and attempts to grow polyarylenes off solid substrates [130]. Also, sp^2/sp^3 and sp^3/sp^3 CC couplings have not yet been really tried for polymer synthesis, although in low molar mass, organic chemistry–related examples are known [131]. As this short list of important yet virtually unexplored field shows, we believe that SPC can be developed further and that polymer chemistry and materials chemistry will witness more exciting findings in the years to come.

Acknowledgments

The authors express there deep thanks to all their coworkers involved over the years for their substantial input. Special thanks go to Prof. Gerhard Wegner, Max Planck Institute for Polymer Research, Mainz; Prof. Norio Miyaura, Hokkaido University, Sapporo; Prof. Martin Heeney, Imperial College, London; and Dr Niels Schulte, Merck, Frankfurt, for their invaluable advice during writing. This work was financially supported over the years by the Max Planck Society, Free University Berlin, Technical University Darmstadt, ETH Zürich, Deutsche Forschungsgemeinschaft, and the Swiss National Science Foundation for which we are thankful.

References

1. For a review, see: de Meijere, A. and Diederich, F. (2005) in *Metal-catalyzed Cross-coupling Reactions, Second Completely Revised and Enlarged Edition*, Wiley-VCH Verlag GmbH.
2. (a) Tamao, K., Sumitani, K., and Kumada, M. (1972) *J. Am. Chem. Soc.*, **94**, 4374–4376; (b) Corriu, R.J.P. and Masse, J.P. (1972) *J. Chem. Soc. Chem. Commun.*, 144.
3. Yamamoto, T., Hayashi, Y., and Yamamoto, Y. (1978) *Bull. Chem. Soc. Jpn.*, **51**, 2091.
4. (a) Miyaura, N., Yamada, K., and Suzuki, A. (1979) *Tetrahedron Lett.*, 3437; (b) Miyaura, N. and Suzuki, A. (1979) *J. Chem. Soc. Chem. Commun.*, 866; (c) Miyaura, N., Yanagi, T., and Suzuki, A. (1981) *Synth. Commun.*, **11**, 513; Also, see: (d) Miller, R.B. and Dugar, S. (1984) *Organometallics*, **3**, 1261.
5. Rehahn, M., Schlüter, A.D., Wegner, G., and Feast, W.J. (1989) *Polymer*, **30**, 1060.
6. (a) Schlüter, A.D. and Wegner, G. (1993) *Acta Polym.*, **44**, 59; (b) Schlüter, A.D. (2001) *J. Polym. Sci., Part A: Polym. Chem.*, **39**, 1533; (c) Sakamoto, J., Rehahn, M., Wegner, G., and Schlüter, A.D. (2009) *Macromol. Rapid Commun.*, **30**, 653.
7. (a) Sangvikar, Y., Sakamoto, J., and Schlüter, A.D. (2008) *Chimia*, **62**, 678 (b) Sangvikar, Y., Fischer, K., Schmidt, M., Schlüter, A.D., and Sakamoto, J. (2009) *Org. Lett*, **11**, 4112.
8. Bo, Z. and Schlüter, A.D. (2000) *Chem. Eur. J.*, **6**, 3235.
9. Onak, T. (1987) *Boron and Oxygen, Gmelin Handbook of Inorganic and Organometallic Chemistry*, vol. 2, B 3rd Supplement, Springer, Berlin, p. 47.

10. Barder, T.E., Walker, S.D., Martinelli, J.R., and Buchwald, S.L. (2005) *J. Am. Chem. Soc.*, **127**, 4685.
11. Kuivila, H.G., Reuwer, J.F., and Mangravite, J.A. (1963) *Can. J. Chem.*, **41**, 3081.
12. Miyaura, N. (2005) in *Metal-Catalyzed Cross-Coupling Reactions, Second Completely Revised and Enlarged Edition*, vol. 1 (eds A. de Meijere and F. Diederich), Wiley-VCH Verlag GmbH, p. 41.
13. Inbasekaran, M., Wu, W., and Woo, E.P. (1999) WO 9920675 A1.
14. Towns, C.R. and O'Dell, R. (2000) WO 0053656 A1.
15. Treacher, K., Stössel, P., Spreitzer, H., Becker, H., and Falcou, A. (2005) US Patent 6,956,095 B2.
16. (a) Towns, C.R. and Wallace, P. (2006) US Patent 7,074,884; (b) Towns, C.R., Wallace, P., Allen, I., Pounds, T., and Murtagh, L. (2007) US Patent 7,173,103.
17. Liu, C., Repolev, A., and Zhou, B. (2008) *J. Polym. Sci., Part A: Polym. Chem.*, **46**, 7268.
18. Murage, J., Eddy, J.W., Zimbalist, J.R., McIntyre, T.B., Wagner, Z.R., and Goodson, F.E. (2008) *Macromolecules*, **41**, 7330.
19. For some other optimization studies, see: (a) Frahn, J., Karakaya, B., Schäfer, A., and Schlüter, A.D. (1997) *Tetrahedron*, **53**, 15459; (b) Kowitz, C. and Wegner, G. (1997) *Tetrahedron*, **53**, 15553; (c) Molina, R., Gómer-Ruiz, S., Montilla, F., Salinas-Cartillo, A., Fernandez-Arroyo, S., del Mar Rauels, M., Micol, V., and Mallavia, R. (2009), *Macromolecules*, **42**, 5471.
20. For details on solvent and base effects for the Suzuki-Miyaura coupling, see Ref. [12].
21. Kandre, R., Feldman, K., Meijer, H.E.H., Smith, P., and Schlüter, A.D. (2007) *Angew. Chem. Int. Ed. Engl.*, **46**, 4956.
22. For an early study on the reduction of catalyst amounts in SPC, see: Rehahn, M. (1990) Metallkatalytisierte kupplungsreaktionen als methode zum aufbau hochmolekularer poly-p-phenylene und anderer polyaromatischer polymere. Thesis, University of Mainz, Mainz.
23. (a) Walker, S.D., Barder, T.E., Martinelli, J.R., and Buchwald, S.L., (2004) *Angew. Chem. Int. Ed. Engl.*, **43**, 1871; (b) Barder, T.E., Walker, S.D., Martinelli, J.R., and Buchwald, S.L. (2005) *J. Am. Chem. Soc.*, **127**, 4685.
24. Arvela, R.K., Leadbeater, N.E., Sangi, M.S., Williams, V.A., Granados, P., and Singer, R.D. (2005) *J. Org. Chem.*, **70**, 161.
25. Adamo, C., Amatore, C., Ciofini, I., Jutand, A., and Lakmini, H. (2006) *J. Am. Chem. Soc.*, **128**, 6829.
26. Homocouplings in Suzuki Miyaura cross-coupling are known: (a) Campi, E.M., Jackson, R., Marcuccio, S., and Naeslund, C.G.M. (1994) *J. Chem. Soc., Chem. Commun.*, 2395; (b) Gillmann, T. and Weeber, T. (1994) *Synlett*, 649; (c) Song, Z.Z. and Wong, H.N.C. (1994) *J. Org. Chem.*, **59**, 33; (d) Moreno-Manas, M., Pérez, M., and Pleixats, R. (1996) *J. Org. Chem.*, **61**, 2346; (e) Aramendia, M.A., Lafont, M., Morena-Manas, M., Pérez, M., and Pleixas, R. (1999) *J. Org. Chem.*, **64**, 3592; Homocouplings have also been reported for SPC: (f) Koch, F. and Heitz, W. (1997) *Macromol. Chem. Phys.*, **198**, 1531.
27. Kappaun, S., Zelzer, M., Bartl, K., Saf, R., Stelzer, F., and Slugovc, C. (2006) *J. Polym. Sci. Polym. Chem. Ed.*, **44**, 2130.
28. (a) Miyaura, N., Yamada, K., Suginome, H., and Suzuki, A. (1985) *J. Am. Chem. Soc.*, **107**, 972; (b) Widenhoefer, R.A., Zhong, H.A., and Buchwald, S.L. (1997) *J. Am. Chem. Soc.*, **119**, 6787.
29. For a careful end group analysis of an SPC product by MALDI-TOF mass spectrometry, for example, see: Jayakannan, M., van Dongen, J.L.J., and Janssen, R.A.J. (2001) *Macromolecules*, **34**, 5386.
30. (a) Konigsveld, R. (1970) *Adv. Polym. Sci.*, **7**, 1; (b) Konigsveld, R. (1985) *Polym. Eng. Sci.*, **25**, 17; (c) Mencer, H.J. (1988) *Polym. Eng. Sci.*, **28**, 8.

31. Nielsen, K.T., Bechgaard, K., and Krebs, F.C. (2005) *Macromolecules*, **38**, 658.
32. Vanhee, S., Rulkens, R., Lehmann, U., Rosenauer, C., Schulze, M., Köhler, W., and Wegner, G. (1996) *Macromolecules*, **29**, 5136.
33. Harre, K. and Wegner, G. (2006) *Polymer*, **47**, 7312.
34. Rehahn, M., Schlüter, A.D., and Wegner, G. (1990) *Makromol. Chem.*, **191**, 1991.
35. Rehahn, M., Schlüter, A.D., and Wegner, G. (1990) *Makromol. Chem., Rapid Commun.*, **11**, 535.
36. Kang, Y.K., Deria, P., Carroll, P.J., and Therien, M.J. (2008) *Org. Lett.*, **10**, 1341.
37. (a) Ishiyama, T., Murata, M., and Miyaura, N. (1995) *J. Org. Chem.*, **60**, 7508; For a recent example of *in situ* borylation at SPC using this method, see: (b) Walczak, R.M., Brookins, R.N., Savage, A.M., van der Aa, E.M., and Reynolds, J.R. (2009) *Macromolecules*, **42**, 1445.
38. Murata, M., Watanabe, S., and Masuda, Y. (1997) *J. Org. Chem.*, **62**, 6458.
39. (a) Nguyen, P., Blom, H.P., Westcott, S.A., Taylor, N.J., and Marder, T.B. (1993) *J. Am. Chem. Soc.*, **115**, 9329; (b) Waltz, K.M. and Hartwig, J.F. (1997) *Science*, **277**, 211; (c) Waltz, K.M., Muhoro, C.N., and Hartwig, J.F. (1999) *Organometallics*, **18**, 3383; (d) Chen, H. and Hartwig, J.F. (1999) *Angew. Chem. Int. Ed. Engl.*, **38**, 3391; (e) Ishiyama, T. and Miyaura, N. (2003) *J. Organomet. Chem.*, **680**, 3; (f) Ishiyama, T. and Miyaura, N. (2004) *Chem. Record*, **3**, 271.
40. (a) Tse, M.K., Cho, J.-Y., and Smith, M.R.III (2001) *Org. Lett.*, **3**, 2831–2833; (b) Ishiyama, T., Takagi, J., Yonekawa, Y., Hartwig, J.F., and Miyaura, N. (2003) *Adv. Synth. Catal.*, **345**, 1103; For SPC, see also: (c) Liversedge, I.A., Higgins, S.J., Giles, M., Heeney, M., and McCulloch, I. (2006) *Tetrahedron Lett.*, **47**, 5143.
41. (a) Ishiyama, T., Matsuda, N., Miyaura, N., and Suzuki, A. (1993) *J. Am. Chem. Soc.*, **115**, 11018; For compound **13** with R = C_3H_7 or Ph, SPC was recently tried with 1,2-dibromoethylene aiming at polyacetylene derivatives. Only strongly colored, mostly insoluble products were obtained: (b) Yamamoto, T., Kobayashi, K., Yasuda, T., Zhou, Z.-H., Yamaguchi, I., Ishikawa, T., and Koshihara, S.-Y. (2004) *Polym. Bull.*, **52**, 315.
42. Hyuga, S., Yamashina, N., Hara, S., and Suzuki, A. (1988) *Chem. Lett.*, 809.
43. (a) Noguchi, H., Hojo, K., and Suginome, M. (2007) *J. Am. Chem. Soc.*, **129**, 758; (b) Noguchi, H., Shioda, T., Chou, C.-M., and Suginome, M. (2008) *Org. Lett.*, **10**, 377.
44. (a) Gillis, E.P. and Burke, M.D. (2007) *J. Am. Chem. Soc.*, **129**, 6716; (b) Lee, S.J., Gray, K.C., Paek, J.S., and Burke, M.D. (2008) *J. Am. Chem. Soc.*, **130**, 466.
45. For other protecting groups for boronic acids, see: (a) Luithle, J.E.A. and Pietruszka, J. (2000) *J. Org. Chem.*, **65**, 9194; (b) Molander, G.A. and Ellis, N. (2007) *Acc. Chem. Res.*, **40**, 275; Note that the protecting groups mentioned here may still be active in Suzuki Miyaura cross-coupling. For a recent study on orthogonal reactivity between organotrifluoro borate and trialkyl borate, see: (c) Molander, G.A. and Sandrock, D.L. (2008) *J. Am. Chem. Soc.*, **130**, 15792.
46. Trimethylsilyl group (TMS) for halides, see for example: (a) Hensel, V. and Schlüter, A.D. (1999) *Chem. Eur. J.*, **5**, 421; (b) Hensel, V. and Schlüter, A.D. (1999) *Eur . J. Org. Chem.*, 451; (c) Triazenes for halides, see for example: Moore, J.S., Weinstein, E.J., and Wu, Z. (1991) *Tetrahedron Lett.*, **32**, 2465; (d) Kimball, D.B. and Haley, M.M. (2002) *Angew. Chem. Int. Ed. Engl.*, **41**, 3338.
47. (a) Negishi, E. (1982) *Acc. Chem. Res.*, **15**, 340; (b) Ciattini, P.G., Morera, E., and Ortar, G. (1992) *Tetrahedron Lett.*, **33**, 4815; (c) Soderquist, J.A., Matos, K., Rane, A., and Ramos, J. (1995) *Tetrahedron Lett.*, **36**, 2401; (d) Fürstner, A. and Seidel, G. (1995) *Tetrahedron*, **51**, 11165; (e) Bumagin,

N.A. and Bykov, V.V. (1997) *Tetrahedron*, **53**, 14437; (f) Cammidge, A.N., Goddard, V.H.M., Gopee, H., Harrison, N.L., Hughes, D.L., Schubert, C.J., Sutton, B.M., Watts, G.L., and Whitehead, A.J. (2006) *Org. Lett.*, **8**, 4071; (g) Molander, G.A. and Ellis, N. (2007) *Acc. Chem. Res.*, **40**, 275.

48. Yamamoto, Y., Takizawa, M., Yu, X.-Q., and Miyaura, N. (2008) *Angew. Chem. Int. Ed. Engl.*, **47**, 928.
49. For example, see: Schlüter, S., Frahn, J., Karakaya, B., and Schlüter, A.D. (2000) *Macromol. Chem. Phys.*, **201**, 139.
50. Kandre, R. and Schlüter, A.D. (2008) *Macromol. Rapid Commun.*, **29**, 1661.
51. For example, see Ref. [15].
52. For example, see: (a) Zhang, W., Chen, C.H.-T., Lu, Y., and Nagashima, T. (2004) *Org. Lett.*, **6**, 1473; (b) Albaneze-Walker, J., Raju, R., Vance, J.A., Goodman, A.J., Reeder, M.R., Liao, J. Maust, M.T., Irish, P.A., Espino P., and Andrews, D.R. (2009) *Org. Lett.*, **11**, 1463, and references therein.
53. For other transition-metal mediated polycondensations using aryltriflate monomers, see for example: (a) Nishihara, Y., Ando, J., Kato, T., Mori, A., and Hiyama, T. (2000) *Macromolecules*, **33**, 2779; (b) Beinhoff, M., Bozano, L.D., Scott, J.C., and Carter, K.R. (2005) *Macromolecules*, **38**, 4147.
54. For a review, see: (a) Littke, A.F. and Fu, G.C. (2002) *Angew. Chem. Int. Ed. Engl.*, **41**, 4176; (b) Martin, R. and Buchwald, S.L. (2008) *Acc. Chem. Res.*, **41**, 1461; For a recent successful example of sterically demanding tetra-*ortho*-substituted biaryl formation by SMC, see: (c) Organ, M.G., Calimsiz, S., Sayah, M., Hoi, K.H., and Lough, A.J. (2009) *Angew. Chem. Int. Ed. Engl.*, **48**, 2383.
55. (a) Walker, S.D., Barder, T.E., Martinelli, J.R., and Buchwald, S.L. (2004) *Angew. Chem. Int. Ed. Engl.*, **43**, 1871; (b) Barder, T.E., Walker, S.D., Martinelli, J.R., and Buchwald, S.L. (2005) *J. Am. Chem. Soc.*, **127**, 4685.
56. Zapf, A., Jackstell, R., Rataboul, F., Riermeier, T., Monsees, A., Fuhrmann, C., Shaikh, N., Dingerdissen, U., and Beller, M. (2004) *Chem. Commun.*, 38.
57. (a) Littke, A.F. and Fu, G.C. (1998) *Angew. Chem. Int. Ed. Engl.*, **37**, 3387; (b) Liu, S.-Y., Choi, M.J., and Fu, G.C. (2001) *Chem. Commun.*, 2408.
58. Stambuli, J.P., Kuwano, R., and Hartwig, J.F. (2002) *Angew. Chem. Int. Ed. Engl.*, **41**, 4746.
59. For Pd/C as catalyst for SMC, see: (a) Marck, G., Villiger, A., and Buchecker, R. (1994) *Tetrahedron Lett.*, **35**, 3277; (b) LeBlond, C.R., Andrews, A.T., Sun, Y., and Sowa, J.R. Jr. (2001) *Org. Lett.*, **3**, 1555; (c) Sakurai, H., Tsukuda, T., and Hirao, T. (2002) *J. Org. Chem.*, **67**, 2721; (d) Heidenreich, R.G., Köhler, K., Krauter, J.G.E., and Pietsch, J. (2002) *Synlett*, 1118; (e) Arcadi, A., Cerichelli, G., Chiarini, M., Correa, M., and Zorzan, D. (2003) *Eur. J. Org. Chem.*, 4080; (f) Tagata, T. and Nishida, M. (2003) *J. Org. Chem.*, **68**, 9412; (g) Arvela, R.K. and Leadbeater, N.E. (2005) *Org. Lett.*, **7**, 2101; (h) Zhang, G. (2005) *Synthesis*, 537; (i) Lysén, M. and Köhler, K. (2006) *Synthesis*, 692; (j) Maegawa, T., Kitamura, Y., Sako, S., Udzu, T., Sakurai, A., Tanaka, A., Kobayashi, Y., Endo, K., Bora, U., Kurita, T., Kozaki, A., Monguchi, Y., and Sajiki, H. (2007) *Chem. Eur. J.*, **13**, 5937; For Pd supported or stabilized by polymers, see: (k) Li, Y., Hong, X.M., Collard, D.M., and El-Sayed, M.A. (2000) *Org. Lett.*, **2**, 2385; (l) McNamara, C.A., Dixon, M.J., and Bradley, M. (2002) *Chem. Rev.*, **102**, 3275; (m) Narayanan, R. and El-Sayed, M.A. (2005) *Langmuir*, **21**, 2027; (n) Gallon, B.J., Kojima, R.W., Kaner, R.B., and Diaconescu, P.L. (2007) *Angew. Chem. Int. Ed. Engl.*, **46**, 7251; (o) Mennecke, K., Cecilia, R., Glasnov, T.N., Gruhl, S., Vogt, C., Feldhoff, A., Larrubia Vargas, M.A., Kappe, C.O., Kunz, U., and Kirschning, A. (2008) *Adv. Synth. Catal.*, **350**, 717; (p) Huang, J., Wang, D., Hou, H., and You, T. (2008) *Adv. Funct. Mater.*, **18**, 441; (q) Lee, D.-H.,

Kim, J.-H., Jun, B.-H., Kang, H., Park, J., and Lee, Y.-S. (2008) *Org. Lett.*, **10**, 1609; For Pd on inorganic supports, see: (r) Augustine, R.L. and O'Leary, S.T. (1992) *J. Mol. Catal.*, **72**, 229; (s) Djakovitch, L. and Köhler, K. (2001) *J. Am. Chem. Soc.*, **123**, 5990; (t) Dams, M., Drijkoningen, L., Pauwels, B., Van Tendeloo, G., De Vos, D.E., and Jacobs, P.A. (2002) *J. Catal.*, **209**, 225; (u) Smith, M.D., Stepan, A.F., Ramarao, C., Brennan, P.E., and Ley, S.V. (2003) *Chem. Commun.*, 2652; (v) Stevens, P.D., Li, G., Fan, J., Yen, M., and Gao, Y. (2005) *Chem. Commun.*, 4435; (w) Bedford, R.B., Singh, U.G., Walton, R.I., Williams, R.T., and Davis, S.A. (2005) *Chem. Mater.*, **17**, 701; (x) Baruwati, B., Guin, D., and Manorama, S.V. (2007) *Org. Lett.*, **9**, 5377; (y) Jana, S., Dutta, B., Bera, R., and Koner, S. (2008) *Inorg. Chem.*, **47**, 5512; (z) Erathodiyil, N., Ooi, S., Seayad, A.M., Han, Y., Lee, S.S., and Ying, J.Y. (2008) *Chem. Eur. J.*, **14**, 3118; For Pd stabilized by ionic liquids, see: (aa) Sasaki, T., Tada, M., Zhong, C., Kume, T., and Iwasawa, Y. (2008) *J. Mol. Catal. A*, **279**, 200; (ab) Yang, X., Fei, Z., Zhao, D., Ang, W.H., Li., Y., and Dyson, P.J. (2008) *Inorg. Chem.*, **47**, 3292; For Pd on graphene, see: (ac) scheuermann, G.M., Rumi, L., Steurer, P., Bannwarth, W., and Mülhaupt, R. (2009). *J. Am. chem. soc.*, **131**, 8262;

60. Liu, C., Repoley, A., and Zhou, B. (2008) *J. Polym. Sci., Part A: Polym. Chem.*, **46**, 7268.
61. Organ, M.G., Çalimsiz, S., Sayah, M., Hoi, K.H., and Lough, A.J. (2009) *Angew. Chem. Int. Ed. Engl.*, **48**, 2383.
62. (a) Bermejo, A., Ros, A., Fernández, R., and Lassaletta, J.M. (2008) *J. Am. Chem. Soc.*, **130**, 15798; (b) Uozumi, Y., Matsuura, Y., Arakawa, T., and Yamada, Y.M.A. (2009) *Angew. Chem. Int. Ed. Engl.*, **48**, 2708.
63. (a) Tobisu, M., Shimasaki, T., and Chatani, N. (2008) *Angew. Chem. Int. Ed. Engl.*, **47**, 4866; (b) Quasdorf, K.W., Tian, X., and Garg, N.K. (2008) *J. Am. Chem. Soc.*, **130**, 14422; (c) Guan, B.-T., Wang, Y., Li, B.-J., Yu, D.-G., and Shi, Z.-J. (2008) *J. Am. Chem. Soc.*, **130**, 14468.
64. (a) Yokoyama, A., Miyakoshi, R., and Yokozawa, T. (2004) *Macromolecules*, **37**, 1169; (b) Miyakoshi, R., Yokoyama, A., and Yokozawa, T. (2004) *Macromol. Rapid Commun.*, **25**, 1663; (c) Miyakoshi, R., Yokoyama, A., and Yokozawa, T. (2005) *J. Am. Chem. Soc.*, **127**, 17542; (d) Adachi, I., Miyakoshi, R., Yokoyama, A., and Yokozawa, T. (2006) *Macromolecules*, **39**, 7793; (e) Miyakoshi, R., Shimono, K., Yokoyama, A., and Yokozawa, T. (2006) *J. Am. Chem. Soc.*, **128**, 16012.
65. (a) Sheina, E.E., Liu, J., Iovu, M.C., Laird, D.W., and McCullough, R.D. (2004) *Macromolecules*, **37**, 3526; (b) Iovu, M.C., Sheina, E.E., Gil, R.R., and McCullough, R.D. (2005) *Macromolecules*, **38**, 8649; (c) Senkovskyy, V., Khanduyeva, N., Komber, H., Oertel, U., Stamm, M., Kuckling, D., and Kiriy, A. (2007) *J. Am. Chem. Soc.*, **129**, 6626; (d) Khanduyeva, N., Senkovskyy, V., Beryozkina, T., Bocharova, V., Simon, F., Nitschke, N., Stamm, M., and Groetzschel, R. (2008) *Macromolecules*, **41**, 7383; (e) Beryozkina, T., Senkovskyy, V., Kaul, E., and Kiriy, A. (2008) *Macromolecules*, **41**, 7817; (f) Lanni, E.L., McNeil, A.J. (2009) *J. Am. chem. soc*, **131**, 16573; For a review, see: (g) Yokoyama, A. and Yokozawa, T. (2007) *Macromolecules*, **40**, 4093.
66. Brown, J.M. and Cooley, N.A. (1988) *J. Chem. Soc., Chem. Commun.*, 1345.
67. Yokoyama, A., Suzuki, H., Kubota, Y., Ohuchi, K., Higashimura, H., and Yokozawa, T. (2007) *J. Am. Chem. Soc.*, **129**, 7236.
68. For a recent report on blockcopolymer formation, see: Yokoyama, A., Kato, A., Miyakoshi, R., and Yokozawa, T. (2008) *Macromolecules*, **41**, 7271.
69. Preferential twofold oxidative addition of dihaloarenes with arylboronic acids is not in contradiction to the sliding hypothesis but does not prove it either: (a) Sinclair, D.J. and Sherburn, M.S. (2005) *J. Org. Chem.*, **70**, 3730; (b) Dong, C.-G. and Hu, Q.-S. (2005) *J. Am. Chem. Soc.*, **127**,

10006; (c) Weber, S.K., Galbrecht, F., and Scherf, U. (2006) *Org. Lett.*, **8**, 4039.

70. (a) Heeney, M., Zhang, W., Duffy, W., McCulloch, I., and Koller, G. WO 2007/059838, Merck Patent GmbH; (b) Koller, G., Falk, B., Weller, C., Giles, M., and McCulloch, I. WO 2005/014691 A2, Merck Patent GmbH.
71. From Ref. 70[a]: . . . .a suitable monomer, for example a 2,5-dibromo-3-alkylthiophene is reacted with an appropriate Grignard reagent, for example methylmagnesium bromide, in the presence of a catalytic amount of a Ni(0) catalyst, for example bis(1,5-cyclooctadiene)nickel (0), and a bidentate ligand, for example a phosphine ligand like diphenylphosphinopropane (dppp). It was surprisingly found that the use of a Ni(0) catalyst, rather than a Ni(II) catalyst, results in a highly reactive catalyst system affording polymers of very high molecular weights and high regioregularity.
72. Hiorns, R.C., de Bettignies, R., Leroy, J., Bailly, S., Firon, M., Sentein, C., Khoukh, A., Preud'homme, H., and Dagron-Lartigau, C. (2006) *Adv. Funct. Mater.*, **16**, 2263; For a related Stille-type route, see: (b) Lère-Porte, J.-P., Moreau, J.J.E., and Torreilles, C. (2001) *Eur. J. Org. Chem.*, 1249.
73. For applications of microwaves in SMC, see: (a) Kappe, C.O. (2004) *Angew. Chem. Int. Ed. Engl.*, **43**, 6250; (b) Loupy, A. (ed.) (2006) *Microwaves in Organic Synthesis*, 2nd edn, Wiley-VCH Verlag GmbH, Weinheim.
74. (a) Nehls, B.S., Füldner, S., Preis, E., Farrell, T., and Scherf, U. (2005) *Macromolecules*, **38**, 687; (b) Galbrecht, F., Bünnagel, T.W., Scherf, U., and Farrell, T. (2007) *Macromol. Rapid Commun.*, **28**, 387.
75. Tsami, X.A., Yang, X.H., Farrell, T., Neher, D., and Holder, E. (2008) *J. Polym. Sci., Part A: Polym. Chem.*, **46**, 7794.
76. (a) Hessel, V., Rothstock, S., Agar, D.W., Jadhavrao, P., Löb, P., and Werner, B. (2007) AlChE Spring National Meeting, Houston, p. 74d; (b) Rothstock, S., Hessel, V., Löb, P., Löwe, H., and Werner, B. (2008) 1stSynTOP: Smart Synthesis and Technologiers for Organic Processes Conference, Potsdam; (c) Rothstock, S., Werner, B., Löb, P., and Hessel, V. (2008) IMRET 10, 10th International Conference on Microreaction Technology, New Orleans, Conference Proceedings, p. 513.
77. For example, see: (a) Heidenhain, S. *http://www.sumation.co.uk/documents/icp_florence_10102006.pdf*; accessed Feb, 2009; (b) *http://www.cintellig.com/res/_overview.htm*, p. 6.
78. *http://www.avsforum.com/avs-vb/showthread.php?t=681125* accessed Feb 2009. Cintellig Ltd. *http://www.cintellig.com/res/_overview.htm* p. 6; accessed Feb, 2009.
79. Yamamoto, T., Mortia, A., Miyazaki, Y., Maruyama, T., Wakayawa, H., Zhou, Z., Nakawura, J., and Kanbara, T.
80. Rulkens, R., Wegner, G., Enkelmann, V., and Schulze, M. (1996) *Ber. Bunsen-Ges. Phys. Chem.*, **100**, 707.
81. Bockstaller, M., Köhler, W., Wegner, G., Vlassopoulos, D., and Fytas, G. (2000) *Macromolecules*, **33**, 3951.
82. Kroeger, A., Deimede, V., Belack, J., Lieberwirth, I., Fytas, G., and Wegner, G. (2007) *Macromolecules*, **40**, 105.
83. (a) Fütterer, T., Hellweg, T., Findenegg, G.H., Frahn, J., Schlüter, A.D., and Böttcher, C. (2003) *Langmuir*, **19**, 6537; (b) Fütterer, T., Hellweg, T., Findenegg, G.H., Frahn, J., and Schlüter, A.D. (2005) *Macromolecules*, **38**, 7451.
84. (a) Baskar, C., Lai, Y.H., and Valiyaveettil, S. (2001) *Macromolecules*, **34**, 6255; (b) Ravindranath, R., Vijila, C., Ajikumar, P.K., Hussain, F.S.J., Ng, K.L., Wang, H., Jin, C.S., Knoll, W., and Valiyaveettil, S. (2006) *J. Phys. Chem. B*, **110**, 25958.
85. (a) Wittemann, M., Kelch, S., Blaul, J., Hickl, P., Guilleaume, B., Brodowski, G., Horvath, A., Ballauff, M., and Rehahn, M. (1999) *Macromol. Chem. Phys., Macromol. Symp.*, **142**, 43; (b) Guilleaume, B., Blaul, J., Wittemann, M., Rehahn, M., and Ballauff, M.

(2000) *J. Phys.: Condens. Matter*, **12**, A245; (c) Blaul, J., Wittemann, M., Ballauff, M., and Rehahn, M. (2000) *J. Phys. Chem. B*, **104**, 7077; (d) Derserno, M., Holm, Ch., Blaul, J., Ballauff, M., and Rehahn, M. (2001) *Eur. Phys. J. E*, **5**, 97; (e) Guilleaume, B., Ballauff, M., Goerigk, G., Wittemann, M., and Rehahn, M. (2001) *Colloid Polym. Sci.*, **279**, 829; (f) Guilleaume, B., Blaul, J., Ballauff, M., Wittemann, M., Rehahn, M., and Goerigk, G. (2002) *Eur. Phys. J.*, **8**, 229; (g) Lachenmayer, K. and Oppermann, W. (2002) *J. Chem. Phys.*, **116**, 392; (h) Bohrisch, J., Eisenbach, C.D., Jaeger, W., Mori, H., Müller, A.E., Rehahn, M., Schaller, Ch., Traser, S., and Wittmeyer, P. (2004) *Adv. Polym. Sci.*, **165**, 1; (i) Holm, Ch., Rehahn, M., Oppermann, W., and Ballauff, M. (2004) *Adv. Polym. Sci.*, **166**, 1; (j) Ballauff, M., Blaul, J., Guilleaume, B., Rehahn, M., Traser, S., Wittemann, M., and Wittmeyer, P. (2004) *Macromol. Symp.*, **211**, 1.

86. Traser, S., Wittmeyer, P., and Rehahn, M. (2002) *e-Polymers*, no. 032.

87. (a) Kandre, R., Kutzner, F., Schlaad, H., and Schlüter, A.D. (2005) *Macromol. Chem. Phys.*, **206**, 1610; (b) Yamashita, Y., Kaneko, Y., and Kadokawa, J.I. (2007) *Eur. Polym. J.*, **43**, 3795; (c) Demirel, A.L., Yurteri, S., Cianga, I., and Yagci, Y. (2005) *Macromolecules*, **38**, 6402.

88. (a) Hwang, W.F., Wiff, D.R., Brenner, C.L., and Helminiak, T.E. (1983) *J. Macromol. Sci., Phys.*, **B22**, 231; (b) Schartel, B. and Wendorff, J.H. (1999) *Polym. Eng. Sci.*, **39**, 128.

89. (a) Bayer, A., Datko, A., and Eisenbach, C.D. (2005) *Polymer*, **46**, 6614; For related studies not involving SPC, see: (b) Winter, D., Eisenbach, C.D., Pople, J.A., and Gast, A.P. (2001) *Macromolecules*, **34**, 5943; (c) Winter, D. and Eisenbach, C.D. (2004) *J. Polym. Sci., Part A: Polym. Chem.*, **42**, 1919; (d) Winter, D. and Eisenbach, C.D. (2004) *Polymer*, **45**, 2507; See also: (e) Eggert, C. and Eisenbach, C.D. (2009) 21st Stuttgarter Kunststoff-kolloquium, Conference Abstract.

90. Burroughes, J.H., Bradley, D.D.C., Brown, A.R., Marks, R.N., Mackay, K., Friend, R.H., Burn, P.L., and Holmes, A.B. (1990) *Nature*, **347**, 539.

91. Ohmori, Y., Uchida, A., Muro K., and Yoshino, K. (1991) *Jpn. J. Appl. Phys.*, **30**, 1941.

92. (a) Towns, C.R., O'Dell, R., and O'Connor, S.J.M. (2005) US Patent 6,861,502; (b) Chen, P., Yang, G., Liu, T., Li, T., Wang, M., and Huang, W. (2006) *Polym. Int.*, **55**, 473.

93. (a) Neher, D. (2001) *Macromol. Rapid Commun.*, **22**, 1365; (b) Leclerc, M. (2001) *J. Polym. Sci., Part A: Polym. Chem.*, **39**, 2867; (c) Scherf, U. and List, E.J.W. (2002) *Adv. Mater.*, **14**, 477; (d) Scherf, U. and Neher, D. (eds) (2008) *Polyfluorenes*, Advances in Polymer Science, Vol. 212 Springer, Berlin; (e) Knaapila, M., Stepanyan, R., Torkkeli, M., Garamus, V.M., Galbrecht, F., Nehls, B.S., Preis, E., Scherf, U., and Monkman, A.P. (2008) *Phys. Rev. E*, **77**, 051803.

94. For short foldamers, see: (a) Nelson, J.C., Saven, J.G., Moore, J.S., and Wolynes, P.G. (1997) *Science*, **277**, 1793; (b) Hill, D.J., Mio, M.J., Prince, R.B., Hughes, T.S., and Moore, J.S. (2001) *Chem. Rev.*, **101**, 3893; For recent polymer approaches to foldamers, see: (c) Hecht, S. and Khan, A. (2003) *Angew. Chem. Int. Ed. Engl.*, **42**, 6021; (d) Ben, T., Goto, H., Miwa, K., Goto, H., Morino, K., Furusho, Y., and Yashima, E. (2008) *Macromolecules*, **41**, 4506.

95. For a graphical representation, see: Zhang, A., Sakamoto, J., and Schlüter, A.D. (2008) *Chimia*, **62**, 776.

96. Schlüter, A.D. and Rabe, J.P. (2000) *Angew. Chem. Int. Ed. Engl.*, **39**, 864.

97. (a) Freudenberger, R., Claussen, W., Schlüter, A.D., and Wallmeier, H. (1994) *Polymer*, **35**, 4496; (b) Claussen, W., Schulte, N., and Schlüter, A.D. (1995) *Macromol. Rapid Commun.*, **16**, 89; (c) Karakaya, B., Claussen, W., Schäfer, A., Lehmann, A., and Schlüter, A.D. (1996) *Acta Polym.*, **47**, 79; (d) Karakaya, B.,

Claussen, W., Gessler, K., Saenger, W., and Schlüter, A.D. (1997) *J. Am. Chem. Soc.*, **119**, 3269; (e) Stocker, W., Karakaya, B., Schürmann, B.L., Rabe, J.P., and Schlüter, A.D. (1998) *J. Am. Chem. Soc.*, **120**, 7691; (f) Bo, Z.S. and Schlüter, A.D. (1999) *Macromol. Rapid Commun.*, **20**, 21; (g) Bo, Z.S., Rabe, J.P., and Schlüter, A.D. (1999) *Angew. Chem. Int. Ed. Engl.*, **38**, 2370; (h) Bo, Z.S., Zhang, C.M., Severin, N., Rabe, J.P., and Schlüter, A.D. (2000) *Macromolecules*, **33**, 2688; (i) See also Ref. [8].

98. For a conceptual description, see: (a) Hecht, S. and Fréchet, J.M.J. (2001) *Angew. Chem. Int. Ed. Engl.*, **40**, 74; (b) Frampton, M.J. and Anderson, H.L. (2007) *Angew. Chem. Int. Ed. Engl.*, **46**, 1028.

99. (a) Jiang, D.L. and Aida, T. (1998) *J. Am. Chem. Soc.*, **120**, 10895; (b) Sato, T., Jiang, D.L., and Aida, T. (1999) *J. Am. Chem. Soc.*, **121**, 10658.

100. (a) Bao, Z.N., Amundson, K.R., and Lovinger, A.J. (1998) *Macromolecules*, **31**, 8647; (b) Tan, Z.A., Tang, R.P., Zhou, E.J., He, Y.J., Xi, F., and Li, Y.F. (2008) *J. Appl. Polym. Sci.*, **107**, 514.

101. Schenning, A.P.H.J., Martin, R.E., Ito, M., Diederich, F., Boudon, C., Gisselbrecht, J.-P., and Gross, M. (1998) *Chem. Commun.*, 1013.

102. (a) Kaneto, T., Horie, T., Asano, M., Aoki, T., and Oikawa, E. (1997) *Macromolecules*, **30**, 3118; (b) Percec, V., Rudick, J.G., Peterca, M., Wagner, M., Obata, M., Mitchell, C.M., Cho, W.D., Balagurusamy, V.S.K., and Heiney, P.A. (2005) *J. Am. Chem. Soc.*, **127**, 15257; (c) Rudick, J.G. and Percec, V. (2007) *New J. Chem.*, **31**, 1083; (d) Percec, V., Peterca, M., Rudick, J.G., Aqad, E., Imam, M.R., and Heiney, P.A. (2007) *Chem. Eur. J.*, **13**, 9572; (e) Percec, V., Rudick, J.G., Peterca, M., Aqad, E., Imam, M.R., and Heiney, P.A. (2007) *J. Polym. Sci., Part A: Polym. Chem.*, **45**, 4974.

103. (a) Malenfant, P.R.L. and Fréchet, J.M.J. (2000) *Macromolecules*, **33**, 3634; (b) Otsubo, T., Ueno, S., Taimiya, K., and Aso, Y. (2004) *Chem. Lett.*, **33**, 1154.

104. (a) Jiang, J., Liu, H.W., Zhao, Y.L., Chen, C.F., and Xi, F. (2002) *J. Polym. Sci., Part A: Polym. Chem.*, **40**, 1167; (b) Fei, Z.P., Han, Y., and Bo, Z.S. (2008) *J. Polym. Sci., Part A: Polym. Chem.*, **46**, 4030.

105. (a) Taylor, P.N., O'Connell, M.J., McNeill, L.A., Hall, M.J., Aplin, R.T., and Anderson, H.L. (2000) *Angew. Chem. Int. Ed. Engl.*, **39**, 3456; (b) Michels, J.J., O'Connell, M.J., Taylor, P.N., Wilson, J.S., Cacialli, F., and Anderson, H.L. (2003) *Chem. Eur. J.*, **9**, 6167; (c) Terao, J., Tang, A., Michels, J.J., Krivokapic, A., and Anderson, H.L. (2004) *Chem. Commun.*, 56.

106. (a) Guillerez, S. and Bidan, G. (1998) *Synth. Met.*, **93**, 123; See also (b) Bidan, G., De Nicola, A., Enée, V., and Guillerez, S. (1998) *Chem. Mater.*, **10**, 1052.

107. (a) Jayakannan, M., van Hel, P.A., and Janssen, R.A.J. (2002) *J. Polym. Sci., Part A: Polym. Chem.*, **40**, 2360; Also, see: (b) Liu, C.L., Tsai, J.H., Lee, W.Y., Chen, W.C., and Jenekhe, S.A. (2008) *Macromolecules*, **41**, 6952.

108. Jayakannan, M., Lou, X., van Dongen, J.L.J., and Janssen, R.A. (2005) *J. Polym. Sci., Part A: Polym. Chem.*, **43**, 1454.

109. Loewe, R.S., Khersonsky, S.M., and McCullough, R.D. (1999) *Adv. Mater.*, **11**, 250.

110. Osaka, I. and McCullough, R.D. (2008) *Acc. Chem. Res.*, **41**, 1202.

111. Li, W.W., Han, Y., Li, B.S., Liu, C.M., and Bo, Z.S. (2008) *J. Polym. Sci., Part A: Polym. Chem.*, **46**, 4556.

112. (a) Knapp, R. and Rehahn, M. (1993) *J. Organomet. Chem.*, **452**, 235; (b) Knapp, R. and Rehahn, M. (1993) *Makromol. Chem., Rapid Commun.*, **14**, 451; (c) Knapp, R., Velten, U., and Rehahn, M. (1998) *Polymer*, **39**, 5827.

113. (a) Kim, Y.H. and Webster, O.W. (1990) *J. Am. Chem. Soc.*, **112**, 4592; (b) Kim, Y.H. and Webster, O.W. (1992) *Macromolecules*, **25**, 5561.

114. (a) Li, J., Sun, M.H., and Bo, Z.S. (2007) *J. Polym. Sci., Part A: Polym. Chem.*, **45**, 1084; For another example based on polyfluorene, see: (b) Xin, Y., Wen, G.-A., Zeng, W.-J., Zhao, L., Zhu,

X.-R., Fen, Q.-L., Feng, J.-C., Wang, L.-H., Wie, W., Peng, B., Cao, Y., and Huang, W. (2005) *Macromolecules*, **38**, 6755.

115. (a) Beyerlein, T. and Tieke, B. (2000) *Macromol. Rapid Commun.*, **21**, 182; (b) Rabindranath, A.R., Zhu, Y., Heim, I., and Tieke, B. (2006) *Macromolecules*, **39**, 8250; (c) Zhu, Y., Rabindranath, A.R., Beyerlein, T., and Tieke, B. (2007) *Macromolecules*, **40**, 6981; (d) Zhang, K. and Tieke, B. (2008) *Macromolecules*, **41**, 7287.

116. Yang, C.D., Schreiber, H., List, E.J.W., Jacob, J., and Müllen, K. (2006) *Macromolecules*, **39**, 5213.

117. Mori, T. and Kijima, M. (2007) *Opt. Mater.*, **30**, 545.

118. Velten, U. and Rehahn, M. (1998) *Macromol. Chem. Phys.*, **199**, 127.

119. (a) Frank, W., Wasgindt, M., Pautzsch, T., and Klemm, E. (2001) *Macromol. Chem. Phys.*, **202**, 980; (b) Frank, W., Pautzsch, T., and Klemm, E. (2001) *Macromol. Chem. Phys.*, **202**, 2535.

120. Hiroki, K. and Kijima, M. (2005) *Chem. Lett.*, **34**, 942.

121. (a) Remmers, M., Schulze, M., and Wegner, G. (1996) *Macromol. Rapid Commun.*, **17**, 239; (b) Yang, C.D., Jacob, J., and Müllen, K. (2006) *Macromol. Chem. Phys.*, **207**, 1107; (c) See also Ref. [36].

122. (a) See Ref [211a]; (b) Hu, Q.-S., Vitharana, D., Liu, G.-Y., Jain, V., Wagaman, M.W., Zhang, L., Lee, T.R., and Pu, L. (1996) *Macromolecules*, **29**, 1082; (c) Feast, W.J., Daik, R., Friend, R.H., and Cacialli, F. (2002) US Patent 6,340,732; (d) Katayama, H., Nagao, M., Nishimura, T., Matsui, Y., Fukuse, Y., Wakioka, M., and Ozawa, F. (2006) *Macromolecules*, **39**, 2039; (e) Feast, W.J., Cacialli, F., Koch, A.T.H., Daik, R., Lartigau, C., Friend, R.H., Beljonne, D., and Brédas, J.-L. (2007) *J. Mater. Sci.*, **17**, 907; (f) Babudri, F., Cardone A., Cassano, T., Farinola, G.M., Naso, F., and Tommasi, R. (2008) *J. Organomet. Chem.*, **693**, 2631.

123. (a) Yang, X., Dou, X., Rouhanipour, A., Räder, H.J., and Müllen, K. (2008) *J. Am. Chem. Soc.*, **130**, 4216; (b) Yang, C., Jacob, J., and Müllen, K. (2006) *Macromolecules*, **39**, 5695.

124. Sakamoto, J., van Heijst, J., Lukin, O., and Schlüter, A.D. (2009) *Angew. Chem. Int. Ed. Engl.*, **48**, 1030.

125. Iwasaki, T., Kohinata, Y., and Nishide, H. (2005) *Org. Lett.*, **7**, 755.

126. Kolb, H.C., Finn, M.G., and Sharpless, K.B. (2001) *Angew. Chem. Int. Ed. Engl.*, **40**, 200.

127. Tobisu, M. and Chatani, N. (2009) *Angew. Chem. Int. Ed. Engl.*, **48**, 2.

128. See for example: (a) Guo, Y., Young, D.J., and Hor, T.S.A. (2008) *Tetrahedron Lett.*, **49**, 5620; (b) Kylmälä, T., Valkonen, A., Rissanen, K., Xu, Y., and Franzén, R. (2008) *Tetrahedron Lett.*, **49**, 6679; Fow a review on iron-catalyzed other cross-couplings, see: (c) Sherry, B.D. and Fürstner, A. (2008) *Acc. Chem. Res.*, **41**, 150.

129. Kudla, C.J., Koenen, N., Pisula, W., and Scherf, U. (2009) *Macromolecules*, **42**, 3483.

130. Beryozkina, T., Boyko, K., Khanduyeva, N., Senkovskyy, V., Horecha, M., Oertel, U., Simon, F., Stamm, M., and Kiriy, A. (2009) *Angew. Chem. Int. Ed. Engl.*, **48**, 2695.

131. For a review, see: (a) Chemler, S.R., Trauner, D., and Danishefsky, S.J. (2001) *Angew. Chem. Int. Ed. Engl.*, **40**, 4544; (b) See also Ref. [12].

3
Advanced Functional Regioregular Polythiophenes

Itaru Osaka and Richard D. McCullough

3.1
Introduction

Polythiophenes (PTs) are the most versatile, important class of π-conjugated polymers and organic electronic materials because of their facile synthesis, processability, and charge carrier transport properties [1–6]. In the past decades an enormous number of these plastic materials have been designed, synthesized, and characterized, and today these materials are aggressively used in a variety of organic electronic devices such as polymer light-emitting diodes (PLEDs) [7–9], organic field-effect transistors (OFETs) [10–14], and organic photovoltaics (OPVs)/solar cells [15–19]. Because desired properties of PTs differ from device to device, the polymer structure should be tailored to various application needs. Creative molecular design and synthesis are therefore critical in controlling material properties as well as device performance. For example, head-to-tail (HT) regioregular poly(3-alkylthiophene)s (rrP3ATs), the most common material for polymeric electronic devices, have been designed to avoid the structural defects in the backbone and are strategically synthesized in a regiochemically controlled fashion [20, 21]. The resulting rrP3ATs can be ordered in three dimensions: conformational ordering along the backbone, π-stacking of planar polymer backbones, and lamellar stacking between adjacent backbones [22]. All of these features lead to the enhanced electrical properties of the material and hence excellent device performance, especially in OFETs and OPVs. In order to control the design and synthesis, it is also important to understand details of structure–property relationships, which allow further optimization and development of new advanced materials.

We have reviewed earlier the synthesis and characterization of regioregular PTs and some of their applications [2–6]. Other earlier reviews on PTs have also focused on synthesis and structures [1], or focused on specific applications [8, 23]. Herein, we focus on a wide variety of regioregular PTs primarily used for electronic applications, systematically describing molecular design, synthesis, and properties. We also attempt to merge current understandings of the relationships between chemical structure, physical properties, and device performance. This chapter is divided into three parts according to the structure of the polymer. The first part,

Design and Synthesis of Conjugated Polymers. Edited by Mario Leclerc and Jean-François Morin

ISBN: 978-3-527-32474-3

focusing on P3ATs, depicts their evolution, beginning with unsubstituted PT (Section 3.2) followed by soluble P3ATs (Section 3.3), and well-defined HT rrP3ATs (Section 3.4). The second part describes regiosymmetric P3ATs (Section 3.5) and regiosymmetric thiophene-based copolymers with various (hetero)aromatic rings (Section 3.6). The third part includes block copolymers containing an rrP3AT segment (Section 3.7).

3.2 Unsubstituted Polythiophene

In 1980s, several years after the discovery of polyacetylene, which is known as the *first π-conjugated (conducting) polymer* [24, 25], chemical synthesis of unsubstituted PT (Table 3.1) was reported by the groups of Yamamoto [26] and of Lin and Dudek [27]. In these reports, mono-Grignard of 2,5-dibromothiophene generated by treatment of magnesium metal was polymerized via metal-catalyzed polycondensation. Elemental analysis indicated that these polymers contained ~3% impurity (mainly Mg or metal from the catalyst). Wudl *et al.* reported later that the polymerization of highly purified 2,5-diiodothiophene by a similar method gave high purity PT with barely 50 ppm of Mg and Ni [28]. Yamamoto *et al.* showed that using a Ni(0) complex, for example, $Ni(cod)_2$ (cod = 1,5-cyclooctadiene), and an appropriate ligand, for example, triphenylphosphine (PPh_3), the polymerization of 2,5-dibromothiophene leads to a quantitative yield of PT [29]. The electrical conductivity of doped PT samples was $\sim 50\ S\ cm^{-1}$. It was also reported that PT could be prepared from thiophene using $FeCl_3$ [30], Wurtz coupling [31], and

Table 3.1 Representative reports for chemical synthesis of unsubstituted polythiophene.

X–(thiophene, S)–X ⟶ –(thiophene, S)–$_n$

X	Conditions	References
Br	(i) Mg/THF (ii) $Ni(bpy)Cl_2$[a]	[26]
Br	(i) Mg/THF (ii) $M(acac)_n$[b]	[27]
I	(i) Mg/ether, reflux (ii) $Ni(dppp)Cl_2$[c]/anisole, 100 °C, 5 h	[28]
Br	$Ni(cod)_2$[d], PPh_3/DMF, 60–80 °C, 16 h	[29]
H	$FeCl_3$/chloroform	[30]

[a] bpy = 2,2′-bipyridine.
[b] M = Pd, Ni, Co, or Fe. acac = acetylacetonate.
[c] dppp = 1,3-dipheylphosphinopropane.
[d] cod = 1,5-cyclooctadiene.

electrochemical polymerization [32, 33]. Problematically, however, these methods allowed the possibility of the linkage at the 3- and/or 4-position of thiophene in the PT polymer chains, resulting in reduced electrical properties because of the limits on the π-electron system.

These early works proved that PT is insoluble even at low molecular weight in any organic solvents and further does not melt; these problems revealed the critical lack of processability necessary for every application. In spite of these problems, its proven environmental stability, thermal stability, and high electrical conductivity made this material attractive to researchers, and led to the development of the chemistry of PTs.

3.3 Poly(3-alkylthiophene)s

Soluble PTs became the next important subject of research starting in the mid-1980s. As it was already known that the attachment of flexible side chains onto the backbone of insoluble polymers can dramatically improve solubility, alkylthiophenes were polymerized to explore soluble and processable PTs (Table 3.2). In this method, essentially the same synthetic methods used for the preparation of PT are applied to synthesize P3ATs. Jen *et al.* reported the first chemical synthesis of P3ATs in [34], using Kumada cross-coupling [35]. These polymers were found to be soluble in common organic solvents such as chloroform, dichloromethane, THF, toluene, and xylene, and could form thin films by being cast from solution. Molecular weights of P3ATs prepared by this method were not high ($M_n = 3–8$ K, PDI = 2), but a later report demonstrated that high-molecular-weight P3ATs could be obtained [36]. Similarly from dihalogenoalkylthiophenes, P3ATs could also be

Table 3.2 Representative reports for chemical synthesis of poly(3-alkylthiophene)s.

X_1, X_2	Conditions	References
$X_1 = X_2 = I$	(i) Mg/THF (ii) $Ni(dppp)Cl_2$	[34, 36]
$X_1 = X_2 = Br, I$	$Ni(cod)_2$, PPh_3/DMF	[29]
$X_1 = X_2 = H$	$FeCl_3$/chloroform	[37–39]
$X_1 = H, X_2 = Cl, Br, I$	$FeCl_3$, $AlCl_3$, and so on/nitrobenzene or chloroform	[40]
$X_1 = X_2 = HgCl$	Cu, $PdCl_2$/pyridine	[41]

synthesized using Ni(0) reagents such as $Ni(cod)_2$, under the same conditions used for the synthesis of PT [29]. The molecular weights of the polymers measured by gel permeation chromatography (GPC) were ~19 K. It was reported that the reaction times were longer for P3ATs as compared to those of PT, and diiodothiophenes were found to be more active monomers than dibromothiophenes.

In 1986, electrochemical synthesis of P3ATs was reported by Kaeriyama *et al.* [42]. Sugimoto *et al.* also reported a simple preparation of P3ATs using $FeCl_3$ in 1986 [37]. P3ATs prepared by this method afforded high molecular weight (M_n = 30–300 K, PDI = 1.3–5) [38, 39]. Hotta *et al.* developed a synthesis that was related to this method, in which 2-halogeno-3-alkylthiophenes were polymerized with anhydrous metal halogenides ($AlCl_3$, $FeCl_3$, etc.) via a dehydrohalogenation reaction (M_w = ~250 K) [40]. Curtis *et al.* reported the preparation of P3ATs using 2,5-bis(chloromercurio)-3-alkylthiophenes as monomers using Cu powder and a catalytic amount of $PdCl_2$ in pyridine, which afforded high-molecular-weight polymers as with a coupling method (M_n = 26 K, PDI = 2.5) [41].

A number of synthetic methodologies have been demonstrated to obtain soluble P3ATs. These processible P3ATs can readily be melt or solution processed into thin films that exhibit reasonably high electrical conductivities of up to 50 S cm^{-1} [1, 43, 44]. Although all these methods produce processible P3ATs, it is important to point out that in these methods the coupling of 3-alkylthiophene occurs with no regiochemical control and hence produces structurally irregular polymers, termed regio*ir*regular poly(3-alkylthiophene)s (irP3ATs) [3, 5]. Since the irregular structure causes a loss of π-conjugation and thereby leads to severely limited conductivity, control of the regioregularity in the polymer backbone is crucial for maximizing polymer performance.

3.4
Head-to-Tail Regioregular Poly(3-alkylthiophene)s (rrP3ATs)

3.4.1
Design and Synthesis of rrP3ATs

An important discovery was the design and synthesis of regioregular HT-coupled P3ATs [20, 21]. Owing to the asymmetric nature of the 3-alkylthiophene molecule, three relative orientations are available (Figure 3.1) when two 3-alkylthiophene rings are coupled between the 2- and 5-positions. The first orientation is 2,5′ or HT coupling, the second is 2,2′ or head-to-head (HH) coupling, and the third is 5,5′ or tail-to-tail (TT) coupling. The orientation becomes even more complicated when it comes to the coupling of three 3-alkylthiophene rings, which leads to four possible chemically distinct triad regioisomers (Figure 3.1) [45]. irP3ATs contain mixtures of these regioisomers, and the ratio of favorable HT coupling, HT regioregularity, in the polymers that are synthesized by the above methods is 50–80% [38, 46, 47]. The loss of regioregularity, that is to say, contamination with HH couplings, causes a sterically twisted structure in the polymer backbone, giving

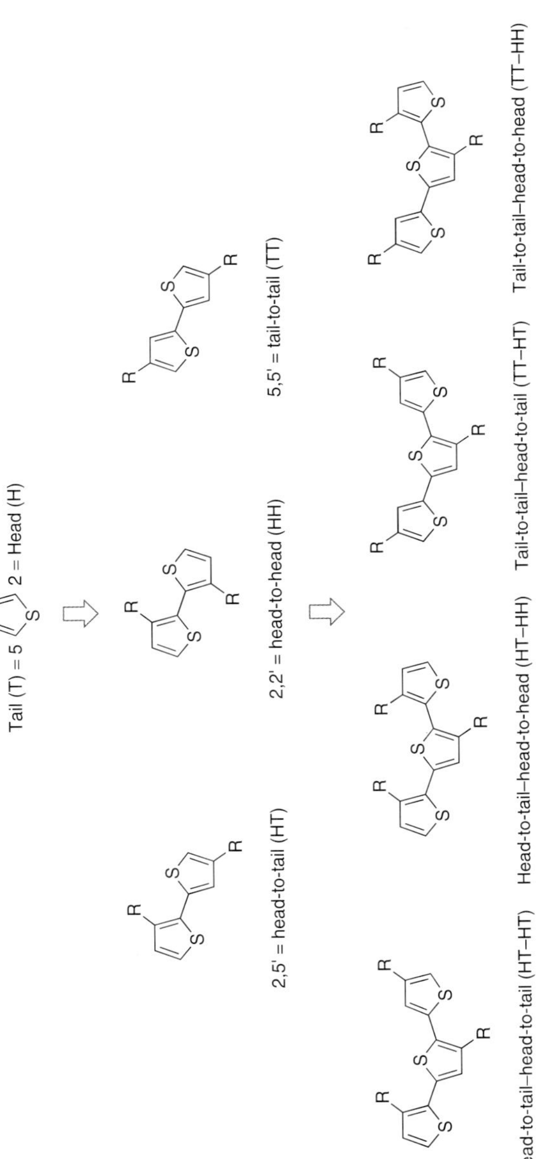

Tail (T) = 5 2 = Head (H)

2,5′ = head-to-tail (HT)

2,2′ = head-to-head (HH)

5,5′ = tail-to-tail (TT)

Head-to-tail–head-to-tail (HT–HT)

Head-to-tail–head-to-head (HT–HH)

Tail-to-tail–head-to-tail (TT–HT)

Tail-to-tail–head-to-head (TT–HH)

Figure 3.1 Possible regiochemical couplings of 3-alkylthiophenes.

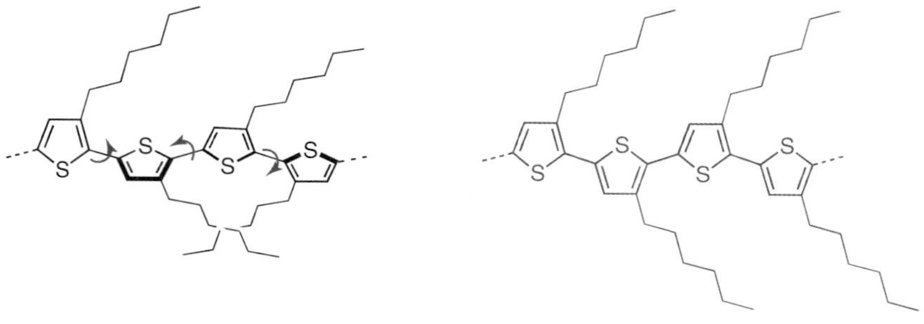

HT–HT–HH

Regio*irr*egular (nonplanar backbone)

HT–HT–HT

Regioregular (planar backbone)

Figure 3.2 (a) Regioirregular P3AT in nonplanar and (b) HT regioregular P3AT (rrP3AT) in planar structure.

rise to a loss of π-conjugation (Figure 3.2) [20, 48, 49]. This is primarily due to the increased repulsive interaction between the alkyl substituent on the 3-position of the thiophene ring and the sp^2 lone pair on sulfur, or perhaps between two alkyl substituents on the alternate thiophene rings. These steric interactions lead to greater bandgaps, with consequent destruction of high conductivity.

The conformational consequence of the four possible regioisomers shown in Figure 3.1 has been modeled in the gas phase by molecular mechanics and *ab initio* methods [48, 50]. The HT–HT triad prefers a trans coplanar orientation and is calculated to have a torsional angle of 7–8° between conjoined rings [48]. In the meantime, introduction of HH coupling (HT–HH triad) dramatically alters the calculated conformation at the defective HH junction. The thiophene rings maintain a trans conformation, but they are severely twisted, approximately 40° from coplanarity. Theoretical studies support the idea that HH coupling should strictly be eliminated in order to build planar backbones.

rrP3ATs, the most well-known regiochemically defined PTs (Figure 3.2), consist of HT arrangement and are almost completely free from HH arrangements. Thus, they can easily access planar polymer backbones that can self-assemble to form well-defined, organized three-dimensional polycrystalline structures [22]. These structures provide efficient intrachain and interchain charge carrier pathways, leading to high conductivity and other desirable properties in PTs.

rrP3ATs are most commonly synthesized by one of the three methods: the McCullough [20], Rieke [21], and Grignard metathesis (GRIM) [51] methods. These methods use nickel-catalyzed cross-coupling and produce comparable rrP3ATs. Since the polymerization is employed *in situ* with 2-monohalogeno- or 2,5-dihalogeno-3-alkylthiophene as the starting monomer, these methods offer an advantage in terms of the accessibility of the material compared to other methods [52, 53] using Stille [54] and Suzuki cross-coupling reactions [55], which use similar starting monomers but require the additional step of isolation before polymerization. However, Stille and Suzuki, in some cases, have an advantage for the synthesis of regioregular PTs with other functional side chains.

3.4.1.1 McCullough Method

The first synthesis of rrP3ATs was reported in 1992 and was called by some as the McCullough method (Scheme 3.1) [20]. The key feature of this method is the regiospecific generation of 2-bromo-5-bromomagnesio-3-alkylthiophene (**2**), which is achieved by treating 2-bromo-3-alkylthiophene (**1**) with LDA (lithium diisopropylamide) at $-40\,^{\circ}C$ followed by the addition of $MgBr_2{\cdot}Et_2O$ (recrystallized from Et_2O in a dry box). Quenching studies performed on intermediate **2** indicate that 98–99% of the desired monomer and less than 1–2% of 2,5-exchanged intermediate **3** are produced [48]. The polymerization is then employed *in situ* by a cross-coupling reaction, the Kumada cross-coupling reaction, using a catalytic amount of $Ni(dppp)Cl_2$ (dppp = 1,3-diphenylphosphinopropane) [35, 56], affording rrP3AT (44–69% yield). The cross-coupling polymerization of these intermediates occurs without any scrambling. The resulting rrP3ATs afford HT–HT regioregularity of 98–100% as seen by NMR study, and the number averaged molecular weights (M_n) are typically 20–40 K with polydispersities (PDI) of around 1.4 [20, 22, 48, 57, 58]. This procedure was later modified by replacing $MgBr_2{\cdot}Et_2O$ with $ZnCl_2$, thus allowing for greater solubility of the reactive intermediate at cryogenic temperatures.

3.4.1.2 Rieke Method

Soon after the report of the McCullough method, the Rieke method (Scheme 3.1) was reported [21, 59–62]. In this method, treating 2,5-dibromo-3-alkylthiophenes (**4**) with highly reactive "Rieke Zinc" (Zn^*) [63, 64] yields a mixture of two isomeric intermediates **2** and **3** at a ratio of 90 : 10, which gives rrP3ATs by *in situ* addition of $Ni(dppe)Cl_2$ (dppe = 1,3-diphenylphosphinoethane) with ~75% yield after purification [21]. Use of a palladium catalyst, $Pd(PPh_3)_4$, yields completely regioirregular P3ATs. Molecular weights of rrP3ATs prepared by this method are M_n=24–34 K (PDI = 1.4).

3.4.1.3 GRIM Method

In 1999, an economical new synthesis method for rrP3ATs, known as the *GRIM* method (Scheme 3.1), was reported [51, 65]. One strong advantage of this method is that the use of both cryogenic temperatures and highly reactive metals is unnecessary; consequently, it offers quick and easy preparation of rrP3ATs and enables the production of kilogram-scale high-molecular-weight rrP3ATs. In this method, 2,5-dibromo-3-alkylthiophene (**4**) is treated with 1 equivalent of any Grignard reagent (R′MgX′) to form a mixture of intermediates **2** and **3** at a ratio of 85 : 15 to 75 : 65. This ratio appears to be independent of reaction time, temperature, or Grignard reagent used. Although the ratio of the desirable to undesirable isomers is higher with the GRIM method than with either the McCullough or Rieke method, this method still affords HT rrP3ATs with a high

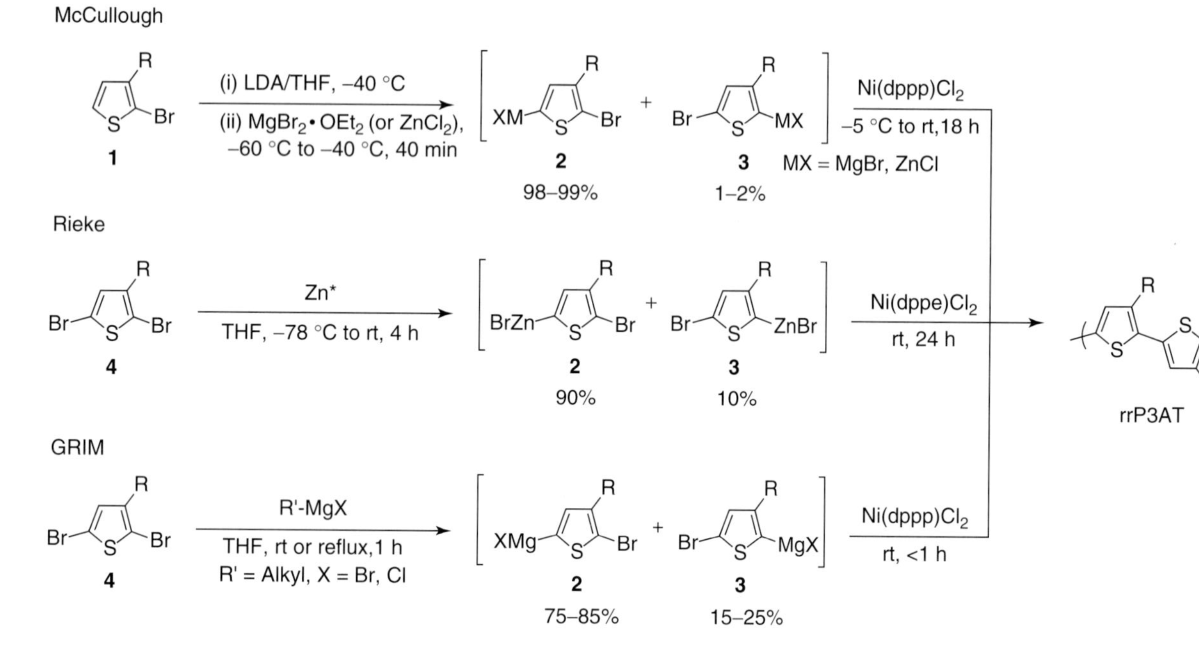

Scheme 3.1 General synthetic methods of rrP3ATs using nickel-catalyzed cross-coupling reaction.

regioregularity of >99% HT couplings. The typical M_n of rrP3ATs synthesized through this method is 20–35 K with very low PDI of 1.2–1.4.

3.4.1.4 Other Methods

Other methods for the synthesis of rrP3ATs using Stille and Suzuki palladium-catalyzed cross-coupling reactions are shown in Scheme 3.2. Iraqi *et al.* investigated the synthesis of rrP3ATs through Stille reaction, using 3-hexyl-2-iodo-5-(tri-*n*-butylstannyl)thiophene (**6**), with a variety of solvents [52]. In all cases, rrP3ATs with greater than 96% of HT couplings were obtained. Molecular weights of rrP3ATs prepared by this method were M_n = 10–16 K with a PDI of 1.2–1.4 after purification. Suzuki reaction using 3-octyl-2-iodo-5-boronatothiophene (**7**) employed by Guillerez gave rrP3ATs with 96–97% HT couplings and a weight-averaged molecular weight (M_w) of 27 K in 51% yield [53]. However, these methods require cryogenic conditions for preparing the corresponding organometallic monomers **6** or **7**, which then must be isolated and purified.

3.4.2 Mechanism of the Nickel-Catalyzed Polymerization

The synthesis of rrP3ATs is based on transition-metal-catalyzed cross-coupling reactions [66], in which the mechanism involves a catalytic cycle of three consecutive steps: oxidative addition, transmetalation, and reductive elimination. Since the nickel-catalyzed polymerization is formally a polycondensation reaction, it is generally accepted that it proceeds via a step-growth mechanism [67]. However, it has been proposed by Yokozawa *et al.* that the nickel-catalyzed cross-coupling polymerization (McCullough and GRIM method) proceeds via chain growth by a catalyst transfer mechanism [68]. Conversion versus M_n and conversion versus M_w/M_n plots of the polymerization of the monomer in the case of the GRIM method have been reported. M_n increased proportionally as monomer conversion increased and M_w/M_n ratios were 1.30–1.39 throughout the polymerization. This new insight is of particular interest because the chain-growth polymerization affords well-defined rrP3ATs with very narrow polydispersities [69, 70].

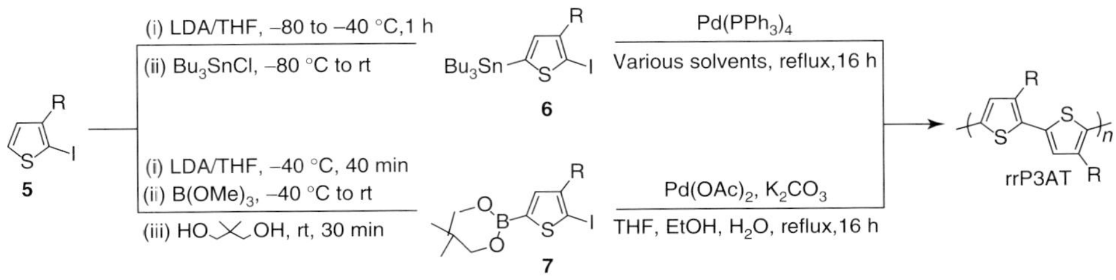

(i) LDA/THF, –80 to –40 °C, 1 h
(ii) Bu_3SnCl, –80 °C to rt
6
$Pd(PPh_3)_4$
Various solvents, reflux, 16 h
5
(i) LDA/THF, –40 °C, 40 min
(ii) $B(OMe)_3$, –40 °C to rt
(iii) HO⨉OH, rt, 30 min
7
$Pd(OAc)_2$, K_2CO_3
THF, EtOH, H_2O, reflux, 16 h
rrP3AT

Scheme 3.2 Synthetic routes of rrP3ATs by Stille (upper) and Suzuki (lower) coupling reactions.

At the same time as the Yokozawa report, the McCullough group also reported that the degree of polymerization of rrP3ATs increased with monomer conversion and can be predicted by the molar ratio of the monomer to the nickel initiator. In addition, they further proposed that this polymerization system is not only a chain growth system but also essentially a living system [71, 72]. This system allows for the control of the molecular weight of the polymer as a function of reaction time and the amount of Ni catalyst. The proposed mechanism for the regioregular polymerization of 3-alkylthiophene is outlined in Scheme 3.3 [5, 6, 71, 72]. The first step is the reaction of 2 equiv. of intermediate **2** with Ni(dppp)Cl_2 affording the organonickel compound (**8**), and reductive elimination immediately occurs to form an associated pair of the 2,2′-dibromo-5,5′-bithiophene (TT coupling) and Ni(0) [**9**·**10**]. Dimer **9** undergoes fast oxidative addition to the nickel center, generating the new organonickel compound **11**. Transmetalation with another **2**, which forms **12**, and reductive elimination gives an associated pair of terthiophene and Ni(0) [**13**·**10**]. Growth of the polymer chain occurs by an insertion of one monomer at a time as shown in the reaction cycle (**14**–**15**–[**16**·**10**]–**14**), where the Ni(dppp) moiety is always incorporated into the polymer chain as an end group via the formation of a π-complex. In this fashion, Ni(dppp)Cl_2 is believed to act as an initiator rather than a catalyst, and therefore this limits polymerization to one end of the polymer chain. The prediction of the mechanism as living is supported by two experimental results: first, the degree of polymerization of rrP3ATs has been found to increase with monomer conversion and can be predicted by the molar ratio of the monomer to the nickel initiator; second, addition of various Grignard reagents (R′MgX) at the end of polymerization results in end capping of rrP3AT with R′ end group [73, 74]. This living nature offers the synthesis of molecular weight–controlled and very pure rrP3AT, which allows systematic studies of structure–property relationships [75]. It also allows the synthesis of multiblock copolymers that may be feasible for electronic devices [72, 76–78]. As a consequence, this synthetic methodology has expanded the research field and hence has led to a better understanding of PT chemistry and physics as well.

3.4.3 End Group Functionalization

End group functionalization of rrP3ATs is expected to result in a number of new uses for these polymers including end group–driven self-assembly onto surfaces and into conducting polymer–assembled networks, as well as the synthesis of reactive end groups that can be used as building blocks for the synthesis of block copolymers. Two approaches have been investigated to alter end group composition: postpolymerization and *in situ* methods.

3.4.3.1 Postpolymerization End Group Functionalization

rrP3ATs synthesized through the McCullough, Rieke, or GRIM route have fairly pure end group composition with a proton at one end and a bromine at the other end (H/Br) [72, 79]. This bromine end can be converted to H by treating the

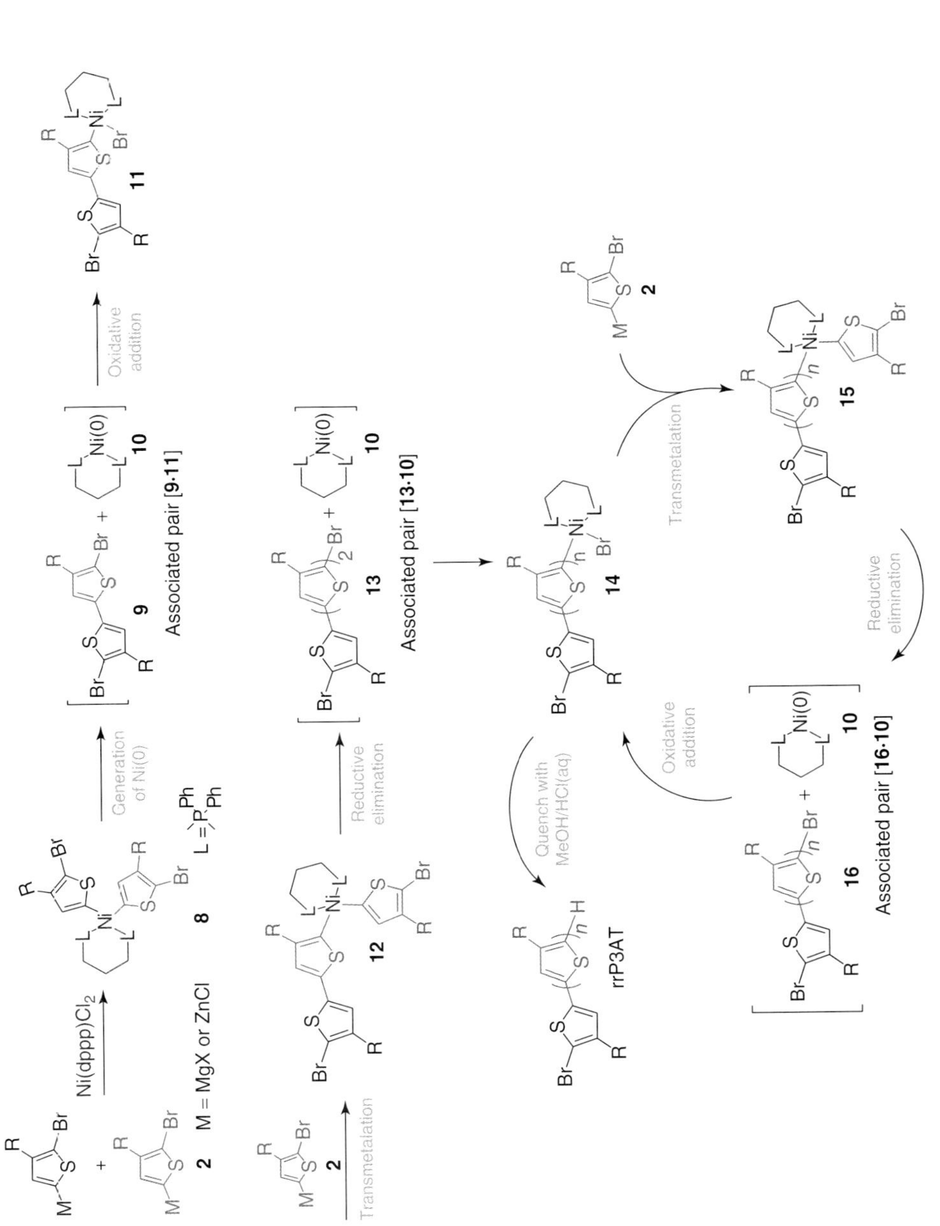

Scheme 3.3 Proposed mechanism of the nickel-initiated cross-coupling polymerization.

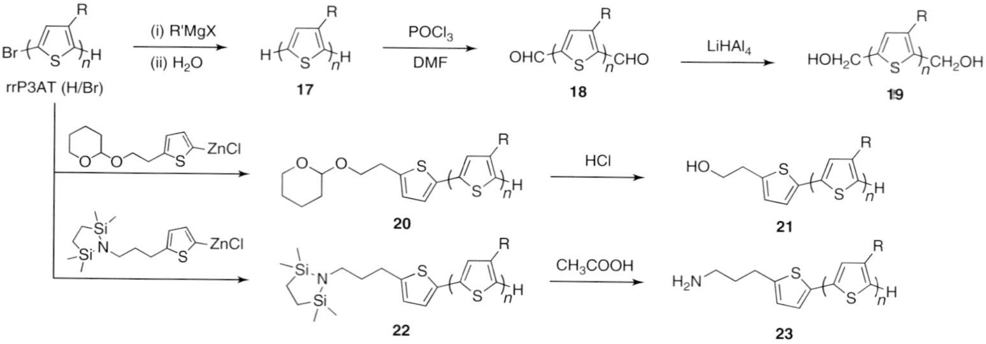

Scheme 3.4 Postpolymerization end group functionalization of rrP3ATs.

polymer with an excess of Grignard reagents and subsequent aqueous workup, yielding an H/H type polymer (**17**), as shown in Scheme 3.4 [79, 80]. Further, both hydrogen end groups can be converted to aldehyde (**18**) by Vilsmeier reaction, and subsequently the aldehyde groups can be reduced to hydroxymethyl groups on both ends (**19**) [80].

On the other hand, in one study the bromine end of H/Br polymer was successfully replaced with a thiophene with tetrahydropyranyl-protected hydroxyl group (**20**) by Negishi coupling reaction [81]. This end group was then deprotected to yield regioregular poly(3-hexylthiophene) (rrP3HT) with hydroxyl end group (**21**). Similarly, a thiophene bearing STABASE [82]-protected amino group was substituted with bromine end group (**22**), and subsequently deprotected to yield rrP3HT with an amino end group (**23**). Further, these end groups can be converted to polyacrylates or polystyrenes (PSs), leading to rrP3AT-based rod–coil block copolymers, as described in Section 3.7.

3.4.3.2 *In situ* End Group Functionalization

The *in situ* method offers an advantage over the postpolymerization method since an end group(s) can be modified by one pot. The first attempt toward *in situ* end group functionalization was reported by Janssen using the McCullough method, in which 2-thienylmagnesium bromide or 5-trimethylsilyl-2-thienylmagnesium bromide was added to the reaction mixture with additional Ni catalyst, to give a mixture of H/H, mono-, and dicapped polymer chains [83].

We have also reported a simple and very versatile method to achieve *in situ* functionalization of rrP3HT using the GRIM method [73, 74]. Since the GRIM method follows a living mechanism, rrP3AT is still bound to the nickel catalyst at the end of the reaction. Therefore, a simple addition of another Grignard reagent effectively terminates the reaction and "end caps" the polymer chains (Scheme 3.5). In this system, a variety of different types of Grignard reagents (alkyl, allyl, vinyl, aryl, etc.) successfully achieved end capping, giving both monocapped (**24**) and dicapped (**25**) rrP3AT. When the end group is allyl, ethynyl, or vinyl, the nickel catalyst is postulated to be bound to the end group through a nickel–π complex

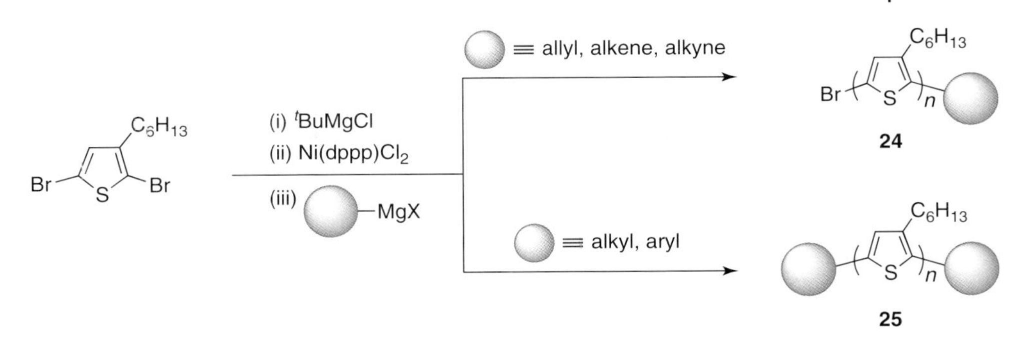

Scheme 3.5 *In situ* end capping of rrP3HT.

to yield monocapped polymers, eliminating the possibility of dicapping. However, when the end groups are alkyl or aryl, the polymers further reacted to yield dicapped polymers. Hydroxyl, formyl, or amino groups were also successfully used to end cap the polymer chains when properly protected. This simple procedure will allow for the synthesis of a library of end-capped polymers that can be used to better understand the effect of end group composition on polymer morphology and self-assembly.

3.4.4 Fundamental Properties of rrP3ATs

3.4.4.1 UV–vis Absorption

Determining the extent of π-conjugation by UV–vis absorption spectroscopy is very important because the π-conjugation implies how far the π-electron or charge carrier can travel along the polymer chain [84]. Maximum absorption (λ_{max}) is regarded as the $\pi-\pi^*$ transition and reflects the extent of the π-electron system and coplanarity of the polymer chains for single chains in solution and for aggregation in film. In studies of UV–vis absorption, rrP3ATs showed a significant redshift in λ_{max} as compared to irP3ATs [3, 20–22, 48, 61]. As an example, the λ_{max} for rrP3HT prepared by the McCullough method with 98% regioregularity in solution was 442 nm, while irP3HT prepared by the $FeCl_3$ method with 70% regioregularity was 436 nm [48]. This redshift indicates that the regioregular polymer has a $\pi-\pi^*$ transition at lower energy and thus a longer π-conjugation. This is because the regioregular polymer chain forms a more planar conformation that facilitates π-orbital overlap [85]. The difference in λ_{max} between regioregular and regioirregular PTs becomes more distinct in the film. rrP3HT shows λ_{max} at 555 nm with well-defined vibronic peaks at 525 and 610 nm, while irP3HT shows λ_{max} at 480 nm. This may be explained by noting that rrP3HT self–assembles to form a three–dimensional superstructure in the film, which cannot be seen in irP3HT because of its sterically twisted structure, and therefore the difference in π-orbital overlap between regioregular and regioirregular polymers is enhanced in the aggregate as compared to the single chain. The λ_{max} in film is dependent on film thickness [3, 86, 87].

Early work on rrP3ATs demonstrated that λ_{max} changed as a function of side chain length [48]. In both solution and film, while irP3ATs showed a similar λ_{max} in all alkyl side chains (butyl, hexyl, octyl, and dodecyl), rrP3ATs with longer side chains gave larger λ_{max}, for example, 500 nm for butyl, 504 nm for hexyl, 520 nm for octyl, and 526 nm for dodecyl (all values in film). This trend may be attributed to the difference in the backbone structure, where the dodecyl polymer is more rigid rod like and butyl polymer is more coil like.

3.4.4.2 Microstructure and Morphology in Thin Films

Supramolecular ordering is one of the most fascinating physical properties of rrP3ATs, one which cannot be seen in irP3ATs. X-ray studies have been used to determine the microstructure. Atomic force microscopy (AFM) is typically used to observe the morphology in film. McCullough *et al.* were the first to show the supramolecular ordering of self-assembled rrP3ATs by X-ray study [22, 49, 57, 88]. X-ray data of an rrP3HT film prepared by simple evaporation of xylene solution on a glass slide showed strong intensity in three small-angle reflections, <100>, <200>, and <300> peaks, corresponding to a well-ordered lamellar structure with an interlayer spacing of 16.0 Å. A narrow single wide-angle reflection, (010) peak, was also observed, corresponding to the π-stacking distance (3.8 Å) of thiophene rings between two polymer chains (Figure 3.3) [22]. Winokur *et al.* also reported a similar structure in stretch-oriented irP3ATs [89]. However, it is important to

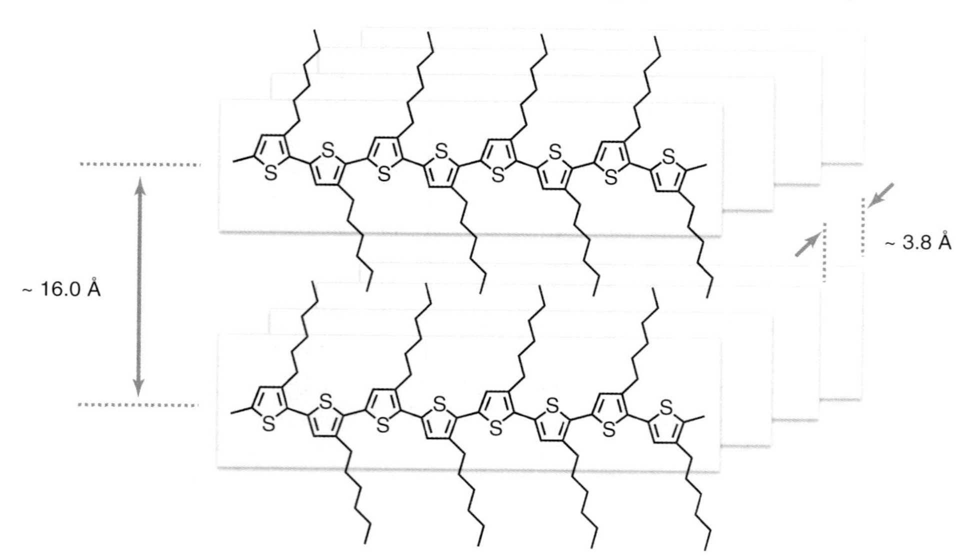

Figure 3.3 Schematic illustrations of self-assembled stacking structure of rrP3HT in thin films as revealed from X-ray study.

point out that all irP3AT samples previously examined show quite broad peaks in the wide-angle region. An irP3HT prepared by the $FeCl_3$ method showed much weaker intensity peaks in the small-angle region, with an interlayer spacing of 17.3 Å, and a very broad amorphous halo centered at 3.8 Å. The observation of a very narrow width in the wide-angle region suggests that along the polymer chain, conformational order gave rise to a single stacking distance rather than a distribution of stacking distances. The narrow widths of all the dominant X-ray features indicated highly ordered crystalline domains in rrP3HT. These results showed an unprecedented structural order for rrP3ATs, indicating that the order is induced by the regiochemical purity afforded by strategic design and synthesis. In addition, interlayer spacing of rrP3ATs is dependent on side chain length, where spacing increases as side chain length increases [90–92]. The crystallinity of rrP3ATs varies depending on their regioregularity [11, 93], molecular weight [92, 94, 95], side chain length [92], and solution processing conditions [92, 95].

Thin films of rrP3ATs show well-organized nanofibril morphologies of differing dimensions. Cho *et al.* have proposed that when an rrP3HT is spin-coated on an SiO_x substrate, the polymer chains in the nanofibrils are packed with their π-system parallel to each other and the alkyl lateral chains perpendicular to the plane of the substrate, which is similar to the case previously discussed here [96]. They also speculated that the extended rrP3HT chains are packed parallel to each other with their long chain axis perpendicular to the long nanofibril axis (Figure 3.4). The Kowalewski and McCullough group also suggested this packing structure of rrP3HT in thin films [75]. Morphologies vary depending on molecular weight [75, 92, 94, 95] and processing conditions [92, 95, 97, 98].

All these structural features play a very important role in device performance. The relationships between chemical structures (regioregularity, molecular weight, and side chain length), solid-state structures, and device properties are discussed in later sections.

3.4.4.3 **Electrical Conductivity**

Electrical conductivity is an important fundamental property in π-conjugated polymers and helps clarify structure–property relationships. Regioregularity is a critical factor in determining polymer conductivity. Owing to their well-defined polymer chain and thus the self-assembled superstructure, rrP3ATs give higher conductivities relative to irP3ATs. In several studies, rrP3ATs gave electrical conductivities of 100–1000 S cm^{-1}, with an average of 600 S cm^{-1}, whereas irP3ATs, under the same conditions, gave conductivities of 0.1–20 S cm^{-1}, with an average of 1 S cm^{-1}, when doped with iodine [20, 22, 48]. Further, in rrP3ATs, side chain length plays an important role in controlling conductivity. While irP3ATs showed roughly the same conductivities regardless of side chain length, rrP3ATs showed a clear trend, in which longer side chains consistently yielded higher conductivities [22]. This result can be understood by recognizing that polymers having longer side chains have more extensive π-electron overlap along the backbone and a larger number of conjugative domains relative to the number found in shorter side chain polymers.

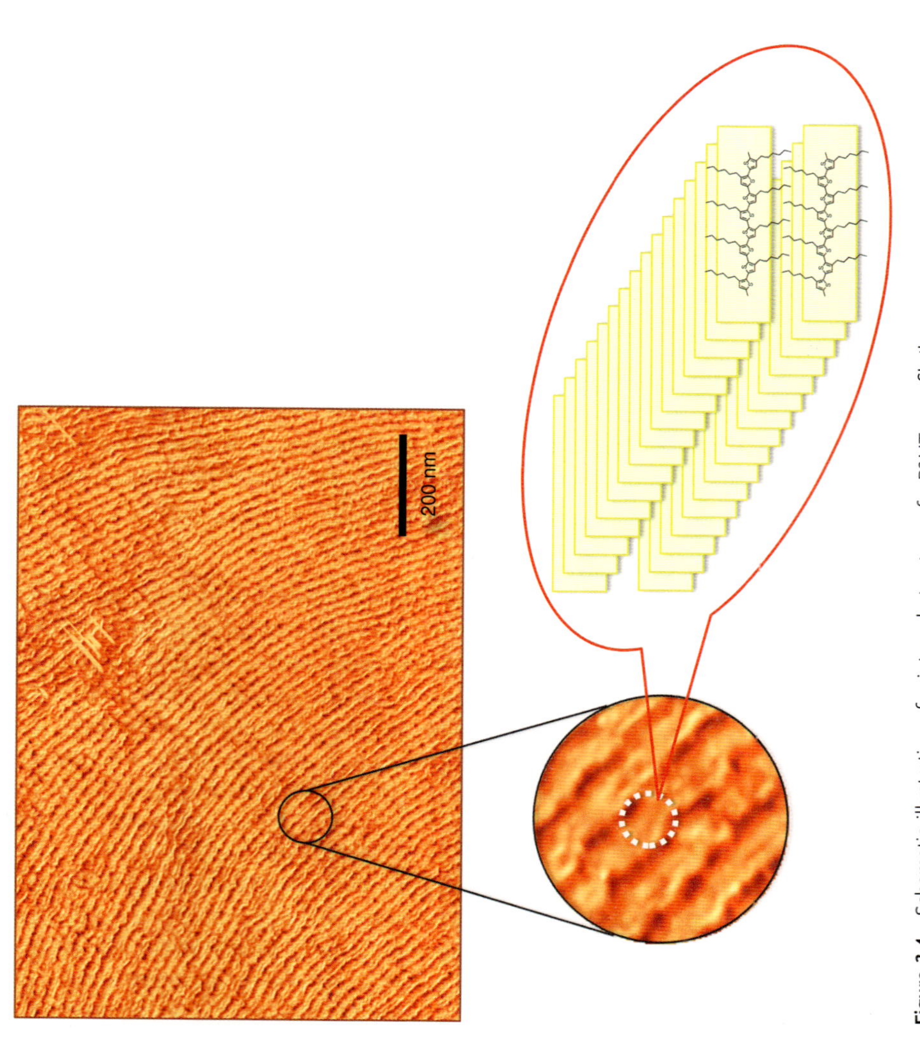

Figure 3.4 Schematic illustrations for internal structure of rrP3HT nanofibrils.

In addition, the conductivity of rrP3AT films cast from the same sample can differ markedly as a result of varying morphology from film to film.

3.4.5
rrP3ATs in Electronic Devices

rrP3ATs are used in various electronic applications, including PLEDs [9], OFETs [13, 14], and OPVs [19]. Of the numerous reports on such devices, we focus on a handful of interesting studies of these devices, which describe structure–property relationships. Such studies have been very important not only for understanding how to achieve the best performance from rrP3AT-based devices but also for understanding design principles that can lead to the future development of new high-performance electronic materials.

3.4.5.1 rrP3ATs in PLEDs

PLEDs are one type of application that attracts researchers in the field of π-conjugated polymers. Although rrP3ATs are less efficient emitters when compared to other π-conjugated polymers such as PPVs and poly(fluorene)s [8, 99], some reports describe interesting findings about the luminescent properties of rrP3ATs, in which effects of structural difference were examined.

An increase in regioregularity (HT ratio) in P3ATs enhances π-conjugation and crystallinity. This, in turn, influences the optical and electrical properties of the polymers and their devices. For instance, Holdcroft and Xu examined the effect of regioregularity in P3HTs on photoluminescence [100]. They showed that increasing the HT ratio from 50 to 80% resulted in a redshift of emission maxima as well as an enhanced fluorescence yield in solution, resulting in a change from 9 to 14%. However, photoluminescent efficiency (Φ_{PL}) in film decreased from 0.8 to 0.2% as the HT ratio increased, perhaps due to the enhanced nonradiative decay caused by an aggregation in high HT-P3HTs. Mehta and McCullough discussed the effect of regioregularity on electroluminescence (EL), in which irP3HT (70% HT) and rrP3HT (>98% HT) were compared [101]. rrP3HT showed a redshifted (30–35 nm) and narrower EL spectrum as compared to irP3HT. rrP3HT exhibited twice as high external efficiency (Φ_{EL}) as irP3HT at a low current (6 mA), 1.5×10^{-4}% for rrP3HT and 7×10^{-5}% for irP3HT. At a high current, the irP3HT device degraded much more rapidly than the rrP3HT device. Turn-on voltage was found to be lower in the case of rrP3HT, reflecting the higher carrier mobility of the polymer. Structural changes in P3HTs affect their luminescence as well as electrical properties, both of which are critical to PLED performance.

3.4.5.2 rrP3ATs in OFETs

OFETs are important devices in understanding charge transport mechanisms as well as charge transport capacities of materials in solid state. The development of the chemistry and physics of rrP3ATs has brought significant improvements in and a deeper understanding of polymer-based OFETs. rrP3ATs show excellent performance in OFETs, where mobilities are usually 0.01–0.1 $cm^2\ (V\ s)^{-1}$ depending on structural features such as regioregularity, molecular weights, and side chain length. These features have a strong impact on thin film structures and strongly influence device performance. Herein, we describe some interesting and important structure–property relationships of rrP3ATs in OFET devices: carrier transport mechanism, effects of regioregularity, molecular weight, and side chain length.

Mechanism of Carrier Transport and Effect of Regioregularity It has already been established that a regioregular structure drastically improves the physical properties of P3ATs [20, 21]. rrP3ATs have also demonstrated significantly higher performance than irP3ATs in OFETs. Careful studies on rrP3AT-based OFETs have manifested the principle of carrier transport in thin films of π-conjugated polymers. Bao *et al.* reported the first study of OFETs using rrP3ATs, in which field-effect mobilities as high as $0.05\,cm^2\,(V\,s)^{-1}$ for rrP3HT were shown [10]. This is in sharp contrast to studies of irP3HTs, where mobilities were only 10^{-5}–10^{-3} $cm^2\,(V\,s)^{-1}$ [102, 103]. They proposed that high mobilities in rrP3HT could be achieved because of the highly ordered crystalline structure. This was supported by an X-ray diffraction study, as described earlier. They also proposed that the preferential edge-on orientation of the polymer lamellae leads to better carrier transport through strong π-stacking between the source and the drain (Figure 3.5). Sirringhaus *et al.* proved this two-dimensional carrier transport capacity in P3HTs with different regioregularities and also showed the effect of regioregularity on the mobility of P3HTs using OFET devices [11]. They found that spin-coated thin films of P3HTs with high regioregularity (>91% HT) and low molecular weight ($M_w = \sim28$ K) adapted edge-on orientation, while spin-coated thin films of P3HTs with low regioregularity (<81% HT) and high molecular weight (Mw = ~175 K) adapted face-on orientation on SiO_2 substrate (Figure 3.5). On the other hand, interestingly, drop-cast films of all P3HT samples formed edge-on orientation. The highest mobilities in this study were observed for P3HTs with 96% HT in both spin-coat and drop-cast films, which were 0.05–$0.1\,cm^2\,(V\,s)^{-1}$. For P3HTs with

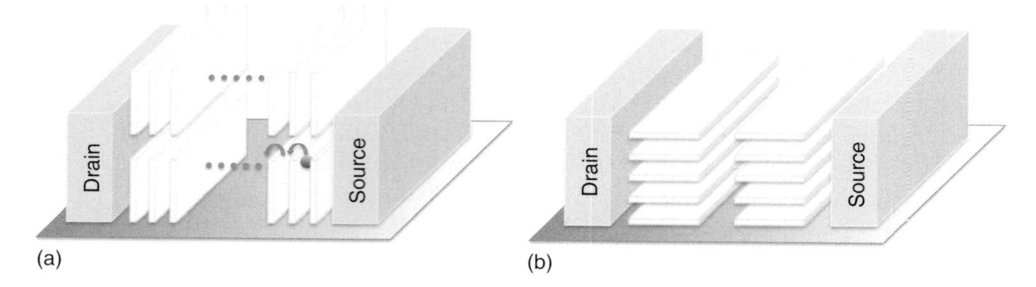

Figure 3.5 Schematic illustrations of two different rrP3AT lamellar orientations with respect to the OFET substrate: (a) edge-on orientation where lamellae pack along the substrate, providing efficient charge carrier (gray sphere) transport between the source and the drain and (b) face-on orientation where lamellae pack perpendicular to the surface.

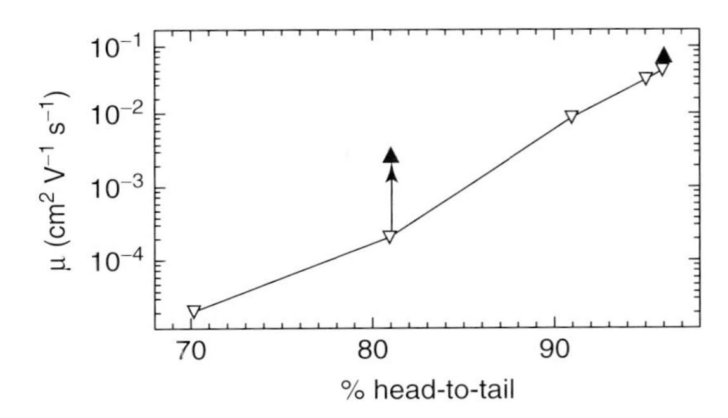

Figure 3.6 Charge carrier mobility of rrP3HT field-effect transistors with different microstructures: dependence of the room-temperature mobility on the regioregularity for spin-coated (downward triangles) and drop-cast (upward triangles) top-contact rrP3HT OFETs. (Adapted from [11].)

81% HT in face-on orientation, mobility was only 2×10^{-4} cm^2 (V s)$^{-1}$, while for P3HTs with 81% HT in edge-on orientation, mobility was more than an order of magnitude higher compared to that seen in the face-on sample (Figure 3.6). This large mobility anisotropy is clear evidence that the carrier transport property in thin films is dominated by orientation of the ordered microcrystalline domains. This, thus, strongly suggests that the majority of the carrier pathway is enabled through π-stacked lamellae, and high mobilities in OFET are achieved when the lamellae are edge-on oriented.

Effect of Molecular Weight on Polymer Structure and Mobility Molecular weight is a very important factor to achieve high mobilities in π-conjugated polymers. Several groups have demonstrated that there is a clear correlation between molecular weights and mobilities in rrP3ATs [75, 94, 95, 104, 105]. Kline *et al.* were the first to show that mobility variation in molecular weights is accompanied by significant changes in film morphology [94, 95]. Mobilities for rrP3HTs have been shown to steadily increase with molecular weights from 1.7×10^{-6} cm^2 (V s)$^{-1}$ at M_n = 3.2 K to 9.4×10^{-3} cm^2 (V s)$^{-1}$ at M_n = 36.5 K. In their work, AFM and X-ray studies suggested that lower molecular weight samples were more crystalline. It is, however, interesting that a sample with a higher degree of crystallinity had significantly lower mobility. AFM data clearly suggested that the low M_n film had more defined grain boundaries than the high M_n film. Grain boundaries have been shown to limit mobility in transistors using polycrystalline molecules [106], and therefore they hypothesized that the higher mobilities of rrP3HTs having higher molecular weight were possibly due to longer polymer chains connecting the ordered regions so as to provide a pathway for charge transport between crystalline domains. They also showed that choice of deposition conditions could lead to a further improvement in mobilities [95].

Neher *et al.* also examined the effect of molecular weight on rrP3HT OFETs [104]. In fact the findings of as-prepared polymer films were in agreement with the earlier work by Kline *et al.* However, further investigation demonstrated that mobilities decreased continuously as temperature increased. This implies that polymer chains adopted a more twisted, disordered conformation at higher temperatures, which in turn gave rise to interchain transport barriers. Based on this, they interpreted these results as meaning that backbone conformation rather than crystallinity is the most crucial parameter controlling charge transport in rrP3HT thin films.

Kowalewski *et al.* have gone one step further. They demonstrated that the width of well-defined nanofibrils (Figure 3.7), which are formed in thin films of rrP3HTs, corresponds very closely to the weight average contour length of polymer chains [75]. Moreover, they showed that mobilities increased exponentially as nanofibril width increased. Initially nanofibril width (w_{AFM}) increased linearly with M_w and then leveled off (Figure 3.8a, red symbols). This dependence was

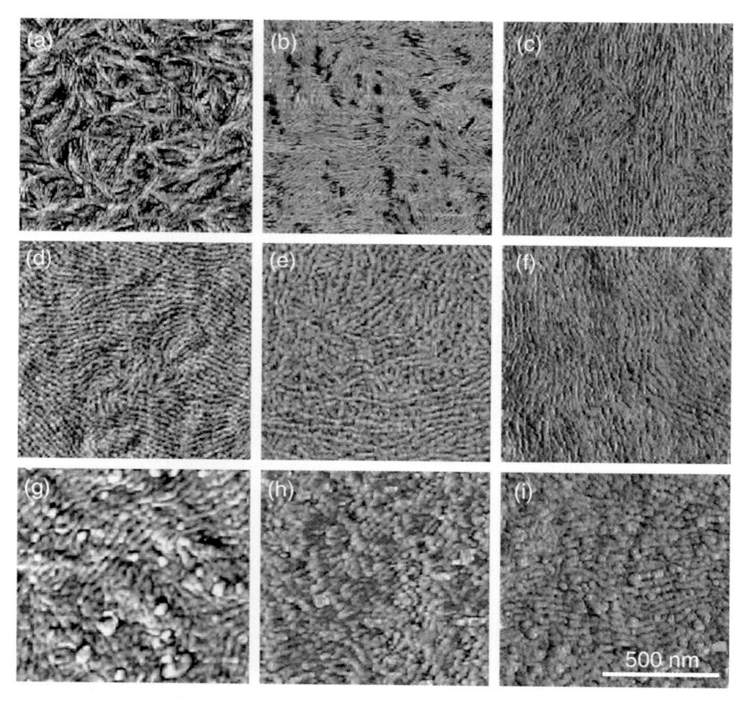

Figure 3.7 Tapping mode AFM images (phase contrast, 1 μm × 1 μm) of thin films of rrP3HTs of various molecular weights in OFET devices prepared by drop casting from toluene. Corrected weight average molecular weights (M_w) in (a–i) were respectively equal to: 2.4, 4.8, 5.1, 7.0, 7.5, 11.8, 15.7, 17.3, and 18.4K. (Adapted from [75].)

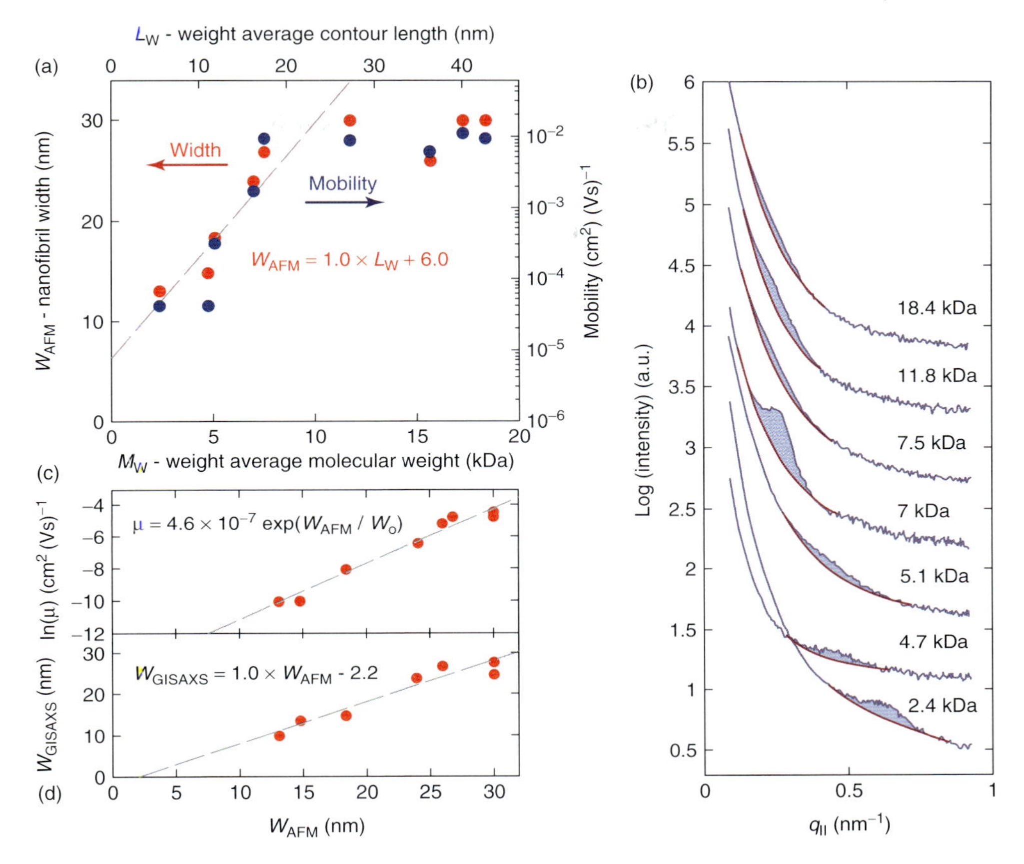

Figure 3.8 (a) Dependence of nanofibril width (w_{AFM}, red) and charge carrier mobility (μ, blue) on weight average molecular weight (M_w, bottom axis) and weight average contour length (L_w, top axis) of rrP3HT. (b) GISAXS scattering profiles for thin films of rrP3HTs with various molecular weights. (c) Data and exponential fit (dash) of μ versus w_{AFM}. (d) Linear correlation between GISAXS periodicities and nanofibril widths (w_{GISAXS} vs w_{AFM}). (Adapted from [75].)

particularly revealing when plotted as a function of weight average contour length (L_w) of rrP3HT (Figure 3.8a). This close, direct relationship between w_{AFM} and L_w is highly suggestive of a structure composed of one-molecule-wide, stacked sheets of rrP3HT, with polymer backbones aligned perpendicular to the nanofibril axis, as shown by Cho *et al.* [96]. Grazing incidence small-angle X-ray scattering (GISAXS) patterns (Figure 3.8b) revealed the presence of periodicity (w_{GISAXS}), which corresponded very well to w_{AFM} (Figure 3.8d). Most interestingly, logarithm of mobility (μ) could be mapped directly against periodicities, L_w (Figure 3.8a, blue symbols). Since nanofibril width, w_{AFM}, tends to directly reflect the contour length of polymer chains, the observed exponential dependence of μ versus w_{AFM} hints at the prominent role of conjugated states in charge transport in π-conjugated polymers. They also suggested that another factor that may explain the exponential dependence of μ versus w_{AFM} is possibly the change in the nature of grain boundaries between adjacent nanofibrils (Figure 3.8c). Another important finding was that both the nanofibril width and the mobility became saturated at higher molecular weight, which may imply a conformational change such as chain folding.

Effect of Side Chain Length on Mobility Early studies of rrP3ATs demonstrated that length of side chain plays an important role in controlling their structure and physical properties [22, 88]. Longer side chain polymers were found to have better solubility and higher conductivity. However, in contrast to the conductivity trend, several reports have shown that longer side chain polymers exhibit lower mobility in OFETs [107–110]. For example, Bao *et al.* reported that in all-printed plastic transistors rrP3HT and rrP3OT (O = octyl) showed similar mobilities of $\sim 10^{-2}$ cm^2 (V s)$^{-1}$, while rrP3DDT (DD = dodecyl) showed much lower mobility, of the order of 10^{-6} cm^2 (V s)$^{-1}$ [107]. Similarly, Kaneto *et al.* reported that mobility decreased continuously as side chain length increased [108]. In particular, they found that the highest mobility was observed in rrP3BT (B = butyl), while the lowest one was observed in rrP3ODT (OD = octadecyl) (Figure 3.9, filled circles). Note that in their study, the conductivity trend agreed with the mobility trend. On the other hand, a report by Babel and Jenekhe indicated that there was an optimal side chain length that would ensure better performances in rrP3AT-based OFETs [109]. More specifically, they showed that mobility increased by an order of magnitude from rrP3BT to rrP3HT, then decreased by almost two orders of magnitude to rrP3OT, and thereafter decreased exponentially to rrP3DDT

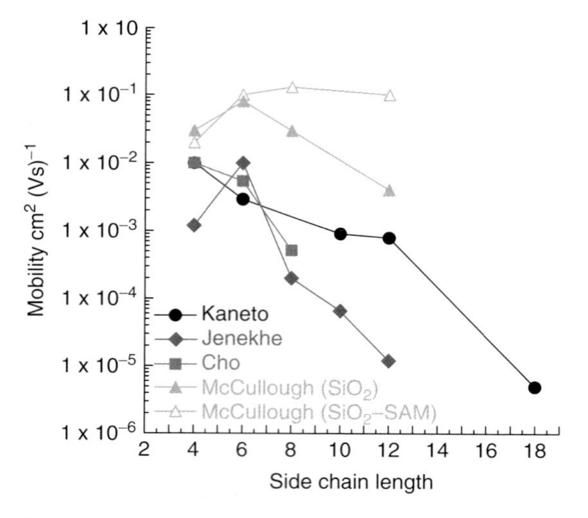

Figure 3.9 Dependence of mobilities on side chain length of rrP3ATs: (●) from [108]; (♦) from [109]; (■) from [110]; (▲, △) unpublished data.

(Figure 3.9, diamonds). Cho *et al.* also reported a similar finding, investigating the relationship between side chain length, structure ordering, and mobility [110]. They found that the mobility of polymer OFETs decreased as side chain increased from butyl to octyl (Figure 3.9, squares). Structural studies of polymer films revealed that shorter side chain polymers tended to have higher crystallinity (X ray), which was consistent with the mobility trend, and narrower nanoribbon structures (AFM).

Our group has also carefully examined how the side chain length of rrP3ATs affects transistor performance. Since the interaction between gate dielectric surface of transistor devices and conjugated polymers will guide self-assembly of polymers at the interface, where the greatest amount of current flows, both bare SiO_2 gate dielectric and octyltrichlorosilane (OTS-8)-treated dielectric were used. It is well known that the treatment of SiO_2 dielectrics with such self-assembled monolayers (SAMs) can improve the orientation of polymer crystalline domains, and this in turn enhances the charge transport property [111, 112]. We found that mobility for polymers on bare SiO_2 had increased from rrP3BT (0.03 cm^2 $(V\ s)^{-1}$) to rrP3HT (0.08 cm^2 $(V\ s)^{-1}$) and subsequently decreased to 0.004 cm^2 $(V\ s)^{-1}$ (for rrP3DDT) as the side chain elongated (Figure 3.9, open and filled triangles); this suggests a trend similar to that reported by Babel and Jenekhe. However, when SiO_2 surfaces were treated with OTS-8, this trend changed dramatically. The OTS-8 treatment had a significant impact on longer side chain polymers: rrP3OT showed the highest average mobility of 0.13 cm^2 $(V\ s)^{-1}$ and rrP3DDT showed a comparable one with rrP3HT, 0.10 cm^2 $(V\ s)^{-1}$ (Figure 3.9, open triangles). The lowest mobility was observed for rrP3BT, 0.02 cm^2 $(V\ s)^{-1}$. While self-assembly of rrP3HT (and probably rrP3BT) is largely governed by π-stacking force (side chains are likely to be disordered and amorphous) [113], rrP3OT and rrP3DDT have been found to self-assemble into two distinct structures, resulting from the competition between backbone and side chain crystallizations [88]. Because of this competition, structural orders in longer side chain polymers might be sensitive to the surface treatment, consequently yielding improved mobilities.

One could surmise that from these results two different trends have been observed in transistors with bare dielectric layers, with one report from Kaneto and Cho, and from Jenekhe and our group. This difference seems to result from the differences in regioregularity and PDI of the samples used, where regioregularity of the samples used in studies by Kaneto and Cho was about 93%, and that in the studies by Jenekhe and our group was >97%. Altogether, for lower regioregular polymers, the butyl side chain may exhibit the best performance and longer side chains may have lower performance, and for higher regioregular polymers the hexyl side chain may be of the optimal length to obtain optimal transistor performance. In addition, when dielectric layers are coated with SAMs, polymers with hexyl to dodecyl side chains may provide comparable transistor performance.

3.4.5.3 rrP3ATs in OPVs

rrP3ATs are the most commonly used and widely studied p-type materials for use in OPVs/solar cells. Despite the nonideal bandgap that does not effectively overlap with the solar spectrum, power conversion efficiencies (PCEs) for rrP3HT-based bulk heterojunction (BHJ) solar cells with [6,6]-phenyl C_{61} butyric acid methyl ester (PCBM) as the n-type material are among the highest reported, reaching ~5% [15, 18, 114, 115]. OPV performance in these systems is, of course, determined by polymer structures, such as regioregularity and molecular weight, which strongly influence their electrical and optical properties, as is the case for other organic devices described previously.

Kim *et al.* investigated a regioregularity effect of P3HTs and found that P3HTs with higher regioregularity had a higher PCE, for example, 95% HT-P3HT exhibited a PCE of 2.4% (reaching 4.4% after optimizing processing conditions), which was almost threefold higher than that for 90% HT-P3HT [116]. This trend can be attributed to enhanced optical absorption and carrier transport resulting from the stronger organization of polymer chains and domains in higher HT-P3HT revealed by GIXS studies. Fréchet *et al.* further investigated the regioregularity effect on P3HT–PCBM solar cells under optimized conditions [117]. The highest PCEs for devices using P3HTs with 86, 90, and 96% HT were 3.9, 3.8, and 3.8%, respectively, showing a quite different trend from the work by Kim *et al.* Long-term thermal stability for these solar cells was also examined. While devices consisting of 96% HT-P3HT decayed quickly to less than 2.5% PCE after 5 hours of annealing (at 150 °C), 86 and 90% HT-P3HT were able to maintain ~3% PCE after more than 10 hours of annealing. This was possible because of less crystallization-induced phase segregation of PCBM.

The influence of the molecular weight of rrP3HT on solar cell performance has been investigated by several groups. Scherf *et al.* used rrP3HT with molecular weights (M_n) of 2.2–19 K (PDI = 1.2~1.9) [118]. Only devices using $M_n > 10$ K rrP3HTs achieved high PCEs (exceeding 2.5%), a result 10- to 20-fold higher than that for devices using low M_n rrP3HT (<5.6 K). They examined the mobilities of rrP3HTs and rrP3HT/PCBM composites and consequently concluded that the significantly higher efficiencies in high-molecular-weight rrP3HTs were caused by distinctly higher mobilities and perhaps higher intermolecular ordering (π-stacking) of the rrP3HT phase as compared to low-molecular-weight ones. Heeger *et al.* later reported similar results: that the longer π-conjugation length and better interconnections within the bicontinuous network in high-molecular-weight rrP3HT afford improved performance in solar cells [119]. Moreover, they showed that much higher performance could be achieved by using rrP3HTs with an optimized ratio between high- (M_n = 62.5 K, PDI = 2.5) and low-molecular-weight (M_n = 13 K, PDI = 2.0) components (1 : 4 in this case), where the morphology comprises highly ordered crystalline regions formed by low-molecular-weight rrP3HT embedded and interconnected by a high-molecular-weight rrP3HT matrix. Ballantyne *et al.* have also studied the effect of molecular weight on rrP3HT/PCBM solar cells as well as charge carrier transport using the time-of-flight method [120]. They then proposed that there was an optimal rrP3HT molecular weight for charge transport and solar cell performance of between 13 and 34 K. World-record lab cells and panels have been produced by Plextonics with efficiencies of nearly 6% for cells.

Several reports investigated the influence of regioregularity and molecular weight on rrP3HT-based solar cells described here. It is apparent that there may be some optimal set of primary structural features of rrP3HT to obtain higher solar cell performance; however, a crucial point might be controlling the microphase structure of polymer/PCBM composites to form a better polymer network with well-blended (less-crystallized) PCBM.

3.5 Regiosymmetric Poly(alkylthiophene)s

PTs with HT orientation are generally accepted as "regioregular" PTs. However, those with HH and TT alternatives as well as complete HH orientations are also regiochemically controlled "regioregular" PTs (Scheme 3.6), which can also be referred to as *regiosymmetric PTs*. Typically, regiosymmetric PTs are prepared by polymerization of a symmetric monomer or polymerization of two different symmetric monomers. However, these structures have huge steric repulsion from HH arrangements. Several regiosymmetric PTs with reduced HH impacts have been, therefore, synthesized and characterized to date (Schemes 3.7–3.10), and have been found to exhibit comparable or even better electrical properties as

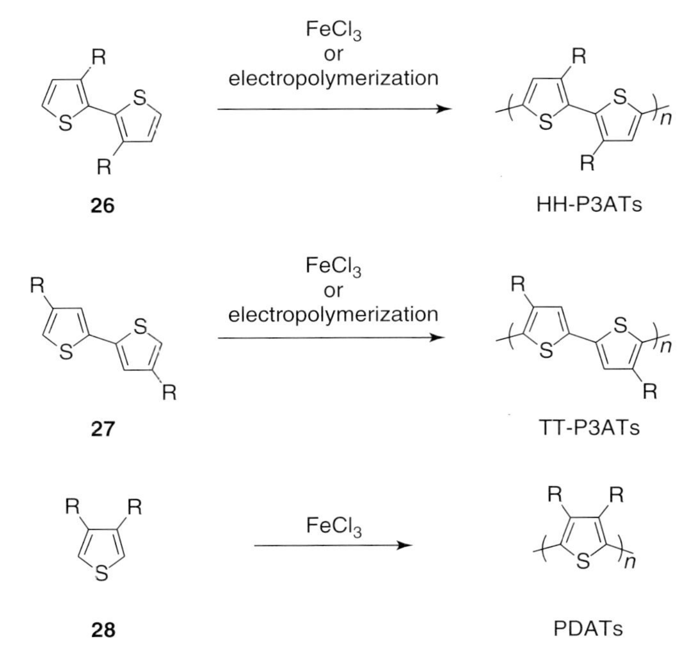

Scheme 3.6 Head-to-head and tail-to-tail coupled regioregular P3ATs (HH-P3ATs and TT-P3ATs) and poly(3,4-dialkylthiophene)s (PDATs).

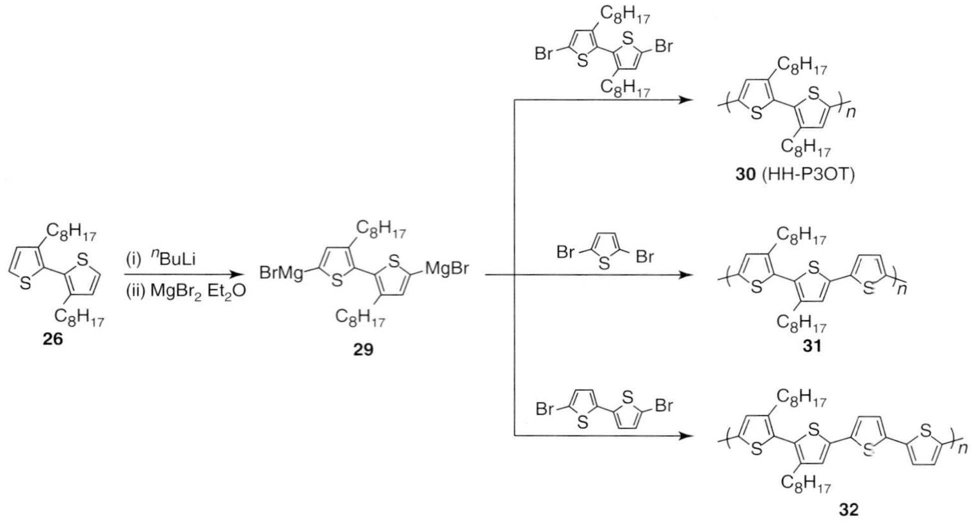

Scheme 3.7 Synthesis of HH-P3OT and its copolymers.

compared to the more familiar rrP3ATs. Herein, we describe the development of regiosymmetric P3ATs.

3.5.1
Head-to-Head–, Tail-to-Tail–Coupled Poly(alkylthiophene)s

The primary structures for regiosymmetric P3ATs are HH- and TT-coupled P3ATs (HH-P3ATs and TT-P3ATs) and poly(3,4-dialkylthiophene)s (PDATs), as shown in Scheme 3.6. In HH-P3ATs or TT-P3ATs, the symmetric monomer, the HH dimer of 3-alkylthiophene (3,3′-dialkyl-2,2′-bithiophene), or the TT dimer of 3-alkylthiophene (4,4′-dialkyl-2,2′-bithiophene) has been polymerized to yield essentially the same polymer. In one such study, Wudl *et al.* prepared poly(3,3′-dihexyl-2,2′-bithiophene) (HH-P3HT) using $FeCl_3$ and electropolymerization [121]. In another, Hadziioannou *et al.* synthesized HH-P3OT (O = octyl) by treating the HH dimer with *n*-butyllithium followed by the addition of $MgBr_2{\cdot}Et_2O$, and then coupling with a dibromo-HH dimer [122]. TT-P3ATs were first synthesized electrochemically by Krische *et al.* [123] and then chemically by both Krische *et al.* [124] and Pron *et al.* [125]. Although the HH-P3ATs are regiochemically defined polymers, these contain 50% of HH couplings, a ratio even higher than that of irP3ATs, and thus backbone coplanarity must be destroyed. The HH-P3AT film gave λ_{max} of 398 nm in UV–vis absorption measurement, whereas regioirregular poly(3-hexylthiophene) (irP3HT) gave 508 nm. The conductivity of HH-P3HT doped with $NOPF_6$ was 4 S cm^{-1}, whereas irP3HT (80% HT couplings) showed 15 S cm^{-1} [121]. Barta *et al.* reported that Φ_{PL} of TT-P3DDT was 11%, that is, one order of magnitude larger than that of rrP3DDT (about 1%) [126]. Since

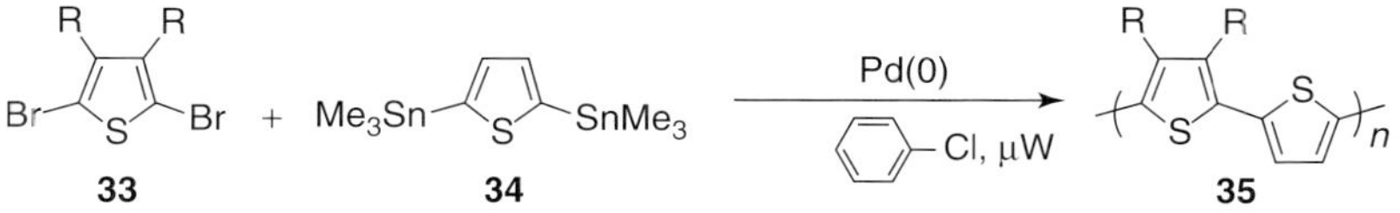

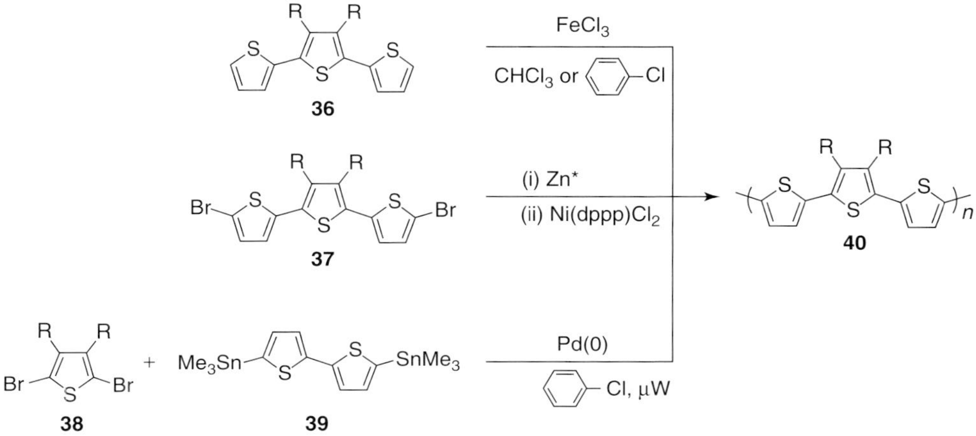

Scheme 3.8 Synthesis of 3,4-dialkylthiophene–(bi)thiophene copolymers.

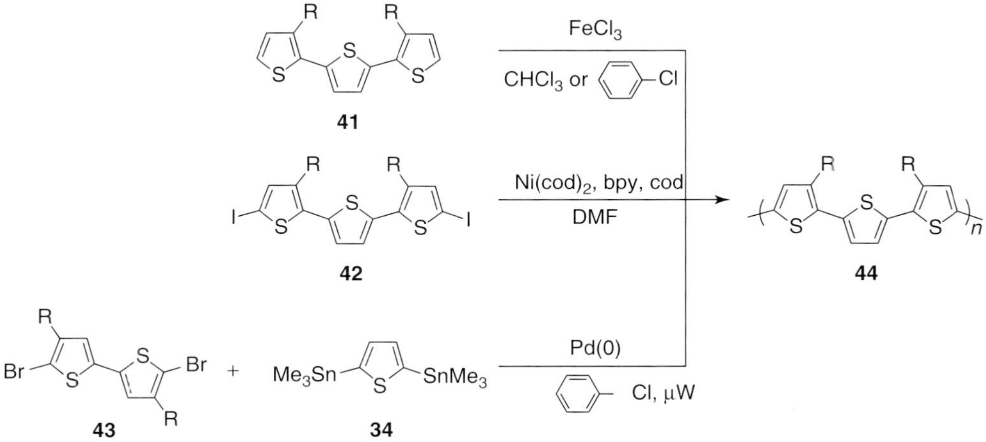

Scheme 3.9 Synthesis of poly(3,3″-dialkylterthiophene)s (**44**).

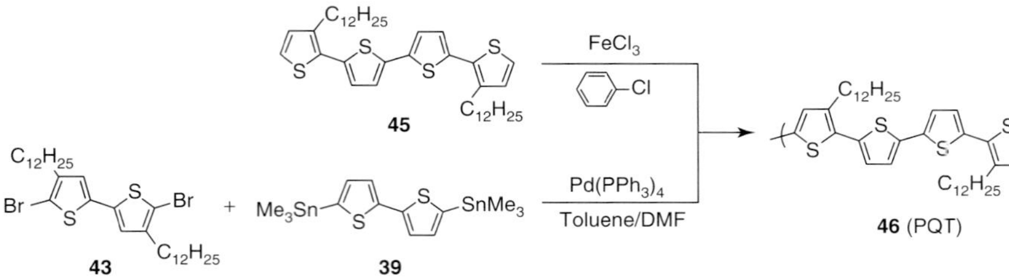

Scheme 3.10 Synthesis of poly(3,3‴-didodecylquaterthiophene) (PQT).

the planar blocks can act as exciton traps [7], the low coplanarity in TT-P3DDT (having lower coplanarity than even that of irP3ATs) reduced the exciton migration to quenching sites and, in turn, exhibited a more efficient luminescent property. Hadziioannou *et al.* synthesized PDAT with an octyl side chain by polymerizing 3,4-dioctylthiophene with $FeCl_3$ [122]. The polymer showed λ_{max} of about 330 nm in film, which is even shorter than that of HH-P3HT. These results suggest that the HH (TT)-coupled poly(alkylthiophene)s, the primary structure for regiosymmetric P3ATs, are not appropriate for electrical properties, but may be effective for obtaining better emitting polymers (with hypsochromic shift).

3.5.2
Regiosymmetric Alkylthiophene–Thiophene Copolymers

To improve the electrical properties of such regiosymmetric P3ATs, the steric impact of the HH arrangement must be reduced. One efficient design strategy for reducing steric impact is to incorporate unsubstituted thiophene rings in the polymer chain. Hadziioannou *et al.* synthesized HH-P3AT with octyl side chains (**30**) and a series of its derivatives having thiophene (**31**) and bithiophene (**32**) rings between the HH dyads, where the di-Grignard of dioctylbithiophene (**29**) was coupled with corresponding 2,5-dibromothiophene and 5,5′-dibromo-2,2′-bithiophene (Scheme 3.7) [122]. The purpose of their work was to tune the conjugation length of PTs for PLEDs. Redshifts were observed in both UV–vis absorption and photoluminescent spectra from polymer **28** to **30** (increasing the number of thiophene rings between the HH dyads), where λ_{max} in absorption was about 390 nm (**30**), 420 nm (**31**), and 450 nm (**32**). However, these polymers still have HH dyads, which cause a conformational defect in the backbone, and the absorption maximum of polymer **32** was approximately 100 nm shorter than that of rrP3ATs. If one expects better physical and electrical properties in PTs, complete absence of HH arrangements is required. From this perspective, incorporating an unsubstituted spacer thiophene(s) between HH-coupled alkylthiophenes is likely to be beneficial.

Both Faïd and Leclerc [127] and Andreani *et al.* [128] synthesized poly(3′,4′-dialkylterthiophene) (**40**) with hexyl and with hexyl and dodecyl side chains, respectively, by polymerizing the corresponding terthiophene monomer (**36**) with $FeCl_3$ (Scheme 3.8). A thin film of polymer **33** with hexyl side chain exhibited λ_{max} of 494 nm in the absorption spectrum. Since this was significantly redshifted ($\Delta\lambda_{max} \approx 160$ nm) from PDAT (with octyl side chain), it suggests an enhanced coplanar backbone [127]. However, λ_{max} of the polymer was shorter than that of rrP3ATs (~560 nm), implying that the backbone was still somewhat twisted. Synthesis of polymer **40** with decyl side chains was also reported by Ong *et al.* [129]. They prepared the polymer by both $FeCl_3$-mediated oxidative coupling using corresponding monomer **36** and Rieke-type coupling using dibromoterthiophene **37**. Tierney *et al.* also synthesized **40** with decyl side chains in a different route [130], where dibromodidecylthiophene **38** and stannylated bithiophene **39** were coupled by Stille cross-coupling reaction under microwave radiation [131]. The advantage of microwave radiation is that it is generally faster and cleaner than conventional heating, reducing reaction times by more than a factor of 10 and yielding products in high yield with fewer side products. In addition, a similar route has also been used to synthesize poly(3,4-dialkylbithiophene) (**35**) with decyl side chain. By optimizing polymerization conditions, they observed that careful choice of reaction solvent, catalyst, ligand, and reaction time could increase the molecular weight. Polymers from these studies had high molecular weights ($M_n = \sim 18\ K$); however, polydispersities were high (about 2), which was much larger than those of rrP3ATs. Mobilities of **40** (decyl side chain) were evaluated using OFET devices [129]. Although the backbone was not as coplanar as that of rrP3ATs, the obtained mobility, $\sim 0.01\ cm^2\ (V\ s)^{-1}$, was quite close to that of rrP3HT. In addition, interestingly, an OFET device with **40** showed better environmental stability as compared to rrP3HT: the **40** device degraded slightly after standing in air for 30 days, while the rrP3HT device lost most OFET activity after only 10 days. This is because the shorter conjugation (wider bandgap) of **40** deepens its HOMO level; in other words, this increases its ionization potential (Ip), and hence the polymer possesses greater resistance to p-doping by atmospheric oxygen.

Using the same strategy, Gallazzi *et al.* synthesized another type of regiosymmetric PT, in which a thiophene ring was inserted between HH-coupled 3-alkylthiophenes (**44**, Scheme 3.9) [132]. The corresponding terthiophene monomer (**41**) with hexyl side chains was polymerized with $FeCl_3$. Several other groups also synthesized polymer **44** with hexyl, octyl, or dodecyl side chains via one of the following methods: the same $FeCl_3$ method [128, 133], Yamamoto polymerization using $Ni(cod)_2$ [134], or Stille coupling reaction (Scheme 3.8) [130]. Molecular weights of the polymers were $M_n = \sim 17$ K (PDI = ~2.5). Electrical conductivity of an iodine-doped pressed pellet was 100 S cm^{-1}, which is comparable to that of rrP3ATs [132]. An XRD study of the polymer thin film revealed a highly ordered lamellar structure similar to that seen in rrP3ATs; moreover, the interlayer spacing was evaluated to be 13.0 Å (for hexyl side chain), which is approximately 3 Å shorter than that of rrP3HT (16.0 Å). This indicates a closer packing of adjacent polymer backbones, which may be due to well-spaced distribution of the side chains [132]. However, another report showed that π-stacking distance appeared to be 4.0 Å, which is about 0.2 Å larger than that of rrP3ATs, indicative of a loosely packed lamellar structure in the π-stacking direction. This is probably because of steric interference between two alkyl side chains on the same repeating unit, causing the polymer backbone to twist [133]. The appearance of λ_{max} at 540 nm in film, which is about 20 nm shorter than in rrP3HT, supports this loose packing structure in the π-stacking direction. Mobilities for **44** with octyl side chains in OFETs were reported to be ~0.03 cm^2 $(V\ s)^{-1}$, which is comparable to those of rrP3HT under the same conditions [130, 133, 135].

Ong *et al.* synthesized this type of regiosymmetric polymer having bithiophene (**46**, poly(3,3‴-didodecylquaterthiophene) (PQT)) via the $FeCl_3$ method, with $M_n = 17$ K and PDI = 1.3 (Scheme 3.10) [136]. The polymer can also be synthesized via Stille coupling reaction [137]. PQT with a dodecyl side chain showed liquid crystalline behavior; annealing at 135 °C (liquid crystalline phase) enhanced its crystalline structure in thin film. PQT exhibited mobilities as high as 0.14 cm^2 $(V\ s)^{-1}$ in OFETs, which is even higher than those of rrP3HT, after annealing. These high mobilities are likely due to the highly organized lamellar structures brought about by its liquid crystalline nature. OFET performance for PQT showed only a slight decrease after being stored under ambient conditions in the dark for one month. This is in sharp contrast to the drastic degradation for rrP3HT, indicating excellent environmental stability.

In the quest to explore regioregular poly(alkylthiophene)s, regiosymmetric P3ATs that eliminated HH couplings were developed soon after the first synthesis of rrP3ATs. Recent studies have revealed that although the extent of π-conjugation in these polymers is not as high as that in rrP3ATs, they show similar lamellar structures in thin films, leading to comparable or even higher charge transport characteristics. They also have wider bandgaps that are detrimental to carrier transport, possibly due to the increased rotational freedom given by incorporation of unsubstituted thiophene(s). However, owing to these wider bandgaps, interestingly, they show large Ip (deep HOMO level), giving rise to higher stability in oxygen doping relative to rrP3ATs. These findings have brought about more recent advances in π-conjugated polymers for organic electronics, which are discussed in the next section.

3.6 Regiosymmetric Polythiophenes with (Hetero)aromatic Rings

Regiosymmetric PTs with (hetero)aromatic rings afford fascinating functional copolymers for electronic devices. Of particular interest today are copolymers incorporating thiophene-based (-related) fused rings, which allow high carrier mobilities in OFETs or high efficiencies in OPVs. Using a design strategy similar to that used for regiosymmetric P3ATs, various kinds of rings have been successfully incorporated in PT backbones. Since the number of copolymers of this type that have been reported to date is too large to cover in depth in this section, we focus here on regiosymmetric thiophene copolymers mainly used for OFETs. Other polymers have been reviewed in other publications [138–140].

Phenylene- (**48**) [141] and naphthalene-substituted (**50**) [135] copolymers were synthesized through oxidative coupling and Suzuki coupling reactions, respectively (Scheme 3.11), giving M_n = ~5.4 K (PDI = ~1.6) and $M_n = 9$ K, respectively. Owing

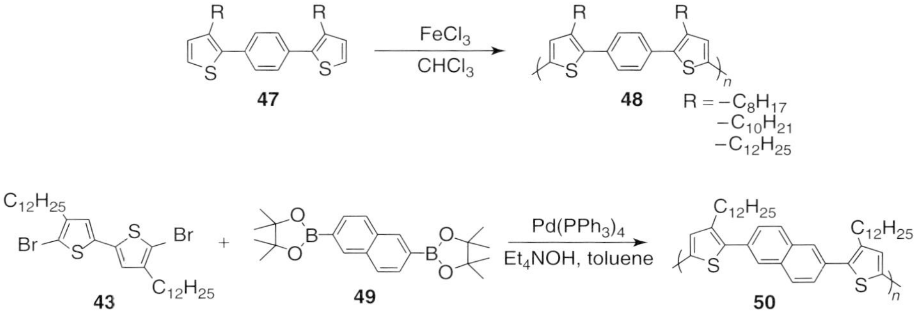

Scheme 3.11 Synthesis of regiosymmetric thiophene copolymers bearing phenylene and naphthalene units.

to their highly twisted structures, the λ_{max} of **48** and **50** was 420–430 nm in thin films and the mobility of **50** was low (in the order of 10^{-4} cm^2 $(V\ s)^{-1}$). Ng *et al.* synthesized pyridine containing copolymers (**53**) via Stille coupling reaction using diiodo (**51**) and ditin (**52**) symmetric monomers consisting of the same building unit ($M_n = \sim 6.7$ K, PDI $= \sim 1.8$) [142]. The λ_{max} of polymer **53** redshifted for about 20 nm from polymer **48**, which is perhaps due to the intramolecular charge transfer driven by the donor–acceptor type backbone, where alkylthiophene is the (weak) donor and pyridine is the acceptor, and/or enhanced coplanarity driven by the absence of C–H bonds. However, the polymers may not be essentially regiosymmetric since the pyridine ring linked at the 2- and 5-positions is not centrosymmetric and the polymerization method does not allow regiochemical control. Yamamoto *et al.* reported on the synthesis of pyridazine containing copolymer (**55**) via Stille coupling reaction using corresponding dibrominated monomer (**54**) and hexamethylditin (Scheme 3.12), with M_n = 32–160 K (PDI = ~ 3.3) [143]. Polymer **55** showed further extension of π-conjugation from polymer **53** with $\Delta\lambda = \sim$ 30 nm and showed mobilities of 3×10^{-3} cm^2 $(V\ s)^{-1}$.

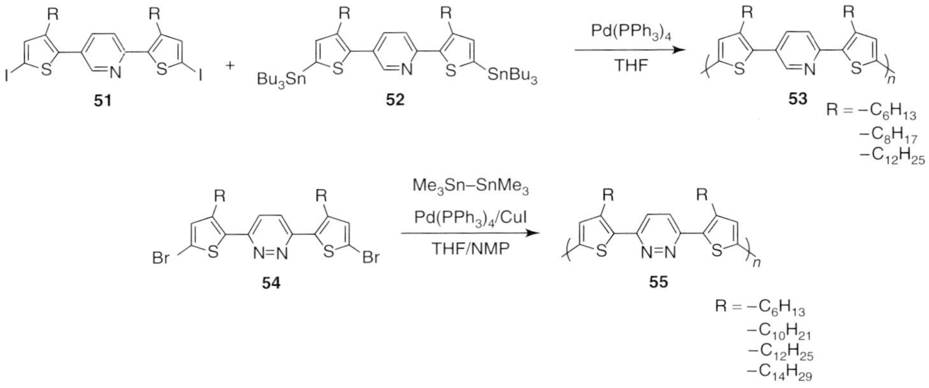

Scheme 3.12 Synthesis of thiophene copolymers bearing pyridine and pyridazine units.

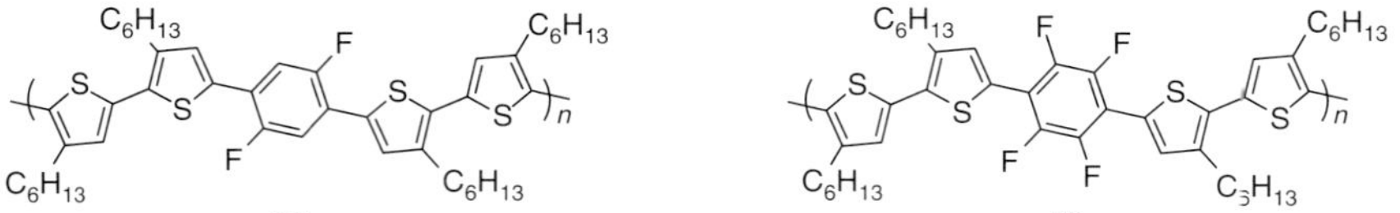

Scheme 3.13 Regiosymmetric thiophene copolymers bearing fluorinated phenylenes.

Fluorinated phenylene rings have also been introduced in PT backbones. Skabara and McCulloch *et al.* reported the synthesis of regiosymmetric PTs having di- and tetrafluorobenzene, for example, polymers **56** and **57**, via the $FeCl_3$ method, in which HH arrangements were used for the polymers (Scheme 3.13) [144]. Incorporation of perfluorinated aromatic rings is particularly interesting because these combinations of fluorinated and nonfluorinated π-electron systems enforce face-to-face π-stacking [145, 146], and this increases electron affinity of the materials, leading to n-type organic semiconductors [147–149]. Fluorine-functionalized polymers **56** and **57** showed increased mobility in OFETs as compared to their parent nonfluorinated polymer. Wang and Watson reported an interesting synthetic methodology to prepare thiophene–perfluoroarene copolymers (**59–61**), in which silylated 3,4-dialkoxythiophene (**58**) was copolymerized with perfluoroarenes using CsF and 18-crown-6 in toluene (Scheme 3.14) [150]. This methodology appears to offer a new design and synthetic strategy for advanced fluorine-functionalized PTs.

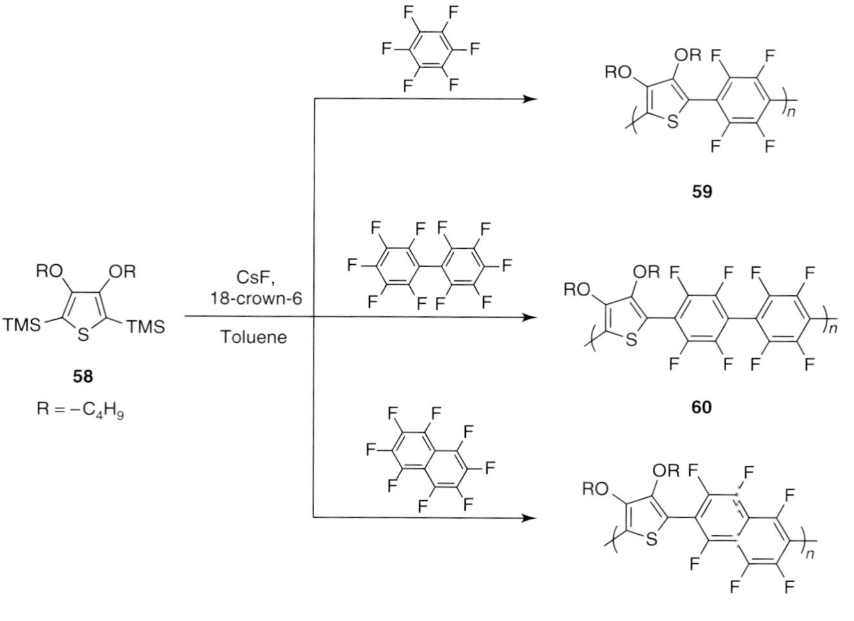

Scheme 3.14 Transition-metal-free synthesis of thiophene–perfluoroarene copolymers.

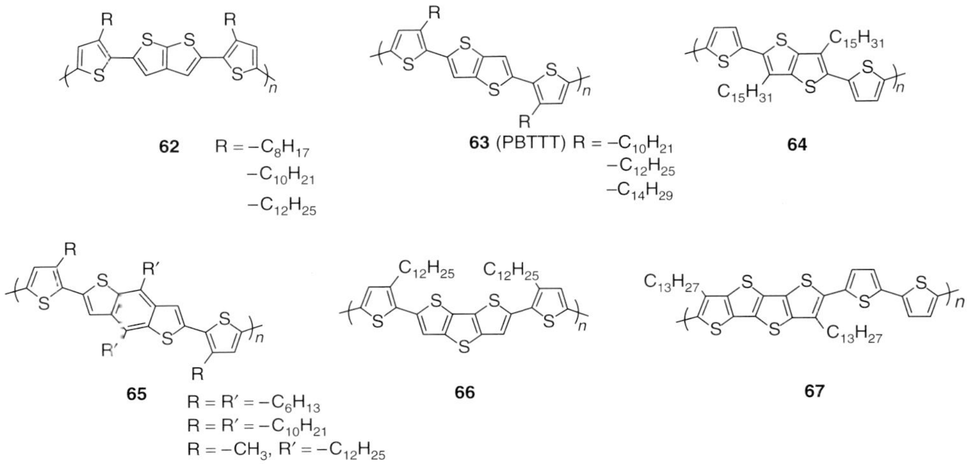

62 R = $-C_8H_{17}$, $-C_{10}H_{21}$, $-C_{12}H_{25}$

63 (PBTTT) R = $-C_{10}H_{21}$, $-C_{12}H_{25}$, $-C_{14}H_{29}$

64 ($C_{15}H_{31}$)

65 R = R′ = $-C_6H_{13}$; R = R′ = $-C_{10}H_{21}$; R = $-CH_3$, R′ = $-C_{12}H_{25}$

66 ($C_{12}H_{25}$)

67 ($C_{13}H_{27}$)

Scheme 3.15 Regiosymmetric polythiophenes with thiophene-based fused rings.

Recently, several interesting thiophene-based regiosymmetric copolymers having a thiophene-based fused ring system have been reported. Heeney and McCulloch *et al.* have reported on copolymers incorporating thieno[2,3-*b*]thiophene (**62**, Scheme 3.15) [151]. Thieno[2,3-*b*]thiophene shows limited π-conjugation of the resulting polymer with a λ_{max} of ~470 nm in film. However, backbone planarity appears to be preserved, and thus the close π-stacking necessary for high carrier transport can be achieved. In fact the polymer exhibited mobilities as high as 0.15 cm^2 $(V\ s)^{-1}$. In addition, the limited π-electron system gave large Ip of 5.3 eV, which was 0.5 eV larger than that observed for rrP3HT; as a consequence, it showed high air stability in OFET performance. Following this work, they introduced thieno[3,2-*b*]thiophene in the PT backbone (**63**/PBTTT, Scheme 3.15) [152]. The polymers were synthesized in a pathway similar to that described in previous reports [130, 135, 144, 151]. Substitution of the aromatic thienothiophene derived, of course, extended π-conjugation (λ_{max} = 547 nm in film), but the bandgap was sufficiently wide to preserve a deep HOMO level (Ip = 5.1 eV). The polymers afforded very high mobilities of up to 0.6 cm^2 $(V\ s)^{-1}$, which are the highest values observed for polymer semiconductors to date, and additionally showed high environmental air stability. These high mobilities can be attributed to their highly crystalline structure, as revealed by AFM and GIXS studies. The polymer thin films were shown to exhibit large domains with terrace structures (Figure 3.10) and provided even more highly ordered lamellar and π-stacking structures [152–154]. On the other hand, Ong *et al.* synthesized a related thiophene-thieno[3,2-*b*]thiophene copolymer by changing side chain placement (**64**). Although this polymer gave a slightly loose π-stacking structure, high mobilities of up to 0.25 cm^2 $(V\ s)^{-1}$ were obtained.

Several other thiophene-based fused rings, consisting of three or four rings, have successfully been introduced in a regiosymmetric PT system. Thiophene

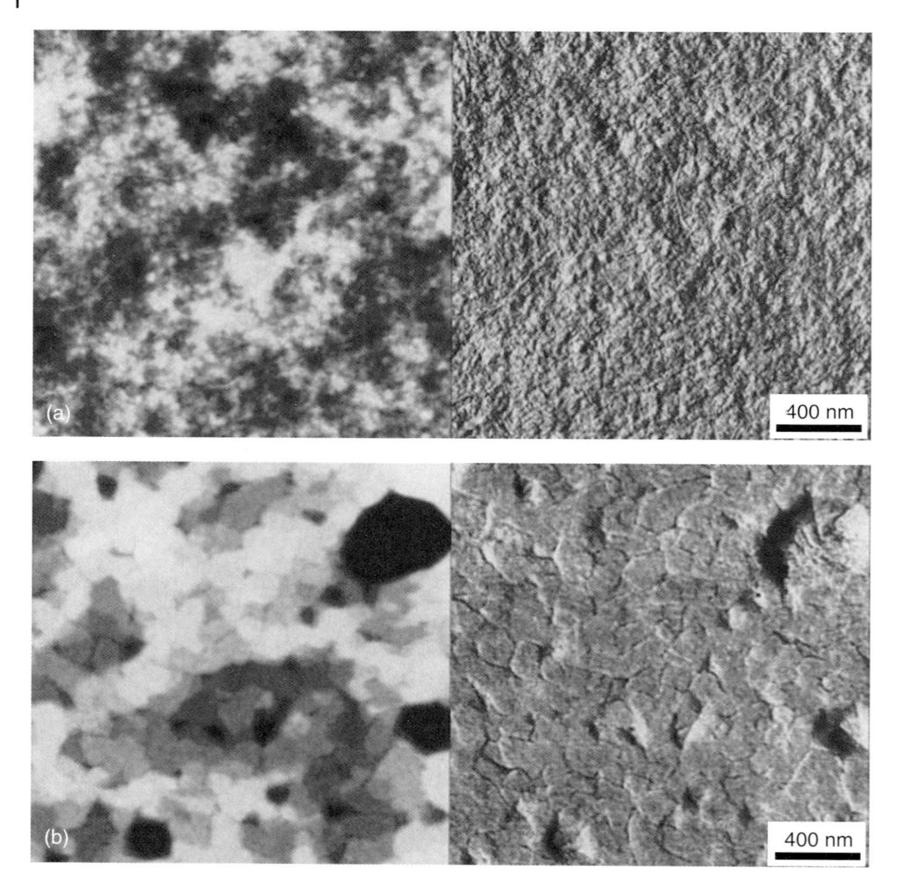

Figure 3.10 AFM images of PBTTT (R = $C_{12}H_{25}$): (a) Before and (b) after annealing (at 180 °C) above the liquid–crystal isotherm. The left images show the topography and the right images show the phase image. The dark spots in the topography of the annealed film are voids where the film partially dewetted during annealing. (Adapted from [152].)

copolymers with benzodithiophene were reported by Ong *et al.* (**65**, Scheme 3.15) [155–157]. Further, high mobilities of up to 0.4 cm^2 $(V\ s)^{-1}$ were obtained without postdeposition thermal annealing [157]. Ong *et al.* reported on the synthesis and OFET performance of a regiosymmetric thiophene copolymer incorporating a dithienothiophene ring (**66**). Devices fabricated from the polymer in chlorobenzene afforded mobilities of ~0.3 cm^2 $(V\ s)^{-1}$ [158]. A further extended fused ring, tetrathienoacene, was introduced in PT backbone by Malliaras *et al.* (**67**) [159]. In this polymer, long alkyl side chains were substituted at the β-position of tetrathienoacenes, rendering the molecule soluble [160]. The polymer afforded a reasonably high crystalline structure and in turn exhibited mobilities as high as 0.33 cm^2 $(V\ s)^{-1}$.

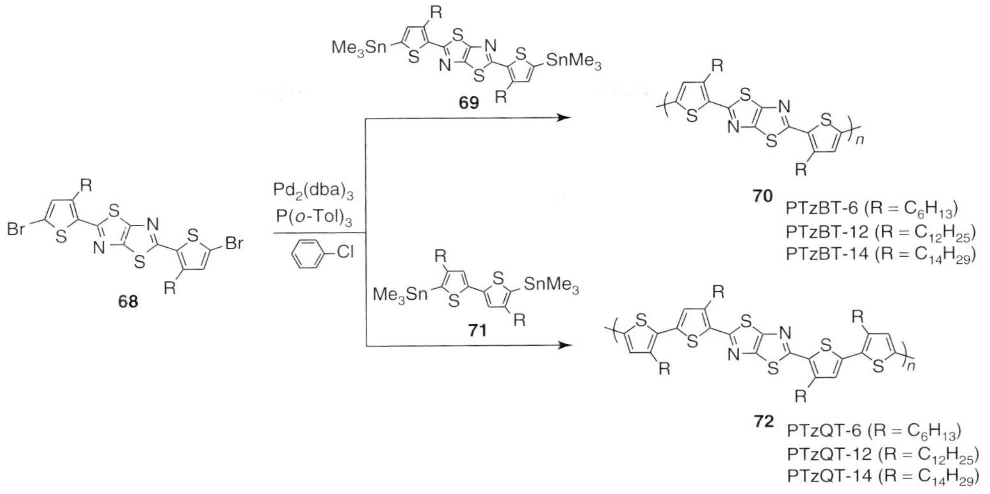

Scheme 3.16 Synthesis of regiosymmetric thiophene–thiazolothiazole copolymers.

We have reported on a related thiophene-based copolymer incorporating a rigid fused thiazolothiazole ring in the backbone (**70**/PTzBTs and **72**/PTzQTs, Scheme 3.16) [161, 162]. Synthesis of PTzBTs was also carried out via oxidative coupling with $FeCl_3$, Yamamoto polymerization, and GRIM; however, these methods gave only low-molecular-weight materials [161, 163]. The use of a thiazolothiazole ring ensures a very rigid and coplanar backbone and thereby produces a highly extended π-electron system and strong π stacking [164–166]. Despite the low molecular weight ($M_n = \sim 8.7$ K), PTzQTs showed high field-effect mobilities of $\sim 0.30\ cm^2\ (V\ s)^{-1}$ (for PTzQT-14). It was also found that in this system, polymers with longer side chains tended to show higher mobilities, for example, $\sim 0.23\ cm^2\ (V\ s)^{-1}$ for PTzQT-12, and $\sim 0.05\ cm^2\ (V\ s)^{-1}$ for PTzQT-6, because of the more highly ordered structure in longer side chains (Figure 3.11). It has been proposed by Kline and Delongchamp *et al.* that PBTTTs and PQT, both of which have low side chain attachment density and uniform side chain arrangement, promote efficient side chain packing and interdigitation, which in turn exhibit a high degree of strong π-stacking and high degree of lamellar ordering [154]. They pointed out that these highly ordered crystalline structures facilitate the high performance of those polymers (PBTTTs and PQT) in OFET devices. Interestingly, despite the lack of side chain interdigitation and ordering, which may be understood as a consequence of uneven side chain placement along the backbone, thus preventing their regular packing, PTzQTs exhibited a very high degree of lamellar ordering, due to the rigid nature of thiazolothiazole rings and perhaps because of the strong intermolecular interactions in the donor–acceptor backbone [167]. Moreover, while PBTTTs tended to form extended smooth, terrace-like structures (Figure 3.10), PTzQTs had characteristics of disordered fractals composed of ill-defined "granular" domains (Figure 3.12). Nevertheless, PTzQT exhibited very high performance

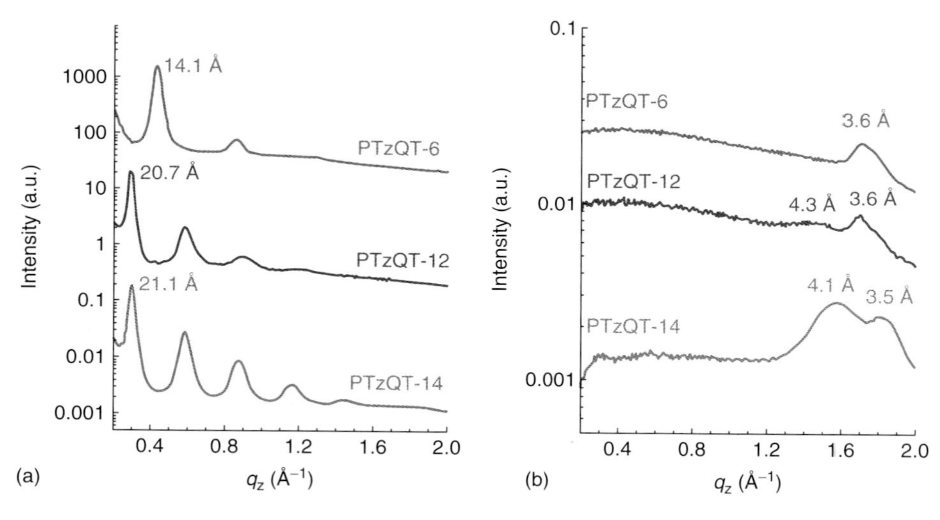

Figure 3.11 GIXS patterns of the polymer thin film cast on OTS-8-treated SiO_2 surfaces: (a) out-of-plane and (b) in-plane patterns of PTzQT-6, -12, and -14 (after annealing at 150 °C for 30 minutes). (Adapted from [162].).

in OFET devices. These results suggest that high mobilities may also be achieved in noninterdigitating systems, where domains in which the polymer lamellae are highly ordered are packed into isotropic amorphous-like superstructures [162].

We have also reported on the synthesis of electroactive and photoactive thiophene copolymers with dithienopyrrole (**74–80**) via Stillle coupling reaction (Scheme 3.17) [168]. These polymers gave high molecular weight (M_n = ~50 K), low bandgap (~1.7 eV), and excellent solubility, all of which are important for polymer semiconductors. Interestingly, carrier mobilities in these polymers were higher for poorly ordered as-cast samples, rather than for more ordered annealed ones. For example, mobilities of **80** were as high as 0.21 cm^2 $(V\ s)^{-1}$ in a disordered and isotropic lamellar system, and they dropped by a factor of 2–3 in a well-ordered lamellar system (Figure 3.13). It should be noted that there was no evidence of appreciable $\pi-\pi$ stacking, even after annealing. This atypical behavior can be tentatively ascribed to enhanced backbone-to-backbone contacts in the less ordered system, perhaps facilitated by sparse placement of alkyl side chains, and thus more effective interchain carrier hopping. These results appear to confirm that disordered, isotropic systems can exhibit enhanced performance.

High OFET performance in a macroscopically disordered and thus amorphous system has also been reported in a cyclopentadithiophene–benzothiadiazole copolymer (**81**, Scheme 3.18) by Müllen *et al.* [169]. The OFET devices using this low bandgap polymer **81** (λ_{max} = 750 nm) exhibited high mobilities of up to 0.17 cm^2 $(V\ s)^{-1}$. X-ray studies on polymer thin films revealed the close π-stacking (3.7 nm) of the polymer backbones; however, relatively diffuse reflections and the lack of higher order ones implied pronounced disorder. They pointed out the importance

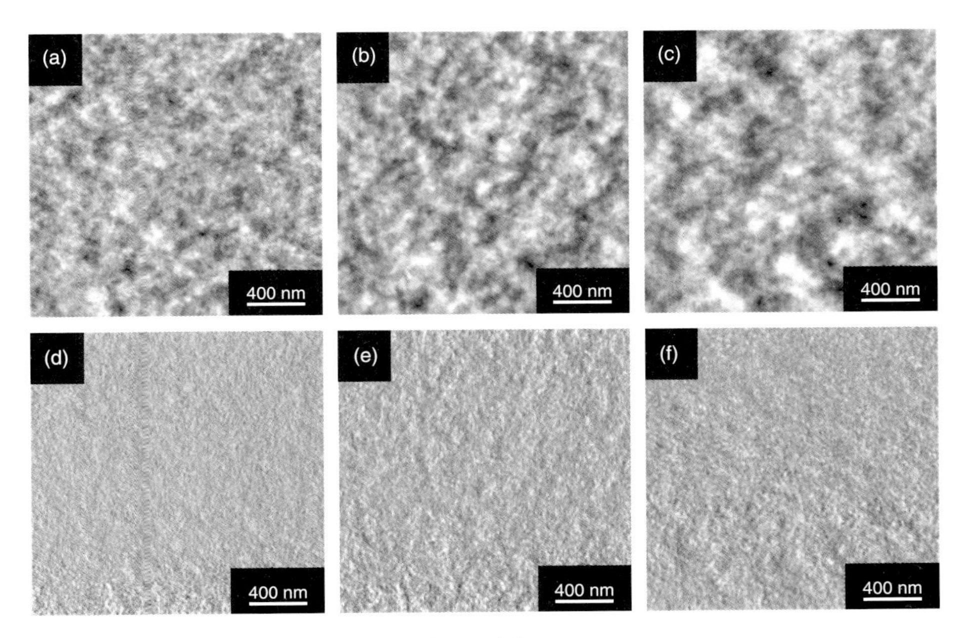

Figure 3.12 Tapping mode AFM images of (a and d) PTzQT-6, (b and e) PTzQT-12, and (c and f) PTzQT-14 on the OFET devices after annealing at 150 °C for 30 minutes. The upper row (a–c) shows topographic images and the lower row (d–f) shows phase contrast images. (Adapted from [162].)

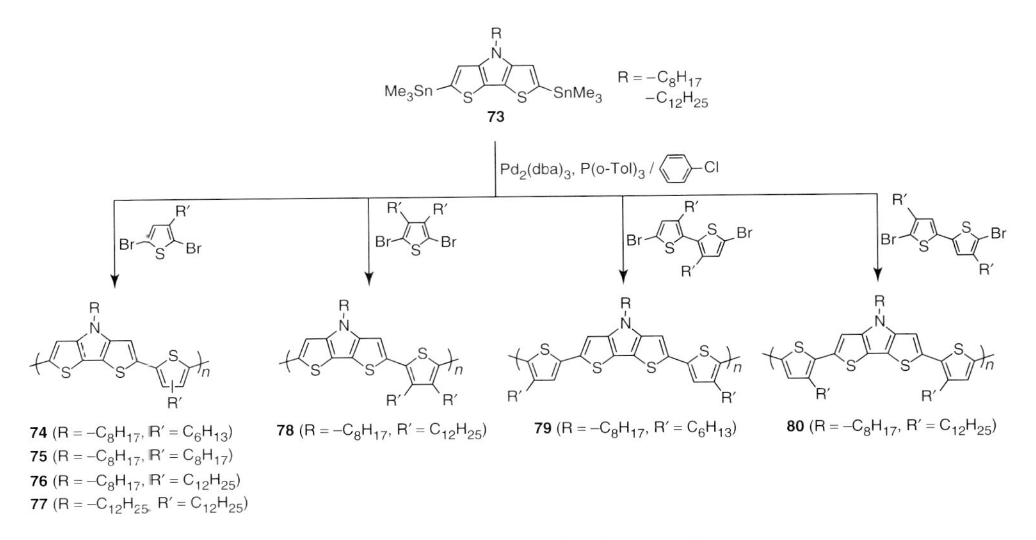

Scheme 3.17 Synthesis of dithienothiophene containing thiophene copolymers.

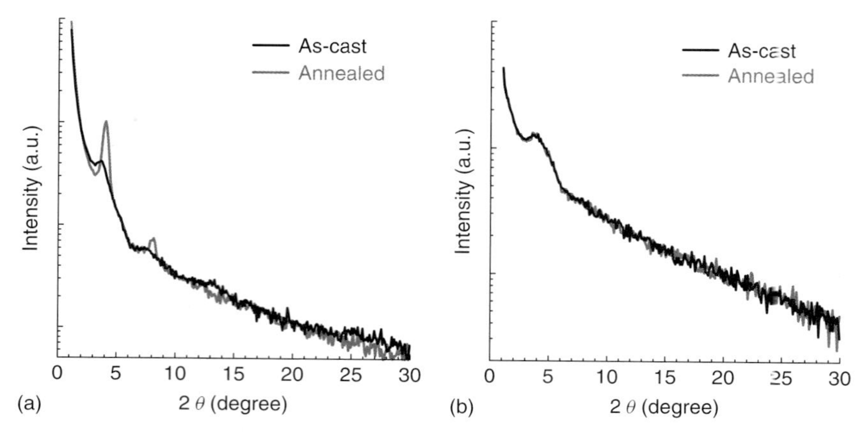

Figure 3.13 XRD profile of **80**: as-cast (black trace) and after annealing at 120 °C (gray trace), on OTS-treated SiO_2/Si substrate:. (a) Out-of-plane and (b) in-plane. The as-cast film gave higher mobility despite the less ordered lamellar structure as compared to the annealed film. (Reproduced from [168].)

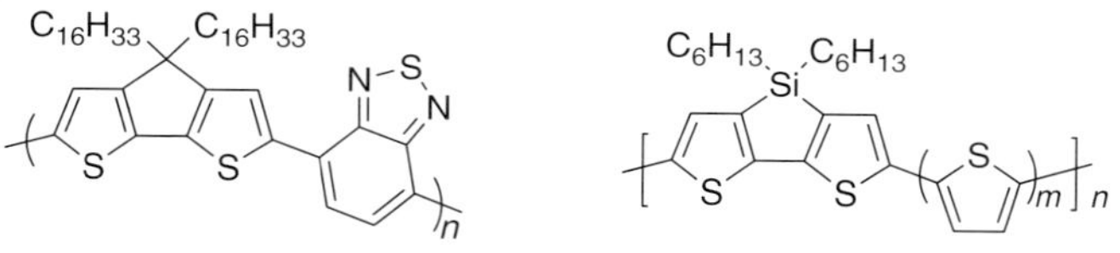

81 **82** ($m = 0,1,2$)

Scheme 3.18 Regiosymmetric copolymers incorporating thiophene-bridged fused rings.

of controlling, thus increasing the intermolecular interaction, and of decreasing the gap between building blocks.

Another analog of a bridged thiophene fused ring system, dithienosilole containing thiophene copolymer (**82**, Scheme 3.18), was reported by Facchetti *et al.* [170, 171]. These polymers were successfully synthesized via Stille coupling reaction ($M_w = \sim41$ K, PDI $= \sim3.0$), while low-molecular-weight oligomers were obtained via Suzuki coupling reaction [171]. These polymers exhibited mobilities of up to $0.08\,cm^2\,(V\,s)^{-1}$.

While most π-conjugated polymers investigated to date are p-type semiconductors, only a few polymers are known to show n-type or ambipolar behavior that can carry both hole and electron [172]. It has already been recognized that the chemical structure is not the only factor that determines whether the material predominantly exhibits p- or n-type behavior (processing conditions, device architectures, etc., are also important factors) [173, 174]. Although knowing that it may thus be inappropriate to speak of p-type, n-type, or ambipolar behavior, we here describe some thiophene-based copolymers that exhibit electron mobilities (Scheme 3.19).

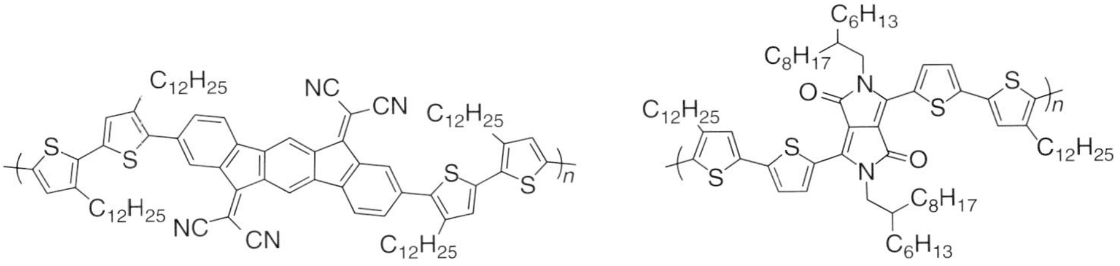

83

84

Scheme 3.19 Regiosymmetric thiophene-based copolymers with n-type behavior.

One necessary design rule making it possible to produce n-type semiconductors is the incorporation of a strong acceptor moiety in the backbone, thus lowering the LUMO level and in turn enhancing the electron affinity of the materials [174]. A copolymer incorporating an indenofluorenebis(dicyanovinylene) core (**83**) synthesized by Stille coupling reaction ($M_n = 6700$, PDI = 3.5) was reported by Facchetti and Marks *et al.* [175]. Polymer **83** was found to have deep LUMO (−4.15 eV) and HOMO (−5.51 eV) levels, indicative of the significant electron-deficient nature necessary for n-type semiconductors. As a result, the polymer exhibited both hole and electron mobilities, both with orders of 10^{-4} cm^2 (V s)$^{-1}$. Another ambipolar polymer semiconductor has been reported by Winnewisser *et al.* [176]. They synthesized a thiophene-based copolymer bearing a diketopyrrolopyrrole moiety (**84**). Polymer **84** gave a high molecular weight of M_n = 28 K (PDI = 2.9) and showed good solubility in toluene, chloroform, and chlorobenzene. Since diketopyrrolopyrrole is a good acceptor component, **84** gave deeper LUMO (−4.0 eV) and HOMO (−5.5 eV) levels. Having a donor–acceptor nature and hence having intra- as well as intermolecular interactions, the polymer showed a low optical bandgap of 1.5 eV. OFET devices using **84** demonstrated high hole mobilities of ~0.11 cm^2 (V s)$^{-1}$ and electron mobilities of ~0.05 cm^2 (V s)$^{-1}$. When the electrode on the devices was changed from gold to barium, giving rise to improved electron injection with its lower work function, the polymer devices showed enhanced electron mobilities of 0.04–0.09 cm^2 (V s)$^{-1}$, still ensuring good hole mobilities of 0.05–0.1 cm^2 (V s)$^{-1}$.

Regiosymmetric PTs with (hetero)aromatic (fused) rings have been shown to demonstrate high performance in electronic devices, in fact even higher than that observed in rrP3ATs. Although the synthesis can be very complicated, with this design strategy one can tailor polymer structures and hence tune their properties, including crystalline to amorphous, wide to narrow bandgaps, and p-type to n-type natures.

3.7 Polythiophene Block Copolymers

Block copolymers composed of two different polymeric segments are fascinating materials that are expected to self-organize and microphase separate [177]. This phase segregation can lead to the formation of nanoscale morphologies, owing to the immiscibility and/or the difference in crystallinity between the covalently connected segments. This may in turn lead to advanced new materials for use as components in nanoelectronic devices. Herein, we describe rrP3AT-based block copolymers comprising all-conjugated segments, as well as conjugated and nonconjugated segments.

3.7.1
All-Conjugated rrP3AT-Based Block Copolymers

Block copolymers consisting of different π-conjugated segments are particularly interesting since different electrical or optical characters from each segment can be wrapped into one polymer chain, possibly giving very unique properties, especially when self-assembled to form nanostructured morphologies [178–181]. To realize highly ordered and nanostructured films, high molecular weights of the individual blocks and narrow polydispersities are preferred. The living nature of the synthetic methodology of rrP3ATs may fulfill these challenging demands, and facilitate the access to those block copolymers by simple chain extension reaction in one pot.

The first synthesis of all-conjugated rrP3AT-based block copolymers was reported by Iovu and McCullough *et al.* In their study, diblock copolymer P3HT-*b*-P3DDT (**85**) and triblock copolymer P3DDT-*b*-P3HT-*b*-P3DDT (**86**) were synthesized by sequential addition of dibrominated monomers of 3-hexylthiophene and 3-dodecylthiophene (Scheme 3.20) [72]. Molecular weights of P3HT-*b*-P3DDT

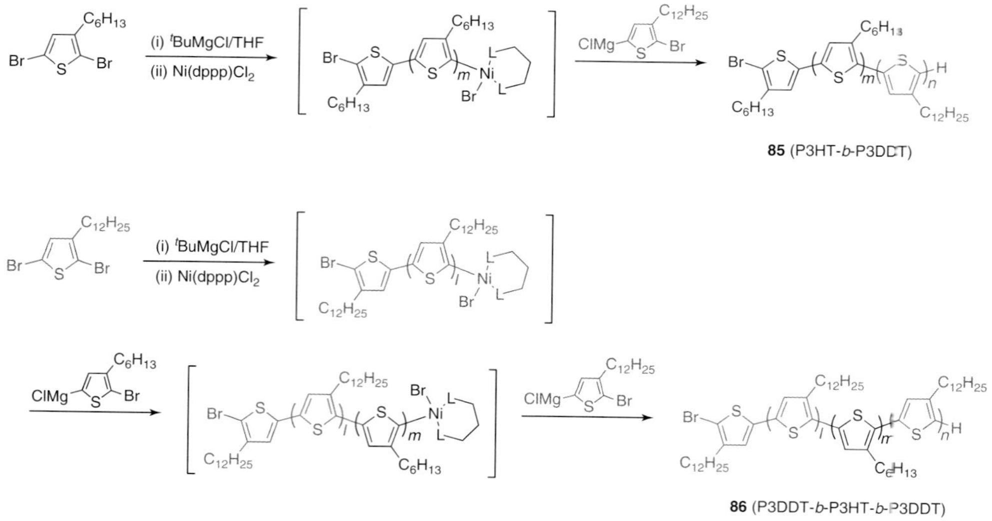

Scheme 3.20 Synthesis of all-conjugated diblock copolymer P3HT-*b*-P3DDT and triblock copolymer P3DDT-*b*-P3HT-*b*-P3DDT by chain extension through sequential monomer addition.

were as high as $M_n = 21$ K, which, of course, can be controlled by reaction time and catalyst concentration, given that the synthesis has a living nature. The PDI of the polymer, 1.44, was slightly higher than that of the homopolymer, which was 1.2, indicating the formation of some dead or inactive chains during the chain extension process. The chain length of P3DDT-*b*-P3HT-*b*-P3DDT was carefully controlled because of its low solubility, which could cause polymer precipitation during the reaction process. Molecular weights of the triblock copolymer were as high as $M_n = 9.8$ K with a relatively wide PDI of 1.53. rrP3AT-based diblock copolymer with hexyl and 2-ethylhexyl side chains can also be synthesized by a similar procedure (**87**, Scheme 3.21) [182].

Yokozawa *et al.* synthesized rrP3AT-based diblock copolymers of hydrophobic 3-hexylthiophene and hydrophilic 3-[2-(2-methoxyethoxy)ethoxy]methylthiophene (**88**) using this methodology [183]. It was found that PDI for these polymers was dependent on the choice of the catalyst due to the difference in reactivity between the two monomers. They also reported on the synthesis of diblock copolymers consisting of an rrP3AT segment and a poly(*p*-phenylene) (PPP) segment (**89**) [184]. While high molecular weight ($M_n = 19.4$ K) with narrow PDI (1.24) block copolymer **89** was successfully obtained when the polymerization was initiated with PPP followed by a chain extension by rrP3ATs, a low molecular weight ($M_n = 5.6$ K) with broad PDI (2.36) block copolymer was obtained when the polymerization was carried out in reverse order. They proposed that in the latter case the Ni catalyst does not smoothly transfer to phenylene rings since thiophene and Ni form a relatively strong π-complex, and thus do not allow efficient postpolymerization of the PPP segment.

A different synthetic methodology was used to prepare a triblock copolymer comprising rrP3HT and polyfluorene (PF) (**91**, Scheme 3.22). Scherf *et al.* first synthesized an end-functionalized PF segment (**90**), and it was then reacted with GRIM type reagent to form an rrP3HT segment at its terminal position [185]. The length of the rrP3HT block was limited to about 6–7 repeating thiophene units. A triblock copolymer P3HT-*b*-CN-PPV-*b*-rrP3HT (CN-PPV = cyano-substituted poly-*p*-phenylenevinylene) (**93**) was also synthesized by the same group using a different methodology [186]. In this case, the Yamamoto method was used to polymerize a cyano-substituted dibromophenylenevinylene monomer, and subsequently the polymerization was terminated by adding the pre-prepared rrP3HT

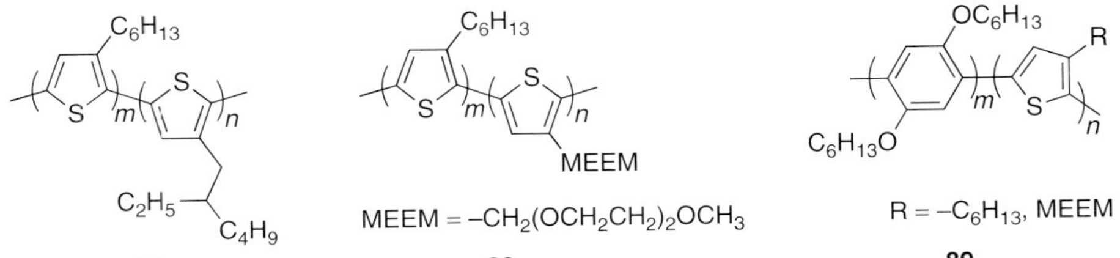

C_6H_{13} C_2H_5 C_4H_9 **87**

C_6H_{13} MEEM; MEEM = $-CH_2(OCH_2CH_2)_2OCH_3$ **88**

OC_6H_{13} $C_6H_{13}O$ R; R = $-C_6H_{13}$, MEEM **89**

Scheme 3.21 All-conjugated diblock copolymers comprising different physical features in each segment.

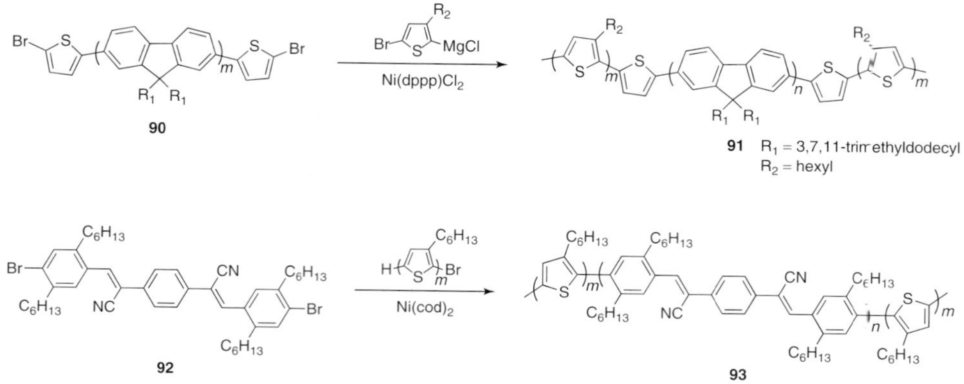

Br S S Br m R1 R1 90 R2 Br S MgCl Ni(dppp)Cl2 R2 S m S n S R2 S m R1 R1 91 R1 = 3,7,11-trimethyldodecyl R2 = hexyl

C6H13 Br C6H13 NC CN C6H13 Br C6H13 92 C6H13 H S m Br Ni(cod)2 C6H13 C6H13 S m C6H13 NC CN C6H13 n S m C6H13 C6H13 93

Scheme 3.22 Synthesis of all-conjugated triblock copolymers, P3HT-*b*-PF-*b*-P3HT and P3HT-*b*-CN-PPV-*b*-P3HT.

with monobromo end group, giving a triblock copolymer with molecular weight of $M_n = \sim 53$ K (PDI = ~1.5). Thin films of the triblock copolymer formed nanosized spherical aggregates with a diameter of 60–90 nm in AFM (Figure 3.14a). Annealing of these thin films (6 hours, 120 °C) led to an increased surface roughness (Figure 3.14b), indicative of an ongoing aggregation and nanostructure formation. On the other hand, in homopolymer rrP3HT with Br end group, nanosized aggregates were also formed (Figure 3.14c), but with a little larger diameter (70–120 nm) and increased dispersity. In contrast, a 2 : 1 blend of rrP3HT and CN-PPV formed irregular nanostructures (Figure 3.14d). The covalent connection of rrP3HT donor and CN-PPV acceptor blocks limited the scale length of nanostructure formation and hence may allow the optimization of OPVs.

Several successful methodologies for the synthesis of all–π-conjugated block copolymers have been shown. As some block copolymers showed nanosized mesostructures, these copolymers seem to be favorable for use in organic electronic devices such as bulk hetero junction solar cells.

3.7.2
Conjugated–Nonconjugated rrP3AT-Based Block Copolymers

Rod–coil block copolymers can produce numerous phase-separated nano- or microstructures that may be of use in various applications [187]. Incorporation of flexible coil-like segments may improve the mechanical properties and processability of rigid rod rrP3ATs, in which high crystallinity may deteriorate reproducibility of the self-assembled structures in films and thus impede device performance. The McCullough group has reported the first conjugated–nonconjugated rrP3AT-based block copolymers [80]. According to this report, end groups of rrP3HTs synthesized via the McCullough method were modified by postpolymerization functionalization [188], and then atom transfer radical polymerization (ATRP) [189, 190] was employed to introduce PS (**94**, **101**) and polymethylacrylate (PMA) segments

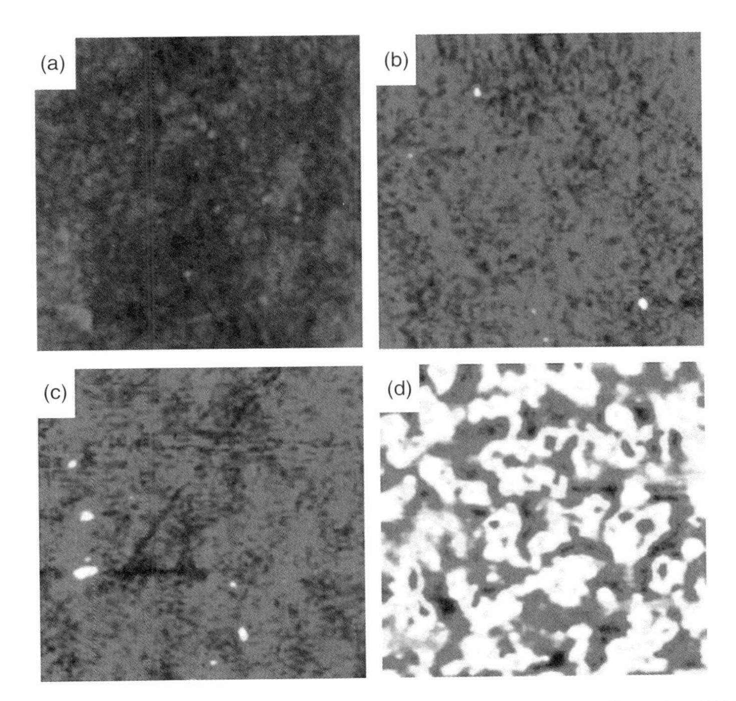

Figure 3.14 Contact-mode atomic force microscopy (AFM) images (5×5 μm) of thin films spin-coated onto SiO_2 from chloroform (concentrated 2 mg ml^{-1}): (a) P3HT-*b*-CN-PPV-*b*-P3HT, **93** (as-prepared); (b) P3HT-*b*-CN-PPV-*b*-P3HT, **93** (annealed for 6 hours at 120 °C); (c) rrP3HT with Br end group (annealed for 6 hours at 120 °C); and (d) a blend of rrP3HT (Br end group) and CN-PPV with a molar ratio of 2 : 1 (annealed for 6 hours at 120 °C). Z range, 10 nm. (Adapted from [186].)

(**95**, **102**), giving di- and triblock copolymers (Scheme 3.23 and 3.24). Weight percentage of the PS and PMA blocks was completely controlled by the feed ratio of the monomers. Conductivities of the block copolymers decreased as the ratio of insulating blocks (PS and PMA) increased, as expected. For example, in rrP3HT–PS diblock copolymer **94**, while 100% rrP3HT (M_n = 16.8 K) had a conductivity of 110 S cm^{-1}, a copolymer with 37% rrP3HT (total M_n = 53.4 K) had a conductivity of 5 S cm^{-1}, and then the conductivity dropped to about 0.1 S cm^{-1} for a copolymer with <22% rrP3HT.

McCullough *et al.* have also reported on a series of rrP3HT-based block copolymers, in which nonconjugated coil blocks, PS, polyacrylates, polymethacrylate, and isoprene were introduced at the *in situ* functionalized end groups of rrP3HTs [76–78, 191, 192]. Various kinds of rrP3HT–PS diblock copolymers were synthesized, where the extension of PS segments was employed by ATRP (**96**) [77], anionic polymerization (**97**) [77], and reversible addition fragmentation chain transfer (RAFT) polymerization [190, 193] (**98**), as shown in Scheme 3.23 [191]. AFM study of thin films of diblock copolymers **96** and **97** showed the formation of nanowire structures (Figure 3.15). These nanowires were sparsely packed relative

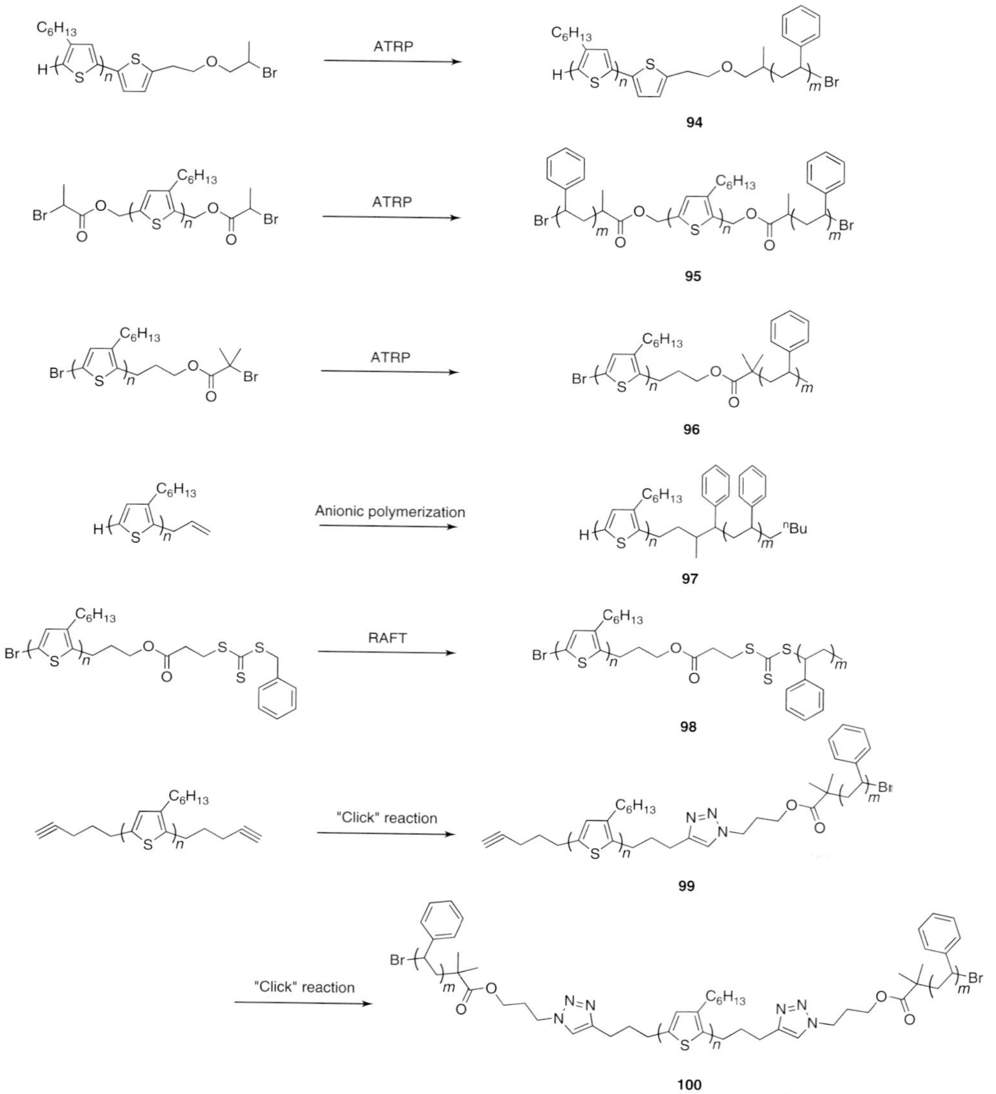

Scheme 3.23 rrP3HT-based di- and triblock copolymers with various polystyrene segments.

to those observed in rrP3HT, which may be due to the presence of the PS coil blocks [77]. In addition, recently, "click" chemistry has also been used to prepare rrP3HT–PS di- (**99**) and triblock (**100**) copolymers by another group (Scheme 3.23) [194].

Diblock copolymers of rrP3HT with polyacrylate and polymethacrylate segments were synthesized by ATRP (**103–108**), as shown in Scheme 3.24 [76, 78, 192]. Tapping mode AFM analysis revealed that the dominating feature of all these diblock

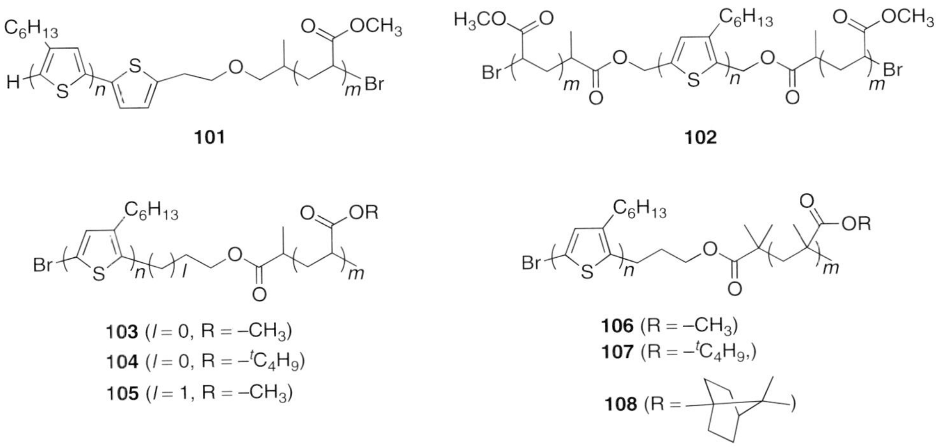

101

102

103 ($l = 0$, R = $-CH_3$)
104 ($l = 0$, R = $-^tC_4H_9$)
105 ($l = 1$, R = $-CH_3$)

106 (R = $-CH_3$)
107 (R = $-^tC_4H_9$,)
108 (R =)

Scheme 3.24 rrP3HT–polyacrylate di- and triblock copolymers (**101**–**105**) and rrP3HT–polymethacrylate diblock copolymers (**106**–**108**) synthesized via ATRP.

copolymers in thin films was the nanofibrillar morphologies that to various degrees resembled those observed in rrP3HT (Figure 3.16). It was found that the nature and the molar ratio of coil segments had a profound impact on nanofibrillar width, length, and distribution, as well as interfibrillar ordering. The phase contrast AFM image of rrP3HT-*b*-P^{t}BuMA (P^{t}BuMA = poly^{t}butylmethacrylate), **107**, containing 15 mol% of the rrP3HT block (Figure 3.16a) clearly shows nanofibrillar morphology with densely packed, elongated, and locally parallel nanofibrils. On the other hand, morphology of rrP3HT-*b*-PIBMA (PIBMA = polyisobornylmethacrylate), **108**, was clearly different (Figure 3.16b). The nanofibrils were shorter, isolated from one another, and randomly oriented throughout the sample. This distinct difference in **108** is apparently attributed to the suppression of nanofibrillar self-assembly by the constraints imposed by the bulkiness and steric hindrance brought in by isobornyl groups [78]. Furthermore, nanofibrillar width of diblock copolymers increased as the size of the coil block increased. The observed dependence of nanofibrillar spacing on copolymer composition points to the interplay between the nanofibrillar self-assembly driven by π-stacking of rrP3HT segments and phase separation, leading to the segregation of coil blocks in interfibrillar spaces. Electrical conductivities of these polymers tended to increase as the rrP3HT segment increased. In addition, rrP3HT diblock copolymers with polyisoprene coil blocks (**109**, Scheme 3.25) can be synthesized using nitroxide-mediated polymerization (NMP) [190]. These polymers also showed nanofibrillar morphologies in thin films [191].

Sauvé and McCullough reported on OFET performances of a series of rrP3HT-*b*-PMA (**105**) with a mole ratio different from that of the PMA segment (0–57%) [192]. They found that polymer devices with bare SiO_2 dielectric surfaces showed good mobilities of $\sim$0.036 cm^2 $(V\ s)^{-1}$ (for rrP3HT = 0% PMA, M_n = 12.5 K), and that mobilities decreased as PMA ratio increased, as observed in

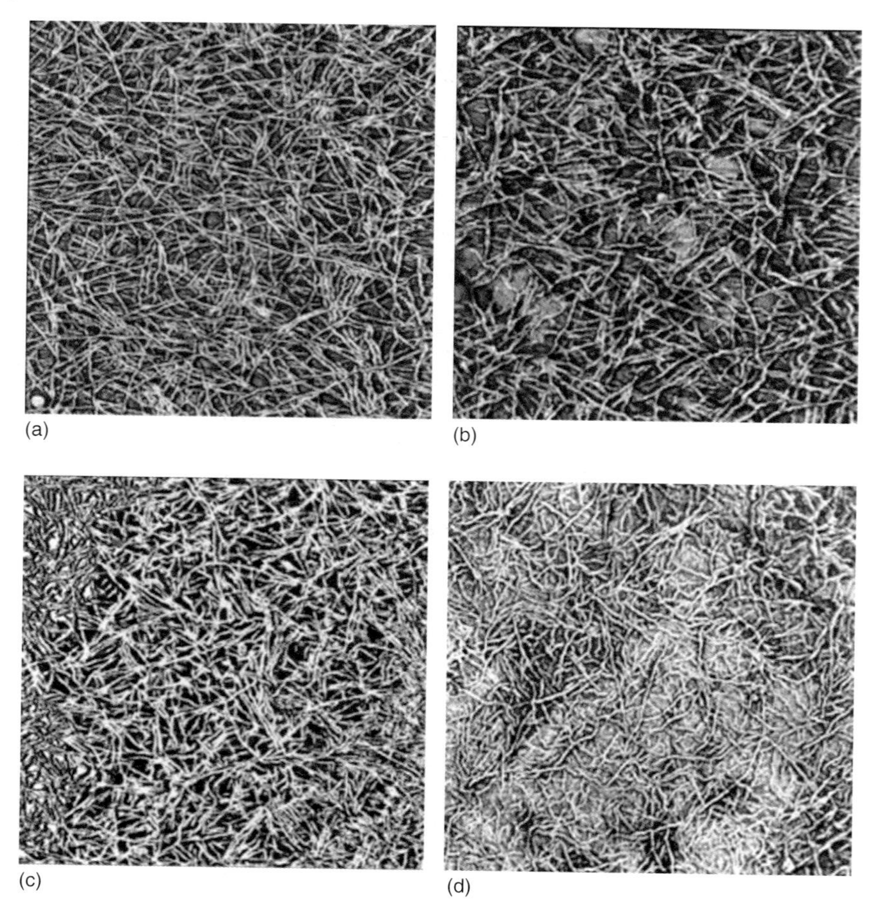

Figure 3.15 AFM phase images (scale: 2 μm): (a) allyl terminated rrP3HT; (b) rrP3HT-*b*-PS synthesized by coupling of "living" polystyrene with allyl terminated rrP3HT (**97**); (c) bromoester terminated rrP3HT, and (d) rrP3HT-*b*-PS synthesized by ATRP (**96**). (Adapted from [77].)

conductivities (Figure 3.17). The lowest value of 0.0055 cm^2 $(V\ s)^{-1}$ was observed for the copolymer with 57 mol% PMA. However, when dielectric surfaces were treated with octyltrichlorosilane to form SAMs, a process that usually promotes edge-on orientation in rrP3HT, the copolymers still had high mobilities at higher PMA content. Initially mobility of rrP3HT decreased from 0.07 to <0.03 cm^2 $(V\ s)^{-1}$ (11 mol% PMA), and then increased as PMA content increased. The highest block copolymer mobility was 0.05 cm^2 $(V\ s)^{-1}$ for the copolymer with 57 mol% PMA. This interesting finding can be explained in terms of the tendency of these block copolymers to self-assemble into conducting nanofibrils of rrP3HT, which are surrounded by insulating polymer segments, near the interface of bottom contact transistors.

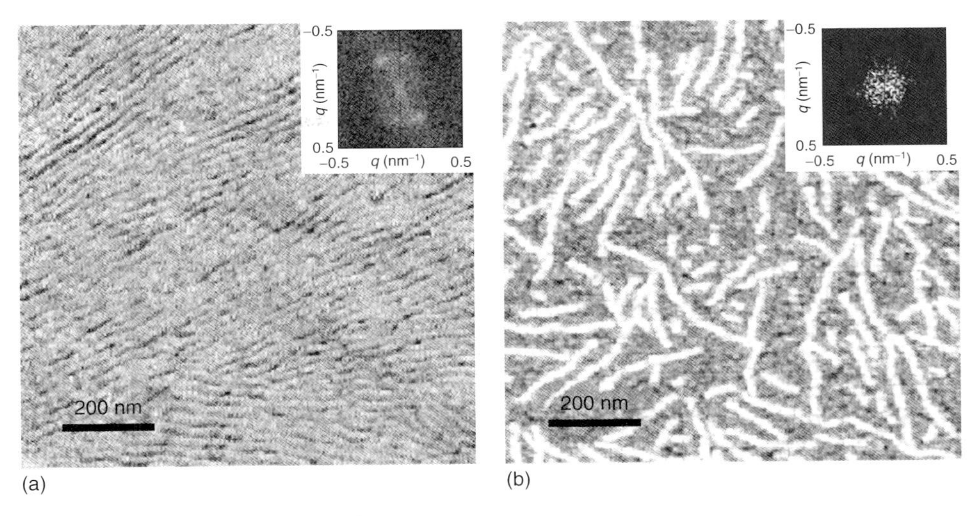

Figure 3.16 Tapping mode atomic force microscopy (TMAFM) phase images and GISAXS patterns of rrP3HT diblock copolymers: (a) rrP3HT-*b*-poly(^{t}BuMA) (**107**, 15 mol% rrP3HT); (b) rrP3HT-*b*-poly(IBMA) (**108**, 15 mol% PHT). Superimposed image: GISAXS scattering profile of (a) **107** and (b) **108**. (Adapted from [78].)

Some other rrP3HT-based diblock copolymers have been reported (Scheme 3.25). Dai *et al.* synthesized rrP3HT-*b*-P2VP (P2VP = poly-2-vinylpyridine), **110**, by anionic polymerization with different composition ratios [195]. The polymers microphase separated and self-assembled into nanostructures of sphere, cylinder, lamellae, and nanofiber structures according to different P2VP volume fractions. A

109

110

111 (R = $-C_6H_{13}$, $-C_{12}H_{25}$)

112

Scheme 3.25 rrP3HT–polyisoprene (**109**), rrP3HT–poly(2-vinylpyridine) (**110**), rrP3HT–polylactide (**111**), and rrP3HT–polyethylene (**112**) diblock copolymers.

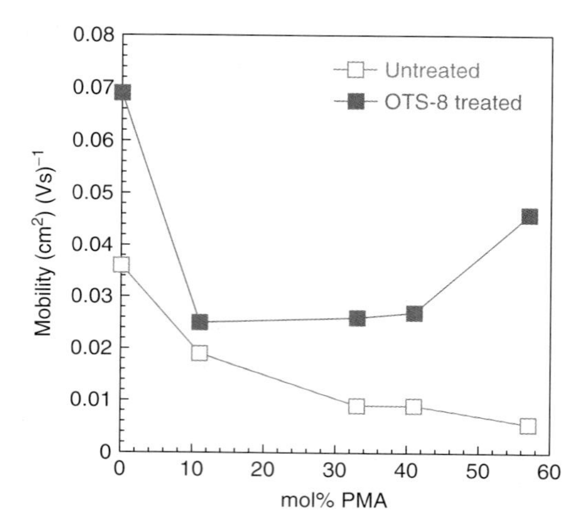

Figure 3.17 Average mobility of rrP3HT-*b*-PMA (**105**) as a function of PMA content. (Reproduced from [192].)

polylactide (PLA) coil block was shown to integrate rrP3AT diblock copolymers (**111**) using ring-opening polymerization in a strategy by Hillmyer *et al.* [196]. Hexyl and dodecyl side chains were introduced on the rrP3AT block in this system. The dodecyl type copolymer had an advantage in terms of processibility (low melting point, low crystallinity), suggesting it would be a better candidate for electronic devices as compared to the hexyl copolymer. Janssen *et al.* synthesized crystalline–crystalline rrP3HT diblock copolymers (**112**) [197] by introducing polyethylene segments via ruthenium-catalyzed ring-opening metathesis polymerization [198]. Grazing incidence wide-angle X-ray scattering (GIWAXS) of copolymer thin films confirmed that the material formed separate domains of crystalline rrP3HT and PE.

A number of rrP3AT-based block copolymers with various kinds of insulating blocks have been synthesized and characterized. Thin films of block copolymers have indeed shown phase-separated nano- or microstructures. Their morphologies have changed as a function of insulating block size, which may lead to better control of electronic device performance. Furthermore, using block copolymers opened up a wide range of possibilities in designing new organic semiconductors by allowing us to modify the material properties with one polymer segment, while preserving the solid-state order of the conjugated polymer segment, both of which are critical for organic electronic devices.

3.8
Conclusion

Molecular design, synthesis, and properties of the regioregular PT family have been reviewed. In particular, HT rrP3ATs can be objectively seen to be one of the most versatile polymer systems in the area of printable electronics. This can apparently be attributed to its well-defined planar PT backbone, which leads to the formation of self-assembled supramolecular structures and hence a dramatic increase in conductivity/mobility. The ease of their synthesis, controllable molecular weights, and narrow polydispersity, due to the living nature of the polymerization, have allowed systematic studies of regioregular PT and their structure–property relationships. These important studies and their conclusions have given rise to deeper understanding of the correlation of chemistry and physics in polymeric electronic materials and to the optimization of device performances as well. On the other hand, a number of new regioregular (regiosymmetric) PTs have recently been designed and synthesized. Some of these polymers have demonstrated even better carrier transport performance relative to rrP3ATs, and some others have shown n-type behavior. All-conjugated and conjugated–nonconjugated block copolymers based on regioregular PTs have also been synthesized. These block copolymers have been shown to exhibit unique nanophase structures that might be feasible for electronic devices. As shown in this chapter, design and synthesis of these new plastic materials have greatly contributed to recent advances in this field. With further innovative molecular design and synthesis, regioregular PTs and analogs will continue to lead the way in printable plastic electronics.

References

1. Roncali, J. (1992) *Chem. Rev.*, **92**, 711.
2. Ewbank, P.C. and McCullough, R.D. (1998) in *Handbook of Conducting Polymers*, 2nd edn (eds T.A.Skotheim, R.L. Elsenbaumer, and J.R. Reynolds), Marcel Dekker, New York, p. 225.
3. McCullough, R.D. (1998) *Adv. Mater.*, **10**, 93.
4. McCullough, R.D. (1999) in *Handbook of Oligo- Polythiophenes* (ed. D.Fichou), Wiley-VCH Verlag GmbH, Weinheim, p. 1.
5. Jeffries-El, M. and McCullough, R.D. (2007) in *Handbook of Conducting Polymers*, Vol. 1 , 3rd edn (eds T.A.Skotheim and J.R. Reynolds), CRC Press, Boca Raton, p. 9.1.
6. Osaka, I. and McCullough, R.D. (2008) *Acc. Chem. Res.*, **41**, 1202.
7. Burn, P.L., Holmes, A.B., Kraft, A., Bradley, D.D.C., Brown, A.R., Friend, R.H., and Gymer, R.W. (1992) *Nature*, **356**, 47.
8. Perepichka, I.F., Perepichka, D.F., Meng, H., and Wudl, F. (2005) *Adv. Mater.*, **17**, 2281.
9. Klaus Müllen, U.S. (2006) *Organic Light Emitting Devices: Synthesis, Properties and Applications*, Wiley-VCH Verlag GmbH & Co. KGaA, Weinheim.
10. Bao, Z., Dodabalapur, A., and Lovinger, A.J. (1996) *Appl. Phys. Lett.*, **69**, 4108.
11. Sirringhaus, H., Brown, P.J., Friend, R.H., Nielsen, M.M., Bechgaard, K., Langeveld-Voss, B.M.W., Spiering, A.J.H., Janssen, R.A.J., Meijer, E.W., Herwig, P., and De Leeuw, D.M. (1999) *Nature*, **401**, 685.
12. Horowitz, G. (2004) *J. Mater. Res.*, **19**, 1946.
13. Klauk, H. (2006) *Organic Electronics*, Wiley-VCH Verlag GmbH & Co. KGaA, Weinheim.
14. Bao, Z. and Locklin, J. (2007) *Organic Field-Effect Transistors*, CRC Press, Boca Raton.
15. Guenes, S., Neugebauer, H., and Sariciftci, N.S. (2007) *Chem. Rev.*, **107**, 1324.
16. Yu, G., Gao, J., Hummelen, J.C., Wudl, F., and Heeger, A.J. (1995) *Science*, **270**, 1789.
17. Hoppe, H. and Sariciftci, N.S. (2004) *J. Mater. Res.*, **19**, 1924.

18. Thompson, B.C. and Frechet, J.M.J. (2008) *Angew. Chem. Int. Ed. Engl.*, **47**, 58.
19. Christoph Brabec, V.D. and Scherf, U. (2008) *Organic Photovoltaics*, Wiley-VCH Verlag GmbH & Co. KGaA, Weinheim.
20. McCullough, R.D. and Lowe, R.D. (1992) *J. Chem. Soc., Chem. Commun.*, 70.
21. Chen, T.A. and Rieke, R.D. (1992) *J. Am. Chem. Soc.*, **114**, 10087.
22. McCullough, R.D., Tristram-Nagle, S., Williams, S.P., Lowe, R.D., and Jayaraman, M. (1993) *J. Am. Chem. Soc.*, **115**, 4910.
23. Ewbank, P.C., Laird, D., and McCullough, R.D. (2008) in *Organic Photovoltaics* (ed. V.D. Christoph Brabec and U. Scherf), p. 3, Wiley-VCH Verlag GmbH & Co. KGaA.
24. Shirakawa, H., Louis, E.J., MacDiarmid, A.G., Chiang, C.K., and Heeger, A.J. (1977) *J. Chem. Soc., Chem. Commun.*, 578.
25. Chiang, C.K., Fincher, C.R. Jr., Park, Y.W., Heeger, A.J., Shirakawa, H., Louis, E.J., Gau, S.C., and MacDiarmid, A.G. (1977) *Phys. Rev. Lett.*, **39**, 1098.
26. Yamamoto, T., Sanechika, K., and Yamamoto, A. (1980) *J. Polym. Sci., Polym. Lett. Ed.*, **18**, 9.
27. Lin, J.W.P. and Dudek, L.P. (1980) *J. Polym. Sci., Polym. Chem. Ed.*, **18**, 2869.
28. Kobayashi, M., Chen, J., Chung, T.C., Moraes, F., Heeger, A.J., and Wudl, F. (1984) *Synth. Met.*, **9**, 77.
29. Yamamoto, T., Morita, A., Miyazaki, Y., Maruyama, T., Wakayama, H., Zhou, Z.H., Nakamura, Y., Kanbara, T., Sasaki, S., and Kubota, K. (1992) *Macromolecules*, **25**, 1214.
30. Yoshino, K., Hayashi, S., and Sugimoto, R. (1984) *Jpn. J. Appl. Phys., Part 2*, **23**, 899.
31. Berlin, A., Pagani, G.A., and Sannicolo, F. (1986) *J. Chem. Soc., Chem. Commun.*, 1663.
32. Tourillon, G. and Garnier, F. (1982) *J. Electroanal. Chem. Interfacial Electrochem.*, **135**, 173.
33. Zotti, G. and Schiavon, G. (1984) *J. Electroanal. Chem. Interfacial Electrochem.*, **163**, 385.
34. Jen, K.Y., Oboodi, R., and Elsenbaumer, R.L. (1985) *Polym. Mater. Sci. Eng.*, **53**, 79.
35. Tamao, K., Sumitani, K., and Kumada, M. (1972) *J. Am. Chem. Soc.*, **94**, 4374.
36. Chen, S.A. and Tsai, C.C. (1993) *Macromolecules*, **26**, 2234.
37. Sugimoto, R., Takeda, S., Gu, H.B., and Yoshino, K. (1986) *Chem. Express*, **1**, 635.
38. Leclerc, M., Martinez Diaz, F., and Wegner, G. (1989) *Makromol. Chem.*, **190**, 3105.
39. Pomerantz, M., Tseng, J.J., Zhu, H., Sproull, S.J., Reynolds, J.R., Uitz, R., Arnott, H.J., and Haider, M.I. (1991) *Synth. Met.*, **41**, 825.
40. Hotta, S., Soga, M., and Sonoda, N. (1988) *Synth. Met.*, **26**, 267.
41. McClain, M.D., Whittington, D.A., Mitchell, D.J., and Curtis, M.D. (1995) *J. Am. Chem. Soc.*, **117**, 3887.
42. Sato, M., Tanaka, S., and Kaeriyama, K. (1986) *J. Chem. Soc., Chem. Commun.*, 873.
43. Jen, K.Y., Miller, G.G., and Elsenbaumer, R.L. (1986) *J. Chem. Soc., Chem. Commun.*, 1346.
44. Elsenbaumer, R.L., Jen, K.Y., and Oboodi, R. (1986) *Synth. Met.*, **15**, 169.
45. Sato, M. and Morii, H. (1991) *Macromolecules*, **24**, 1196.
46. Mao, H., Xu, B., and Holdcroft, S. (1993) *Macromolecules*, **26**, 1163.
47. Stein, P.C., Botta, C., Bolognesi, A., and Catellani, M. (1995) *Synth. Met.*, **69**, 305.
48. McCullough, R.D., Lowe, R.D., Jayaraman, M., and Anderson, D.L. (1993) *J. Org. Chem.*, **58**, 904.
49. McCullough, R.D., Williams, S.P., Jayaraman, M., Reddinger, J., Miller, L., and Tristram-Nagle, S. (1994) *Mater. Res. Soc. Symp. Proc.*, **328**, 215.
50. Barbarella, G., Bongini, A., and Zambianchi, M. (1994) *Macromolecules*, **27**, 3039.
51. Loewe, R.S., Khersonsky, S.M., and McCullough, R.D. (1999) *Adv. Mater.*, **11**, 250.

52. Iraqi, A. and Barker, G.W. (1998) *J. Mater. Chem.*, **8**, 25.
53. Guillerez, S. and Bidan, G. (1998) *Synth. Met.*, **93**, 123.
54. Stille, J.K. (1986) *Angew. Chem.*, **98**, 504.
55. Miyaura, N. and Suzuki, A. (1995) *Chem. Rev.*, **95**, 2457.
56. Tamao, K., Sumitani, K., Kiso, Y., Zembayashi, M., Fujioka, A., Kodama, S., Nakajima, I., Minato, A., and Kumada, M. (1976) *Bull. Chem. Soc. Jpn.*, **49**, 1958.
57. McCullough, R.D., Williams, S.P., Tristram-Nagle, S., Jayaraman, M., Ewbank, P.C., and Miller, L. (1995) *Synth. Met.*, **69**, 279.
58. McCullough, R.D., Lowe, R.D., Jayaraman, M., Ewbank, P.C., Anderson, D.L., and Tristram-Nagle, S. (1993) *Synth. Met.*, **55**, 1198.
59. Chen, T.-A. and Rieke, R.D. (1993) *Synth. Met.*, **60**, 175.
60. Chen, T.A., O'Brien, R.A., and Rieke, R.D. (1993) *Macromolecules*, **26**, 3462.
61. Chen, T.-A., Wu, X., and Rieke, R.D. (1995) *J. Am. Chem. Soc.*, **117**, 233.
62. Wu, X., Chen, T.-A., and Rieke, R.D. (1995) *Macromolecules*, **28**, 2101.
63. Rieke, R.D. (1977) *Acc. Chem. Res.*, **10**, 301.
64. Rieke, R.D., Li, P.T.-J., Burns, T.P., and Uhm, S.T. (1981) *J. Org. Chem.*, **46**, 4323.
65. Loewe, R.S., Ewbank, P.C., Liu, J., Zhai, L., and McCullough, R.D. (2001) *Macromolecules*, **34**, 4324.
66. de Meijere., A. and Diederich, F. (eds) (2004) in *Metal-Catalyzed Cross-Coupling Reactions, Second, Completely Revised and Enlarged Edition*, Wiley-VCH Verlag GmbH & Co. KGaA, Weinheim.
67. Odian, G. (2004) *Principles of Polymerization*, 4th edn, Wiley-Interscience, New York.
68. Yokoyama, A., Miyakoshi, R., and Yokozawa, T. (2004) *Macromolecules*, **37**, 1169.
69. Miyakoshi, R., Yokoyama, A., and Yokozawa, T. (2005) *J. Am. Chem. Soc.*, **127**, 17542.
70. Miyakoshi, R., Yokoyama, A., and Yokozawa, T. (2004) *Macromol. Rapid Commun.*, **25**, 1663.
71. Sheina, E.E., Liu, J., Iovu, M.C., Laird, D.W., and McCullough, R.D. (2004) *Macromolecules*, **37**, 3526.
72. Iovu, M.C., Sheina, E.E., Gil, R.R., and McCullough, R.D. (2005) *Macromolecules*, **38**, 8649.
73. Jeffries-El, M., Sauve, G., and McCullough, R.D. (2004) *Adv. Mater.*, **16**, 1017.
74. Jeffries-El, M., Sauve, G., and McCullough, R.D. (2005) *Macromolecules*, **38**, 10346.
75. Zhang, R., Li, B., Iovu, M.C., Jeffries-El, M., Sauve, G., Cooper, J., Jia, S., Tristram-Nagle, S., Smilgies, D.M., Lambeth, D.N., McCullough, R.D., and Kowalewski, T. (2006) *J. Am. Chem. Soc.*, **128**, 3480.
76. Iovu, M.C., Jeffries-El, M., Sheina, E.E., Cooper, J.R., and McCullough, R.D. (2005) *Polymer*, **46**, 8582.
77. Iovu, M.C., Jeffries-El, M., Zhang, R., Kowalewski, T., and McCullough, R.D. (2006) *J. Macromol. Sci., Part A Pure Appl. Chem.*, **43**, 1991.
78. Iovu, M.C., Zhang, R., Cooper, J.R., Smilgies, D.M., Javier, A.E., Sheina, E.E., Kowalewski, T., and McCullough, R.D. (2007) *Macromol. Rapid Commun.*, **28**, 1816.
79. Liu, J., Loewe, R.S., and McCullough, R.D. (1999) *Macromolecules*, **32**, 5777.
80. Liu, J., Sheina, E., Kowalewski, T., and McCullough, R.D. (2002) *Angew. Chem. Int. Ed. Engl.*, **41**, 329.
81. Negishi, E. (1982) *Acc. Chem. Res.*, **15**, 340.
82. Djuric, S., Venit, J., and Magnus, P. (1981) *Tetrahedron Lett.*, **22**, 1787.
83. Langeveld-Voss, B.M.W., Janssen, R.A.J., Spiering, A.J.H., van Dongen, J.L.J., Vonk, E.C., and Claessens, H.A. (2000) *Chem. Commun. (Camb.)*, 81.
84. Patil, A.O., Heeger, A.J., and Wudl, F. (1988) *Chem. Rev.*, **88**, 183.
85. Rughooputh, S.D.D.V., Hotta, S., Heeger, A.J., and Wudl, F. (1987) *J. Polym. Sci., Part B: Polym. Phys.*, **25**, 1071.
86. Roncali, J., Yassar, A., and Garnier, F. (1988) *J. Chem. Soc., Chem. Commun.*, 581.
87. Yassar, A., Roncali, J., and Garnier, F. (1989) *Macromolecules*, **22**, 804.

88. Prosa, T.J., Winokur, M.J., and McCullough, R.D. (1996) *Macromolecules*, **29**, 3654.
89. Prosa, T.J., Winokur, M.J., Moulton, J., Smith, P., and Heeger, A.J. (1992) *Macromolecules*, **25**, 4364.
90. Aasmundtveit, K.E., Samuelsen, E.J., Guldstein, M., Steinsland, C., Flornes, O., Fagermo, C., Seeberg, T.M., Pettersson, L.A.A., Inganaes, O., Feidenhans'l, R., and Ferrer, S. (2000) *Macromolecules*, **33**, 3120.
91. Yamamoto, T. and Kokubo, H. (2000) *J. Polym. Sci., Part B: Polym. Phys.*, **38**, 84.
92. Yang, H. (2007) in *Organic Field-Effect Transistors* (eds Z. Bao and J. Locklin), CRC Press, Boca Raton, p. 371.
93. Aasmundtveit, K.E., Samuelsen, E.J., Hoffmann, K., Bakken, E., and Carlsen, P.H.J. (2000) *Synth. Met.*, **113**, 7.
94. Kline, R.J., McGehee, M.D., Kadnikova, E.N., Liu, J., and Frechet, J.M.J. (2003) *Adv. Mater.*, **15**, 1519.
95. Kline, R.J., McGehee, M.D., Kadnikova, E.N., Liu, J., Frechet, J.M.J., and Toney, M.F. (2005) *Macromolecules*, **38**, 3312.
96. Kim, D.H., Park, Y.D., Jang, Y., Kim, S., and Cho, K. (2005) *Macromol. Rapid Commun.*, **26**, 834.
97. Chang, J.-F., Sun, B., Breiby, D.W., Nielsen, M.M., Soelling, T.I., Giles, M., McCulloch, I., and Sirringhaus, H. (2004) *Chem. Mater.*, **16**, 4772.
98. Yang, H., Shin, T.J., Yang, L., Cho, K., Ryu, C.Y., and Bao, Z. (2005) *Adv. Funct. Mater.*, **15**, 671.
99. Greenham, N.C., Samuel, I.D.W., Hayes, G.R., Phillips, R.T., Kessener, Y.A.R.R., Moratti, S.C., Holmes, A.B., and Friend, R.H. (1995) *Chem. Phys. Lett.*, **241**, 89.
100. Xu, B. and Holdcroft, S. (1993) *Macromolecules*, **26**, 4457.
101. Chen, F., Mehta, P.G., Takiff, L., and McCullough, R.D. (1996) *J. Mater. Chem.*, **6**, 1763.
102. Assadi, A., Svensson, C., Willander, M., and Inganäs, O. (1988) *Appl. Phys. Lett.*, **53**, 195.
103. Yoshino, K., Takahashi, H., Muro, K., Ohmori, Y., and Sugimoto, R. (1991) *J. Appl. Phys.*, **70**, 5035.
104. Zen, A., Pflaum, J., Hirschmann, S., Zhuang, W., Jaiser, F., Asawapirom, U., Rabe, J.P., Scherf, U., and Neher, D. (2004) *Adv. Funct. Mater.*, **14**, 757.
105. Goh, C., Kline, R.J., McGehee, M.D., Kadnikova, E.N., and Frechet, J.M.J. (2005) *Appl. Phys. Lett.*, **86**, 122110/1.
106. Horowitz, G., Hajlaoui, M.E., and Hajlaoui, R. (2000) *J. Appl. Phys.*, **87**, 4456.
107. Bao, Z., Feng, Y., Dodabalapur, A., Raju, V.R., and Lovinger, A.J. (1997) *Chem. Mater.*, **9**, 1299.
108. Kaneto, K., Lim, W.Y., Takashima, W., Endo, T., and Rikukawa, M. (2000) *Jpn. J. Appl. Phys., Part 2*, **39**, L872.
109. Babel, A. and Jenekhe, S.A. (2005) *Synth. Met.*, **148**, 169.
110. Park, Y.D., Kim, D.H., Jang, Y., Cho, J.H., Hwang, M., Lee, H.S., Lim, J.A., and Cho, K. (2006) *Org. Electron.*, **7**, 514.
111. Salleo, A., Chabinyc, M.L., Yang, M.S., and Street, R.A. (2002) *Appl. Phys. Lett.*, **81**, 4383.
112. Veres, J., Ogier, S., Lloyd, G., and de Leeuw, D. (2004) *Chem. Mater.*, **16**, 4543.
113. Gurau, M.C., Delongchamp, D.M., Vogel, B.M., Lin, E.K., Fischer, D.A., Sambasivan, S., and Richter, L.J. (2007) *Langmuir*, **23**, 834.
114. Li, G., Shrotriya, V., Huang, J., Yao, Y., Moriarty, T., Emery, K., and Yang, Y. (2005) *Nat. Mater.*, **4**, 864.
115. Ma, W., Yang, C., Gong, X., Lee, K., and Heeger, A.J. (2005) *Adv. Funct. Mater.*, **15**, 1617.
116. Kim, Y., Cook, S., Tuladhar, S.M., Choulis, S.A., Nelson, J., Durrant, J.R., Bradley, D.D.C., Giles, M., McCulloch, I., Ha, C.-S., and Ree, M. (2006) *Nat. Mater.*, **5**, 197.
117. Woo, C.H., Thompson, B.C., Kim, B.J., Toney, M.F., and Frechet, J.M.J. (2008) *J. Am. Chem. Soc.*, **130**, 16324.
118. Schilinsky, P., Asawapirom, U., Scherf, U., Biele, M., and Brabec, C.J. (2005) *Chem. Mater.*, **17**, 2175.
119. Ma, W., Kim, J.Y., Lee, K., and Heeger, A.J. (2007) *Macromol. Rapid Commun.*, **28**, 1776.

120. Ballantyne, A.M., Chen, L., Dane, J., Hammant, T., Braun, F.M., Heeney, M., Duffy, W., McCulloch, I., Bradley, D.D.C., and Nelson, J. (2008) *Adv. Funct. Mater.*, **18**, 2373.
121. Maior, R.M.S., Hinkelmann, K., Eckert, H., and Wudl, F. (1990) *Macromolecules*, **23**, 1268.
122. Gill, R.E., Malliaras, G.G., Wildeman, J., and Hadziioannou, G. (1994) *Adv. Mater.*, **6**, 132.
123. Krische, B., Hellberg, J., and Lilja, C. (1987) *J. Chem. Soc., Chem. Commun.*, 1476.
124. Zagorska, M. and Krische, B. (1990) *Polymer*, **31**, 1379.
125. Zagorska, M., Kulszewicz-Bajer, I., Pron, A., Firlej, L., Bernier, P., and Galtier, M. (1991) *Synth. Met.*, **45**, 385.
126. Barta, P., Cacialli, F., Friend, R.H., and Zagorska, M. (1998) *J. Appl. Phys.*, **84**, 6279.
127. Faid, K. and Leclerc, M. (1993) *J. Chem. Soc., Chem. Commun.*, 962.
128. Andreani, F., Salatelli, E., and Lanzi, M. (1996) *Polymer*, **37**, 661.
129. Ong, B., Wu, Y., Jiang, L., Liu, P., and Murti, K. (2004) *Synth. Met.*, **142**, 49.
130. Tierney, S., Heeney, M., and McCulloch, I. (2005) *Synth. Met.*, **148**, 195.
131. Lidstrom, P., Tierney, J., Wathey, B., and Westman, J. (2001) *Tetrahedron*, **57**, 9225.
132. Gallazzi, M.C., Castellani, L., Marin, R.A., and Zerbi, G. (1993) *J. Polym. Sci., Part A Polym. Chem.*, **31**, 3339.
133. Wu, Y., Liu, P., Gardner, S., and Ong, B.S. (2005) *Chem. Mater.*, **17**, 221.
134. Kokubo, H. and Yamamoto, T. (2001) *Macromol. Chem. Phys.*, **202**, 1031.
135. McCulloch, I., Bailey, C., Giles, M., Heeney, M., Love, I., Shkunov, M., Sparrowe, D., and Tierney, S. (2005) *Chem. Mater.*, **17**, 1381.
136. Ong, B.S., Wu, Y., Liu, P., and Gardner, S. (2004) *J. Am. Chem. Soc.*, **126**, 3378.
137. Thompson, B.C., Kim, B.J., Kavulak, D.F., Sivula, K., Mauldin, C., and Frechet, J.M.J. (2007) *Macromolecules*, **40**, 7425.
138. Zhu, Z., Waller, D., and Brabec, C.J. (2008) in *Organic Photovoltaics* (ed. V.D. Christoph Brabec and U. Scherf), p. 129, Wiley-VCH Verlag GmbH & Co. KGaA.
139. Rasmussen, S.C. and Pomerantz, M. (2007) in *Handbook of Conducting Polymers*, 3rd edn, Vol. 1 (eds T.A. Skotheim and J.R. Reynolds), p. 12.1, Marcel Dekker.
140. Blanchard, P., Leriche, P., Frere, P., and Roncali, J. (2007) in *Handbook of Conducting Polymers*, 3rd edn, Vol. 1 (eds T.A. Skotheim and J.R. Reynolds), p. 13.1, Marcel Dekker.
141. Ng, S.C., Xu, J.M., and Chan, H.S.O. (2000) *Macromolecules*, **33**, 7349.
142. Lu, H.-F., Chan, H.S.O., and Ng, S.-C. (2003) *Macromolecules*, **36**, 1543.
143. Yasuda, T., Sakai, Y., Aramaki, S., and Yamamoto, T. (2005) *Chem. Mater.*, **17**, 6060.
144. Crouch, D.J., Skabara, P.J., Lohr, J.E., McDouall, J.J.W., Heeney, M., McCulloch, I., Sparrowe, D., Shkunov, M., Coles, S.J., Horton, P.N., and Hursthouse, M.B. (2005) *Chem. Mater.*, **17**, 6567.
145. Weck, M., Dunn, A.R., Matsumoto, K., Coates, G.W., Lobkovsky, E.B., and Grubbs, R.H. (1999) *Angew. Chem. Int. Ed. Engl.*, **38**, 2741.
146. Ponzini, F., Zagha, R., Hardcastle, K., and Siegel, J.S. (2000) *Angew. Chem. Int. Ed. Engl.*, **39**, 2323.
147. Facchetti, A., Yoon, M.-H., Stern, C.L., Katz, H.E., and Marks, T.J. (2003) *Angew. Chem. Int. Ed. Engl.*, **42**, 3900.
148. Sakamoto, Y., Suzuki, T., Kobayashi, M., Gao, Y., Fukai, Y., Inoue, Y., Sato, F., and Tokito, S. (2004) *J. Am. Chem. Soc.*, **126**, 8138.
149. Tang, M.L., Reichardt, A.D., Miyaki, N., Stoltenberg, R.M., and Bao, Z. (2008) *J. Am. Chem. Soc.*, **130**, 6064.
150. Wang, Y. and Watson, M.D. (2006) *J. Am. Chem. Soc.*, **128**, 2536.
151. Heeney, M., Bailey, C., Genevicius, K., Shkunov, M., Sparrowe, D., Tierney, S., and McCulloch, I. (2005) *J. Am. Chem. Soc.*, **127**, 1078.
152. McCulloch, I., Heeney, M., Bailey, C., Genevicius, K., MacDonald, I., Shkunov, M., Sparrowe, D., Tierney,

S., Wagner, R., Zhang, W., Chabinyc, M.L., Kline, R.J., McGehee, M.D., and Toney, M.F. (2006) *Nat. Mater.*, **5**, 328.
153. Chabinyc, M.L., Toney, M.F., Kline, R.J., McCulloch, I., and Heeney, M. (2007) *J. Am. Chem. Soc.*, **129**, 3226.
154. Kline, R.J., DeLongchamp, D.M., Fischer, D.A., Lin, E.K., Richter, L.J., Chabinyc, M.L., Toney, M.F., Heeney, M., and McCulloch, I. (2007) *Macromolecules*, **40**, 7960.
155. Pan, H., Li, Y., Wu, Y., Liu, P., Ong, B.S., Zhu, S., and Xu, G. (2007) *J. Am. Chem. Soc.*, **129**, 4112.
156. Pan, H., Wu, Y., Li, Y., Liu, P., Ong, B.S., Zhu, S., and Xu, G. (2007) *Adv. Funct. Mater.*, **17**, 3574.
157. Ong, B.S., Wu, Y., Li, Y., Liu, P., and Pan, H. (2008) *Chem. Eur. J.*, **14**, 4766.
158. Li, J., Qin, F., Li, C.M., Bao, Q., Chan-Park, M.B., Zhang, W., Qin, J., and Ong, B.S. (2008) *Chem. Mater.*, **20**, 2057.
159. Fong, H.H., Pozdin, V.A., Amassian, A., Malliaras, G.G., Smilgies, D.-M., He, M., Gasper, S., Zhang, F., and Sorensen, M. (2008) *J. Am. Chem. Soc.*, **130**, 13202.
160. He, M. and Zhang, F. (2007) *J. Org. Chem.*, **72**, 442.
161. Osaka, I., Sauve, G., Zhang, R., Kowalewski, T., and McCullough, R.D. (2007) *Adv. Mater.*, **19**, 4160.
162. Osaka, I., Zhang, R., Sauve, G., Smilgies, D.-M., Kowalewski, T., and McCullough, R.D. (2009) *J. Am. Chem. Soc.*, **131**, 2521.
163. Narasos, W.F. (2008) *Macromolecules*, **41**, 3169.
164. Ando, S., Nishida, J.-I., Inoue, Y., Tokito, S., and Yamashita, Y. (2004) *J. Mater. Chem.*, **14**, 1787.
165. Ando, S., Nishida, J., Fujiwara, E., Tada, H., Inoue, Y., Tokito, S., and Yamashita, Y. (2004) *Chem. Lett.*, **33**, 1170.
166. Ando, S., Nishida, J.-I., Tada, H., Inoue, Y., Tokito, S., and Yamashita, Y. (2005) *J. Am. Chem. Soc.*, **127**, 5336.
167. Yamamoto, T., Kokubo, H., Kobashi, M., and Sakai, Y. (2004) *Chem. Mater.*, **16**, 4616.
168. Liu, J., Zhang, R., Sauve, G., Kowalewski, T., and McCullough, R.D. (2008) *J. Am. Chem. Soc.*, **130**, 13167.
169. Zhang, M., Tsao Hoi, N., Pisula, W., Yang, C., Mishra Ashok, K., and Mullen, K. (2007) *J. Am. Chem. Soc.*, **129**, 3472.
170. Usta, H., Lu, G., Facchetti, A., and Marks, T.J. (2006) *J. Am. Chem. Soc.*, **128**, 9034.
171. Lu, G., Usta, H., Risko, C., Wang, L., Facchetti, A., Ratner, M.A., and Marks, T.J. (2008) *J. Am. Chem. Soc.*, **130**, 7670.
172. Mallik, A.B., Locklin, J., Mannsfeld, S.C.B., Reese, C., Roberts, M.E., Senatore, M.L., Zi, H., and Bao, Z. (2007) in *Organic Field-Effect Transistors*, Vol. 128 (eds Z. Bao and J. Locklin), p. 159, CRC Press.
173. Chua, L.-L., Zaumseil, J., Chang, J.-F., Ou, E.C.W., Ho, P.K.H., Sirringhaus, H., and Friend, R.H. (2005) *Nature*, **434**, 194.
174. Zaumseil, J. and Sirringhaus, H. (2007) *Chem. Rev.*, **107**, 1296.
175. Usta, H., Facchetti, A., and Marks, T.J. (2008) *J. Am. Chem. Soc.*, **130**, 8580.
176. Burgi, L., Turbiez, M., Pfeiffer, R., Bienewald, F., Kirner, H.-J., and Winnewisser, C. (2008) *Adv. Mater.*, **20**, 2217.
177. Massimo Lazzari, G.L. and Lecommandoux, S. (2006) *Block Copolymers in Nanoscience*, Wiley-VCH Verlag GmbH & Co. KGaA, Weinheim.
178. Schmitt, C., Nothofer, H.-G., Falcou, A., and Scherf, U. (2001) *Macromol. Rapid Commun.*, **22**, 624.
179. Sun, S., Fan, Z., Wang, Y., Haliburton, J., Taft, C., Maaref, S., Seo, K., and Bonner, C.E. (2003) *Synth. Met.*, **137**, 883.
180. Sun, S.-S. (2003) *Sol. Energy Mater. Sol. Cells*, **79**, 257.
181. Scherf, U., Gutacker, A., and Koenen, N. (2008) *Acc. Chem. Res.*, **41**, 1086.
182. Zhang, Y., Tajima, K., Hirota, K., and Hashimoto, K. (2008) *J. Am. Chem. Soc.*, **130**, 7812.
183. Yokozawa, T., Adachi, I., Miyakoshi, R., and Yokoyama, A. (2007) *High Perform. Polym.*, **19**, 684.

184. Miyakoshi, R., Yokoyama, A., and Yokozawa, T. (2008) *Chem. Lett.*, **37**, 1022.
185. Asawapirom, U., Guentner, R., Forster, M., and Scherf, U. (2005) *Thin Solid Films*, **477**, 48.
186. Tu, G., Li, H., Forster, M., Heiderhoff, R., Balk, L.J., and Scherf, U. (2006) *Macromolecules*, **39**, 4327.
187. Jenekhe, S.A. and Chen, X.L. (1998) *Science*, **279**, 1903.
188. Liu, J. and McCullough, R.D. (2002) *Macromolecules*, **35**, 9882.
189. Matyjaszewski, K. and Xia, J. (2001) *Chem. Rev.*, **101**, 2921.
190. Matyjaszewski, K. (2000) *Controlled/Living Radical Polymerization: Progress in ATRP, NMP and RAFT*, Vol. 768 , American Chemical Society, Washington, DC.
191. Iovu, M.C., Craley, C.R., Jeffries-El, M., Krankowski, A.B., Zhang, R., Kowalewski, T., and McCullough, R.D. (2007) *Macromolecules*, **40**, 4733.
192. Sauve, G. and McCullough, R.D. (2007) *Adv. Mater.*, **19**, 1822.
193. Chiefari, J., Chong, Y.K., Ercole, F., Krstina, J., Jeffery, J., Le, T.P.T., Mayadunne, R.T.A., Meijs, G.F., Moad, C.L., Moad, G., Rizzardo, E., and Thang, S.H. (1998) *Macromolecules*, **31**, 5559.
194. Urien, M., Erothu, H., Cloutet, E., Hiorns, R.C., Vignau, L., and Cramail, H. (2008) *Macromolecules*, **41**, 7033.
195. Dai, C.-A., Yen, W.-C., Lee, Y.-H., Ho, C.-C., and Su, W.-F. (2007) *J. Am. Chem. Soc.*, **129**, 11036.
196. Boudouris, B.W., Frisbie, C.D., and Hillmyer, M.A. (2008) *Macromolecules*, **41**, 67.
197. Radano, C.P., Scherman, O.A., Stingelin-Stutzmann, N., Mueller, C., Breiby, D.W., Smith, P., Janssen, R.A.J., and Meijer, E.W. (2005) *J. Am. Chem. Soc.*, **127**, 12502.
198. Trnka, T.M. and Grubbs, R.H. (2001) *Acc. Chem. Res.*, **34**, 18.

4
Poly(phenylenevinylenes)

Yi Pang

4.1
Introduction

Poly(*para*-phenylenevinylene)s (PPVs) represent one of the most intensively investigated classes of π-conjugated materials, which have applications in light-emitting diodes (LEDs) [1–4], solar cells [5, 6], and chemical sensors [7]. Since the demonstration of electroluminescence from PPV in 1990s [8, 9], considerable effort has been directed to further improving the synthesis of PPV and its derivatives, and investigating their physical properties. The aromatic rings in the backbone of PPV are connected by vinylene bonds. Effective conjugation between the phenylenevinylene (PV) segments is affected by substituents (on the phenyl ring), various chain defects, vinylene bond geometry, overall polymer conformation, and morphology. The combination of these factors gives each PPV derivatives a distinctive physical and optical property. With regard to the optical properties, for improved device performance, the introduction of donor and acceptor groups along the PPV backbone alters the charge mobility and luminescence characteristics of the respective materials. Furthermore, the display applications require the material to provide emission in the specific color region of interest, which is usually achievable through modification of polymer structure.

An overview[1)] of research activity on "PPV" shows that the material has been receiving increasing attentions in the field since the demonstration of its electroluminescence in 1990s [8, 9]. Various methodologies have been developed in the past for synthesis of PPVs, and earlier examples can be found in the previous reviews [10, 11]. The aromatic rings in the backbone of PPV are connected by vinylene bonds. The construction of PPVs can thus be achieved via formation of new C=C bonds, which includes the Gilch route, Wittig condensation, and Wessling route (Scheme 4.1, disconnection 1). Another commonly used methodology is to couple an olefinic fragment with an aromatic halide by using the Heck coupling (disconnection 2). As is the case in most highly conjugated materials, PPVs have

1) On the basis of the hits from "PPV" in SciFinder data base, the number of publications on PPV has risen steadily to about 1100 in 2008.

Design and Synthesis of Conjugated Polymers. Edited by Mario Leclerc and Jean-François Morin

ISBN: 978-3-527-32474-3

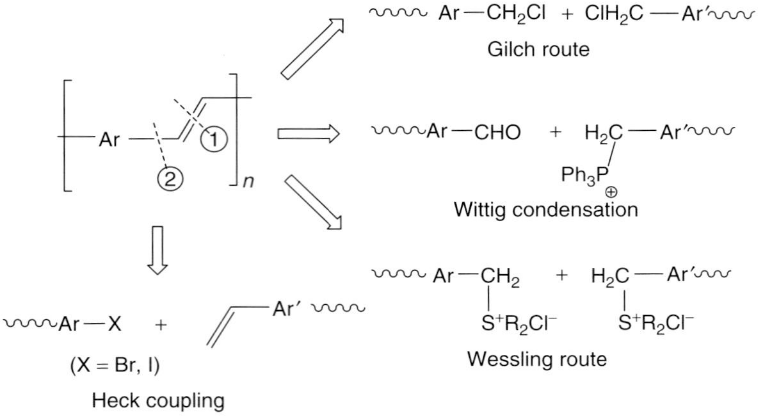

Scheme 4.1 Synthesis of PPV.

poor solubility in organic solvents. In addition, device performance requires the presence of various functional groups to facilitate, for example, electron- and hole-transporting across the materials. Various structural modifications have been made to overcome these difficulties. This chapter focuses on the recent progress in the synthesis of PPV materials.

4.2 Poly(*p*-phenylenevinylene)s via Polymerization Methods

4.2.1 Gilch Approach

The polymerization of 1,4-bis(halomethyl)benzenes **1** to PPVs in the presence of excess potassium *t*-butoxide is referred to as the *Gilch route* [12]. The advantages of the reaction include easily accessible starting materials, mild reaction conditions, formation of film-forming products, and potential for large-scale production.

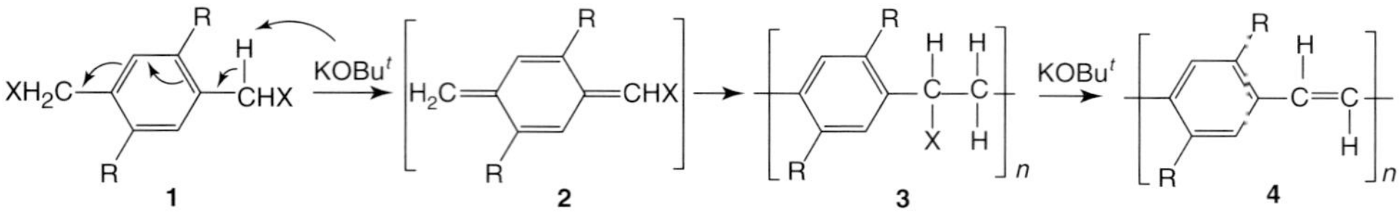

Despite its early discovery [12], little attention has been paid to understanding the mechanistic aspect of the polymerization until recently [13–16]. The first step of the reaction is believed to be 1,6-dehydrohalogenation of **1** via an E2 type 1,6-elimination to give the monomer **2**, which then polymerizes to give the corresponding polymer

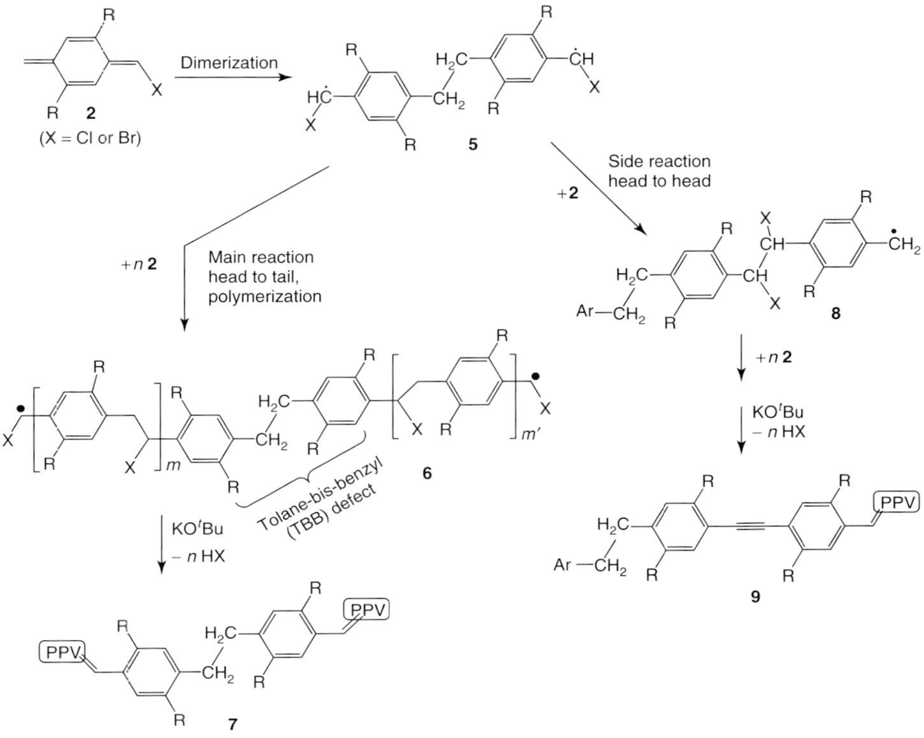

Scheme 4.2 The radical mechanism in the Gilch polymerization.

precursor **3**. The reactive monomer **2**, α-chloro-*p*-quinodimethane, can be formed at low temperature ($T < -70\,^{\circ}\mathrm{C}$), and has been observed by using low-temperature NMR spectroscopy [16, 17]. The current mechanistic investigation suggests that the polymerization from **2** to **3** proceeds predominantly via a radical [18, 19] rather than anionic growth [20] mechanism. The initial step in the radical mechanism is thought to form α,ω-biradical **5** via dimerization of **2** (Scheme 4.2). Addition of the monomer **2** to one end of **5** in the head-to-tail fashion will give the PPV polymer precursor **6**, which is then converted to **7** via dehydrohalogenation reaction. In the proposed mechanism, the growing species are α,ω-macro-biradicals, whose recombination does not cause chain termination as in conventional radical polymerization [21]. The assumption is consistent with the characteristics of Gilch polymerization, which usually produces PPVs of high molecular weight and often leads to gelation [17]. The radical characteristics are reported to be useful for molecular weight control by using oxygen [22].

Although the radical addition to the monomer **2** primarily occurs in the head-to-tail fashion, the head-to-head addition may also happen and lead to chain defects along the PPV backbone. The possible chain defects include saturated ethylene bridges (i.e., tolane-bis-benzyl (TBB) defects), rodlike diphenyl ethynylene subunits **9**, and halogenated chain ends [23, 24]. It should be noted that the TBB defect formed from the biradical initiator **5** is negligible. These TBB defects, which form along with the Ph–C≡C–Ph subunit as a result of head-to-head addition, will be accumulated in significant amount. These structural defects limit the application of the Gilch route, since the performance of devices such as LEDs require the materials to be not only perfect in design but also free of constitutional defects and impurities [23–26].

Some examples of PPV homopolymers obtained by using the Gilch reaction are shown in Table 4.1. Polymerization of the 2-decyloxy-5-(4′-*tert*-butylphenyl)-1,4- bis(bromomethyl)benzene produces PPV **10** in ~62% yield (entry 2) [27]. Presence of the bulky *tert*-butylphenyl substituent promotes the desirable head-to-tail addition of monomer **2**, thereby minimizing the structural defects. The content of TBB defects is undetectable in the ^{1}H NMR of PPV **10**, whereas it is significant in the less hindered PPV **11**. The LED device using **10** shows a lower turn-on voltage (at 8.4 V) and a longer electroluminescent (EL) emission wavelength ($\lambda_{em} = 546$ nm) than that using **11** (turn on at 9.8 V, $\lambda_{em} = 527$ nm), illustrating the importance of eliminating structural defects. Another useful feature of the Gilch process is that the reaction can be used to polymerize monomers with up to four substituents on the phenyl ring (entries 3 and 4, monomers). Thus, PPV **12** with tetrakis(ethylhexyloxy) groups has also been synthesized in ~50% yield as green fiber [28]. The solution of **12** exhibits absorption $\lambda_{max} = 419$ nm and photoluminescence $\lambda_{em} = 475$ nm, which is notably blueshifted from that of less substituted MEH–PPV ($\lambda_{max} = 485$ nm and $\lambda_{em} = 555$ nm in solution) due to the steric hindrance between the vinylene and substituents on phenylene. Similarly, the trialkoxyalkyl-substituted PPV (**13**) exhibits blueshifted electrolumescence at 493 nm (entry 4) [29]. An LED device using **12** (TEH–PPV) gives green electrolumescence at 505 and 542 nm. Interestingly, reaction of 1,4-bis(chloromethyl)-2-methoxy-5-octyloxybenzene with 1,3,5-trichloromethyl-2,4,6-tributoxybenzene gives a soluble hyperbranched copolymer **14** (entry 5) [30]. Some additional examples with aryl [31, 32], porphrin [33], and bulky tricyclodecane (TCD) [34] substitution can be found in literatures.

The Gilch reaction has also been successfully used to prepare copolymers, in which two or more different PV units are incorporated randomly into a single polymer chain. As an example, the copolymers **15** are prepared from 1,4-bis(chloromethyl)-2-(2-ethylhexyloxy)-5-methoxybenzene and the polyhedral oligomeric silsesquioxanes (POSSs)-substituted 1,4-bis(chloromethyl)benzene [38]. The ratio of MEH–PPV to POSSs–PPV in the resulting copolymers is directly correlated to the feed ratio of the two different starting materials. Similarly the $x:y$ ratio between the feed and product composition indicates that the polymerization is relatively insensitive to the attached bulky substituent, thereby allowing to produce the copolymer with a predictable composition. An LED device using the copolymer of 5% POSS–PPV and 95% MEH–PPV exhibits enhanced electroluminescence by a factor of 6.4 (with maximum brightness of 11 000 cd m^{-2} at 14.3 V).

Table 4.1 Examples of Gilch polymerization.

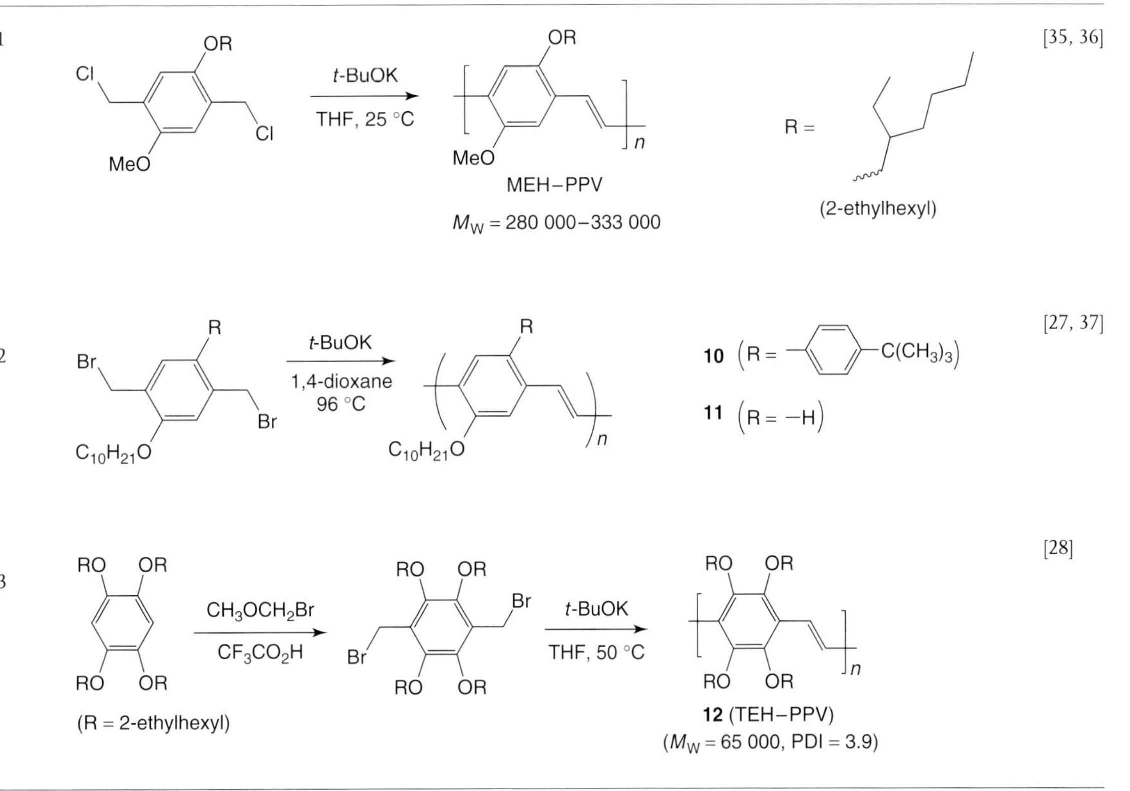

(continued overleaf)

Table 4.1 *(continued)*

4	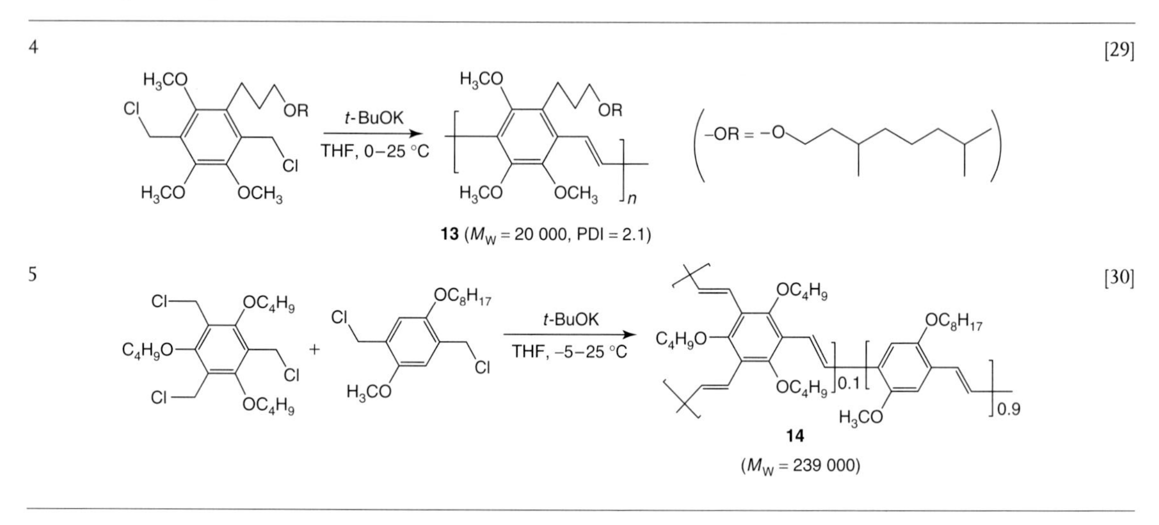	[29]
5		[30]

T-BuOK, THF, 25 °C

$x:y$	M_W ($\times 10^4$)
100 : 0	1.9
95 : 5	21.5
75 : 25	10.8
0 : 100	24.5

15

Various functional groups can be incorporated into the PPVs by using the Gilch process. A recent example is to incorporate fluoro group on the vinylene to increase the electron affinity [39]. In the resulting copolymer **17**, the content of (poly)*p*-phenylenedifluorovinylene (PPDFV) fragments significantly influences the optical properties of the respective materials. Other PPV materials, which have triphenylamine [40] and oxadiazole [41] side chains, have also been prepared. In addition, PPVs wrapped with 1,3,5-phenylene-based rigid dendrons have also been prepared, whose EL properties are shown to be strongly affected by the size of the phenylene Dendron groups [42].

16 — *t*-BuOK, THF, 40 °C to 25 °C — **17**

Feed ratio $x : y = 1 : 9$, $= 5 : 5$, $= 9 : 1$

copolymer absorption λ_{max} = 372 nm, = 449 nm, = 490 nm

M_W = 11 000–43 000 (PDI = 1.4–2.6)

One of the challenging problems for the Gilch polymerization is that most reaction mixtures change from homogeneous solutions into gels within the first few minutes of the conversion. Continual stirring for hours or days is often necessary, sometimes combined with refluxing, to redissolve the gels, in most cases, to give homogeneous solutions. Wudl and Srdanov first found that the troublesome gelation can be minimized by slow or intermittent addition of a great excess (~6 equiv) of the base [43]. Later, Hsieh and coworkers observed that addition of monofunctional benzylhalogenides (about 1–3% *tert*-butylbenzyl chloride) can reliably prevent gel formation during the Gilch reactions [31, 35]. Introduction of poly(ethylene glycol) (PEG), a complexing agent, has also been shown to suppress the gelation formation [20]. Recent studies from Rehahn and coworkers [17] show that addition of monofunctional benzylhalogenides in the Hsieh's approach does not influence the molecular weight of PPV products as originally thought. The results suggest that the additives may act as cosolvents that influence the polymer chain conformation and segment–segment interactions, thereby affecting the chain-entanglement and gel-formation process.

4.2.2
The Wessling Method

The Wessling method [44], developed by Wessling and Zimmerman in 1968, represents a polymerization route via a water-soluble and processible polyelectrolyte precursor. The general process involves conversion of 1,4-bis (dialkysulfoniomethyl)-benzenedihalides **18** to the reactive intermediate **19** via a base-promoted E2 type 1,6-elimination. The intermediate **19**, which is the actual monomer and is formed *in situ*, is polymerized to give the precursor **20**, followed by thermal treatment to provide the final PPV products. Alkoxy, alkyl, and aryl substituents R do not greatly influence the ability to generate and polymerize the intermediate **19** [10, 45, 46]. The charged sulfonium groups solubilize the polymer **20** and are removed in the subsequent thermal conversion step. The polymerization of **19** is believed to proceed according to a chain growth mechanism, based on the facts that high molecular weight is formed very quickly, within the first minutes of the reaction. Although both radical and anionic mechanisms have been proposed, experiment evidence appears to favor the radical chain growth as suggested from radical-trapping [47] and ESR [48] experiments.

$Cl^- R_2^+S$ … $S^+R_2Cl^-$ (**18**) $\xrightarrow{OH^-}$ [… $S^+R_2Cl^-$] (**19**) → [… $S^+R_2Cl^-$]$_n$ (**20**) $\xrightarrow{200-300\ °C}$ PPV

Nanoparticles of functional polymers are key materials in nanotechnology, and the Wessling method has found a special role. With the aid of ionic liquid, which serves as a poor solvent and high-temperature reaction media, the water-soluble high-molecular precursor can be converted to nanoparticles upon slow evaporation of water (a self-organized precipitation process) (Figure 4.1). Thermal treatment of the formed nanoparticles provides spherical PPV particles of about 340±70 nm [49].

The sulfonium dication **18** can interact with the cucurbit[*n*]uril hosts CB[*n*], which have a pumpkinlike molecular shape and tend to form stable inclusion complexes with positively charged organic molecules [50]. The highly stable guest–host complex **21** is reported to polymerize to give polymeric precursor **22**, which is then thermally converted to PPV threading in multiple CB[*n*] rings [51]. The content of CB[*n*] in **23** is dependent on the size of the host. When CB[7] is used as host, the ratio of $m : n$ in **23** is determined to be 6 : 1 by using chemical analysis data based on the N content (5.75%). When CB[6] is used, however, no CB[6] is found in the final product **23**. It appears that the smaller size of CB[6] shields the reaction site in **21**, thereby preventing its incorporation into polymer structure **22**. Preparation of **22** can also be achieved by treating water-soluble **20** (R=H) with CB[7], though with less threading efficiency [52]. Conversion of the CB[7]-treated **22** to PPV occurs more readily and at lower temperature (~110 °C) than that of **20**. The lifetime of fluorescence is considerably shorter (0.47 ns) for PPV **23** (with CB[7]) than the emission of pure PPV films (1.23 ns), as the PPV chain in the former is protected from fluorescence quenchers [51].

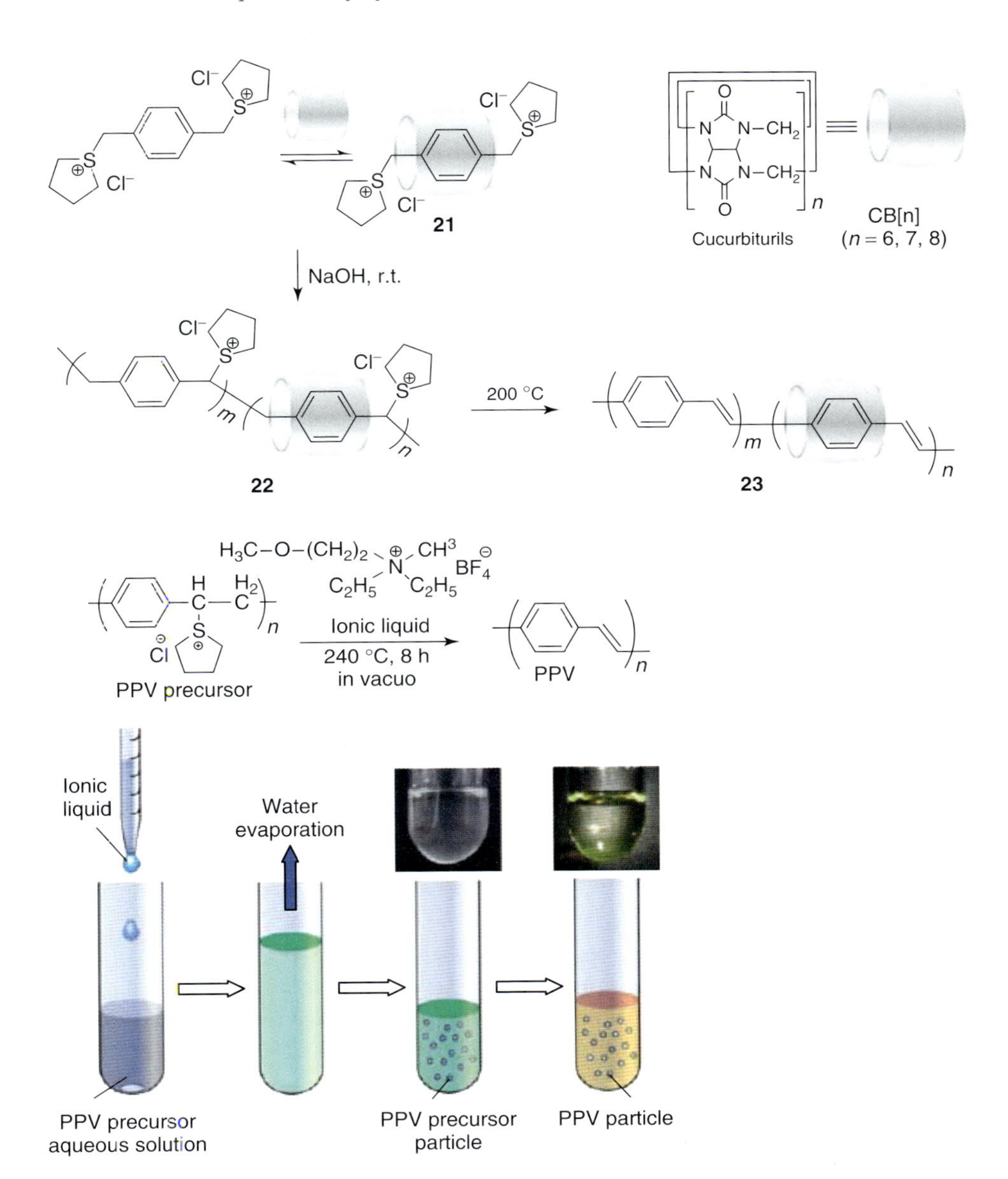

Figure 4.1 Schematic illustration of PPV nanoparticle formation, the reaction process of PPV precursor conversion to PPV, and photographs of nanoparticle dispersion in the ionic liquid before and after thermal annealing. (Reprinted with permission from [49].)

An alternative precursor route [53] to PPV uses the sodium dithiocarbamate to treat α,α'-dichloro-p-xylene to give monomer **24a**, which is then polymerized by using a strong base. The resulting precursor **24b** has a high molecular weight (up to $M_w = 250\,000$) and relative lower polydispersity index (PDI = 1.4–2.1) Soluble PPVs can be obtained by heating the precursor polymer **24b** in 1,2-dichlorobenzene at 175 °C. Photoluminescence of PPV films suggests that the PPV from this dithiocarbamate precursor route may have slightly more defects than that from the sulfinyl precursor **20**. The dithiocarbamate precursor route is also useful for preparation of segmented PPV derivatives via controlled elimination [54].

4.3 Poly(*p*-phenylenevinylene)s via Polycondensation

The polymerization methods to PPVs described in the previous section involve 1,6-polymerization of a 1,4-xylylene intermediate formed during the reaction. The reaction is limited to incorporate 1,4-phenylene along the PPV chain. In addition, the vinylene bonds generated in the Gilch process are primarily in trans configuration. The vinylene in PPVs can also be generated by using an olefination reaction such as Wittig, which provides synthetic flexibility to construct *cis*-vinylene, and to incorporate not only 1,4-phenylene but also 1,3-, and 1,2-phenylene units in the PPV materials.

4.3.1 Wittig and Horner–Wadsworth–Emmons Reaction

The Wittig reaction is a step-growth polycondensation method that involves AA/BB type reaction between an aromatic dialdehyde and bisphosphonium ylide monomers. The reaction is easy to carry out and proceeds under mild conditions, which gives PPV chains free of structural defects. The condensation, however, leads to products of moderate molecular weight (degree of polymerization, DP $\approx$ 10–20), thereby leaving residual terminal groups in the products, which are often detectable in ^{1}H NMR. The olefinic C=C bond formed in the reaction can be either in *cis* or *trans* geometry, depending on the phosphonium ion, monomer structure, and reaction conditions [55]. If the Wittig reaction (phosphonium ion = Ph_3P^+) is used, significant *cis*-vinylene will be found in the PPV products.

A valuable alternative makes use of phosphonate ester (-PO(OEt)$_2$), which is a modified Wittig reaction and known as the *Horner–Wadsworth–Emmons* reaction [56]. The Horner–Wadsworth–Emmons reaction is generally superior to the Wittig reaction with resonance-stabilized phosphonium ylides, and is widely employed in the preparation of conjugated systems. The reaction often gives better yields than the Wittig reaction, and olefin bond in the products are primarily in trans configuration. It also has the practical advantage that the phosphate by-product is water soluble and easily removed from the PPV product.

OHC–C$_6$H$_2$R$^1_2$–CHO + R$_3$P$^+$CH$_2$–C$_6$H$_2$R$^2_2$–CH$_2$P$^+$R$_3$ 2X$^\ominus$ (**25**) $\xrightarrow{\text{Base}}$ [–C$_6$H$_2$R$^1_2$–CH=CH–C$_6$H$_2$R$^2_2$–CH=CH–]$_n$ (**26**)

(R^1, R^2 = H, alkyl, alkoxy)

	Phosphonium ion	Vinylene bond geometry
Wittig	R$_3$P$^+$ = Ph$_3$P$^+$	A mixture of *cis*- and *trans*-CH=CH
	Bu$_3$P$^+$	*trans*-CH=CH
Horner–Wadsworth–Emmons	(EtO)$_2$P(=O)	*trans*-CH=CH

Incorporation of a triphenylamino (TPA) functional group into the main chain leads to PPVs **27–29** (Table 4.2), which illustrates the impact of different synthetic routes on the percentage of *trans*-CH=CH content (entry 1) [57]. Replacement of the phenyl on the phosphorus atom with a less-bulky butyl group modifies the Wittig reaction, leading to drastically increased *trans*-olefin content and improved molecular weight. The Wittig reaction also allows incorporation of 1,2-phenylene [58] and 1,3-phenylene [59–61] building blocks into PPV main chains. Useful functions associated with the *meta*-phenylene linkage include improved polymer solubility and defined conjugation length for PPVs. Effective conjugation length control can be illustrated from PPVs **31**, **32**, and **33**, which give respectively blue, orange, and green electroluminescence. In these polymers, a block of *para*-phenylenevinylene is sandwiched between two adjacent *meta*-phenylenes, thereby providing a uniform chromophore for each polymer. Conjugation length control via *meta*-phenylene can be understood by considering the resonance interaction between the two substituents "A" and "D" on the phenyl ring (Figure 4.2). The electronic connection between the two substituents at meta position is significantly weakened, owing to the absence of the resonance interaction.

It should be noted that the EL peaks of **31–33** are redshifted from their respective solution fluorescence by ~50 nm [59, 62, 63], which is notably smaller than ~100 nm observed from the oligo(*p*-phenylenevinylene)s of comparable conjugation length [64]. It appears that the chromophores in the *meta*-phenylene-containing PPVs are randomly aligned, in comparison with the oligo PPVs, thereby alleviating the aggregation and excimer formation. The green-emitting **33a** gives electrolumescence

Table 4.2 Synthesis of PPVs by using the Wittig condensations.

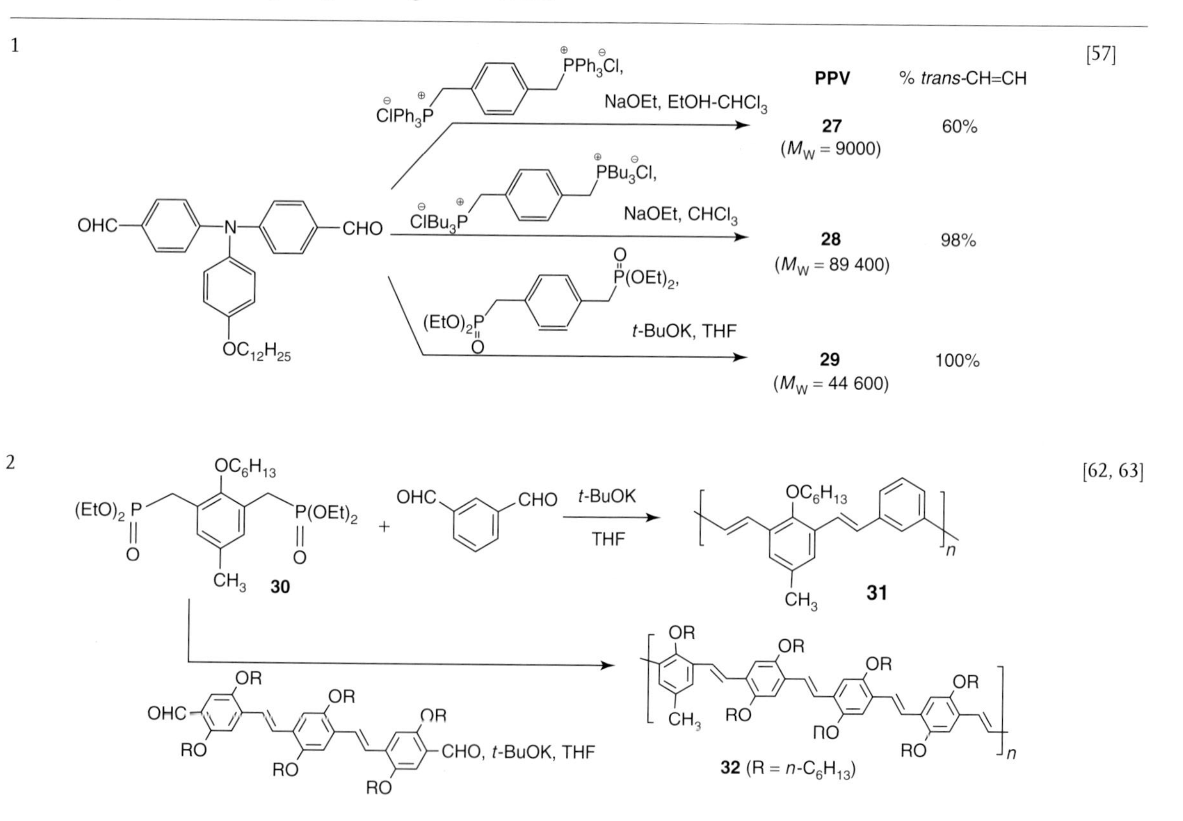

3	OHC, CHO, R + $(EtO)_2P(O)$, OC_6H_{13}, $C_6H_{13}O$, $P(OEt)_2$, O $\xrightarrow[\text{THF}]{t\text{-BuOK}}$ **33** (M_W ~ 45 000) **a**: R = –H; **b**: R = –NPh_2	[59, 65]
4	OHC, OMe, MeO, O–$(CH_2)_6$–O, CHO + Cl^- $Ph_3\overset{+}{P}$, $\overset{+}{P}Ph_3$ Cl^-, X; EtONa, $CHCl_3$–EtOH; X = H, Cl, or Br $\rightarrow$ **34**	[67, 68]
5	OHC, O–Me, Br, R–O $\xrightarrow[150\ °C]{P(OEt)_3}$ **35** $P(OEt)_2$ $\xrightarrow[\text{THF}]{t\text{-BuOK}}$ **36** R = 2′-ethylhexyl — M_W (PDI) 13 000 (1.8) R = 3′,7′-dimethyloctyl — 21 000 (2.3)	[69]

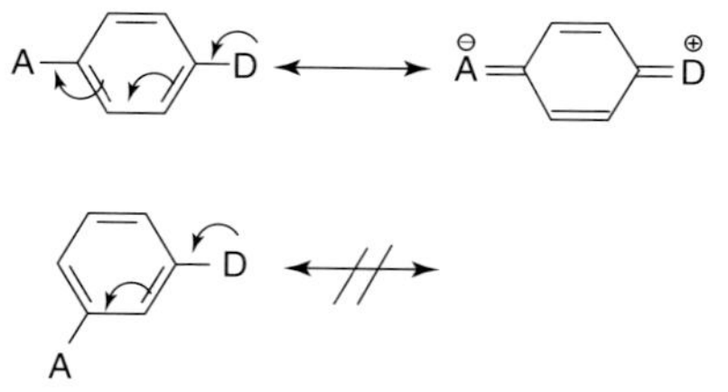

Figure 4.2 Resonance interaction is allowed when two substituents "A" and "D" are placed in para position, but not allowed when they are in meta position.

with a balanced charge injection [59]. Addition of a hole-injecting diphenylamino group, which is at meta position relative to both vinylene bonds shown in **33b**, improves the charge injection without significant shift in emission wavelength and raises the EL brightness by an order of magnitude to ~7 000 cd m^{-2} [65]. Recent development [66] in the synthesis of *meta*-phenylene-containing PPVs includes the use of renewable resource starting materials 3-pentadecylphenol, which is an industrial by-product from the cashew-nut-processing industry.

An interesting example is the synthesis of regioregular PPV **36** [69], where the vinylene bond is formed in a pure head-to-tail manner, which is not likely achieved in the Gilch process. The methodology relies on the synthesis of pure asymmetrically functionalized monomer **35**, which fixes the aldehyde and phosphate groups in the 1 : 1 ratio. The regioregular PPV **36** exhibits higher crystallinity in comparison with regiorandom PPVs, as a result of the highly ordered substitution pattern along the polymer backbone.

Owing to the chain defects, the chromophores in the fully conjugated PPV can be in variable conjugation lengths. The emission occurs usually from the more highly conjugated segments as a result of Forster energy transfer, and the emission spectrum is broad [4]. This raises challenge for structure–property correlation, and for effective control of pure color emission (especially blue). Using a saturated carbon chain to isolate chromophore has been proven to be a reliable strategy to confine the conjugated segment into a well-defined length as well as morphology, leading to **34** (X=H) as one of the best blue-emitting materials [67]. In analogy, silicon [70], germanium [71], and sulfur [72] have been incorporated into the main chain of PPV to confine the conjugated segments and to influence the emission properties of the respective materials. The Wittig reaction has also been used to incorporate the *ortho*-phenylene linkage in PPVs [58, 73].

An alternative synthesis route to **33** (R=H) utilizes the thermal isomerization of *cis*-CH=CH in **37** in the presence of a catalytic amount of iodine [59]. The resulting **33** contains a small percentage of iodine contamination (about 1 per 240 PV units), which leads to a notable decrease in EL efficiency (by ~4.4 times). The example supports the prevailing view that it is structural disorder or impurity, which is very hard to eliminate in polymeric materials, that prevents realization of optimum device performance.

4.3.2
PPVs with *cis*-Vinylene: a Useful Tool to Modify Physical Properties

The geometry of vinylene bond in PPV can adopt a cis or trans configuration. Earlier study shows that presence of a small percentage of *cis*-vinylene introduces disorder and improves the morphology, thereby leading to amorphous PPV for increased EL efficiency [74]. The polymer solution property study shows that **37** in THF exhibits a semirigid conformation with the Mark–Houwink constant $\alpha \approx 0.85$, which is notably smaller than its *trans*-vinylene isomer **33a** ($\alpha \approx 1.0$) [60]. The *cis*-vinylene bonds therefore lower the chain rigidity in PPV, which translates into improved polymer solubility and permits the side chain for soluble and EL **37** to be as short as ethyl (Table 4.3) [75, 76]. In the *meta*-phenylene-bridged PPVs, the influence of *cis*-vinylene linkage on the fluorescence quantum efficiency ϕ_{fl} appears to be dependent on the effective conjugation length of PPVs. For example, the **37** exhibits nearly the same ϕ_{fl}($\approx$0.70) as its isomer **33** [60, 75, 77, 78], while the presence of ~40% *cis*-CH=CH in the less conjugated **31** notably decreases the ϕ_{fl} values (e.g., from $\phi_{\text{fl}} \approx 0.55$ for the trans to $\phi_{\text{fl}} \approx 0.35$ for the cis isomer) [77].

With bulky substituents on the phenyl ring, the *cis*-CH=CH content can be increased to ~80% as shown in **38** [79], which is about the limit one can expect from the Wittig condensation as the reaction is not stereospecific. PPVs with pure *cis*-vinylene **39** have been synthesized by using Suzuki–Miyaura coupling of 2,5-dioctyloxy-1,4-benzenediboronic acid with pure (*cis,cis*)-bis(2-bromoethenyl)benzenes [80]. The *cis*-CH=CH in the soluble **39** can undergo photoisomerization to the corresponding *trans*-CH=CH under photoirradiation, thereby transforming the soluble precursor into an insoluble isomer. This cis to trans isomerization is demonstrated to be useful for direct patterning of PPVs on a micrometer scale. Further development has led to soluble PPV **40a**, which has no alkyl substituents and gives strong blue emission ($\lambda_{\text{em}} = 485$ nm, $\phi_{\text{fl}} = 0.78$) in film state [82].

4.3.3
Knoevenagel Polycondensation

Most of the EL polymers exhibit suitable hole-injecting and transporting properties. The electron-withdrawing cyano group may be attached to PPVs to improve the electron affinity, which is desirable to achieve the balanced charge injection in the device operation. Knoevenagel condensation of 1,4-xylylene dinitriles and dialdehydes gives a convenient access to cyano-substituted PPVs, and the earlier examples can be found in previous reviews [4, 11]. In the Knoevenagel condensation, most vinylene bonds in CN–PPVs are in trans configuration, but the *cis*-vinylene content can be significant, which has an adverse effect on the fluorescence quantum yield. For example, the ratio of *trans*- to *cis*-vinylene in CN–PPV **41a** is estimated to be 85 : 15 [84]. The pendant TPA group in **41b** exhibits the higher HOMO energy level than that of the PPV backbone, which permits the hole injection from electrode (ITO) into TPA, then from TPA into PPV [85]. The energetically favorable pathway leads to the effective hole injection and transport in the respective device.

Table 4.3 Synthesis of PPVs with *cis*-vinylene.

1	CHO, CHO + $BrPh_3P$, OR, RO, PPh_3Br → NaOEt, THF/EtOH → OR, RO, *n* R = *n*-hexyl, *n*-butyl, *n*-propyl, ethyl; 40 – 60% *cis*-olefin; ϕ_{fl} = 0.55–0.8 **37**	[60, 75, 78]
2	OR^1, $\overset{+}{P}Ph_3Cl^-$, $Cl^-Ph_3\overset{+}{P}$, R^1O + OR^2, OHC, CHO, R^2O → EtONa, $CHCl_3$–EtOH → OR^1, OR^2, R^1O, R^2O, C=C (H, H), *n* $R^1 = Et_2NCH_2CH_2$– R^2 = alkyl– **38** (~80% *cis*-CH=CH; M_W ~ 10 000; ϕ_{fl} ~0.5)	[79]

3	Br–CH=CH–Ar–CH=CH–Br + $(HO)_2B$–$C_6H_2(OC_8H_{17})_2$–$B(OH)_2$ $\xrightarrow{Pd(PPh_3)_4,\ K_3PO_4,\ Bu_4NBr,\ toluene,\ 80\ °C}$ **39** Ar = **a** (p-phenylene), **b** (m-phenylene)	[80, 81]
4	$BrPh_3P$–CH_2–$C_6H_2R_2$–CH_2–PPh_3Br + OHC–$C_6H_2Ph_2$–CHO $\xrightarrow{t\text{-}BuOK,\ THF}$ –[CH=CH–$C_6H_2R_2$–CH=CH–$C_6H_2Ph_2$]$_n$– **40** **a**: R = Ph–; ~50% *cis*-CH=CH **b**: R = $C_6H_{13}O$–; 87% *cis*-CH=CH	[82, 83]

The yne-containing CN–PPV **41c** is similarly synthesized [86], which exhibits higher film fluorescence but lower EL efficiency than the corresponding CN-free polymer.

OHC—Ar—CHO + [NC–CH₂–(R′O)(OR)C₆H₂–CH₂–CN] →(Base, e.g., t-BuOK) [Ar–CH=C(CN)–(R′O)(OR)C₆H₂–C(CN)=CH]ₙ 4-

Ar = (a) 2-OR-5-R′O-phenylene, (b) 3-(O(CH₂)₆O-biphenyl-NPh₂)phenylene, (c) OC_8H_{17}/$C_8H_{17}O$-substituted bis(phenylene)ethyne

4.4 Palladium-Catalyzed Cross-Coupling (Heck-, Suzuki-, and Stille Reactions)

The synthetic procedures described in previous sections involve the generation of the olefinic double bond during the PPV construction. Modern synthetic reactions, such as Heck or Suzuki coupling, provide alternative and powerful methodologies to form the aryl–vinyl single bond in the presence of a palladium catalyst.

4.4.1 The Heck Coupling

In the Heck coupling reaction [87, 88], the aryl halides react with an alkene in the presence of a palladium catalyst, leading to direct formation of the aryl–vinyl bond. The reaction gives products that formally result from the substitution of a hydrogen atom in the alkene-coupling partner. The mild conditions associated with the Heck coupling allow the presence of various functional groups, making it an attractive tool in the preparation of various functional PPVs (Table 4.4). For example, the water-soluble **44** has been prepared, which can form multilayer self-assembly films with hexa(sulfobutyl)fullerences [89]. Resulting from the interaction between the donor (PPV) and acceptor (fullerence) layers, the assembled multilayer films exhibit a steady and rapid photocurrent response upon light irradiation (at ~430 nm). In addition to copolymerization, which uses two different monomers, polymerization of **45** leads to regioregular PPVs **46** with alkylthio side chains [90]. The polymer is EL, and its blend with [6,6]-phenyl-C_{61} butyric acid methyl ester exhibits photovoltaic properties with the power conversion of 1.4% under the AM 1.5 illumination.

The distinctly different reaction conditions of the Heck coupling from the other synthetic methodologies allow one to build various functional materials. For example, the Ar–X bond is not affected under the Wittig condensation. Thus, the branched or comblike PPV **49** has been synthesized by using the Wittig

Table 4.4 Synthesis of PPV by using the Heck-coupling reactions.

1	**42** ($OC_{12}H_{25}$, $C_{12}H_{25}O$) + Br—Ar—Br $\xrightarrow[\text{DMF, } Bu_3N]{Pd(OAc)_2}$ **43**; –Ar– = **a** (NC, CN; M_W = 8000, PDI = 1.25; EL at 646 nm), **b** (NPh_2; N–N; (Mn = 8200, PDI = 1.7)), **c** (OC_8H_{17}, $C_8H_{17}O$; (Mn = 1.4×10^4, PDI = 4.7; EL at 563 nm))	[91–94]
2	$Br^{\ominus}$, $N^{\oplus}$, I, O, OH, HO + ... $\xrightarrow[\text{DMF, 80 °C}]{Pd(OAc)_2/Et_3N/(o\text{-tolyl})_3P}$ **44** (M_W = 13 789; PDI = 1.08)	[89]

(continued overleaf)

Table 4.4 (*Continued*)

3	**45** (R^1 = n-C_8H_{17} or 2-methylhexyl; R^2 = H– or CH_3O–) → $Pd(OAc)_3/P(o\text{-tol})_3$, Et_3N, 80 °C → **46** (regioregular PPV) ($Mn = 25 \times 10^3$, PDI = 2.5; EL at 563 nm)	[90]
4	Br, OC_6H_{13}, N–N, Br + **42** ($Pd(PPh_3)_4$; Br, $OC_{12}H_{25}$, Br, $C_{12}H_{25}O$ + $SnBu_3$) → $Pd(OAc)_2$, tri(o-tolyl) phosphine, DMF, Bu_3N → **47** (M_W = 15 600; PDI = 1.9)	[95]

reaction to construct the PV side chain in the monomer **48**, which is subsequently polymerized by using Heck coupling to build the PV main chain [96]. The synthetic sequence allows one to incorporate various PV fragments along the main chain, and the respective copolymers exhibit a broader absorption range (300–500 nm) than the corresponding homopolymers. The strategy is also used in the synthesis of hyperbranched PPV **51**, where the vinylene in the AB_2 type monomer **50** is obtained by using the Wittig reaction. The molecular weights of **51** increases with the reaction time, which exhibits slight influence on the fluorescence maxima (λ_{em} = 524–564 nm) of the film [97]. Similarly, the blue-emitting copolymer **52**, which bears oxadiazole and carbazole pendent groups, is synthesized by coupling of 1,4-divinylbenzene with the corresponding dibromide monomer [98]. Additional examples include attachment of PPV-based oligomer on the side chain of helical peptide scaffolds [99], and preparation of bipolar PPV bearing electron-donating triphenylamine and electron-accepting quinoxaline moieties [100].

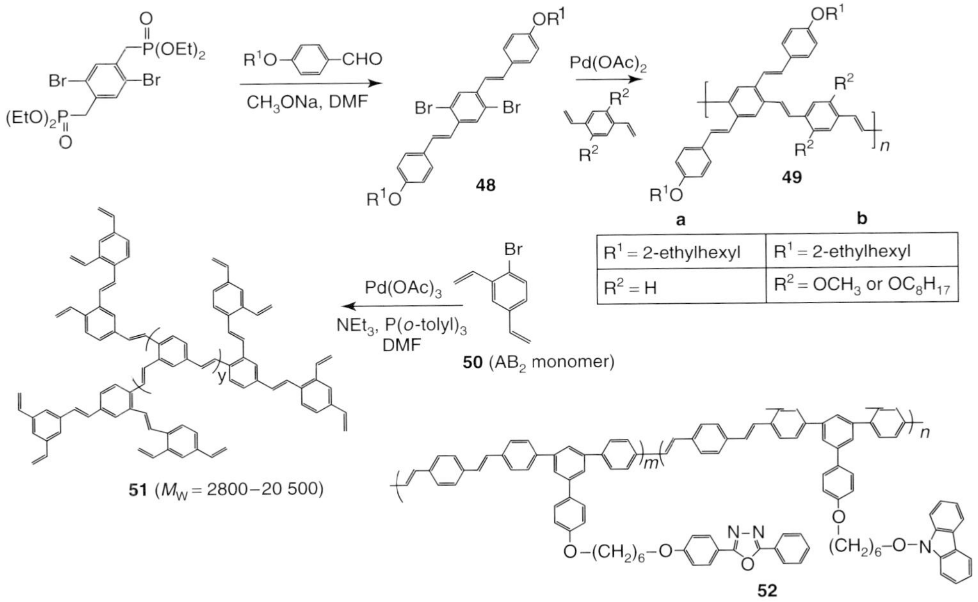

a	**b**
R^1 = 2-ethylhexyl	R^1 = 2-ethylhexyl
R^2 = H	R^2 = OCH_3 or OC_8H_{17}

50 (AB_2 monomer)

51 (M_W = 2800–20 500)

52

It should be noted that the Heck reaction is regioselective but not regiospecific, although *trans*-vinylene is usually the major product [101]. For example, the PPV **46** contains a mixture of *trans-/cis*-CH=CH (≈98 : 2), depending on the reaction conditions [90].

4.4.2
Stille and Suzuki Coupling

To eliminate *cis*-CH=CH structural unit, one may use the Stille reaction [102], whose application in PPV synthesis has been explored only in recent years. The

Stilly reaction refers to the coupling reaction between aryl halide (Ar–X, X=Br or I) and alkenyl stannanes in the presence of a palladium(0) catalyst. As an example, copolymers **56** have been prepared by reaction of aryl dibromides **53** and **54** with *trans*-1,2-bis(tributylstannyl)ethylene **55** [103]. The copolymer composition is consistent with the monomer feed ratio ($x:y = 100:0, 99.5:0.5, 99:1, 95:5, 85:15, 70:30$, and $50:50$). Presence of the low bandgap unit, 2,1,3-benzothiadiazole, systematically tunes the electroluminescence emission of copolymer **56** to longer wavelength from EL $\lambda_{max} = 591$ nm ($x:y = 100:0$, homopolymer) to EL $\lambda_{max} = 723$ nm ($x:y = 1:1$).

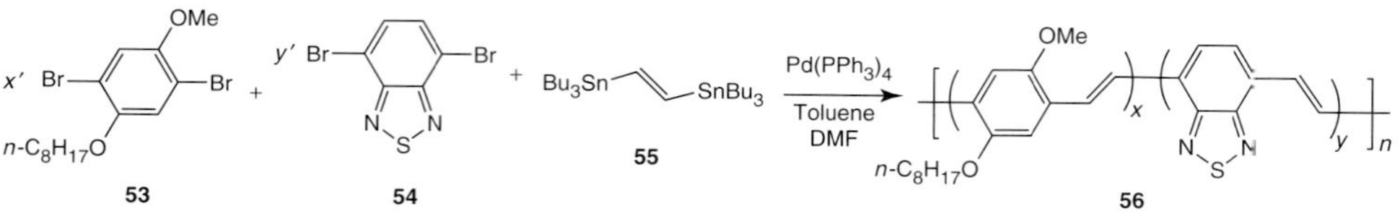

The synthesis of **59** is achieved by coupling the aryl dibromide **58** with *trans*-1,2-bis(tributylstannyl)ethylene [104]. The PV arms in the monomer **58** is obtained by bromination of 5,8-dibromo-2,3-dimethylquinoxaline (**57**), followed by the Wittig reaction. The polymer film **59** exhibits a broad absorption band (300–700 nm), a useful feature to harvest solar energy, is attributed to a combination of PPV backbone and donor–acceptor interaction in the PV-substituted quinoxaline arm.

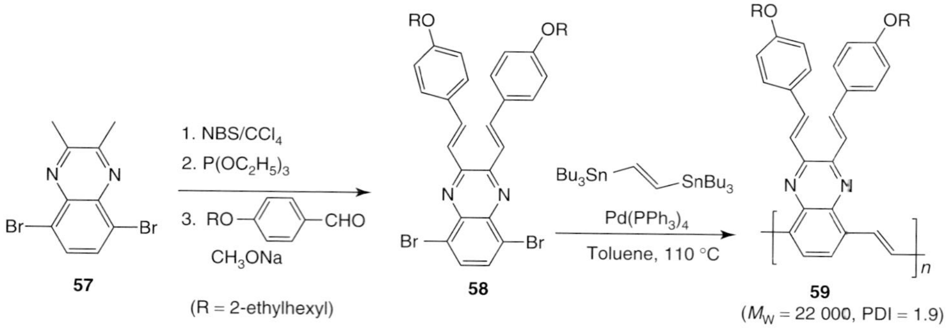

Coupling of 1,4-diiodo-2,3,5,6-tetrafluorobenzene with **55** provide highly fluorinated PPV **60** [105], which exhibits large refractive nonlinearity ($n_2 = -10 \times 10^{-12}\,\text{cm}^{-2}\cdot\text{W}^{-1}$). PPV derivatives with fluorinated vinylene units **63** are prepared similarly by reaction of diiodoarene with (*E*)-(1,2-difluoro-1,2- ethenediyl) bis(tributylstannane) **62** [106].

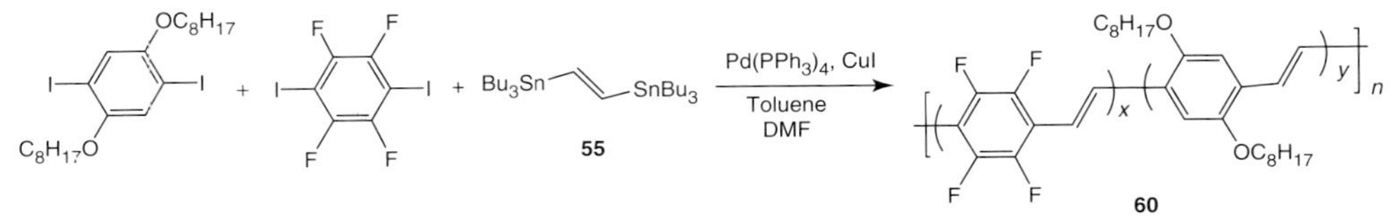

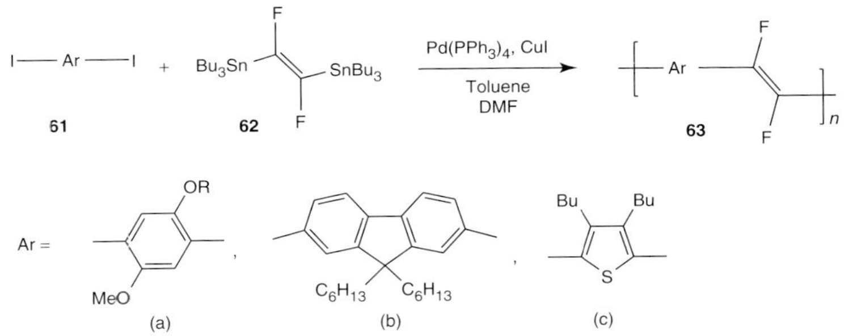

The Suzuki-coupling reaction [107] provides another useful tool in the synthesis of PPVs, due to its milder reaction conditions and easily accessible borane compounds. The reaction does not involve toxic tin, and can be used for all the aryl, heteroaryl, vinyl, benzyl, and other groups that Heck and Stille use. In addition, the Suzuki coupling is regiospecific, as the positions for coupling are clearly marked, one with a boronic acid and the other with a halide. Thus, the following reaction leads to poly(fluorenevinylene-*co*-phenylenevinylene) **66** [108], in which the vinylene bonds are in trans configuration as defined in the monomer **64**. Similarly, the *p*-terphenyl PPV derivative **69** is prepared by coupling of **68** with dibromide **67** [109]. Purification of **67** allows to control the *Z*-/*E*-vinylene ratio (from 5 : 95 to 94 : 6) in the polymer structure. The polymers emit 438–444 nm in solution, and 447–450 nm in film states. The fluorescence quantum yields of **69** do not appear to be affected significantly by the *Z*-/*E*-ratio of the vinylene bonds.

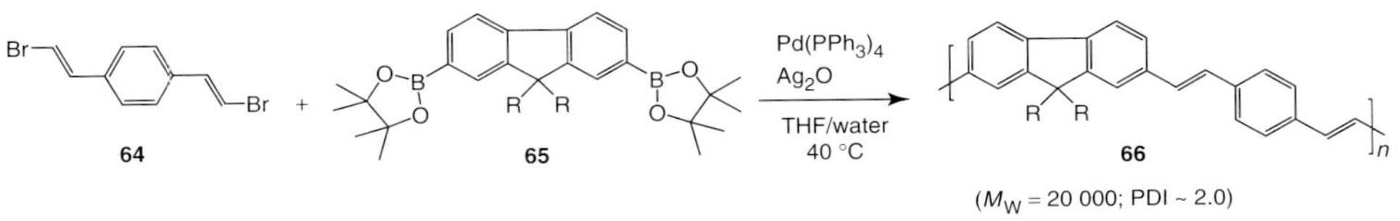

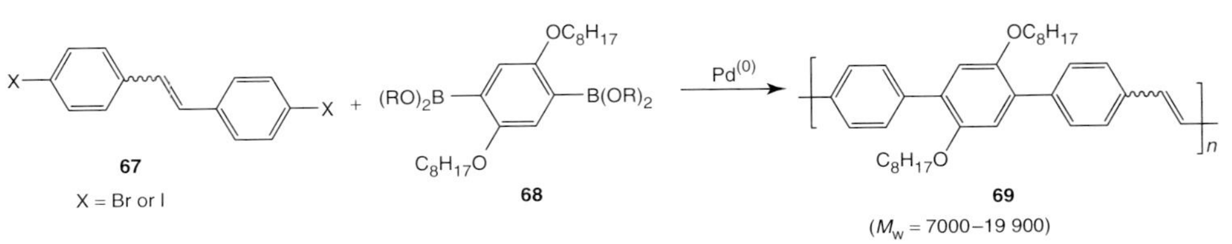

(M_W = 20 000; PDI ~ 2.0)

X = Br or I

(M_w = 7000–19 900)

In the synthesis of **71** [110], the cyano group is assembled in the crystalline monomer **70** via Knoevenagel condensation, which is then polymerized by using

Suzuki-coupling reaction. Resulting from the biphenyl linkage along the PPV backbone, the polymer film exhibits the photoluminescent (PL) emission peak at 480 nm, which is significantly blueshifted from that of CN–PPV (at 550 nm) [111]. The electrolumescence of the ITO/PPV/**71**/Al device gives a maximum peak at 491 nm, low turn-on voltage (2.5 V), and maximum brightness of about 5200 cd m^{-2} at 8 V. A notable feature for the Suzuki coupling, which has been used to prepare the PPV-threaded rotaxanes, is that the reaction can be carried out under aqueous conditions. [112].

4.5 Conclusion

Over the years, significant progress has been made in understanding the reaction mechanisms (such as in Gilch reaction), in controlling the vinylene bond geometry, in assembling functional groups, and in achieving specific nano structures in PPV materials. Synthesis of new, structurally defined PPVs will continue to play an essential role in achieving the processible materials with improved physical properties. To establish reliable structure–property correlation, synthetic chemists are constantly challenged by providing structural defect-free polymers with desirable properties. Demands of new functional materials with unique nano structures also require us to come up with new and intelligent approaches. Synthetic elimination of the structural disorder or impurity in the PPV materials, which is believed to prevent realization of optimum device performance, requires the chemists to continue exploring new synthetic tools and conditions.

References

1. Dini, D. (2005) *Chem. Mater.*, **17**, 1933–1945.
2. (2003) *Handbook of Luminescence, Display Materials, and Devices: Organic Light-emitting Diodes*, (eds H.S. Nalwa, L.S. Rohwer), American Scientific Publishers, Stevenson Ranch.
3. Kraft, A., Grimsdale, A.C., and Holmes, A.B. (1998) *Angew. Chem. Int. Ed. Engl.*, **37**, 402–428.
4. Akcelrud, L. (2003) *Prog. Polym. Sci.*, **28**, 875–962.
5. Segura, J.L., Martin, N., and Guldi, D.M. (2005) *Chem. Soc. Rev.*, **34**, 31–47.

6. Gunes, S., Neugebauer, H., and Sariciftci, N.S. (2007) *Chem. Rev.*, **107**, 1324–1338.
7. Thomas, S.W., Joly, G.D., and Swager, T.M. (2007) *Chem. Rev.*, **107**, 1339–1386.
8. Burroughes, J.H., Bradley, D.D.C., Brown, A.R., Marks, R.N., MacKay, K., Friend, R.H., Burn, P.L., and Holmes, A.B. (1990) *Nature*, **347**, 539–541.
9. Greenham, N.C. and Friend, R.H. (2004) in *Organic Light-emitting Devices: A Survey* (ed. J. Shinar.), Springer, New York, pp. 127–153.
10. Scherf, U. (1999) in *Carbon Rich Compounds II* (ed. A.D. Meijere), Springer, Berlin, pp. 163–222.
11. Denton, F.R. and Lahti, P.M. (1998) in *Photonic Polymer Systems: Fundamentals, Methods, and Applications* (eds D.L. Wise, G.E. Wnek, D.J. Trantolo, T.M. Cooper, and J.D. Gresser), Marcel Dekker, New York, pp. 61–102.
12. Gilch, H.G. and Wheelwright, W.L. (1966) *J. Polym. Sci., Part A: Polym. Chem.*, **4**, 1337–1346.
13. Wiesecke, J. and Rehahn, M. (2003) *Angew. Chem. Int. Ed. Engl.*, **42**, 567–570.
14. Schwalm, T. and Rehahn, M. (2007) *Macromolecules*, **40**, 3921–3928.
15. Wiesecke, J. and Rehahn, M. (2007) *Macromol. Rapid Commun.*, **28**, 78–83.
16. Wiesecke, J. and Rehahn, M. (2007) *Macromol. Rapid Commun.*, **28**, 188–193.
17. Schwalm, T. and Rehahn, M. (2008) *Macromol. Rapid Commun.*, **29**, 33–38.
18. Schwalm, T., Wiesecke, J., Immel, S., and Rehahn, M. (2007) *Macromolecules*, **40**, 8842–8854.
19. Hontis, L., Vrindts, V., Vanderzande, D., and Lutsen, L. (2003) *Macromolecules*, **36**, 3035–3044.
20. Yin, C. and Yang, C.-Z. (2001) *J. Appl. Polym. Sci.*, **82**, 263–268.
21. Odian, G. (2004) *Principles of Polymerization*, John Wiley & Sons, Ltd, pp. 74–80.
22. Schwalm, T. and Mattias, R. (2008) *Macromol. Rapid Commun.*, **29**, 207–213.
23. Becker, H., Spreitzer, H., Ibrom, K., and Kreuder, W. (1999) *Macromolecules*, **32**, 4925–4932.
24. Becker, H., Spreitzer, H., Kreuder, W., Kluge, E., Schenk, H., Parker, I., and Cao, Y. (2000) *Adv. Mater.*, **12**, 42–48.
25. Roex, H., Adriaensens, P., Vanderzande, D., and Gelan, J. (2003) *Macromolecules*, **36**, 5613–5622.
26. Johansson, D.M., Wang, X., Johansson, T., Inganas, O., Yu, G., Srdanov, G., and Andersson, M.R. (2002) *Macromolecules*, **35**, 4997–5003.
27. Chang, H.-T., Lee, H.-T., Chang, E.-C., and Yeh, M.-Y. (2007) *Polym. Eng. Sci.*, **47**, 1380–1387.
28. Jin, Y., Kim, J.Y., Song, S., Xia, Y., Kim, J., Woo, H.Y., Lee, K., and Suh, H. (2008) *Polymer*, **49**, 467–473.
29. Lee, J.-H. and Hwang, D.-H. (2008) *Synth. Met.*, **158**, 273–277.
30. Wang, H.L., Sun, Y.M., Qi, Z.J., Kong, F., Ha, Y.Q., Yin, S.G., and Lin, S. (2008) *Macromolecules*, **41**, 3537–3542.
31. Hsieh, B.R., Yu, Y., Forsythe, E.W., Schaaf, G.M., and Feld, W.A. (1998) *J. Am. Chem. Soc.*, **120**, 231–232.
32. Jin, Y., Song, S., Park, S.H., Park, J.-A., Kim, J., Woo, H.Y., Lee, K., and Suh, H. (2008) *Polymer*, **49**, 4559–4568.
33. Feng, J.C., Zhang, Q.H., Li, W., Li, Y., Yang, M.J., and Cao, Y. (2008) *J. Appl. Polym. Sci.*, **109**, 2283–2290.
34. Amrutha, S.R. and Jayakannan, M. (2007) *Macromolecules*, **40**, 2380–2391.
35. Hsieh, B.R., Yu, Y., VanLaeken, A.C., and Lee, H. (1997) *Macromolecules*, **30**, 8094–8095.
36. Lin, K.-F., Fan, Y.-L., and Chow, H.-L. (2006) *Polym. Int.*, **55**, 938–944.
37. Chang, H.-T., Chang, E.-C., Yeh, M.-Y., and Lee, H.-T. (2007) *J. Macromol. Sci., Part A: Pure Appl. Chem.*, **44**, 831–837.
38. Kang, J.-M., Cho, H.-J., Lee, J., Lee, J., Lee, S.-K., Cho, N.-S., Hwang, D.-H., and Shim, H.-K. (2006) *Macromolecules*, **39**, 4999–5008.
39. Jin, Y., Jee, J., Kim, K., Kim, J., Song, S., Park, S.H., Lee, K., and Suh, H. (2007) *Polymer*, **48**, 1541–1549.
40. Tan, Z., Tang, R., Xi, F., and Li, Y.F. (2007) *Polym. Adv. Technol.*, **18**, 963–970.

41. Kim, S.-C., Park, S.-M., Park, J.S., Lee, S.J., and Jin, S.-H. (2007) *J. Polym. Sci., Part A: Polym. Chem.*, **46**, 1098–1110.
42. Kimura, M., Sato, M., Adachi, N., Fukawa, T., Kanbe, E., and Shirai, H. (2007) *Chem. Mater.*, **19**, 2809–2815.
43. Wudl, F. and Srdanov, G. (1993) US Patent 5 189 136.
44. Wessling, R.A. (1985) *J. Polym. Sci., Polym. Symp.*, **72**, 55–66.
45. McCoy, R.K., Karasz, F.E., Sarker, A., and Lahti, P.M. (1991) *Chem. Mater.*, **3**, 941–947.
46. Jang, M.-S., Song, S.-Y., and Shim, H.-K. (2000) *Polymer*, **41**, 5675–5679.
47. Denton, F.R., Lahti, P.M., and Karasz, F.E. (1992) *J. Polym. Sci., Part A: Polym. Chem.*, **30**, 2223–2231.
48. Kok, M.M., Nguyen, T.P., Molinie, P., van Breemen, A.J.J.M., Vanderzande, D.J., and Gelan, J.M. (1999) *Synth. Met.*, **102**, 949–950.
49. Yabu, H., Tajima, A., Higuchi, T., and Shimomura, M. (2008) *Chem. Commun.*, 4588–4589.
50. Lee, J.W., Samal, S., Selvapalam, N., Kim, H.J., and Kim, K. (2003) *Acc. Chem. Res.*, **36**, 621–630.
51. Corma, A., Garcia, H., and Montes-Navajas, P. (2007) *Tetrahedron Lett.*, **48**, 4613–4617.
52. Ling, Y. and Kaifer, A.E. (2006) *Chem. Mater.*, **18**, 5944–5949.
53. Henckens, A., Duyssens, I., Lutsen, L., Vanderzande, D., and Cleij, T.J. (2006) *Polymer*, **47**, 123–131.
54. Padmanaban, G., Nagesh, K., and Bamakrishnan, S. (2003) *J. Polym. Sci., Part A: Polym. Chem.*, **41**, 3929–3940.
55. Lawrence, N.J. (1996) in *Preparation of Alkenes: A Practical Approach* (ed. J.M.J. Williams), Oxford University Press, Oxford, pp. 19–58.
56. Maryanoff, B.E. and Reitz, A.B. (1989) *Chem. Rev.*, **89**, 863–927.
57. Li, H., Wang, L., Jing, X., and Wang, F. (2004) *Tetrahedron Lett.*, **45**, 2823–2826.
58. Liao, L., Pang, Y., Ding, L., and Karasz, F.E. (2003) *J. Polym. Sci., Part A: Polym. Chem.*, **41**, 2650–2658.
59. Liao, L., Pang, Y., Ding, L., and Karasz, F.E. (2002) *Macromolecules*, **35**, 6055–6059.
60. Pang, Y., Li, J., Hu, B., and Karasz, F.E. (1999) *Macromolecules*, **32**, 3946–3950.
61. Drury, A., Maier, S., Rüther, M., and Blau, W.J. (2003) *J. Mater. Chem.*, **13**, 485–490.
62. Liao, L., Pang, Y., Ding, L., and Karasz, F.E. (2001) *Macromolecules*, **34**, 7300–7305.
63. Liao, L., Pang, Y., Ding, L., Karasz, F.E., Smith, P.R., and Meador, M.A. (2004) *J. Polym. Sci., Part A: Polym. Chem.*, **42**, 5853–5862.
64. Stalmach, U., Detert, H., Meier, H., Gebhardt, V., Haarer, D., Bacher, A., and Schmidt, H.W. (1998) *Opt. Mater.*, **9**, 77–81.
65. Liao, L., Pang, Y., Ding, L., and Karasz, F.E. (2004) *Macromolecules*, **37**, 3970–3972.
66. Cyriac, A., Amrutha, S.R., and Jayakannan, M. (2008) *J. Polym. Sci., Part A: Polym. Chem.*, **46**, 3241–3256.
67. Yang, Z., Sokolik, I., and Karasz, F.E. (1993) *Macromolecules*, **26**, 1188–1190.
68. Peres, L.O., Fernandes, M.R., Garcia, J.R., Wang, S.H., and Nart, F.C. (2006) *Synth. Met.*, **156**, 529–536.
69. Suzuki, Y., Hashimoto, K., and Tajima, K. (2007) *Macromolecules*, **40**, 6521–6528.
70. Paik, K.L., Baek, N.S., Kim, H.K., Lee, J.-H., and Lee, Y. (2002) *Macromolecules*, **35**, 6782–6791.
71. Cho, H.J., Hwang, D.H., Lee, J.D., Cho, N.S., Lee, S.K., Lee, J., Jung, Y.K., and Shim, H.K. (2008) *J. Polym. Sci., Part A: Polym. Chem.*, **46**, 979–988.
72. Jaballah, N., Majdoub, M., Fave, J.L., Barthou, C., Jouini, M., and Tanguy, J. (2008) *Eur. Polym. J.*, **44**, 2886–2892.
73. Wu, T.Y., Sheu, R.B., and Chen, Y. (2004) *Macromolecules*, **37**, 725–733.
74. Son, S., Dodabalapur, A., Lovinger, A.J., and Galvin, M.E. (1995) *Science*, **269**, 376–378.
75. Liao, L. and Pang, Y. (2004) *Synth. Met.*, **144**, 271–277.
76. Liao, L., Pang, Y., Ding, L., and Karasz, F.E. (2005) *Thin Solid Films*, **479**, 249–253.
77. Liao, L., Ding, L., Karasz, F.E., and Pang, Y. (2004) *J. Polym. Sci., Part A: Polym. Chem.*, **42**, 303–316

78. Liao, L., Pang, Y., Ding, L., and Karasz, F.E. (2001) *Macromolecules*, **34**, 6756–6760.
79. Fan, Q.L., Lu, S., Lai, Y.H., Hou, X.Y., and Huang, W. (2003) *Macromolecules*, **36**, 6976–6984.
80. Katayama, H., Nagao, M., Nishimura, T., Matsui, Y., Umeda, K., Akamatsu, K., Tsuruoka, T., Nawafune, H., and Ozawa, F. (2005) *J. Am. Chem. Soc.*, **127**, 4350–4353.
81. Katayama, H., Nagao, M., Nashimura, T., Matsui, Y., Fukuse, Y., Wakioka, M., and Ozawa, F. (2006) *Macromolecules*, **39**, 2039–2048.
82. Wang, F.F., He, F., Xie, Z.Q., Li, M., Hanif, M., Gu, X., Yang, B., Zhang, H.Y., Lu, P., and Ma, Y.G. (2008) *J. Polym. Sci., Part A: Polym. Chem.*, **46**, 5242–5250.
83. Wang, F.F., He, F., Xie, Z.Q., Li, Y.P., Hanif, M., Li, M., and Ma, Y.G. (2008) *Macromol. Chem. Phys.*, **209**, 1381–1388.
84. Li, Y., Xu, H., Wu, L., He, F., Shen, F., Liu, L., Yang, B., and Ma, Y. (2008) *J. Polym. Sci., Part B: Polym. Phys.*, **46**, 1105–1113.
85. Liang, F., Pu, Y.-J., Kurata, T., Kido, J., and Nishide, H. (2005) *Polymer*, **46**, 3767–3775.
86. Egbe, D.A.M., Kietzke, T., Carbonnier, B., Muhlbacher, D., Horhold, H.H., Neher, D., and Pakula, T. (2004) *Macromolecules*, **37**, 8863–8873.
87. Beletskaya, I.P. and Cheprakov, A.V. (2000) *Chem. Rev.*, **100**, 3009–3066.
88. Nicolaou, K.C., Bulger, P.G., and Sarlah, D. (2005) *Angew. Chem. Int. Ed. Engl.*, **44**, 4442–4489.
89. Li, H., Li, Y., Zhai, J., Cui, G., Liu, H., Xiao, S., Liu, Y., Lu, F., Jiang, L., and Zhu, D. (2003) *Chem. Eur. J.*, **9**, 6031–6038.
90. Hou, J., Fan, B., Huo, C., He, C., Yang, C., and Li, Y. (2006) *J. Polym. Sci., Part A: Polym. Chem.*, **44**, 1279–1290.
91. Kim, J.H. and Lee, H. (2002) *Chem. Mater.*, **14**, 2270–2275.
92. Kim, J.H., Park, J.H., and Lee, H. (2003) *Chem. Mater.* **15**, 3414–3416.
93. Barberis, V.P. and Mikroyannidis, J.A. (2006) *J. Polym. Sci., Part A: Polym. Chem.*, **44**, 3556–3566.
94. Kim, J.H. and Lee, H.S. (2007) *Synth. Met.*, **157**, 1040–1045.
95. Peng, Q., Li, M., Tang, X., Lu, S., Peng, J., and Cao, Y. (2007) *J. Polym. Sci., Part A: Polym. Chem.*, **45**, 1632–1640.
96. Huo, L.J., Hou, J.H., Zhou, Y., Han, M.F., and Li, Y.F. (2008) *J. Appl. Polym. Sci.*, **110**, 1002–1008.
97. Lim, S.J., Seok, D.Y., An, B.K., Jung, S.D., and Park, S.Y. (2006) *Macromolecules*, **39**, 9–11.
98. Mikroyannidis, J.A. (2006) *J. Appl. Polym. Sci.*, **101**, 3842–3849.
99. Kas, O.Y., Charati, M.B., Rothberg, L.J., Galvin, M.E., and Kiick, K.L. (2008) *J. Mater. Chem.*, **18**, 3847–3854.
100. Karastatiris, P., Mikroyannidis, J.A., and Spiliopoulos, I.K. (2008) *J. Polym. Sci., Part A: Polym. Chem.*, **46**, 2367–2378.
101. Martelock, H., Greiner, A., and Heitz, W. (1991) *Makromol. Chem.*, **192**, 967–979.
102. Mitchell, T.N. (1992) *Synthesis*, **1992**, 803–815.
103. Li, X., Zhang, Y., Yang, R., Huang, J., Yang, W., and Cao, Y. (2005) *J. Polym. Sci., Part A: Polym. Chem.*, **43**, 2325–2336.
104. Huo, L.J., Tan, Z.A., Zhou, Y., Zhou, E.J., Han, M.F., and Li, Y.F. (2007) *Macromol. Chem. Phys.*, **208**, 1294–1300.
105. Babudri, F., Cardone, A., Farinola, G.M., Naso, F., Cassano, T., Chiavarona, L., and Rommasi, R. (2003) *Macromol. Chem. Phys.*, **204**, 1621–1627.
106. Babudri, F., Cardone, A., Farinola, G.M., Martinelli, C., Mendichi, R., Naso, F., and Striccoli, M. (2008) *Eur. J. Org. Chem.*, 1977–1982.
107. Miyaura, N. and Suzuki, A. (1995) *Chem. Rev.*, **95**, 2457–2483.
108. Lopez, L.C., Strohriegl, P., and Stübinger, T. (2002) *Macromol. Chem. Phys.*, **203**, 1926–1930.
109. Babudri, F., Cardone, A., Cassano, T., Farinola, G.M., Naso, F., and

Tommasi, R. (2008) *J. Organomet. Chem.*, **693**, 2631–2636.

110. Kim, Y.-H., Shin, D.-C., and Kwon, S.-K. (2005) *Polymer*, **46**, 4647–4653.

111. Greenham, N.C., Moratti, S.C., Bradley, D.D.C., Friend, R.H., and Holmes, A.B. (1993) *Nature*, **365**, 628–630.

112. Terao, J., Tang, A., Michela, J J., Krivokapic, A., and Anderson, H.L. (2004) *Chem. Commun.*, 56–57.

5
Poly(aryleneethynylene)s

Brett VanVeller and Timothy M. Swager

5.1
Introduction

Poly(aryleneethynylene)s (PAEs) are a class of conjugated polymers which are structurally similar to poly(arylenevinylene)s (PAVs) except that triple bonds exist in place of the double bonds. PAEs are robust, rigid materials that are generally luminescent in both solution and solid state. The spectral and electronic properties of the polymer, such as HOMO and LUMO energy levels and hence bandgap, can be tuned by the identity of the arylene unit or through attachment of side chains. This structural flexibility has allowed for application of PAEs in sensing materials [1–3], supramolecular assemblies [4], and molecular wires [5, 6]. While this chapter mainly discusses PAEs in the context of phenylene as the arylene unit, the arylene's identity can be diverse, and thus, a large library of structurally distinct PAE polymers is possible. This chapter gives an overview of some current synthetic methodologies and topologies of PAEs, and connections between synthesis and polymer properties are established where appropriate. In addition, a brief mention of macrocycle synthesis will be included to illustrate the strengths and weaknesses of palladium and metathesis methods. Finally, before we begin, it should be noted that two reviews were published in the early part of this decade comprising extensive tables of various monomer scaffolds and their associated polymerization conditions [7, 8]. This invaluable cache of information should not be overlooked.

5.2
Palladium-Catalyzed Polymerizations

The most popular method for the synthesis of PAEs makes use of the Sonogashira reaction (Eq. (5.1)) [7–15]. The reaction involves a palladium-mediated cross-coupling between an sp^2-hybridized carbon (**1**) and an sp-hybridized carbon (**2**) to create a single bond and a conjugated bridge. Its popularity can be attributed to its mild reaction conditions, remarkable functional group tolerance, and high,

Design and Synthesis of Conjugated Polymers. Edited by Mario Leclerc and Jean-François Morin

ISBN: 978-3-527-32474-3

often quantitative, yields. The reaction is catalytic in palladium and copper and requires a stoichiometric equivalent of base.

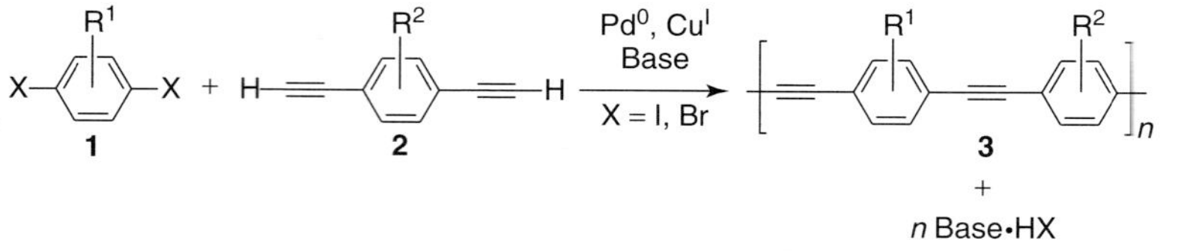

$$X\text{-Ar}(R^1)\text{-}X\ (\mathbf{1}) + H\text{-}\equiv\text{-Ar}(R^2)\text{-}\equiv\text{-}H\ (\mathbf{2}) \xrightarrow[X = I, Br]{Pd^0, Cu^I, Base} [\equiv\text{-Ar}(R^1)\text{-}\equiv\text{-Ar}(R^2)]_n\ (\mathbf{3}) + n\ Base{\cdot}HX \quad (5.1)$$

The mechanism of the reaction, following Scheme 5.1, begins with a Pd^0 species undergoing oxidative addition to the aryl–X bond of **B** to give **C**. Transmetallation with the putative Cu^I acetylide **D** leads to **G**, which undergoes reductive elimination to yield the product, **H**, and regenerating **A**.

5.2.1
The Palladium, A

A palladium catalyst loading of 1–5% is typically sufficient, and a variety of palladium sources have been used in PAE synthesis. The most common are $Pd(PPh_3)_4$, $(PPh_3)_2PdCl_2$, and Pd_2dba_3 plus added phosphine. While Pd^{II} precatalysts are desirable because of their ease of handling, they will consume two equivalents of alkyne to generate **A**, producing diyne **I** according to Scheme 5.2. During small molecule synthesis this is not a problem; however, these diyne linkages will be incorporated into the polymer. Theoretically, using a Pd^0 catalyst at the outset should prevent these occurrences, but even trace amounts of

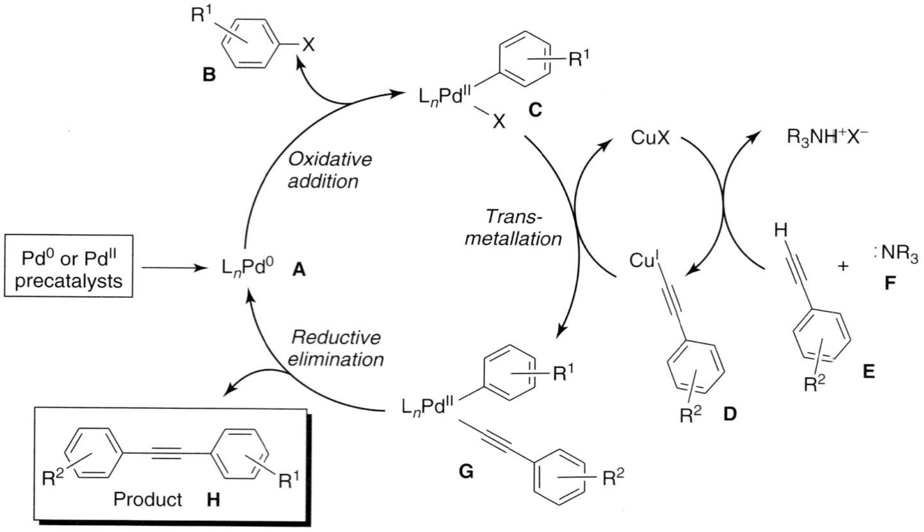

Scheme 5.1 Catalytic cycle.

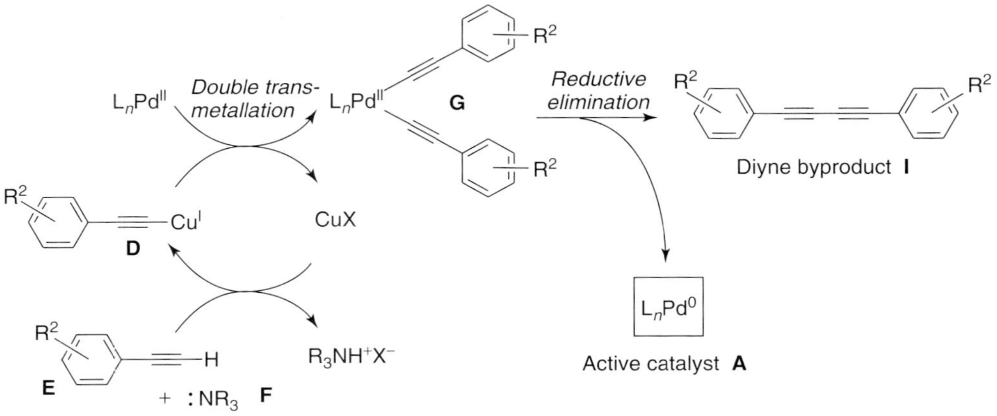

Scheme 5.2 Catalyst activation and/or reactivation.

oxygen are able to oxidize **A**. Additionally, Cu^{II} impurities in the CuI cocatalyst may also effect palladium oxidation. Thus, small diyne impurities often result, and the highest molecular weights are obtained when a small excess (small percentage) of the bis-alkyne monomer is employed to compensate for this side reaction [16]. By far, the most popular palladium ligand used is triphenylphosphine; however, some reports indicate the use of $^t Bu_3P$ and $(furyl)_3P$ (*vide infra*). The last decade has witnessed many advances in phosphine ligands for the improvement of cross-couplings [9], but these ligands have not found use in PAE synthesis.

5.2.2
The Aryl Halide, B

The identity of X can play an important role in terms of reaction rate. Aryl iodides undergo oxidative addition much faster than the corresponding bromides; however, both halides have proven successful for polymerizations. The bromides often require higher reaction temperatures (>80 °C), whereas the iodides can react at room temperature. This makes iodides a preferred choice as the milder conditions can limit side reactions (see Section 5.8).

5.2.3
The Amine, F, Solvent, and Copper

The choice of amine for these couplings can have a big effect; however, it is sometimes difficult to predict the best amine for the reaction. With aryl iodides, a good option to start with is diisopropylamine, but pyrrolidine, piperidine, and morpholine have all found use. Triethylamine or Hunig's base appears to be a suitable base for aryl bromides [8]. The reaction can be run in neat amine or a suitable cosolvent (THF, Toluene, DMF, DCM, etc.). Again there are no hard and fast rules and often some degree of experimentation is wise. The copper cocatalyst, CuI, is believed to facilitate alkyne deprotonation [17]. Reaction progress is not harmed by its presence and almost all polymerization reactions make use of it.

5.2.4
Substituents, R^1 and R^2

The substituent(s) on the aryl ring (Scheme 5.1, **B**–$\mathbf{R^1}$) can greatly affect the reaction rate. Generation of intermediate **C** requires oxidation of **A**; thus, the more electrophilic (or electron poor) the Ar–X bond, the more facile the oxidative addition. In general, electron-withdrawing substituents on **B** increase the rate of the reaction and hence molecular weight substantially [16], relative to electron-donating groups like alkoxyalkanes. The substituents on the bis-alkyne **E** do not appear to have as great an effect.

Ring substituents are extremely important for polymer solubility. PAEs are like rigid rods and are poorly soluble [15] unless flexible side chains are present to entropically drive solubility. Alternatively, ionic or hydrogen-bonding groups that exhibit strong enthalpic interactions with solvent (e.g., water) can also be used. Further, the polymer's spectral and electronic properties can depend greatly on substituent identity. For example, a PAE bandgap depends on the energy difference between its HOMO and its LUMO. Electron-donating substituents act to raise the HOMO level relative to the LUMO, making the polymer more prone to oxidation, while electron-withdrawing groups lower the LUMO relative to the HOMO, making reduction more facile. The result in both situations is a decrease in the band-gap energy and generally a redshift in absorbance.

5.3
Different Palladium Schemes

Numerous examples of PAE polymerizations employ monomers with the reactivity pattern shown in Eq. (5.1), where a monomer with functional group pattern A–A reacts with another monomer B–B. This reactivity scheme is quite common and leads to alternating copolymers; however several variations have been reported.

5.3.1
A–B Monomers

The use of an A–B type monomer (Eq. (5.2)) involves homopolymerization of a single monomer [18, 19]. It is not evident that this approach yields polymers with improved properties (i.e., MW, PDI, etc.), and this may explain why this variation has found less application. However, one advantage of the A–B scheme is the direct control over head (h) to tail (t) coupling.

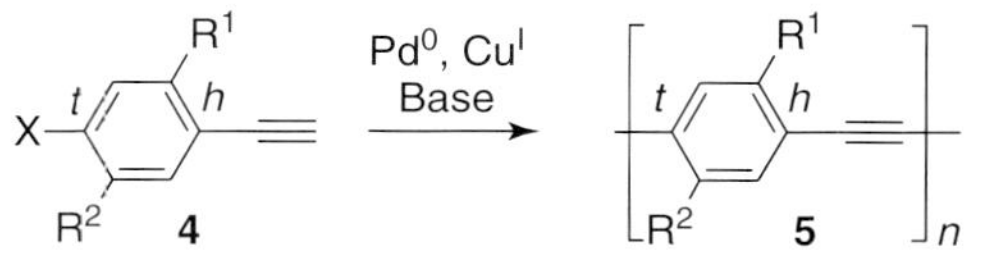

(5.2)

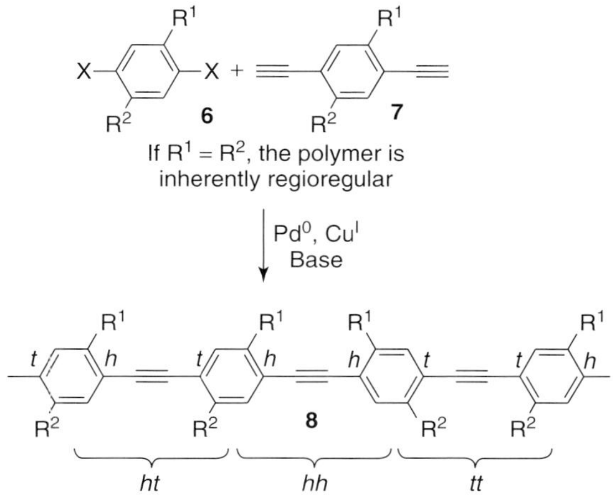

If $R^1 = R^2$, the polymer is inherently regioregular

(5.3)

Unless desired for specific applications [20–22], most examples of PAEs employ symmetrical side chains ($R^1 = R^2$), so the polymer is inherently regioregular (i.e., only *ht*) because of C_2 symmetry across the monomer (Eq. (5.3)). Alternatively, if a monomer has dissimilar side chains ($R^1 \neq R^2$), then polymerization using an A–A/B–B scheme will introduce regio-isomerism along the polymer backbone (i.e., Eq. (5.3), mixture of *ht*, *hh*, and *tt*). However, when employing a monomer bearing an A–B reactivity pattern and dissimilar side chains, regioregularity is necessitated. Recently, this effect was probed by Collard and coworkers using dissimilar linear alkoxy side chains and found to influence solid-state electronic structure and molecular packing [23]. It still remains to be seen whether regioregularity will be as important to PAEs as it has been to polythiophene (refer to Chapter 3). Finally, the A–B reactivity pattern may still suffer regioisomeric defects due to alkyne homocoupling detailed in Scheme 5.2.

5.3.2
Acetylene as a Monomer

Dihalo arenes can also be polymerized directly with acetylene gas, and the best results have been reported by Bunz and coworkers using low catalyst loadings of $(PPh_3)_2PdCl_2$ (Eq. (5.4)) [24]. The resultant polymers can be isolated in excellent yield and purity with degree of polymerization (DP) as high as 316 for R = ethylhexyloxy.

9

Toluene/piperidine
R = alkyl, alkyloxy

$Pd(PPh_3)_2Cl_2$ (0.1–0.2%)
CuI (0.4%)

10

a R = ethylhexyl
b R = octyl
c R = hexyloxy
d R = ethylhexyloxy

(5.4)

Several uses of protected acetylene have also been reported where the protecting group is removed by *in situ* hydrolysis (Eq. (5.5)) [25, 26]. The advantage of this strategy is that **12** is a liquid and easier to handle than a gas reagent; however, it is much more expensive than acetylene gas. The results show that acetylene deprotection is rate limiting and that coupling of the iodide and acetylene occurs initially, followed by deprotection. This finding has been used to limit the occurrence of diyne defects [27].

11 12 (TMS)

H_2O, DBU, solvent | $Pd(PPh_3)_4CuI$

13

(5.5)

5.4
Ortho and *Meta* PAEs

The polymerization of *ortho*- and *meta*-substituted dihalo and bis-alkynyl arenes produces "*ortho*-linked" (*o*-) and "*meta*-linked" (*m*-) PAEs (Figure 5.1). The polymerization procedures are very similar to those of *para*-PAEs, but *o,m*-PAEs are considerably more soluble because the chain now contains kinks. Consequently, these materials are not as rigid and have a blue-shifted emission relative to *p*-PAEs.

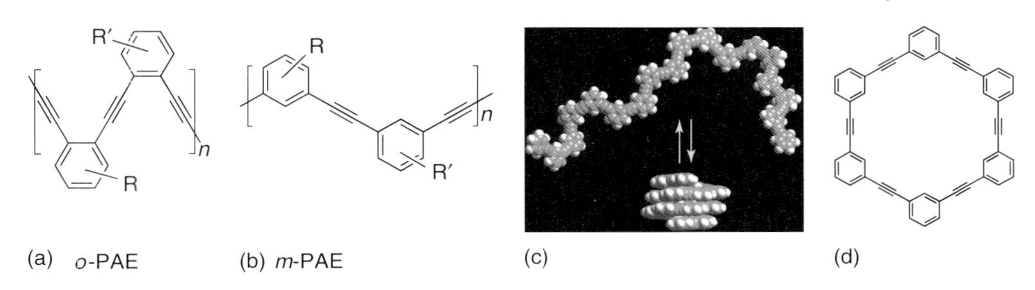

Figure 5.1 (a) *ortho*-linked PAE, (b) *meta*-linked PAE, (c) *meta*-linked PAE foldamer, and (d) *m*-PAE macrocycle.

In addition, the molecular weights are usually not as high as those for *p*-PAEs, and special consideration must sometimes be given to monomer synthesis in order to achieve the desired substitution pattern.

One application of *o,m*-PAEs has been in the area of foldamers [28–30], polymers that have a strong tendency to adopt a specific compact conformation. In the case of *o,m*-PAEs, the main chain is able to twist into a helix (Figure 5.1c) and this structure has been investigated in the context of supramolecular organization [31–33]. Solvophobic interactions drive helix formation, leading to a dynamic structure in which stability is controlled by solvent composition, temperature, oligomer length, and secondary interactions [30, 34–37]. Further, the chirality of the helix can be controlled by chiral side chains [38, 39] or wrapping around chiral guest molecules [40–42]. Moore investigated the well-defined core of the foldamer as a "reactive sieve" [43], similar to the amino acid sieving mechanism of tRNA synthetase enzyme [44, 45], where the reactivity of substrates is differentiated through size discrimination. A pyridine unit was placed in the middle of a *meta*-PAE oligomer and hence in the middle of the folded structure (**14**) [46]. Methylating agents (**15**) of different sizes and shapes were tested and evaluated according to the rate of the reaction (Scheme 5.3). Unfortunately, no sieving was observed as all substrates underwent reaction, perhaps as a result of the flexible nature of the foldamer. However, a wide variety of rate enhancements were observed (45- to 1600-fold increase in rate relative to a reference trimer), and the system was able to differentiate subtle differences in substrate structure, perhaps highlighting the possibility of supramolecular catalysts possessing substrate recognition ability.

5.5 Macrocycles: an Introduction

Conceivably, it should be possible to make macrocycles based on *ortho*- and *meta*-substituted monomers (Figure 5.1d). In fact, macrocycles are commonly encountered as byproducts during the polymerization reaction. The yields of such

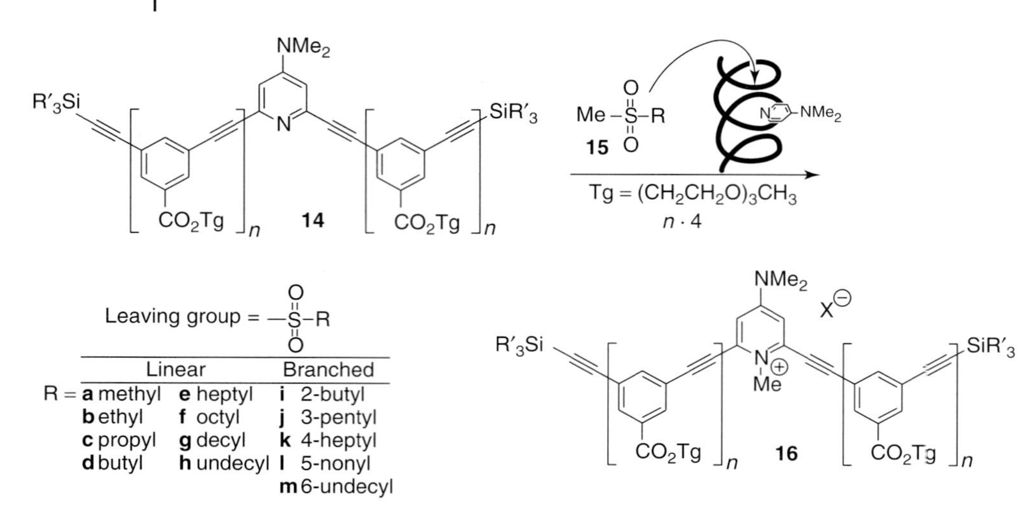

Scheme 5.3 Reactive sieving of a *m*-PAE foldamer.

materials using the Pd methods discussed are low, owing to oligomer lengthening that results in "overshooting" the length required to form the macrocycle. The yield therefore suffers as these oligomers, too long to form the desired cycle, cannot be shortened and are effectively wasted. Typically, yields rarely exceed 40% unless some form of template effect is present; further, yield decreases quickly with increasing cycle size. Section 5.10 discusses how alkyne metathesis has been brought to bear on this problem and how it is exquisitely adept at overcoming these shortfalls. However, Pd methods are still used and yields have been improved by using monodisperse precursors (Section 5.6) [47–54]. Macrocycles are of great interest because of their stacking behavior and ability to form porous supramolecular aggregates.

5.6
Synthesis of Monodisperse and Sequence-Specific PAEs

Monodisperse *o,m,p*-PAE oligomers with sequence-defined arylene units have received great attention [5, 6, 55, 56] for application in organic molecular wires [57–59], nonlinear optical materials [60–62], improving macrocycle yields [63], and dendrimer synthesis [51, 52]. Further, it has allowed for the careful study of chain-length dependence relationships for understanding foldamer behavior [36]. Monodisperse PAEs can be synthesized using solution or solid support methods.

Numerous variations of solution phase syntheses of such molecules have been reported. Strategies can involve (i) an excess of either the dihalo or bis-alkynyl moiety relative to the other [64], (ii) rate discrimination between two reactive aryl halides [65], or (iii) orthogonal protecting groups in an iterative deprotection–coupling sequence [58]. Eq. (5.6) shows a generalized example of a deprotection–activation–Sonogashira coupling triad used by both Gong [37] and Tew [66]. The iodine atom can be masked as the diethyltriazene [67] while the TMS-protected acetylene can be easily revealed with either TBAF or basic methanol [68]. Iterations of the triad shown will lengthen the oligomer by one unit. A significant disadvantage of solution methods is that each iteration often requires chromatographic separation of the desired product, meaning that long oligomers require substantial time investment. For this reason solid-phase techniques have largely superseded solution methods.

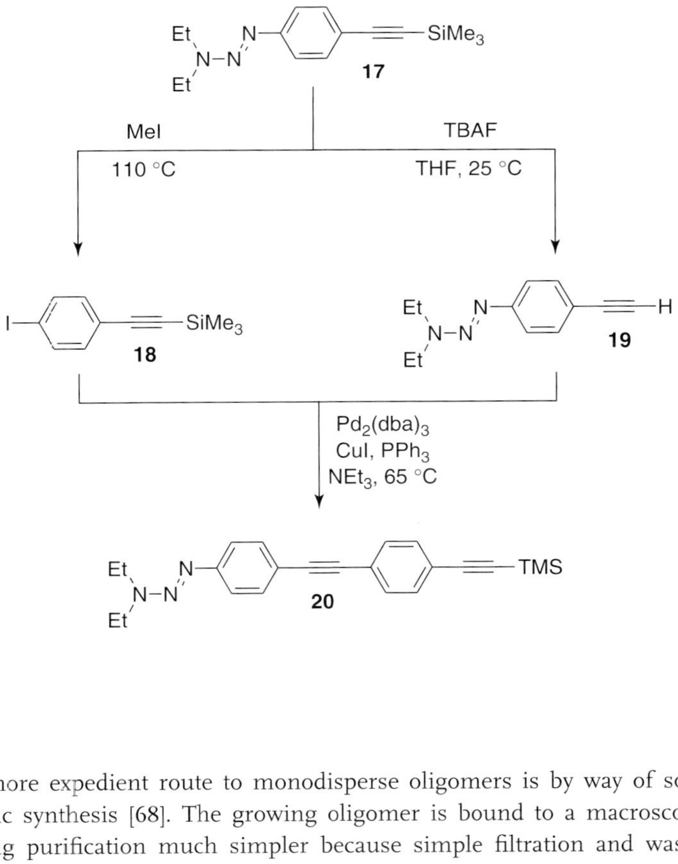

(5.6)

A more expedient route to monodisperse oligomers is by way of solid-phase organic synthesis [68]. The growing oligomer is bound to a macroscopic bead, making purification much simpler because simple filtration and washing can remove the excess reagents and column chromatography is avoided. Despite this ease of purification, lengthy reaction times still prolong the process. Moore developed a solid-phase procedure where addition of a subsequent monomer could be completed in 2 hours at room temperature [69].

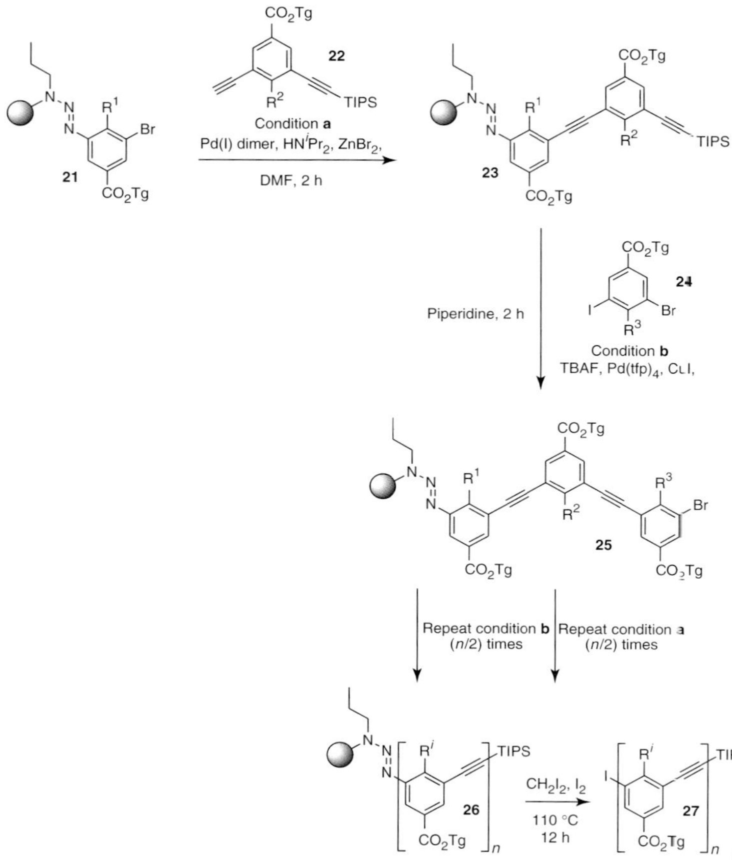

(5.7)

Bound to the solid-phase support by a triazene linker, **21** is coupled to **22** in the usual way. After appropriate washing, chain extension is achieved by *in situ* deprotection using TBAF and subsequent cross-coupling of **23** and **24** using the iodide-selective catalyst shown. These steps are alternated to lengthen the oligomer, which can be cleaved from the resin (**26**→**27**). The catalyst systems employed here can generate products much faster than traditional solid-phase methods (hexamer in 48 hours as opposed to six days).

Recently, Sleiman and Lo communicated an interesting nucleobase-templated polymerization to yield products that – while not strictly mouodiopere – were of low polydispersity (Eq. (5.8)) [70]. Ring-opening metathesis polymerization (ROMP) was used to synthesize thymine-containing **28**, with narrowly defined molecular weight. The templated synthesis was achieved by alignment of the complementary adenine-bound monomer **29**, and the resulting polymer **30** was found to have PDIs and DP similar to template **28**. Further, the molecular weight of

30 was much higher than control polymerizations run in the absence of template, suggesting the template's ability for programming polymer length and structure.

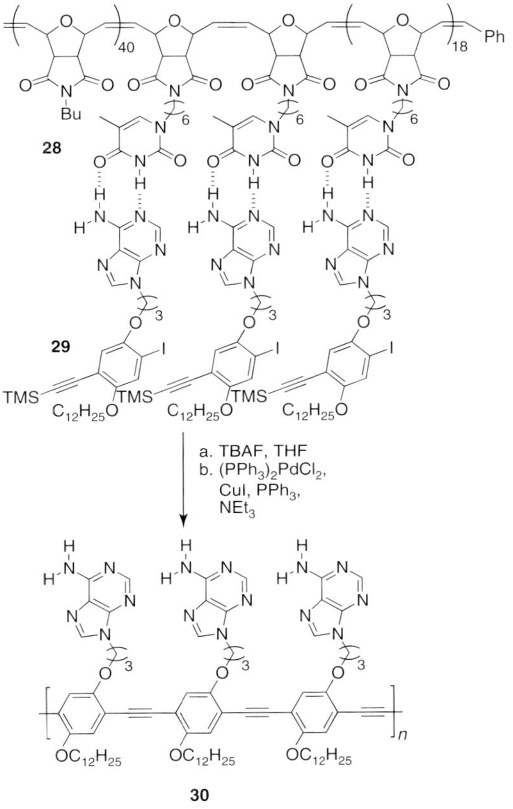

(5.8)

5.7
Synthesis of Poly(phenylenebutadiynylenes) PPBs

The synthesis of a related polymer called poly(arylenebutadiynylene) (PAB) [71] makes use of the facile way in which butadiynes (**I**) are formed in Scheme 5.2. Scheme 5.4 shows the mechanism of how these butadiynes can be applied to polymer synthesis. The reaction uses Pd^{II}, **J**, as the active catalyst instead of Pd^0 and begins with a double transmetallation step to form **G** followed by reductive elimination to produce **A** and **K**, similar to palladium activation in Scheme 5.2. However, if there is an oxidant present, **A** can be oxidized back to the Pd^{II} species **J**, and the cycle repeats. In contrast to the Sonogashira reaction where the amine base is required in stoichiometric quantities, here the amine acts as a proton transfer catalyst.

Kijima and coworkers first reported the synthesis of short polymers with this palladium-mediated cycle using I_2 as an oxidant [72, 73]. However, Williams and

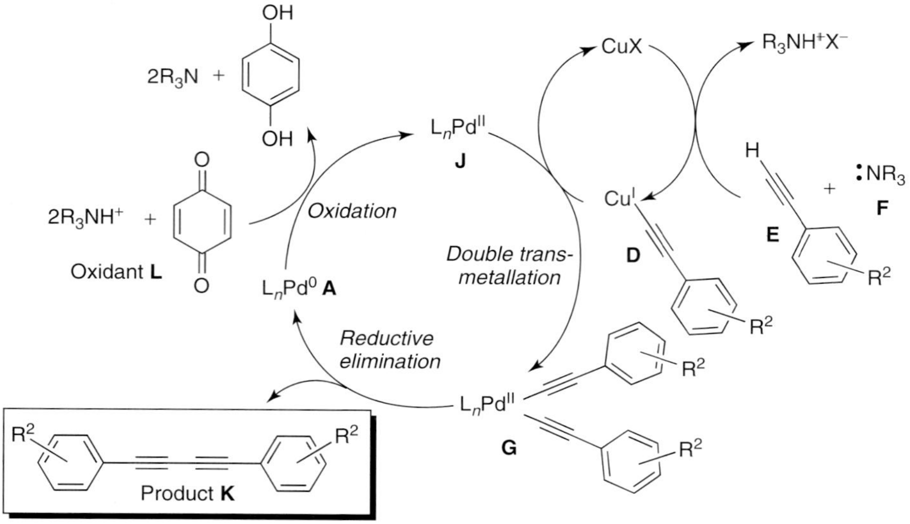

Scheme 5.4 Oxidative palladium acetylene homocoupling (OPAH).

Swager showed benzoquinone to be a more effective oxidant as it is mild and does not react with the acetylene moiety [71].[1] The functional group tolerance matches that of Sonogashira methods and can provide even higher molecular weight polymers by comparison (Eq. (5.9)). In addition, the optical and conductive properties of PABs are very similar to those of their PAE cousins.

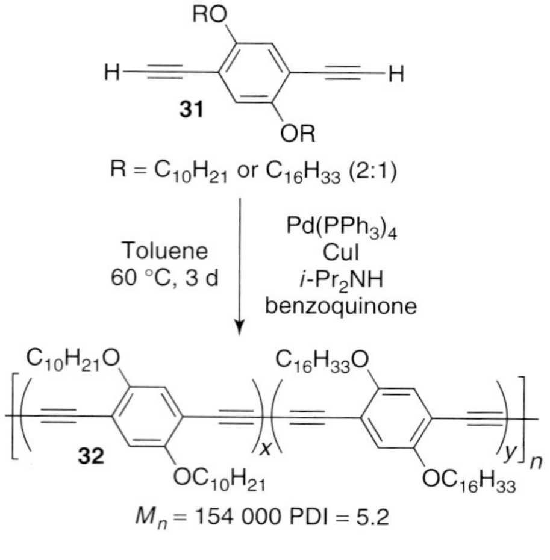

R = $C_{10}H_{21}$ or $C_{16}H_{33}$ (2:1)

Toluene, 60 °C, 3 d; $Pd(PPh_3)_4$, CuI, i-Pr_2NH, benzoquinone

M_n = 154 000 PDI = 5.2 (5.9)

1) Unpublished results show that the best results are obtained when the benzoquinone is purified through sublimation.

5.8
Palladium-Mediated Synthesis: Limitations and Conclusions

The greatest strength of palladium-mediated polymerization methods is their broad functional group tolerance. Indeed, even radical-bearing monomers can be successfully polymerized [8, 74]. However, except for a few select cases, DP remains low (20–50). Further, the unfortunate incorporation of diyne structural defects has been estimated to be as high as 10%. In addition, polymers made in this way typically have ill-defined end groups due to proto-dehalogenation and substitution of iodides by triarylphosphines [75]. Finally, removal of the ligand and metal catalyst reagents from the final polymer can be challenging [7, 8].

5.9
Metathesis Polymerizations

Alkene metathesis has had tremendous impact on organic synthesis and polymer chemistry. Similarly, the analogous alkyne metathesis, while less developed, has shown great promise as a synthesis technique [76, 77]; thus it should not be surprising to see that it has been applied to PAE synthesis [77, 78]. The mechanism, proposed by Katz [79] and experimentally established by Schrock [80, 81], involves an equilibrium of metallacyclobutadienes formed between a metal alkylidyne complex and a disubstituted acetylene (Scheme 5.5a); however, reaction efficiency

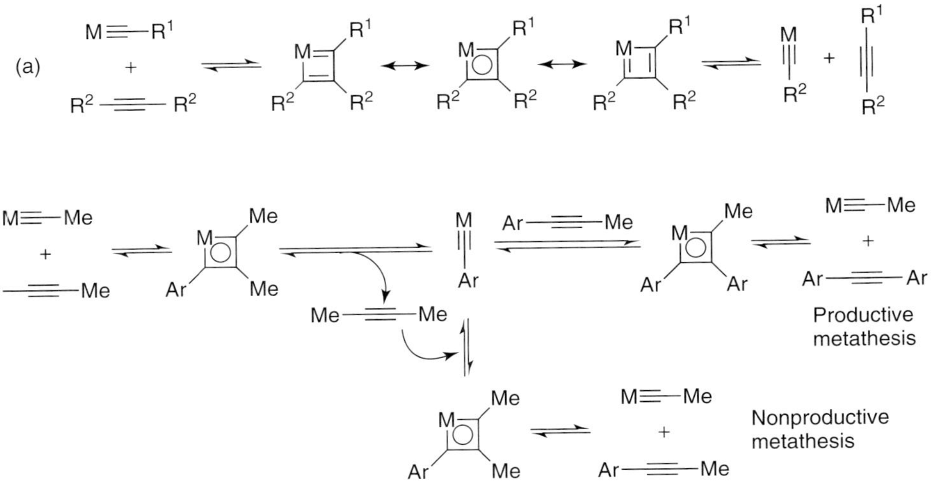

Scheme 5.5 (a) Mechanism of metathesis. (b) Productive and nonproductive metathesis using ethylidyne catalyst and propynyl-substituted aromatic substrate as examples.

depends on the position of this equilibrium. Consider Scheme 5.5b, which shows two possible outcomes after a molecule of starting material reacts. The catalyst has a choice of reacting with another molecule of the starting material, or instead with the 2-butyne byproduct. The former results in a productive metathesis reaction where product is formed, and the latter nonproductive step is essentially the reverse reaction. Thus, the 2-butyne byproduct should be removed to drive the reaction forward. Further, metathesis catalysts are known to react faster with alkyl-substituted alkynes than with aryl-substituted alkynes. If the byproduct is not removed, the catalyst will remain preoccupied with nonproductive metathesis (often referred to as "*pseudo-poisoning*"). Finally, another possible catalyst side reaction is the polymerization of the 2-butyne byproduct [82]. The take-home message is that fruitful yields can only be achieved through removal of the 2-butyne byproduct.

The first synthesis of PAE by alkyne metathesis was reported by Weiss and coworkers [83] using Schrock's tungsten alkylidyne catalyst **34** (Eq. (5.10)) [84, 85]. The step-growth process of polymerization is known as acyclic diyne metathesis (ADIMET). Dipropynyl monomer **33** was reacted at 80 °C in 1,2,4-trichlorobenzene under strict exclusion of air and water to yield polymer **35** in good yield and high degree of polymerization (DP = 100). The reaction was carried out under vacuum to remove the volatile 2-butyne byproduct and drive the equilibrium forward. The polymer structure and propynyl end groups were confirmed by NMR and **34** showed optical and spectroscopic properties identical to those of PAEs synthesized by cross-coupling means. However, the Schrock catalyst requires a relatively demanding synthesis, is highly sensitive to air and moisture, and is incompatible with amines or polyether chains [86–88].

$$\text{Me}{-}{\equiv}{-}C_6H_2(C_6H_{13})_2{-}{\equiv}{-}\text{Me}\ (\mathbf{33}) \xrightarrow[80\ ^\circ C]{(Me_3CO)_3W{\equiv}C{-}C(Me)_3\ (\mathbf{34})} \text{Me}{-}[{\equiv}{-}C_6H_2(C_6H_{13})_2{-}]_n{\equiv}{-}\text{Me}\ (\mathbf{35},\ n \sim 100) + (n-1)\text{Me}{-}{\equiv}{-}\text{Me}\uparrow \quad (5.10)$$

Bunz and coworkers have used the Mortreux $Mo(CO)_6$/phenol catalyst system [89] for the synthesis of PAEs (Eq. (5.11)) [90–92]. The active catalyst forms *in situ* from commercially available precursors at temperatures above 140 °C. Further, the reaction can be run in "off the shelf" solvents under a stream of nitrogen to drive off the 2-butyne (however, the authors note that the high reaction temperature and nitrogen lead to an anhydrous environment) [93]. Polymers **33a–e** were obtained in excellent yields, but the DP was strongly dependent on the identity of the side chains with heteroatom-containing chains giving the lowest results. Bunz also reported the ADIMET polymerization of carbazole-based monomers and found that the DP of copolymers increased with decreasing carbazole content [94]. Thus, functional group restrictions and high-temperature requirements make this system best suitable for alkyl-substituted PAEs, and bis-propynyl monomers bearing ethyl–hexyl chains were reported to have DPs above 1200 by gel permeation chromatography (GPC) [90]. Although these results seem to exceed Pd methods, the tendency of Mo catalysts to give oligomers of polyacetylene [82] suggests that these materials may be branched (branched architectures greatly overestimate MW when determined by GPC).

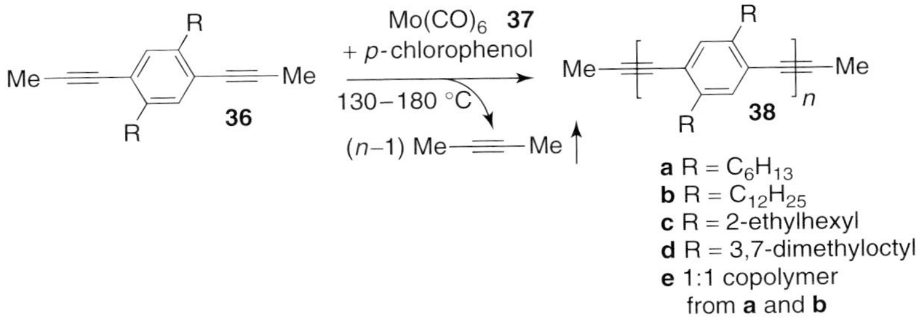

$Mo(CO)_6$ **37** + *p*-chlorophenol, 130–180 °C; **36** → **38** + (*n*–1) Me–≡–Me ↑

a R = C_6H_{13}
b R = $C_{12}H_{25}$
c R = 2-ethylhexyl
d R = 3,7-dimethyloctyl
e 1:1 copolymer from **a** and **b**

(5.11)

Recently, Zhang and Moore applied a molybdenum alkylidyne catalyst to the synthesis of poly(thienylene ethynylene) (Eq. (5.12)) [95]. Thiophene was known to be incompatible with alkyne metathesis [78]; however, this system produced **41** in excellent yield and DP. The inactive catalyst precursor **40** must be synthesized [96], but the mild conditions and broad functional group tolerance [82] of this system make it particularly attractive.

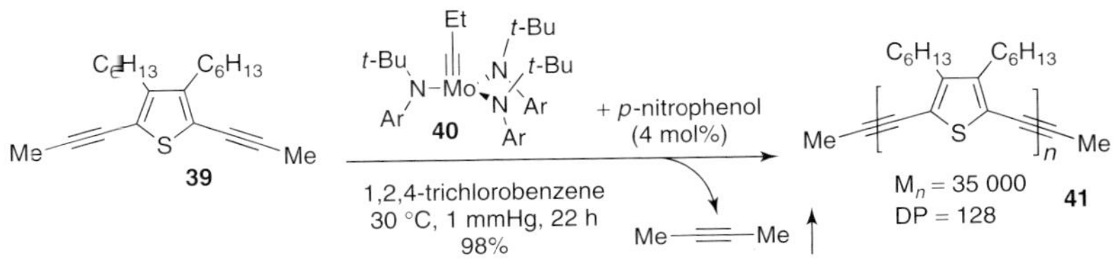

39 → **41**: **40** + *p*-nitrophenol (4 mol%), 1,2,4-trichlorobenzene, 30 °C, 1 mmHg, 22 h, 98%; Me–≡–Me ↑

M_n = 35 000, DP = 128

(5.12)

5.10 Macrocycles: the Continued Story

In Section 5.5, a brief discussion of the synthesis of macrocycles using the Pd method revealed that product yield suffers because a significant proportion of oligomers grow beyond the required length to form the macrocycle. This implies that, under Pd conditions, the macrocycle is formed as the kinetic product. In Section 5.9, the mechanism of alkyne metathesis was shown to involve an equilibrium reaction. Thus, the reversibility of metathesis may allow for oligomers that are too long (i.e., kinetic products) to be reversibly shortened. Under these conditions, the macrocycle would be the thermodynamic product.

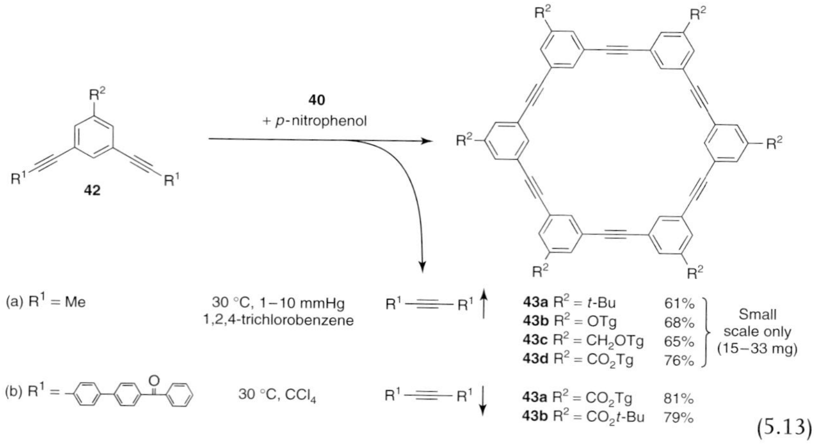

42

40 + *p*-nitrophenol

(a) R^1 = Me — 30 °C, 1–10 mmHg, 1,2,4-trichlorobenzene — R^1–≡–R^1 ↑

43a R^2 = *t*-Bu 61%
43b R^2 = OTg 68%
43c R^2 = CH_2OTg 65%
43d R^2 = CO_2Tg 76%
Small scale only (15–33 mg)

(b) R^1 = (benzoylbiphenyl) — 30 °C, CCl_4 — R^1–≡–R^1 ↓

43a R^2 = CO_2Tg 81%
43b R^2 = CO_2t-Bu 79%

(5.13)

Initial results using the Schrock and Mortreux catalyst systems were inefficient. However, the mild reaction conditions of the Moore catalyst system were much more suitable, and good yields of hexacycles (**43**) could be obtained on small scale (Eq. (5.13a)) [54, 77]. Unfortunately, upon scale up, these yields could not be realized. Moore reasoned that vacuum-driven removal of 2-butyne was insufficient on larger scale, leading to pseudo-poisoning effects discussed in Section 5.9. In an alternative approach, a precipitation-driven removal of the alkyne was developed using the poorly soluble benzoylbiphenyl group (Eq. (5.13b)). With the advent of this technique, multigram synthesis of macrocycles was achieved (unfortunately, atom economy suffers because of the high-molecular-weight bis(benzoylbiphenyl)acetylene byproduct (MW = 538)). The scope of the reaction was further demonstrated by the synthesis of tetracarbazole **46** (Eq. (5.14)), a molecule that self-assembles into nanofibrils [97] and shows sensitivity to explosives [98]. Finally, a solid-supported catalyst version of **40** has also been developed (**45**) [99].

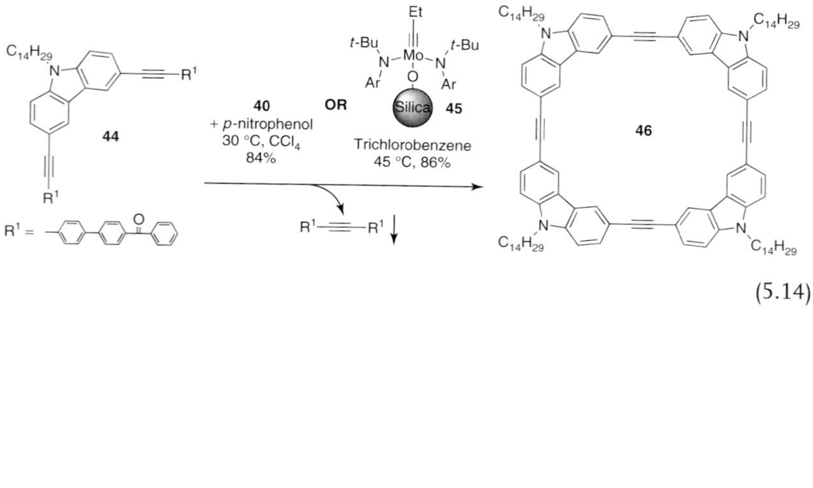

44 ($C_{14}H_{29}$); R^1 = (benzoylbiphenyl)

40 + *p*-nitrophenol, 30 °C, CCl_4, 84% OR **45** (Silica), Trichlorobenzene, 45 °C, 86%

R^1–≡–R^1 ↓

46

(5.14)

5.11
Metathesis: Concluding Remarks

In the case of *ortho*- and *meta*-substituted bis-alkynyl monomers, alkyne metathesis is aptly suited for macrocyclic synthesis. Numerous studies have experimentally established this dynamic equilibrium and probed the nature of the thermodynamic products (i.e., preferred ring size) [54, 77]. With regard to alkyl- and alkoxy-substituted *p*-PAE synthesis, the molecular weights obtained using metathesis are much higher than those obtained by Sonogashira polymerizations. Further, diyne defects are necessarily eliminated and end group identity is better defined. However, at this point, alkyne metathesis cannot match palladium cross-coupling methods in terms of functional group tolerance.

5.12
Transition-Metal-Free Polymerizations

It would seem that all viable methods for the synthesis of PAEs require a transition metal. However, Watson and coworkers reported the synthesis of an alternating aryl-perfluoro PAE (**49**) using catalytic fluoride (Eq. (5.15)) by making use of perfluorobenzene's propensity to undergo regioselective nucleophilic aromatic substitution at the 1,4-positions [100, 101].

47 + **48** $\xrightarrow[-2n\,(CH_3)_3SiF]{\text{Cat. } F^{\ominus}}$ **49** (5.15)

The catalytic fluoride ion generates a nucleophilic pentacoordinate silicate ion from **47**, which attacks **48**, forming the new bond and ejecting a new fluoride ion to repeat the cycle. High-molecular-weight polymers can be prepared using a variety of fluoride sources (i.e., TBAF, CsF, TMAF), but TMAF (tetramethylammonium fluoride) appears to give the best results (MW = 153 000; PDI = 1.54). The polymers can be easily purified with aqueous treatment to remove the catalytic fluoride, and the TMSF byproduct is a gas. The authors note that this method may be most beneficial for polymerization of alkynyl monomers, which are unstable after desilylation prior to Pd-mediated polymerization.

5.13
More Complex Side-Chain Effects

Section 5.2 discussed how side chains could be used to modify the spectral and electronic properties of the polymer in addition to solubility effects. As an extension of this principle, the Swager lab has extensively investigated the effect of three-dimensional shape as a design element in numerous material science applications [102]. Iptycenes are molecules built upon [2,2,2]-ring systems in which the bridges are aromatic rings, and their three-dimensional shape creates interstitial free volume around the molecules (Figure 5.2, **54**) [102]. A few select examples of this effect will be discussed in brief.

When PAE polymers are spread into thin films, interchain $\pi-\pi$ interactions can lead to self-quenching and reduced luminescence intensity. The rigid structure of the pent-iptycene in **50** prevents $\pi-\pi$ interactions and additionally creates a film with molecular level porosity able to allow small molecules to penetrate the film.

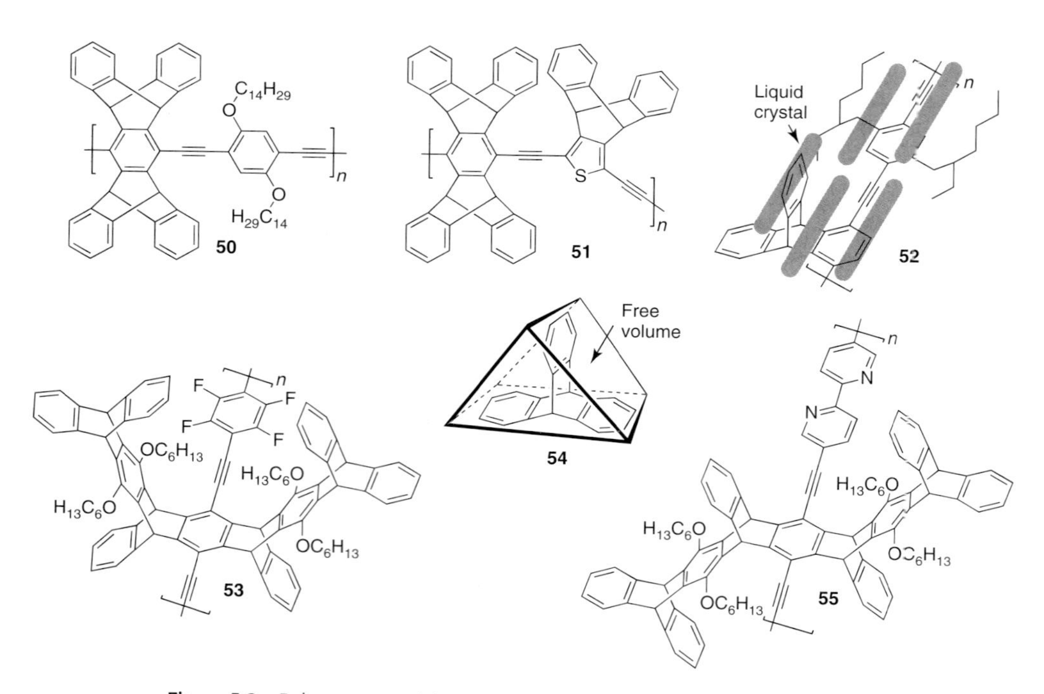

Figure 5.2 Polymers containing iptycene molecules.

This porosity and increased luminescence efficiency has led to films of **50** in the sensing of nitroaromatic explosives like TNT [103, 104].

Another interesting result of iptycene incorporation is the effect on solubility. Section 5.2 mentioned that side chains are necessary to entropically drive solubility, yet the noncompliant structure of iptycenes has been found to increase solubility quite dramatically. These effects are thought to arise from the avoidance of strong $\pi-\pi$ associations. Polymer **50** enjoys greater solubility than PAEs with more than twice the number of side chains. Further, **51** lacks flexible side chains yet shows good solubility in methylene chloride [105]. More elaborate structures like **53** and **55** are even more effective in preventing interchain interactions and show no evidence of self-quenching [106, 107].

Finally, polymer **52**, containing a triptycene moiety, has been shown to adopt a highly aligned ordering when placed in a liquid crystal phase. The rod-like liquid crystals can lie in the free space clefts created by the triptycene and assist alignment of the polymers along the long axis of the liquid crystals. This results in an increase in conjugation and hence exiton migration [108, 109].

Morisaki and Chujo have investigated through-space $\pi-\pi$ interactions as a new kind of polymer conjugation [110], by Sonogashira polymerization of [2.2]paracyclophane **56** to yield flexible polymer **58** (Eq. (5.16)). While not conjugated in the traditional sense, **58** still maintains communication between segments via the cyclophane-enforced $\pi-\pi$ interactions and shows red-shifted absorption characteristic of extended conjugation [111, 112].

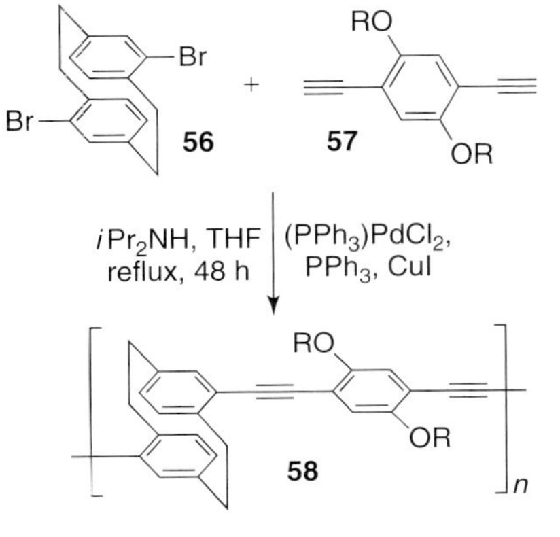

(5.16)

Swager extended the effect of enforced $\pi-\pi$ interactions by synthesizing an arene sandwich [113]. It is well documented that $\pi-\pi$ interactions in stacked arenes can lead to lower oxidation potentials by raising the HOMO energy level [114]. Thus, this would provide an interesting through-space means by which to tune polymer electronic properties. The arene sandwich was synthesized by a remarkably selective [4 + 2] cycloaddition of anthracene onto **59** (Eq. (5.17)). After some functional group manipulation and oxidative palladium acetylene homocoupling (OPAH) polymerization (Section 5.7), highly emissive polymer **61** was synthesized in modest yield (45%, $M_n = 17\,000$, DP = 16). Monomer **60** showed a reduced oxidation potential and red-shifted absorption, providing strong evidence for a raised HOMO level and narrowed HOMO–LUMO gap. However, the polymer results were less conclusive. The authors suggest that the effect of cofacial arenes may be attenuated by conjugation length.

(5.17)

5.14
Post-Polymerization Modification

Functionalization of polymers after polymerization can be challenging, but offers the opportunity to generate novel polymers without moving back to the monomer stage. In effect, a variety of PAE structures could be synthesized late in the synthetic process from a common precursor.

5.14.1
Modification of the Main Chain

Bunz has reported the catalytic reduction of the triple bonds along a PAE backbone using hydrogen gas under relatively forceful conditions (Eq. (5.18), **62**→**63**) [115, 116]. Alternatively, Weder found that *p*-toluenesulfonylhydrazine and tripropyl amine could be used in toluene at much lower temperatures (Eq. (5.18), **64**→**63**) [117]. As expected, the conversion of the rigid PAE backbone to a more flexible coil structure resulted in a decrease in the molecular weight by a factor of 1.42 as determined by GPC (see Section 5.17) [116].

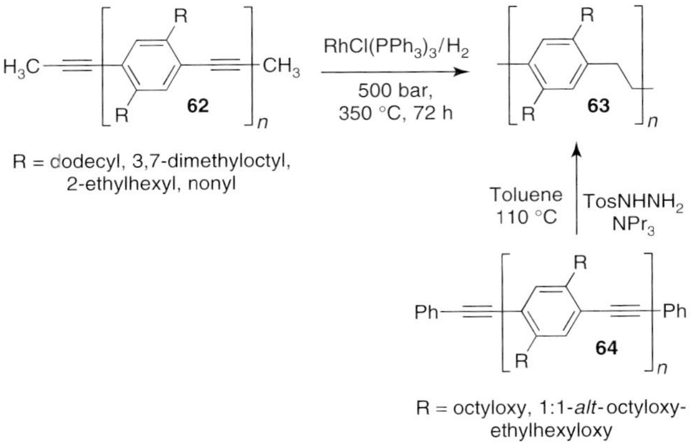

H_3C — R — R — CH_3 **62** *n*

R = dodecyl, 3,7-dimethyloctyl, 2-ethylhexyl, nonyl

$RhCl(PPh_3)_3/H_2$

500 bar, 350 °C, 72 h

R — R **63** *n*

Toluene 110 °C | $TosNHNH_2$ NPr_3

Ph — R — R — Ph **64** *n*

R = octyloxy, 1:1-*alt*-octyloxy-ethylhexyloxy

(5.18)

Moving in the other direction, Taylor and Swager reported a mild strategy to generate a conjugated PAE from a nonconjugated polymer precursor (Eq. (5.19)) [118]. After polymerization of **65** using the OPAH method outlined in Section 5.7, facile aromatization using tin(II) chloride and 1 M HCl produced the desired poly(anthrylenebutadiynylene) **67** in excellent yield. The anthrylene units imbue the polymer with increased quinoid resonance character that decreases polymer bandgap to 1.5 eV, remarkably low for a polymer with an all-carbon backbone. This reductive-aromatization strategy should be useful for arylene units that are unstable in their monomer forms [119]. Several older reports of synthesizing PAE backbone structure after polymerization can be found in [120, 121].

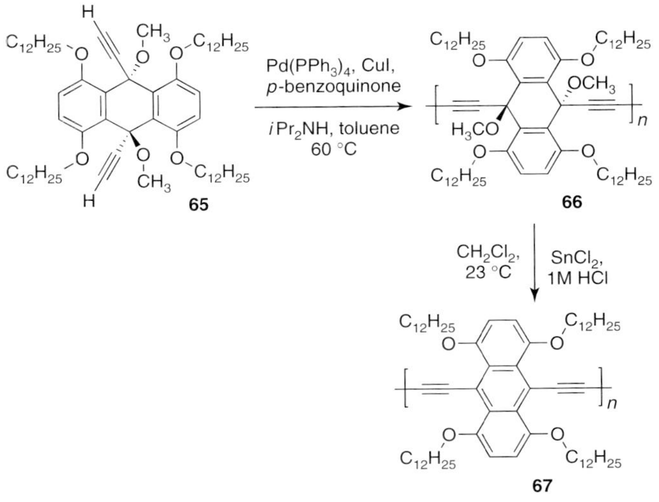

H, $C_{12}H_{25}$, O, CH_3, O, O, $C_{12}H_{25}$, $C_{12}H_{25}$, O, O, O, CH_3, $C_{12}H_{25}$, H **65**

$Pd(PPh_3)_4$, CuI, *p*-benzoquinone

iPr_2NH, toluene 60 °C

$C_{12}H_{25}$, O, O, $C_{12}H_{25}$, OCH_3, H_3CO, O, O, $C_{12}H_{25}$, $C_{12}H_{25}$ *n* **66**

CH_2Cl_2, 23 °C | $SnCl_2$, 1M HCl

$C_{12}H_{25}$, O, O, $C_{12}H_{25}$, O, O, $C_{12}H_{25}$, $C_{12}H_{25}$ *n* **67**

(5.19)

Finally, Swager and coworkers showed the partial reduction of triple bonds to the *all-cis*-PAV (Eq. (5.20)) [122]. The transformation, initially described by Sato [123], is stoichiometric in both titanium and magnesium, but the reduction is quantitative and completely *cis*-selective. Polymer **69** was further converted to the *trans*-PAV under UV radiation.

68: M_n = 103 kDa, PDI = 5.3, Φ = 0.82; λ_{abs} = 449 nm λ_{em} = 472 nm

1. Ti(O*i*-Pr)$_4$, rt to −78 °C
2. *i*-PrMgCl, −78 to −42 °C
3. H_2O, −42 °C to rt
toluene
86%

69: M_n = 61 kDa, PDI = 4.0, Φ = 0.29; λ_{abs} = 419 nm λ_{em} = 533 nm

(5.20)

5.14.2 Side-Chain Manipulation

Swager and coworkers showed that polystyrene grafts could be grown from the PAE backbone using atom transfer radical polymerization (ATRP) (Eq. (5.21)) [124], and the molecular weight and PDI of the grafts can be analyzed by hydrolysis of the ester linkages. The diversity in length and structure of ATRP-compatible monomers gives this method a modular approach to tuning polymer structure and properties [125–127].

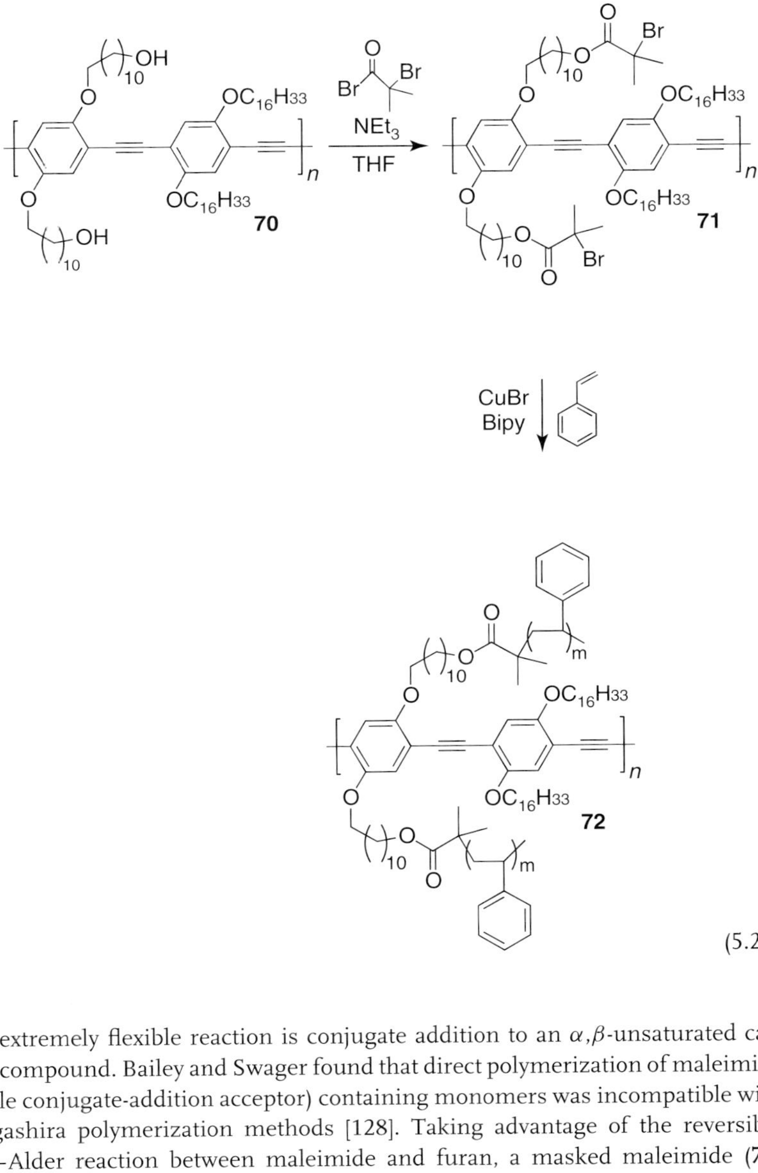

(5.21)

An extremely flexible reaction is conjugate addition to an α,β-unsaturated carbonyl compound. Bailey and Swager found that direct polymerization of maleimide (a facile conjugate-addition acceptor) containing monomers was incompatible with Sonogashira polymerization methods [128]. Taking advantage of the reversible Diels–Alder reaction between maleimide and furan, a masked maleimide (**73**) was successfully polymerized (M_n = 11 000) (Eq. (5.22)). Subsequently, the retro Diels–Alder was effected under mild heating to unveil the maleimide which could be reacted with nucleophiles like thiols to covalently attach a carboxy-X-rhodamine (ROX) dye (**75**). The maleimide group should also be reactive as a diencphile for subsequent Diels–Alder modification.

(5.22)

Bunz developed a strategy to synthesize hyperbranched conjugated polymers with unreactive aryl iodides (17.5% iodide by combustion analysis) (Eq. (5.23)) [129]. These reactive sites were further modified with different alkynes by subsequent Sonogashira reaction. This post-functionalization strategy generates a variety of fluorescent hyperbranched PAEs, with optical properties greatly dependent on the alkyne that is selected.

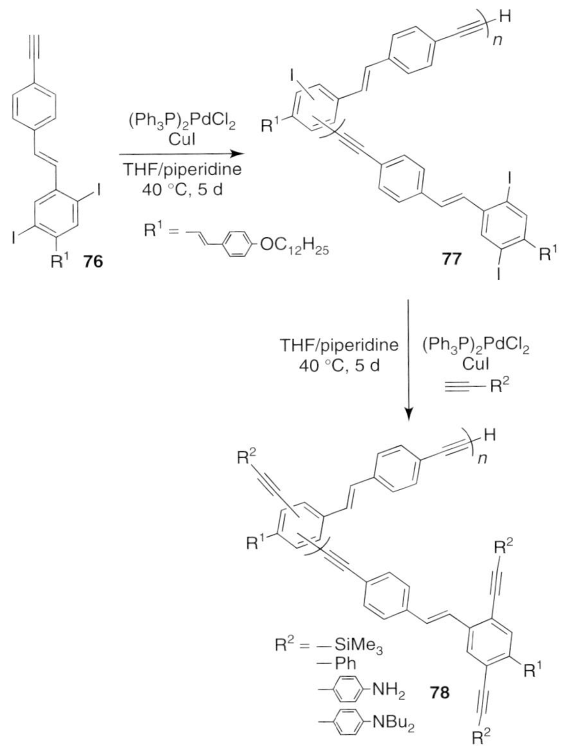

(5.23)

5.15 Characterization of Poly(aryleneethynylene)

All of the methods of PAE synthesis discussed here are step-growth polycondensations. In contrast to chain-growth polymerization, this means that high-molecular-weight polymers cannot be realized unless very high monomer conversion is achieved [130]. To date, no examples of chain-growth polymerizations have been reported.

The most common method for determining PAE molecular weight is GPC. GPC calibration is typically performed with a flexible coiled polymer standard (e.g., polystyrene); because PAEs behave like rigid rods, their molecular weights are overestimated by a factor of ~2. However, Cotts and Swager used light-scattering measurements to demonstrate that higher molecular weight PAEs are in fact more flexible and worm like with a persistence length of 15 nm [131]. MALDI–TOF analysis has been used but appears to be unreliable. Structural characterization can be achieved by traditional NMR techniques (e.g., ^1H, ^{13}C, etc.) and infrared spectroscopy has been found especially useful in solid-phase synthesis [68]. Finally, PAEs are often fluorescent, so the reaction progress can be tracked by naked eye by illuminating the reaction flask with a black light.

5.16 Conclusion

This chapter has discussed the means by which to synthesize and modify PAEs. Like in all of science, we continually observe the emergence of complex systems that are greater than the sum of their parts. Thus, it should not be surprising that through the use of simple building blocks and simple reactions to link them together, numerous novel materials can be realized.

References

1. Thomas, S.W., Joly, G.D., and Swager, T.M. (2007) *Chem. Rev.*, **107**, 1339–1386.
2. Zheng, J. and Swager, T.M. (2005) *Adv. Polym. Sci.*, **177**, 151–179.
3. Swager, T.M. (1998) *Acc. Chem. Res.*, **31**, 201–207.
4. Hoeben, F.J.M., Jonkheijm, P., Meijer, E.W., and Schenning, A.P.H. (2005) *J. Chem. Res.*, **105**, 1491–1546.
5. James, D.K. and Tour, J.M. (2005) *Top. Curr. Chem.*, **257**, 33–62.
6. Tour, J.M. (2000) *Acc. Chem. Res.*, **33**, 791–804.
7. Bunz, U.H.F. (2005) *Adv. Polym. Sci.*, **177**, 1–52.
8. Bunz, U.H.F. (2000) *Chem. Rev.*, **100**, 1605–1644.
9. Doucet, H. and Hierso, J.C. (2007) *Angew. Chem. Int. Ed.*, **46**, 834–871.
10. Negishi, E. and Anastasia, L. (2003) *Chem. Rev.*, **103**, 1979–2017.
11. Sonogashira, K., Tohda, Y., and Hagihara, N. (1975) *Tetrahedron Lett.*, 4467–4470.
12. Yamamoto, T. (1999) *Bull. Chem. Soc. Jpn.*, **72**, 621–638.
13. Yamamoto, T. (2003) *Synlett*, 425–450.
14. Schwab, P.F.H., Levin, M.D., and Michl, J. (1999) *Chem. Rev.*, **99**, 1863–1933.
15. Giesa, R. (1996) *J. M. S.–Rev. Macromol. Chem. Phys.*, **36**, 631–670.
16. One of the highest molecular weights reported by palladium methods utilized a bisamide: Zhou, Q. and Swager, T.M. (1995) *J. Am. Chem. Soc.*, **117**, 12593–12602.
17. Osakada, K., Sakata, R., and Yamamoto, T. (1997) *Organometallics*, **16**, 5354–5364.
18. Kovalev, A.I., Takeuchi, K., Barzykin, A.V., Asai, M., Ueda, M., and Rusanov, A.L. (2005) *Macromol. Chem. Phys.*, **206**, 2112–2121.
19. Francke, V., Mangel, T., and Mullen, K. (1998) *Macromolecules*, **31**, 2447–2453.
20. Kim, J. and Swager, T.M. (2001) *Nature*, **411**, 1030–1034.
21. Clark, A.P.Z., Shen, K.F., Rubin, Y.F., and Tolbert, S.H. (2005) *Nano Lett.*, **5**, 1647–1652.
22. Arnt, L. and Tew, G.N. (2003) *Langmuir*, **19**, 2404–2408.
23. Nambiar, R., Woody, K.B., Ochocki, J.D., Brizius, G.L., and Collard, D.M. (2009) *Macromolecules*, **42**, 43–51.
24. Wilson, J.N., Waybright, S.M., McAlpine, K., and Bunz, U.H.F. (2002) *Macromolecules*, **35**, 3799–3800.
25. Khan, A., Muller, S., and Hecht, S. (2005) *Chem. Commun.*, 584–586.
26. Khan, A. and Hecht, S. (2004) *Chem. Commun.*, 300–301.
27. Hager, H. and Heitz, W. (1998) *Macromol. Chem. Phys.*, **199**, 1821–1826.

28. Gellman, S.H. (1998) *Acc. Chem. Res.*, **31**, 173–180.
29. Hill, D.J., Mio, M.J., Prince, R.B., Hughes, T.S., and Moore, J.S. (2001) *Chem. Rev.*, **101**, 3893–4011.
30. (a) Ray, C.R. and Moore, J.S. (2005) *Adv. Polym. Sci.*, **177**, 91–149; (b) Moore, J.S. (2007) in *Foldamers: Structure, Properties, and Applications* (eds S. Hecht and I. Huc), Wiley-VCH Verlag GmbH; For *o*-PAE foldamers see: (c) Jones, T.V., Slutsky, M.M., Laos, R., de Greef, T.F.A., and Tew, G.N. (2005) *J. Am. Chem. Soc.*, **127**, 17235–17240; (d) Jones, T.V., Blatchly, R.A., and Tew, G.N. (2003) *Org. Lett.*, **5**, 3297–3299.
31. Mio, M.J., Prince, R.B., Moore, J.S., Kuebel, C., and Martin, D.C. (2000) *J. Am. Chem. Soc.*, **122**, 6134–6135.
32. Nelson, J.C., Saven, J.G., Moore, J.S., and Wolynes, P.G. (1997) *Science*, **277**, 1793–1796.
33. Prest, P.J., Prince, R.B., and Moore, J.S. (1999) *J. Am. Chem. Soc.*, **121**, 5933–5939.
34. Prince, R.B., Okada, T., and Moore, J.S. (1999) *Angew. Chem. Int. Ed.*, **38**, 233–236.
35. Smaldone, R.A. and Moore, J.S. (2008) *Chem. Eur. J.*, **14**, 2650–2657.
36. Stone, M.T., Heemstra, J.M., and Moore, J.S. (2006) *Acc. Chem. Res.*, **39**, 11–20.
37. Yang, X.W., Yuan, L.H., Yamamoto, K., Brown, A.L., Feng, W., Furukawa, M., Zeng, X.C., and Gong, B. (2004) *J. Am. Chem. Soc.*, **126**, 3148–3162.
38. Prince, R.B., Brunsveld, L., Meijer, E.W., and Moore, J.S. (2000) *Angew. Chem. Int. Ed.*, **39**, 228–230.
39. Prince, R.B., Moore, J.S., Brunsveld, L., and Meijer, E.W. (2001) *Chem. Eur. J.*, **7**, 4150–4154.
40. Prince, R.B., Barnes, S.A., and Moore, J.S. (2000) *J. Am. Chem. Soc.*, **122**, 2758–2762.
41. Tanatani, A., Hughes, T.S., and Moore, J.S. (2001) *Angew. Chem. Int. Ed.*, **41**, 325–328.
42. Tanatani, A., Mio, M.J., and Moore, J.S. (2001) *J. Am. Chem. Soc.*, **123**, 1792–1793.
43. Smaldone, R.A. and Moore, J.S. (2007) *J. Am. Chem. Soc.*, **129**, 5444–5450.
44. Nureki, O., Vassylyev, D.G., Tateno, M., Shimada, A., Nakama, T., Fukai, S., Konno, M., Hendrickson, T.L., Schimmel, P., and Yokoyama, S. (1998) *Science*, **280**, 578–582.
45. Fukunaga, R., Fukai, S., Ishitani, R., Nureki, O., and Yokoyama, S. (2004) *J. Biol. Chem.*, **279**, 8396–8402.
46. Smaldone, R.A. and Moore, J.S. (2008) *Chem. Commun.*, 1011–1013.
47. Bunz, U.H.F., Rubin, Y., and Tobe, Y. (1999) *Chem. Soc. Rev.*, **28**, 107–119.
48. Faust, R. (1998) *Angew. Chem. Int. Ed. Engl.*, **37**, 2825–2828.
49. Haley, M.M., Pak, J.J., and Brand, S.C. (1999) *Top. Curr. Chem.*, **208**, 81–130.
50. Hoger, S. (1999) *J. Polym. Sci., Part A: Polym. Chem.*, **37**, 2685–2698.
51. Moore, J.S. (1997) *Acc. Chem. Res.*, **30**, 402–413.
52. Young, J.K. and Moore, J.S. (1995) in *Modern Acetylene Chemistry* (eds F.Diederich and P.J. Stang), VCH Weinheim, New York, pp. 415–441.
53. Zhao, D.H. and Moore, J.S. (2003) *Chem. Commun.*, 807–818.
54. Zhang, W. and Moore, J.S. (2006) *Angew. Chem. Int. Ed.*, **45**, 4416–4439.
55. Martin, R.E. and Diederich, F. (1999) *Angew. Chem. Int. Ed.*, **38**, 1350–1377.
56. Tour, J.M. (1996) *Chem. Rev.*, **96**, 537–553.
57. Tam, I.W., Yan, J.M., and Breslow, R. (2006) *Org. Lett.*, **8**, 183–185.
58. Wang, C.S., Batsanov, A.S., and Bryce, M.R. (2006) *J. Org. Chem.*, **71**, 108–116.
59. Wang, C.S., Batsanov, A.S., Bryce, M.R., and Sage, I. (2004) *Org. Lett.*, **6**, 2181–2184.
60. Armitt, D.J. and Crisp, G.T. (2006) *Tetrahedron*, **62**, 1485–1493.
61. Meier, H., Ickenroth, D., Stalmach, U., Koynov, K., Bahtiar, A., and Bubeck, C. (2001) *Eur. J. Org. Chem.*, 4431–4443.
62. Wautelet, P., Moroni, M., Oswald, L., LeMoigne, J., Pham, A., Bigot, J.Y., and Luzzati, S. (1996) *Macromolecules*, **29**, 446–455.
63. Moore, J.S. and Zhang, J. (1992) *Angew. Chem. Int. Ed. Engl.*, **1992**, 922–924.

64. Abe, H., Masuda, N., Waki, M., and Inouye, M. (2005) *J. Am. Chem. Soc.*, **127**, 16189–16196.
65. Huang, S. and Tour, J.M. (1999) *Tetrahedron Lett.*, **40**, 3447–3350.
66. Jones, T.V., Blatchly, R.A., and Tew, G.N. (2003) *Org. Lett.*, **5**, 3297–3299.
67. Moore, J.S., Weinstein, E.J., and Wu, Z.Y. (1991) *Tetrahedron Lett.*, **32**, 2465–2466.
68. Moore, J.S., Hill, D.J., and Mio, M.J. (2000) in *Solid Phase Organic Synthesis* (ed. K. Burgess), John Wiley & Sons, Inc., New York, pp. 119–147.
69. Elliott, E.L., Ray, C.R., Kraft, S., Atkins, J.R., and Moore, J.S. (2006) *J. Org. Chem.*, **71**, 5282–5290.
70. Lo, P.K. and Sleiman, H.F. (2009) *J. Am. Chem. Soc.*, **131**, 4182–4183.
71. Williams, V.E. and Swager, T.M. (2000) *J. Polym. Sci., Part A: Polym. Chem.*, **38**, 4669–4676, and references therein.
72. Kijima, M., Kinoshita, I., Hattori, T., and Shirakawa, H. (1998) *J. Mater. Chem.*, **8**, 2165–2166.
73. Kijima, M., Kinoshita, I., Hattori, T., and Shirakawa, H. (1999) *Synth. Met.*, **100**, 61–69.
74. Matsuda, K., Stone, M.T., and Moore, J.S. (2002) *J. Am. Chem. Soc.*, **124**, 11836–11837.
75. Goodson, F.E., Wallow, T.I., and Novak, B.M. (1997) *J. Am. Chem. Soc.*, **119**, 12441–12453.
76. Furstner, A. and Davies, P.W. (2005) *Chem. Commun.*, 2307–2320.
77. Zhang, W. and Moore, J.S. (2007) *Adv. Synth. Catal.*, **349**, 93–120.
78. Bunz, U.H.F. (2001) *Acc. Chem. Res.*, **34**, 998–1010.
79. Katz, T.J. and Mcginnis, J. (1975) *J. Am. Chem. Soc.*, **97**, 1592–1594.
80. Pedersen, S.F., Schrock, R.R., Churchill, M.R., and Wasserman, H.J. (1982) *J. Am. Chem. Soc.*, **104**, 6808–6809.
81. Wengrovius, J.H., Sancho, J., and Schrock, R.R. (1981) *J. Am. Chem. Soc.*, **103**, 3932–3934.
82. Zhang, W., Kraft, S., and Moore, J.S. (2004) *J. Am. Chem. Soc.*, **126**, 329–335.
83. Weiss, K., Michel, A., Auth, E.M., Bunz, U.H.F., Mangel, T., and Mullen, K. (1997) *Angew. Chem. Int. Ed. Engl.*, **36**, 506–509.
84. Listemann, M.L. and Schrock, R.R. (1985) *Organometallics*, **4**, 74–83.
85. Schrock, R.R., Clark, D.N., Sancho, J., Wengrovius, J.H., Rocklage, S.M., and Pedersen, S.F. (1982) *Organometallics*, **1**, 1645–1651.
86. Furstner, A., Ackermann, L., Gabor, B., Goddard, R., Lehmann, C.W. Mynott, R., Stelzer, F., and Thiel, O.R. (2001) *Chem. Eur. J.*, **7**, 3236–3253.
87. Furstner, A., Guth, O., Rumbo, A., and Seidel, G. (1999) *J. Am. Chem. Soc.*, **121**, 11108–11113.
88. Furstner, A., Mathes, C., and Lehmann, C.W. (1999) *J. Am. Chem. Soc.*, **121**, 9453–9454.
89. Mortreux, A. and Blanchar, M. (1974) *J. Chem. Soc., Chem. Commun.*, 786–787.
90. Kloppenburg, L., Jones, D., and Bunz, U.H.F. (1999) *Macromolecules*, **32**, 4194–4203.
91. Kloppenburg, L., Song, D., and Bunz, U.H.F. (1998) *J. Am. Chem. Soc.*, **120**, 7973–7974.
92. Pschirer, N.G., Vaughn, M.E., Dong, Y.B., zur Loye, H.C., and Bunz, U.H.F. (2000) *Chem. Commun.*, 85–86.
93. Pschirer, N.G. and Bunz, U.H.F. (1999) *Tetrahedron Lett.*, **40**, 2481–2484.
94. Brizius, G., Kroth, S., and Bunz, U.H.F. (2002) *Macromolecules*, **35**, 5317–5319.
95. Zhang, W. and Moore, J.S. (2004) *Macromolecules*, **37**, 3973–3975.
96. Zhang, W., Kraft, S., and Moore, J.S. (2003) *Chem. Commun.*, 832–833.
97. Balakrishnan, K., Datar, A., Zhang, W., Yang, X.M., Naddo, T., Huang, J.L., Zuo, J.M., Yen, M., Moore, J.S., and Zang, L. (2006) *J. Am. Chem. Soc.*, **128**, 6576–6577.
98. Naddo, T., Che, Y., Zhang, W., Balakrishnan, K., Yang, X., Yen, M., Zhao, J., Moore, J.S., and Zang, L. (2007) *J. Am. Chem. Soc.*, **129**, 6978–6979.
99. Cho, H.M., Weissman, H., and Moore, J.S. (2008) *J. Org. Chem.*, **73**, 4256–4258.
100. Dutta, T., Woody, K.B., and Watson, M.D. (2008) *J. Am. Chem. Soc.*, **130**, 452–445.

101. Woody, K.B., Bullock, J.E., Parkin, S.R., and Watson, M.D. (2007) *Macromolecules*, **40**, 4470–4473.
102. Swager, T.M. (2008) *Acc. Chem. Res.*, **41**, 1181–1189.
103. Yang, J.S. and Swager, T.M. (1998) *J. Am. Chem. Soc.*, **120**, 11864–11873.
104. Yang, J.S. and Swager, T.M. (1998) *J. Am. Chem. Soc.*, **120**, 5321–5322.
105. Williams, V.E. and Swager, T.M. (2000) *Macromolecules*, **33**, 4069–4073.
106. Zhao, D. and Swager, T.M. (2005) *Macromolecules*, **38**, 9377–9384.
107. Zhao, D.H. and Swager, T.M. (2005) *Org. Lett.*, **7**, 4357–4360.
108. Zhu, Z.G. and Swager, T.M. (2002) *J. Am. Chem. Soc.*, **124**, 9670–9671.
109. Nesterov, E.E., Zhu, Z.G., and Swager, T.M. (2005) *J. Am. Chem. Soc.*, **127**, 10083–10088.
110. Morisaki, Y. and Chujo, Y. (2008) *Prog. Polym. Sci.*, **33**, 346–364.
111. Morisaki, Y., Ishida, T., Tanaka, H., and Chujo, Y. (2004) *J. Polym. Sci., Part A: Polym. Chem.*, **42**, 5891–5899.
112. Morisaki, Y., Ishida, T., and Chujo, Y. (2002) *Macromolecules*, **35**, 7872–7877.
113. McNeil, A.J., Muller, P., Whitten, J.E., and Swager, T.M. (2006) *J. Am. Chem. Soc.*, **128**, 12426–12427.
114. See references discussed in reference 113.
115. Marshall, A.R. and Bunz, U.H.F. (2001) *Macromolecules*, **34**, 4688–4690.
116. Ricks, H.L., Choudry, U.H., Marshall, A.R., and Bunz, U.H.F. (2003) *Macromolecules*, **36**, 1424–1425.
117. Beck, J.B., Kokil, A., Ray, D., Rowan, S.J., and Weder, C. (2002) *Macromolecules*, **35**, 590–593.
118. Taylor, M.S. and Swager, T.M. (2007) *Angew. Chem. Int. Ed.*, **46**, 8480–8483.
119. (a) Khan, M.S., Al-Mandhary, M.R.A., Al-Suti, M.K., Al-Battashi, F.R., Al-Saadi, S., Ahrens, B., Bjernemose, J.K., Mahon, M.F., Raithby, P.R., Younus, M., Chawdhury, N., Kohler, A., Marseglia, E.A., Tedesco, E., Feeder, N., and Teat, S.J. (2004) *Dalton Trans.*, 2377–2385; (b) However, a recent report has found otherwise: Ju, J.U., Chung, D.S., Kim, S.O., Jung, S.O., Park, C.E., Kim, Y.H., and Kwon, S.K. (2009) *J. Polym. Sci., Part A: Polym. Chem.*, **47**, 1609–1616.
120. Chmil, K. and Scherf, U. (1993) *Makromolekulare Chem.-Macromol. Chem. Phys.*, **194**, 1377–1386.
121. Goodson, F.E. and Novak, B.M. (1997) *Macromolecules*, **30**, 6047–6055.
122. Moslin, R.M., Espino, C.G., and Swager, T.M. (2009) *Macromolecules*, **42**, 452–454.
123. Sato, F. and Okamoto, S. (2001) *Adv. Synth. Catal.*, **343**, 759–784.
124. Breen, C.A., Deng, T., Breiner, T., Thomas, E.L., and Swager, T.M. (2003) *J. Am. Chem. Soc.*, **125**, 9942–9943.
125. Breen, C.A., Rifai, S., Bulovic, V., and Swager, T.M. (2005) *Nano Lett.*, **5**, 1597–1601.
126. Breen, C.A., Tischler, J.R., Bulovic, V., and Swager, T.M. (2005) *Adv. Mater.*, **17**, 1981–1985.
127. Deng, T., Breen, C., Breiner, T., Swager, T.M., and Thomas, E.L. (2005) *Polymer*, **46**, 10113–10118.
128. Bailey, G.C. and Swager, T.M. (2006) *Macromolecules*, **39**, 2815–2818.
129. Tolosa, J., Kub, C., and Bunz, U.H.F. (2009) *Angew. Chem. Int. Ed.*
130. Odian, G. (2004) *Principles of Polymerization*, 4th edn, John Wiley & Sons, Inc.
131. Cotts, P.M., Swager, T.M., and Zhou, Q. (1996) *Macromolecules*, **29**, 7323–7328.

6
Synthesis of Poly(2,7-carbazole)s and Derivatives

Pierre-Luc T. Boudreault, Jean-François Morin, and Mario Leclerc

6.1
Introduction

Since their discovery in 1977, conducting (conjugated) polymers (CPs) have been extensively studied [1–4]. The remarkable electrical and optical properties of these materials pushed many academic and industrial researchers to further investigate this class of macromolecules. There are numerous foreseen applications for CPs: optical transducers [5], sensors [6, 7], light-emitting diodes (LEDs) [8], field-effect transistors (FETs) [9, 10], photovoltaic cells (PCs) [11–13], and so forth. Early developments have shown the emergence of well-defined conjugated polymers such as poly(2,5-thiophene)s [14], poly(1,4-phenylene)s [15], and poly(2,7-fluorene)s [16, 17]. In parallel, very interesting results have been obtained with nitrogen-containing conjugated polymers, such as polyaniline (PAni) [18] and polypyrrole (PPy) [19].

PAni has been studied for a long time mainly because aniline is a very cheap monomer and a very stable polymeric material. One of the most interesting features of this polymer is that it can be readily synthesized from a simple oxidation of the monomer. Moreover, as shown in Scheme 6.1, a wide variety of PAni can be obtained by simply oxidizing or reducing the resulting polymer. However, it has been shown that only a few stable states can be obtained: $y = 1$ or 0.5. Only a mixture of the two can be observed between these two values [20]. These oxidation states have a significant influence on its electrical properties. For example, by doping the emeraldine state ($y = 0.5$) with a protonic acid, for example, 1 M HCl, the conductivity increases by 9–10 orders of magnitude giving the fully protonated emeraldine hydrochloric salt with an electrical conductivity of about 1–10 S cm^{-1} [21, 22].

Just like PAni, PPy is still studied by many research groups [2]. It has attracted great attention mainly because of its high electrical conductivity and good environmental stability. Many scientists see PPy as a very promising candidate in applications like electronic devices, electrodes for rechargeable batteries and supercapacitors, sensors, corrosion-protecting materials, actuators, and electrochromic devices. In order to obtain well-defined PPys, it is first important to realize that

Design and Synthesis of Conjugated Polymers. Edited by Mario Leclerc and Jean-François Morin

ISBN: 978-3-527-32474-3

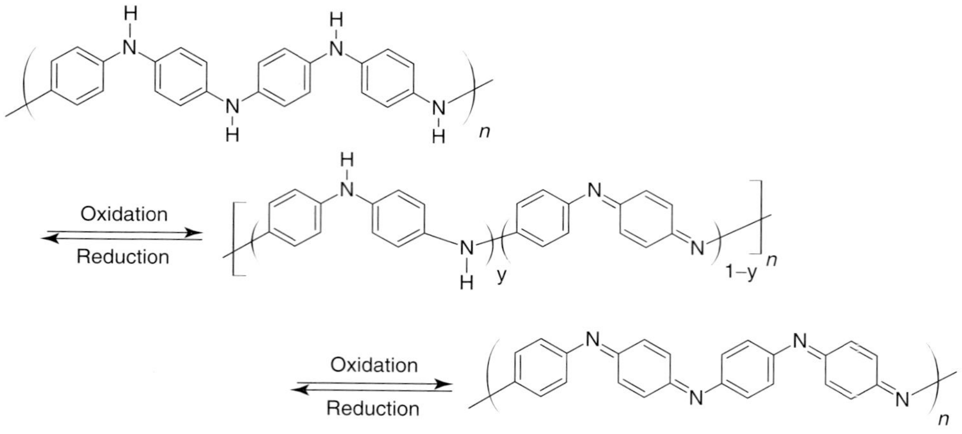

Scheme 6.1 Oxidation of PAni from the fully reduced state to the fully oxidized state.

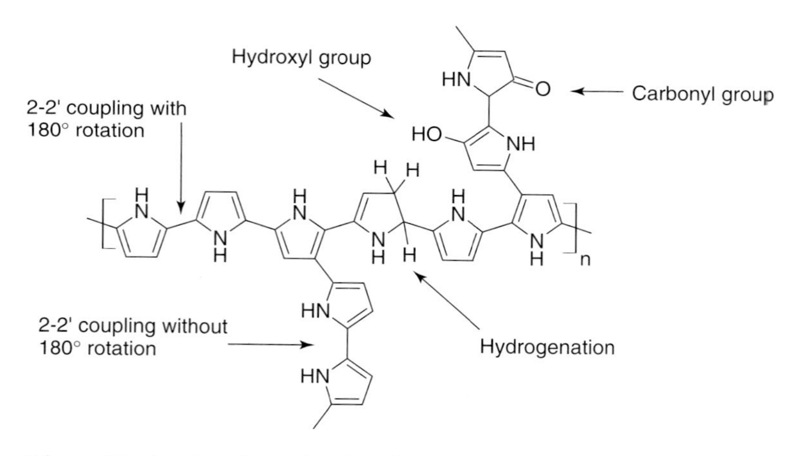

Scheme 6.2 Possible chemical and conformational defects in PPy.

many side reactions can occur during the synthesis of PPy. The more common defects are shown in Scheme 6.2 [28].

Some of these structural defects are caused by a pyrrole unit reacting at the 3-position instead of the 2-position, whereas overoxidation will add oxygen within the backbone in the form of hydroxyl or carbonyl groups. Structural defects between two pyrrole units can also occur if the rotation does not happen and the sterically more hindered conformation prevails. All these structural defects lead to nonplanar conformations and to a decrease of the performances of the polymer (conductivity, field-effect mobility, etc.).

To solve these problems, there have been many new polymers based on 3,4-disubstituted pyrroles. The most common 3,4-disubstituted PPys are poly(3,4-dimethoxypyrrole)s and poly(3,4-diethylenedioxypyrrole)s. These polymers can be readily obtained from electrochemical polymerization [36].

Many other poly(3,4-dialkylenedioxypyrrole)s have been prepared using the same electrochemical method [37].

The electrical and optical properties of these two classes of polymers inspired the development of another class of nitrogen-containing polymers, namely polycarbazoles. The carbazole moiety brings many interesting properties that explain why there is still plenty of research focused on this particular molecule. First, 9*H*-carbazole is a very cheap starting material and its fully aromatic unit leads to a good environmental stability. Moreover, a wide variety of functional groups can be added onto the nitrogen atom to increase the solubility and to modulate the optical and electrical properties [23]. Throughout this chapter, we present the synthesis of polymers based on the carbazole unit, and especially recent studies performed on poly(2,7-carbazole)s and derivatives [24]. In addition, we also discuss the synthesis of the monomers since it is vital to afford it at a very low cost to enable the commercialization of the polymers [25, 26].

6.2 Polycarbazoles

Polycarbazoles have now been studied for more than 25 years [23]. At first, the remarkable photoconductivity of poly(*N*-vinylcarbazole) (PVK) attracted a lot of attention [27, 38–40]. Recent reports have shown that PVK could be very useful as a host material in white-organic light-emitting diodes (WOLEDs) [29, 41], in combination with other layers, because of its high energy blue-emissive singlet excited state and the absence of low energy triplet state [30]. Moreover, PVK has also been used in dye-sensitized solar cells blended with TiO_2 as the electron donor and hole-transporting material [31–34, 42]. The interest has grown up recently because of new promising performances in PCs, FETs, and LEDs for carbazole derivatives and related polymers [35]. Furthermore, carbazole-based polymers have also been utilized in applications such as adaptive camouflage [43], nonlinear optics (NLOs) [44], and even self-assembly of nanoscale π-systems [45]. The scope of these materials is almost unlimited and we describe in the following sections some exciting research developments that have been made over the last few years.

6.2.1 Poly(3,6-carbazole)s

After the synthesis and commercialization of PVK, poly(3,6-carbazole)s rapidly emerged. Indeed, the synthesis of processible poly(3,6-carbazole)s is straightforward and can be easily performed. From the commercially available 9*H*-carbazole, a side chain can be added onto the nitrogen atom. Afterward, a polymerization reaction can be performed directly without functionalizing or protecting any positions because of the high reactivity of the 3- and 6-positions, as shown by Ambrose *et al.* [46, 47]. For instance, Siove *et al.* successfully synthesized poly(3,6-carbazole)s from simple oxidative polymerization [48, 49]. It was found that films cast from homopolymers and copolymers based on the 3,6-carbazole framework presented redox processes that were dependent on the monomer composition. These redox processes were also accompanied by color changes from yellow (neutral state) to blue (oxidized state) [50, 51].

The reductive polymerization of 3,6-dihalocarbazoles from Grignard [52], electrochemical [53–55], and chemical palladium- or nickel-catalyzed [56, 57] coupling reactions have also been developed to obtain poly(3,6-carbazole)s. Depending on the side chain added onto the nitrogen atom, these methods led to more or less soluble polymers with low molecular weights (M_w around 10^4). Higher molecular weights (M_w of 10^5) were obtained by using the well-known Ni(0)-catalyzed Yamamoto reductive homocoupling [58].

Despite extensive efforts from different research groups in the past 20 years, it is still difficult to obtain high-molecular-weight poly(3,6-carbazole)s with good mechanical properties. Moreover, the poor electronic conjugation along the polymer backbone is another factor explaining the limited performances of 3,6-carbazole-based polymers. As shown in Scheme 6.3, it has been demonstrated that a positive charge in a 3,6-carbazole homopolymer is delocalized only on two monomeric units [46, 47]. Consequently, it is difficult to prepare polymers having good photoactive properties in the visible range with such a small effective conjugation length. Nevertheless, 3,6-carbazole-based polymers have been used mainly in LEDs because of their very good hole-transporting ability which is a crucial criteria to obtain high-performance LEDs. The first device using a poly(3,6-carbazole) derivative was fabricated in 1996. Only a few studies have focused on the synthesis of yellow, green, and red light-emitting polymers because of the difficulty to functionalize at different positions. Still, electron-withdrawing units were copolymerized to lower the bandgap of the polymers and therefore to redshift the light-emitting properties of the resulting materials [60]. Lately some research groups have focused on the synthesis of new materials for WOLEDs because of the need for lightweight illumination devices in aircrafts, space shuttles, and liquid crystal displays [61, 62]. Some interesting results have been published using 3,6-carbazole as hole-transporting layer [63] and as electroluminescent polymer [64].

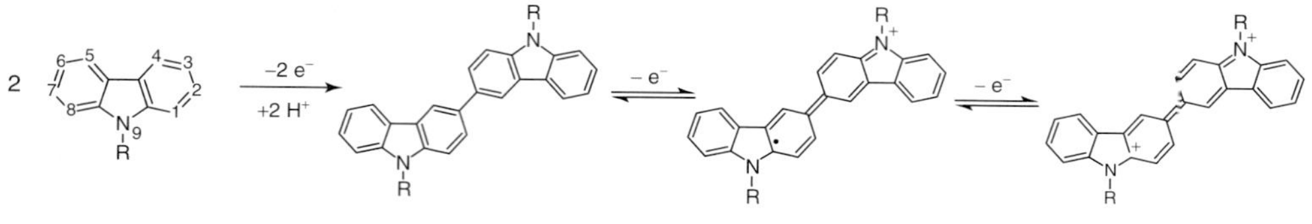

2 [carbazole, positions 1–9, N-R] $\xrightarrow[+2\,H^+]{-2\,e^-}$ [dimer] $\underset{}{\overset{-\,e^-}{\rightleftharpoons}}$ [radical cation] $\overset{-\,e^-}{\rightleftharpoons}$ [dication]

Scheme 6.3 Oxidation reactions within a carbazole dimer substituted at the 3-position.

6.2.2
Synthesis of 2,7-Disubstituted Carbazoles

The synthesis of 2,7-disubstituted carbazoles is a lot trickier than the synthesis of 3,6-disubstituted carbazoles. As shown in Scheme 6.4, there are several ways to

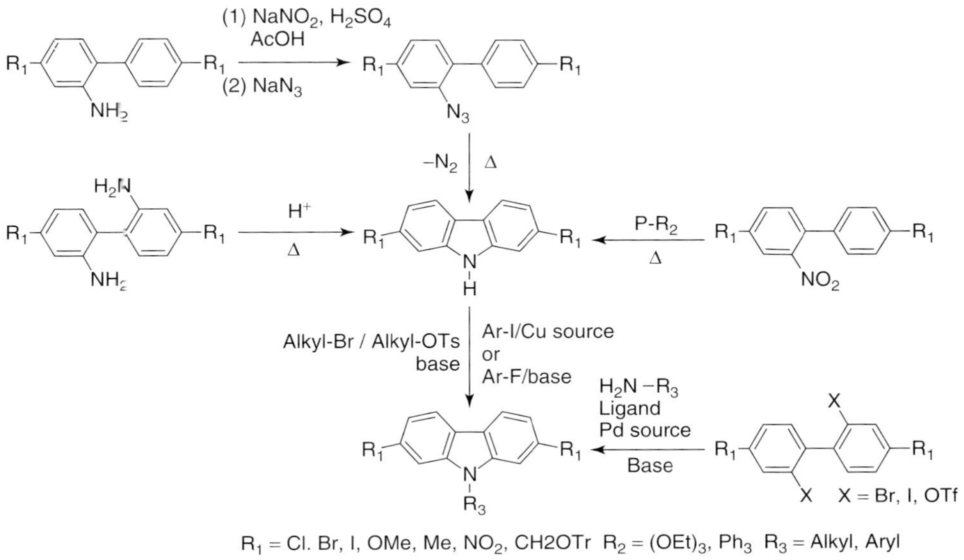

R_1 = Cl. Br, I, OMe, Me, NO_2, CH2OTr R_2 = $(OEt)_3$, Ph_3 R_3 = Alkyl, Aryl

Scheme 6.4 Synthetic pathways leading to 2,7-functionalized carbazoles.

synthesize the carbazole monomer. We discuss the methods developed over the years to obtain efficiently 2,7-disubstituted carbazoles and later on the different methods to synthesize carbazole-based polymers.

Smith and Brown first developed a method for an efficient synthesis of 2,7-disubstituted carbazoles [65, 66]. The key step in this synthesis is the conversion of the amine group into an azide function through a diazonium salt. Afterward, the ring-closure reaction is performed through a nitrene intermediate by simply heating the azide derivative in a high boiling point solvent. Soon after, Heinrich developed a ring-closure reaction from the 2,2′-diaminobiphenyl [67]. This reaction allows to obtain 2,7-functionalized carbazoles by simply treating the starting product in phosphoric acid that acts as a proton source at 200 °C. Later on, other proton sources such as Nafion H were used, which allowed this ring closure with much better yields of 87–93% [68].

In 1965, Cadogan developed an even more efficient reaction to obtain the desired carbazole framework [69, 70]. The starting compound is a functionalized 2-nitrobiphenyl, which is treated in triethylphosphite ($P(OEt)_3$) at high temperatures. This reaction provides the carbazole unit in 50–60% yields. The relatively low yields for this reaction are attributed to a side reaction that leads to the formation of the *N*-ethylcarbazole. New Cadogan ring closure has been developed recently to avoid this side product. The reaction can be carried out in triphenylphosphine (PPh_3) instead of $P(OEt)_3$ [71] or by using microwaves [72], the yield increases, in this case, up to 70–90%.

In 2001, Morin and Leclerc applied the azide strategy to synthesize the first 2,7-carbazole-based polymers (Scheme 6.5) [24]. Starting from a

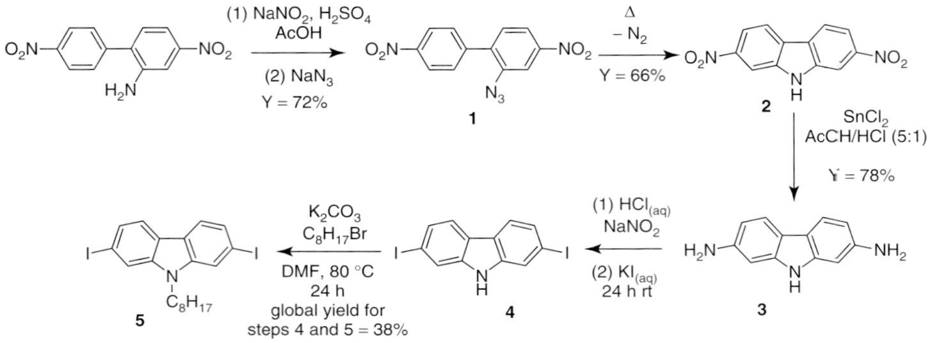

Scheme 6.5 Synthesis of 2,7-diiodocarbazole through the azide strategy.

dinitroaminobiphenyl derivative, they were able to prepare in five steps a soluble 2,7-dihalogenocarbazole derivative (**5**), a very useful derivative for the synthesis of conjugated polymers by palladium-catalyzed cross-coupling reaction.

In order to increase the overall efficiency of the synthetic process used to prepare 2,7-dihalogenocarbazole, Morin and Leclerc proposed to use the Cadogan ring-closing reaction to prepare a reactive 2,7-carbazole monomer. Using the pathway depicted in Scheme 6.6, they were able to synthesize a 2,7-carbazole monomer on a multigram scale. Although the 2,7-dichlorocarbazole monomer (**7**) is very useful for the synthesis of the homopolymer by using Yamamoto coupling, it is less practical for Suzuki, Stille, and other palladium-catalyzed cross-coupling reactions since aromatic carbon atoms bearing a chlorine atom are weakly reactive in the oxidative addition step. Thus, a more reactive monomer with triflate groups at the 2- and 7-positions was prepared (Scheme 6.7) [73].

The greatest advantage of this new synthetic pathway is that it allows the preparation of the very useful 2,7-diboronatecarbazole derivatives. However, this strategy suffers from poor scalability because of the hazardous materials used along the synthesis making it unattractive for industrial applications. Then, Müllen *et al.* proposed an efficient and inexpensive three-step method for the synthesis of 2,7-dibromocarbazole (**15**) [74] from which 2,7-diboronatecarbazole derivatives can be obtained. As shown in Scheme 6.8, the synthesis of 2,7-dibromocarbazole is performed in two simple steps from commercially available starting material, that is, 4,4′-dibromobiphenyl [24, 74, 75]. For now, it is still the most efficient way to synthesize 2,7-functionalized carbazoles.

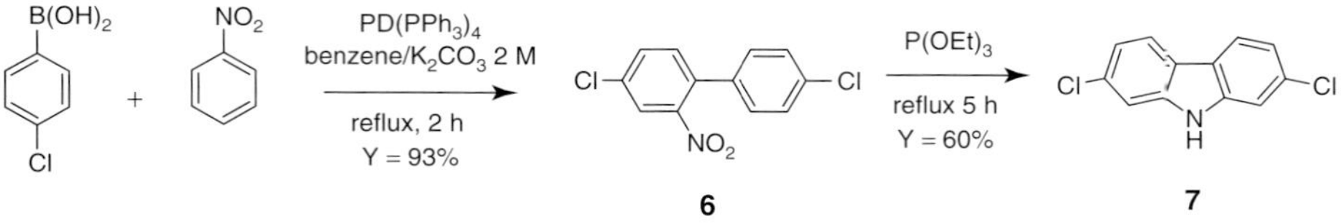

Scheme 6.6 First synthesis of 2,7-dichlorocarbazole from the Cadogan ring-closure reaction.

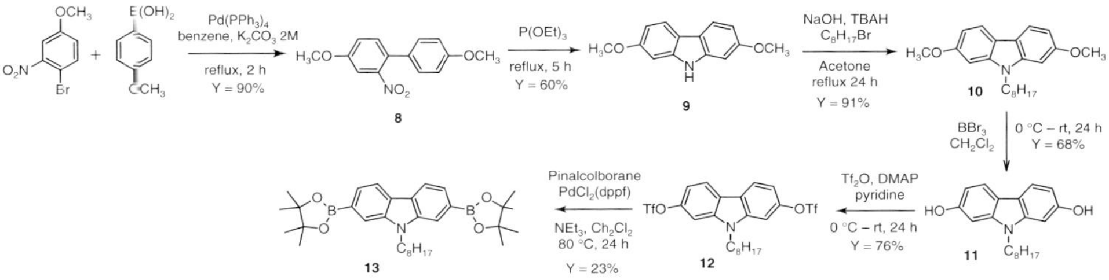

Scheme 6.7 Synthesis of 2,7-disubstituted carbazole unit bearing reactive functional groups.

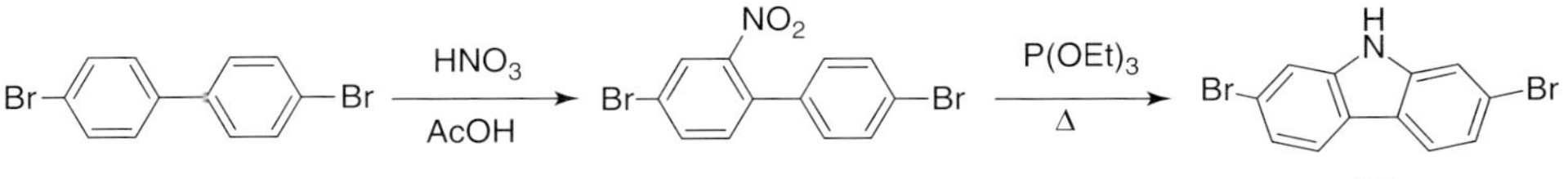

Scheme 6.8 Simple and rapid synthesis of 2,7-dibromocarbazole.

In order to insure a good solubility for the carbazole monomers and related polymers, it is important to perform N-alkylation or N-arylation. Initially, a simple alkyl chain (hexyl, octyl, etc.) was added to the carbazole unit before performing the polymerization reaction [24, 76]. These kinds of side chains were suitable for some applications such as LEDs and FETs, but to obtain polymers with better mechanical properties and even higher molecular weights, new side chains were developed. They were synthesized to mimic the three-dimensional organization of 9,9-dialkylfluorene [77, 78]. In addition to the fine-tuning of the side chain on the nitrogen atom, several functional groups can be added at the 3,6- or 1,8-positions to avoid degradation, to improve the organization, or even to modify the optical properties of the synthesized oligomers or polymers. Functional groups such as halogen, alkyl, nitrile, nitro, ketone, or silane can easily be introduced onto the carbazole backbone [74, 79–81]. Finally, novel methodologies that can afford 2,7-functionalized carbazoles were obtained through modified Hartwig and Buchwald palladium cross-coupling reactions (Scheme 6.4) [82–84]. Although these methods are very promising and lead to a carbazole unit that is already arylated, it has limitations in terms of functional groups that can be used at the 2- and 7-positions. The arylation reaction does not allow the presence of halogens or other reactive groups and this will lead to additional steps after the formation of the carbazole unit. Also, additional steps before the key step of this synthetic pathway are required to prepare the functionalized 4,4′-biphenyl.

6.2.3
Poly(2,7-carbazole)s for Light-Emitting Diodes

Chemical oxidation of the carbazole-based monomers is the most commonly used reaction to afford poly(3,6-carbazole)s. But, as mentioned earlier, the same reaction cannot be used to obtain poly(2,7-carbazole)s because of the low reactivity of these two positions. Poly(2,7-carbazole)s and derivatives can therefore be obtained from Yamamoto [85], Suzuki [86], and Stille coupling [87] reactions. Of these three types of polymerization, Yamamoto and Stille have issues that are difficult to overcome: the former is strictly used for the synthesis of homopolymers and random copolymers and in the case of the latter, purification of organotin compounds is very difficult. Consequently, the Suzuki coupling has been extensively utilized. Over the years, a great variety of polymers have been synthesized for numerous applications. Homopolymers have shown great potential in LED devices. However, truly interesting developments have been accomplished during the past few years by copolymerizing carbazole with suitable comonomers depending on the targeted application [88].

As mentioned above, the first 2,7-carbazole-based polymers were reported by Morin and Leclerc in 2001 who showed a straightforward approach to synthesize homopolymers and copolymers [24]. As expected, the UV–visible absorption spectra of the homopolymers are redshifted ($\lambda_{max} = 380$ nm) compared to the 3,6-carbazole homologs ($\lambda_{max} = 300$ nm). Most importantly, these polymers emit blue light (415–440 nm) with high quantum yields in solution [24, 89, 90]. Also, most of the 2,7-carbazole-based polymers emit light in the solid state making 2,7-carbazole derivatives an interesting alternative to the unstable fluorene unit for blue LED applications [89]. Table 6.1 presents some promising examples of light-emitting polymers that have been synthesized over the last 10 years for LED applications.

At first, the research focused on the synthesis of blue light-emitting homopolymers. The strategy used to improve the electroluminescence properties was the modification of the substituents on the nitrogen atom. A wide variety of alkyl chains [76, 90] and aryl groups [95, 96, 105–107] have been tested to obtain highly performing materials. Recently, a new carbazole monomer bearing a sulfonate group at the end of the alkyl chain was synthesized [108]. This led to the first 2,7-carbazole-based polymers soluble in water and these polymers show an intense blue light emission. Furthermore, copolymerization has also helped in obtaining more soluble polymers and higher molecular weights [109, 110]. One of the major problems of poly(2,7-carbazole)s is the degradation at the 3- and 6-positions that are more reactive, as described earlier. A solution to this problem has been given by the addition of methyl groups onto these positions. After the synthesis of many blue light-emitting polymers, research was focused on the synthesis of lower bandgap polymers that should provide green and red light emission [76]. Since the carbazole is an electron-rich comonomer, insertion of an electron-poor comonomer is necessary to lower the bandgap, by internal charge transfer (ICT), of the resulting copolymer. A wide variety of polymers have been synthesized, as shown in Scheme 6.9 [111, 112].

A green light-emitting polymer was synthesized by copolymerizing 2,7-carbazole with quinoxaline to give **32**, while a red light-emitting polymer was synthesized from thiophene dioxide-based polymer **33**. Both polymers were synthesized using Suzuki cross-coupling with $Pd(PPh_3)_4$ as catalyst and 2 M K_2CO_3 as the base in THF. In order to further decrease the bandgap and thus obtain red light emission,

Table 6.1 Some of the most promising carbazole-based polymers for LED applications.

Polymer	λ_{em} (nm)		References
	Solution	Thin film	
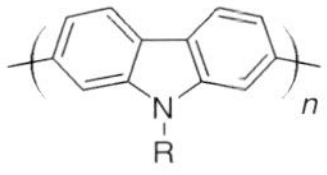 **16**, 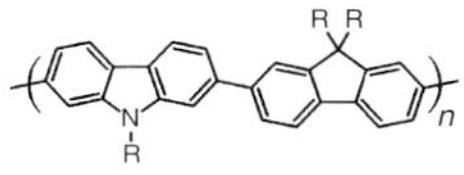 **17**	**16** – 417, 439 **17** – 417, 440	**16** – 423, 447 **17** – 426, 448	[89]
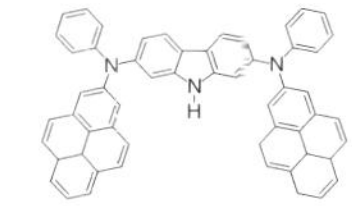 **18**	**18** – 550	**18** – 530, 532	[91]
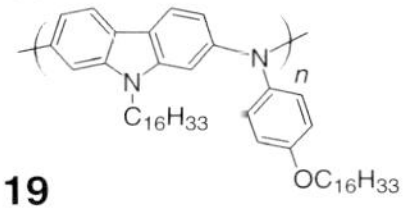 **19**, 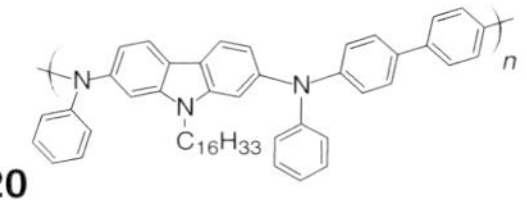 **20**	**19** – 439 **20** – 429	**19** – 373, 420 **20** – N.A.	[92, 93]
21, 22, 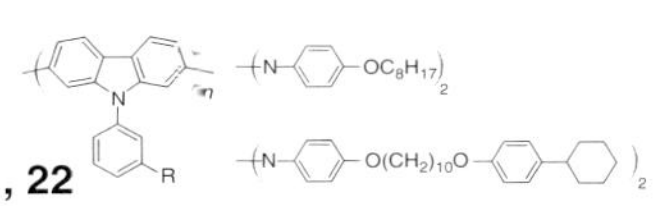 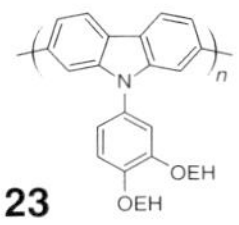 **23**	**21** – N.A. **22** – 412, 442 **23** – 417	**21** – 450 **22** – 428, 450 **23** – 430	[94–96]
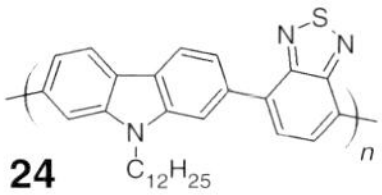 **24**, 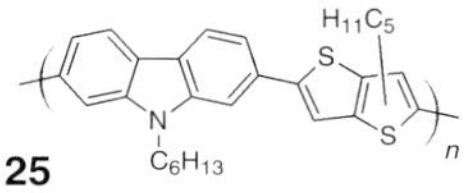 **25**	**24** – 549 **25** – 454	**24** – 577 **25** – 475	[97, 98]
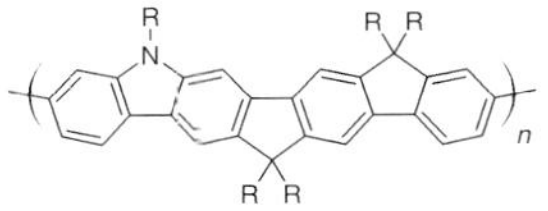 **26**	**26** – 442	**26** – 445	[99]
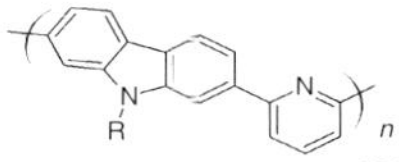 **27**	**27** – 411, 428	**27** – 436	[100]
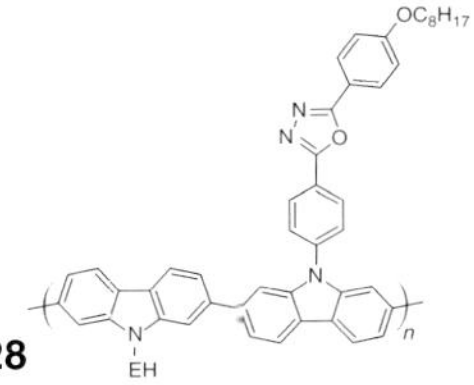 **28**, 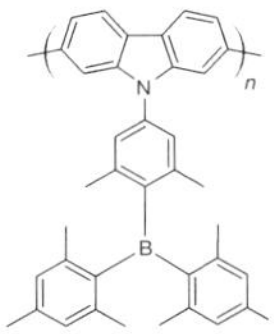 **29**	**28** – 419 **29** – 407	**28** – 437 **29** – 418	[101, 102]

(continued overleaf)

Table 6.1 (*Continued*)

Polymer	λ_{em} (nm)		References
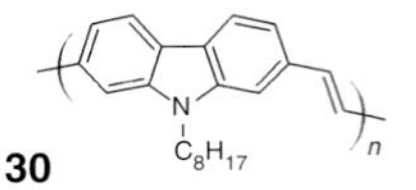 **30**	30 – 470	30 – N.A.	[103]
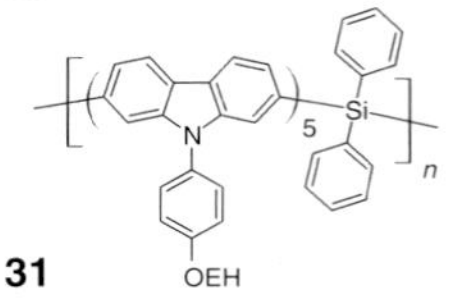 **31**	31 – 412	31 – 425	[104]

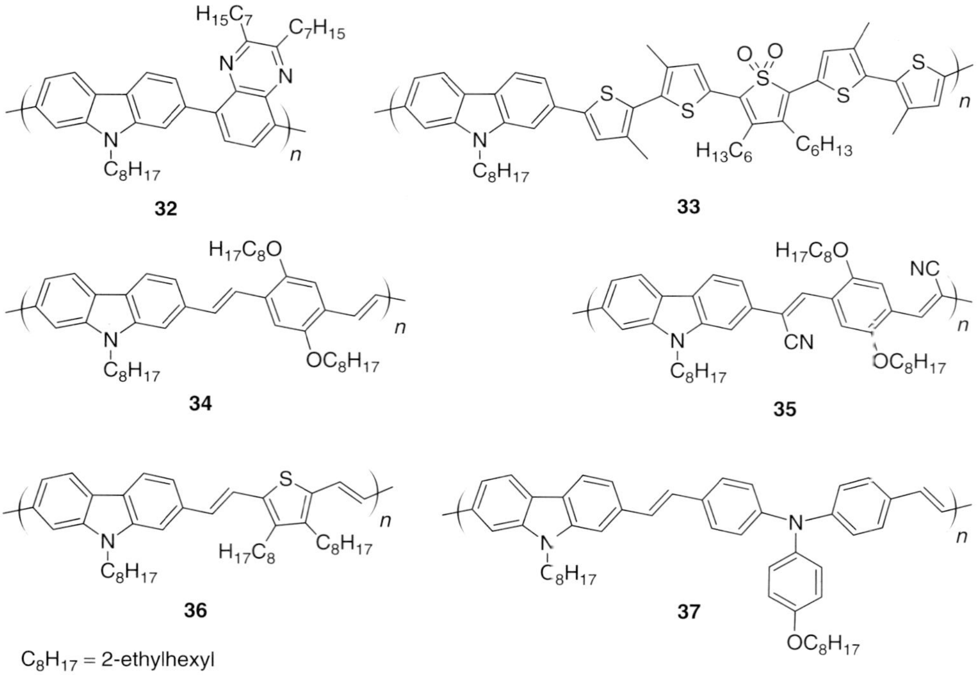
32
33
34
35
36
37
C_8H_{17} = 2-ethylhexyl

Scheme 6.9 2,7-Carbazole-based copolymers for red and green light-emitting diodes.

vinylene unit has been introduced into the poly(2,7-carbazole) backbone. Medium to low bandgap (2.37–2.15 eV) polymers have been obtained (**34–37**) [75]. These polymers were synthesized by Horner–Emmons and Knoevenagel polymerization reactions. These methods lead to highly pure polymers because these two methods do not involve any metal catalyst during the synthesis [113]. The lowest bandgap from this class of polymers was obtained with **35** that showed a strong orange-red luminescence (656 nm). For the other three polymers, it is believed that the fluorescence quenching may be due to strong solid state interactions [114].

6.2.4 Poly(2,7-carbazole)s for Conducting Devices

For the sake of comparison with their 3,6-carbazole-based analogs, several 2,7-carbazole-based homopolymers and copolymers have been synthesized to study their conductivity properties (Scheme 6.10) [73]. Interestingly, poly(2,7-carbazole) derivatives do not usually exhibit much better conductivity values (10^{-2} S cm^{-1}) than their 3,6-carbazole counterparts. Electrochemical characterization shows that the charge created upon electrochemical doping is still pinned on the nitrogen atom of the carbazole unit, thus limiting the conjugation along the polymer backbone. This has been proved by the higher conductivity values obtained for the derivatives bearing an electron-withdrawing group on the nitrogen atom of the carbazole unit that makes the oxidation of the lone pair more difficult.

For example, the addition of 4-hexylbenzoyl moieties (**42**, **43**) onto the nitrogen atom of the carbazole allowed conductivities of about one order of magnitude higher (10^{-1} S cm^{-1}) than a carbazole with a regular alkyl chain [73, 81]. However, the conductivities are still low compared to polythiophenes. Another strategy that has been used to improve the electrical conductivity in these devices is the addition of double alkyl chains onto the nitrogen atom that allows the synthesis of polymers with higher molecular weights (>20 kDa), which led to materials

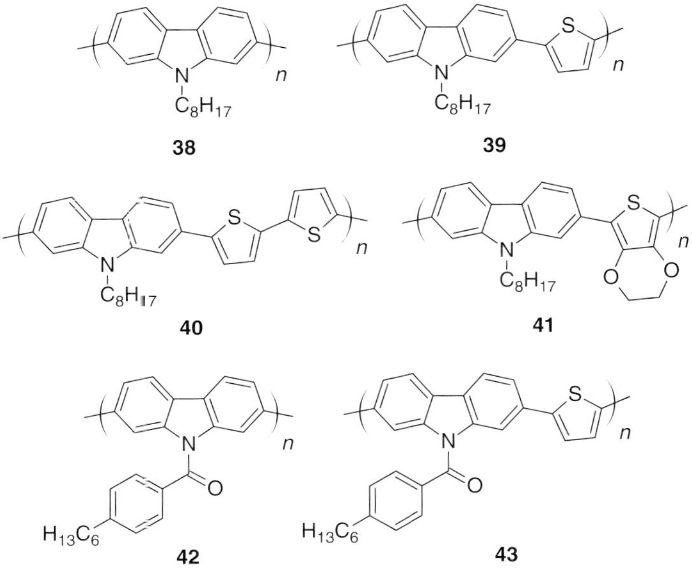

N, C_8H_{17}, n, **38**
S, N, C_8H_{17}, n, **39**
S, S, N, C_8H_{17}, n, **40**
S, N, C_8H_{17}, O, O, n, **41**
N, O, $H_{13}C_6$, n, **42**
S, N, O, $H_{13}C_6$, n, **43**

Scheme 6.10 Poly(2,7-carbazole)s designed specifically for conducting devices.

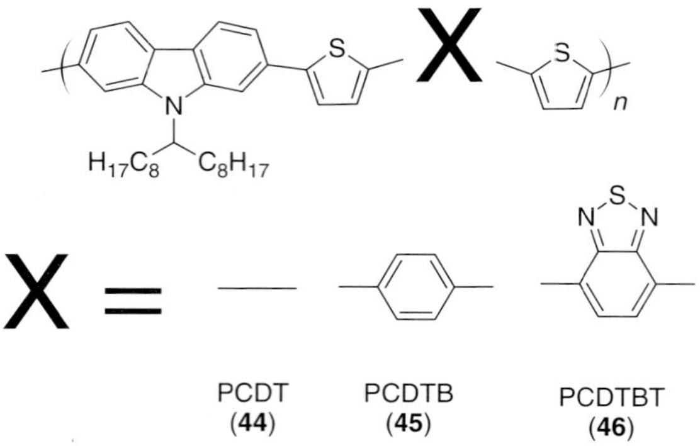

X = — PCDT (**44**), PCDTB (**45**), PCDTBT (**46**)

Scheme 6.11 High-molecular-weight poly(2,7-carbazole) derivatives with higher electrical conductivity.

with better film-forming properties. Alternating and highly conjugated copolymers synthesized by Leclerc *et al.* are shown in Scheme 6.11, which allowed electrical conductivities up to 500 S cm^{-1} when doped with iron trichloride.

This high conductivity for doped PCDTBT (poly[*N*-9″-hepta-decanyl-2,7-carbazole-alt-5,5-(4′,7′-di-2-thienyl-2′,1′,3′-benzothiadiazole)], **46**) combined with the good Seebeck coefficient (70 μV K^{-1}) led to a power factor value of 19 μW m^{-1} K^{-2}, which is among the best ever reported thermoelectric parameters without any thermal and mechanical treatment prior to testing the devices.

6.2.5 Poly(2,7-carbazole)s for Field-Effect Transistors

Unlike LED applications, FETs usually require a high degree of organization within the material. Recently, poly(2,7-carbazole)s have received some attention in this research field. For instance, Leclerc *et al.* designed some poly(2,7-carbazole) derivatives (shown in Scheme 6.12) for FET applications [105]. To date, only modest hole mobilities have been obtained with these polymers (10^{-4}–10^{-3} cm^2 (V s)$^{-1}$). However, interesting DSC and XRD data showed good structural organization and the devices fabricated from these molecules showed good stability under ambient conditions. The polymers were synthesized by Stille coupling using Pd_2dba_3 as the palladium(0) source and P(*o*-tol)$_3$ as the ligand in toluene. This pathway gave high molecular weight for all the new derivatives (10–27 kDa). PCDTBT did also show some interesting results [78]. Indeed, recent studies showed that the hole mobility can be as high as 0.02 cm^2 (V s)$^{-1}$ with a remarkable stability [115].

6.2.6 Poly(2,7-carbazole)s for Photovoltaic Devices

Another interesting application of poly(2,7-carbazole) derivatives is PCs. As mentioned earlier, one of the great advantages of carbazole is clearly its versatility

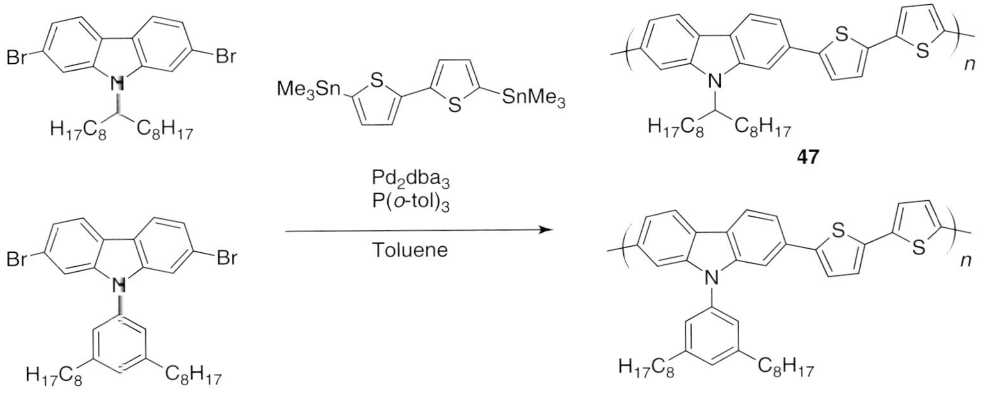

Scheme 6.12 Synthetic pathways to poly(2,7-carbazole) derivatives for field-effect transistors.

and stability [116]. Not only numerous kinds of alkyl or aryl groups can be added onto the nitrogen atom but simple substitution at the 2- and 7-positions can also produce various functional groups. The first poly(2,7-carbazole)s that were used in solar cells were synthesized by Müllen *et al.* and they used a bulk heterojunction (BHJ) architecture [11, 12]. They used a homopolymer as the electron donor and perylene tetracarboxydiimide as the electron acceptor [117, 118]. The very low power conversion efficiency (PCE) is probably due to the poor solar spectrum match of this polymer. This eventually allowed the synthesis of a wide variety of carbazole-based polymers for PC applications. Several copolymers were synthesized by Leclerc *et al.* based on a poly(2,7-carbazolenevinylene) configuration (Scheme 6.13) [119]. Polymers **49–52** were synthesized by a Horner–Emmons reaction. For **53**, a Stille cross-coupling was performed using $AsPh_3$ as the ligand and Pd_2dba_3 as the palladium(0) source. All these syntheses gave relatively low molecular weights, ranging from 3 to 5 kDa, and a PCE up to 0.8% was achieved. A new pathway to synthesize poly(*N*-alkylcarbazole-2,7-vinylene) derivatives has recently been developed by Nomura *et al.* for LED applications [103]. This new method allows the synthesis of **49** through acyclic diene metathesis (ADMET) polymerization using ruthenium–carbene complex catalysts. The molecular weight is relatively high (20–30 kDa) compared to the pathways presented earlier. This technique could be useful for the synthesis of new low bandgap polycarbazole derivatives.

These promising results stimulated the research of new low bandgap polymers, and other new polymers were synthesized leading to high PCE [120, 121]. These new polymers had benzodithiazole-like units in common to lower the bandgap and to obtain, at the same time, good field-effect mobility that is important for solar cell applications. By using similar units, a wide variety of poly(2,7-carbazole) derivatives were synthesized with the structures shown in Scheme 6.14.

These polymers were all synthesized using the same pathway, a Suzuki cross-coupling with Pd_2dba_3 and $P(o\text{-tol})_3$ catalyst system. The reaction worked

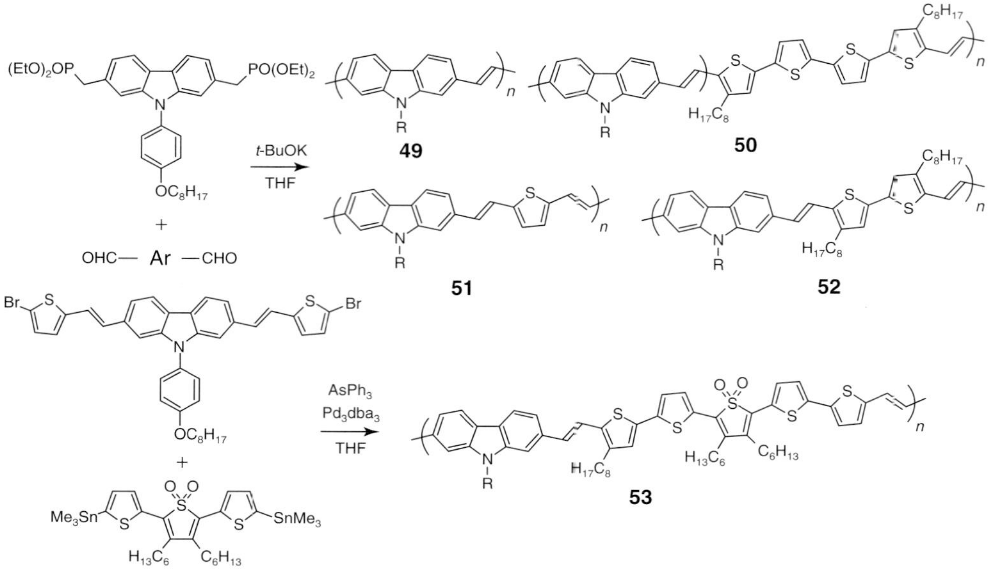

Scheme 6.13 Synthetic pathways to poly(2,7-carbazolenevinylene) derivatives for photovoltaic cell applications [119].

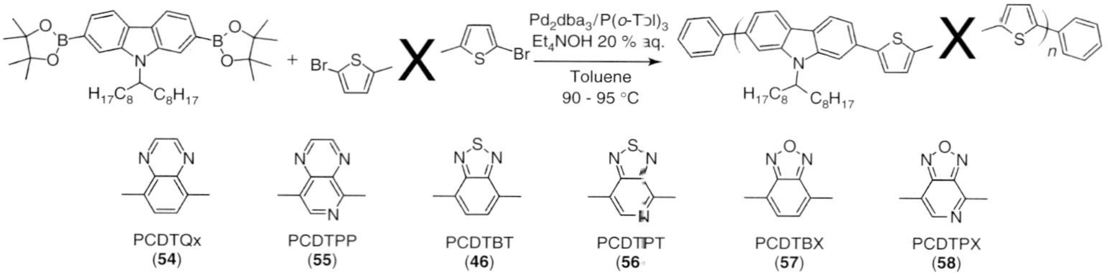

Scheme 6.14 Structure and synthesis of the 2,7-carbazole-based low bandgap polymers.

well in terms of the molecular weight for PCDTQx (**54**), PCDTPP (**55**), PCDTBT (**46**), and PCDTBX (**57**) with M_n of 9.0–36 kDa but was not effective for PCDTPT (**56**) and PCDTPX (**58**) (M_n = 4.0–4.5 kDa) [78]. The authors suggested that this may be due to the asymmetry in the polymer created by the additional nitrogen atom in the backbone or may be due to the complexation of the palladium atoms caused again by the nitrogen atoms in the backbone of the polymer. Among the high-molecular-weight polymers, PCDTBT has gained much attention because of the great performances observed [77]. The PCE of this polymer reached 3.6% and Heeger *et al.* showed that by the optimization of the device it was possible to reach 6.1% [122].

6.3
Other Carbazole Derivatives

Many other carbazole derivatives have been synthesized over the last years for various applications [75, 123]. Indolo[3,2-*b*]carbazole (IC) is very promising because of the possibilities to easily functionalize both ends of the molecule, making it appropriate for the synthesis of new polymers [35]. Moreover, other ladder-like rigid carbazole derivatives, such as diindolo[3,2-*b*:2′,3′-*h*]carbazole (DIC) [79], have been synthesized to enhance the packing and three-dimensional organization of the materials. In this section, we discuss recent progress with respect to monomer and polymer syntheses of IC and DIC derivatives.

6.3.1
Indolo[3,2-b]carbazoles and Poly(indolo[3,2-b]carbazole)s

The synthesis of IC-based monomers has evolved in the last few years [124, 125]. But, despite these advances, still the Robinson mechanism based on a Fischer indolization is still used widely [126]. The yields of the reaction are not very high (20–50%), but disubstituted ICs can be separated from other regioisomers (shown in Scheme 6.15) by simple recrystallization. This synthetic pathway allowed the synthesis of numerous new IC-based oligomers [127–130] and monomers for further polymerization, which are described later. IC-based materials are mainly used for FET applications, but promising performances have also been shown in organic PCs [35, 123].

Over the last years, only a few IC-based monomers have been used and they are generally 2,8-(**59**) or 3,9-dihaloindolo[3,2-*b*]carbazoles (**60**). However, numerous side chains have been added onto the nitrogen atoms to increase the solubility [125, 131, 132], to lower the short contact distance between separate molecules [127, 128], or to completely change the thin film organization [129, 130, 133, 134]. The first successful synthesis of IC-based polymers was reported in 2006 [125]. A Ni(0)-catalyzed Yamamoto homocoupling from 2,8- or 3,9-dichloro ICs allowed the synthesis of the desired polymers. In both cases, the molecular weights were quite low, ranging from 3.0 to 6.0 kDa, probably due to the very low solubility

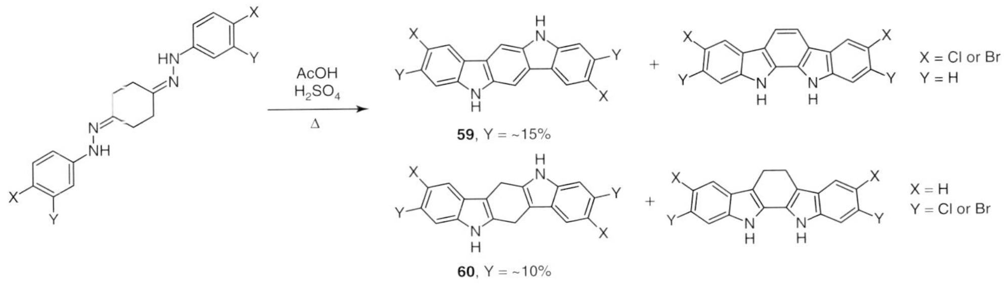

Scheme 6.15 Synthesis of ICs from Fischer indolization reaction.

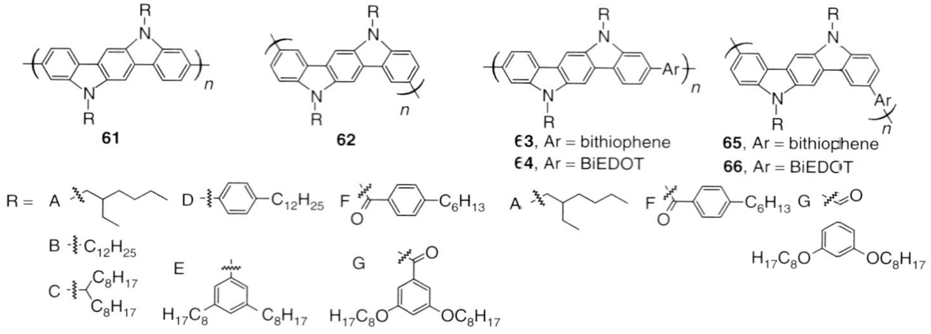

61 62 63, Ar = bithiophene 64, Ar = BiEDOT 65, Ar = bithiophene 66, Ar = BiEDOT

R = A B C D E F G

Scheme 6.16 Chemical structure of some IC-based polymers [35].

of the indolocarbazole unit. To raise the molecular weights of the polymers, copolymerization was performed [135]. IC monomers were polymerized with bithiophene through a Stille cross-coupling or with bis(3,4-ethylenedioxythiophene) through Suzuki cross-coupling. Unfortunately, the molecular weights were still very low leading to irrelevant properties. It was found that by changing the side chains on the nitrogen atoms, the solubility of the monomers could be raised to acceptable levels. More recently, Ong *et al.* were able to obtain homopolymers (**61**, **62**) with molecular weights of 9.0–11.0 kDa via Zn-mediated dehalogenative coupling polymerization and $FeCl_3$-mediated oxidative coupling polymerization [132]. These polymers were tested in FET devices as p-type semiconductors and afforded hole mobilities of almost 0.01 cm^2 $(V\ s)^{-1}$. As shown in Scheme 6.16, a wide variety of IC-based polymers have been synthesized usually showing good charge transport and thermoelectric properties [81].

Recently, Lu *et al.* have shown that it was possible to efficiently use an IC-based copolymer in photovoltaic applications [136]. They were able to polymerize, via Suzuki cross-coupling, the IC unit with 4,7-bis(3,4′-dioctyl-2,2′-bithiophen-5-yl)-2,1,3-benzothiadiazole to obtain a copolymer with a molecular weight of 19.5 kDa (**67** in Scheme 6.17). In order to increase the molecular weight, they were forced to use long double side chains. Best performances were obtained with an IC-based polymer with a PCE of 3.6%. Moreover, new polymers for FET and PC applications have been synthesized by Tsai *et al.*, but the fabricated devices suffer from the low molecular weights of their polymer. The structure is shown in Scheme 6.17 (**68**). The highest mobility they obtained was 10^{-4} cm^2 $(V\ s)^{-1}$ and the best PCE was 1.4% [137].

6.3.2
Diindolo[3,2-b:2′,3′-h]carbazoles

In the quest of extending the π-conjugated unit as much as possible, a new ladder derivative was developed by Leclerc *et al.* in 2004 [79]. The synthesis involved a seven-step synthetic pathway from 1,3,6,8-tetrabromo-2,7-dimethoxy-*N*-octylcarbazole (**69**) (Scheme 6.18, pathway 1). Since the Cadogan ring closure in

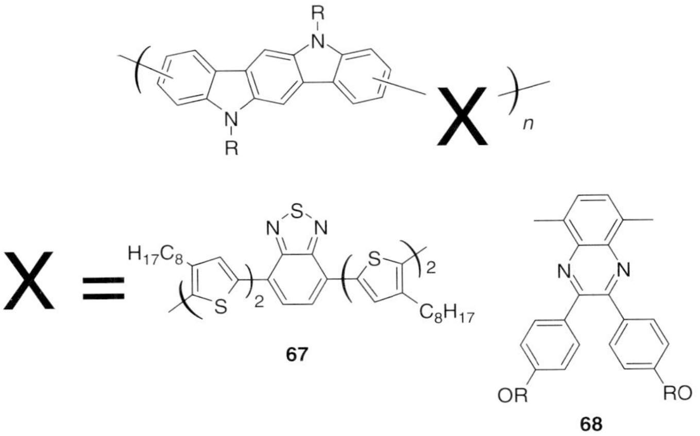

Scheme 6.17 Indolo[3,2-*b*]carbazole-based low bandgap polymers.

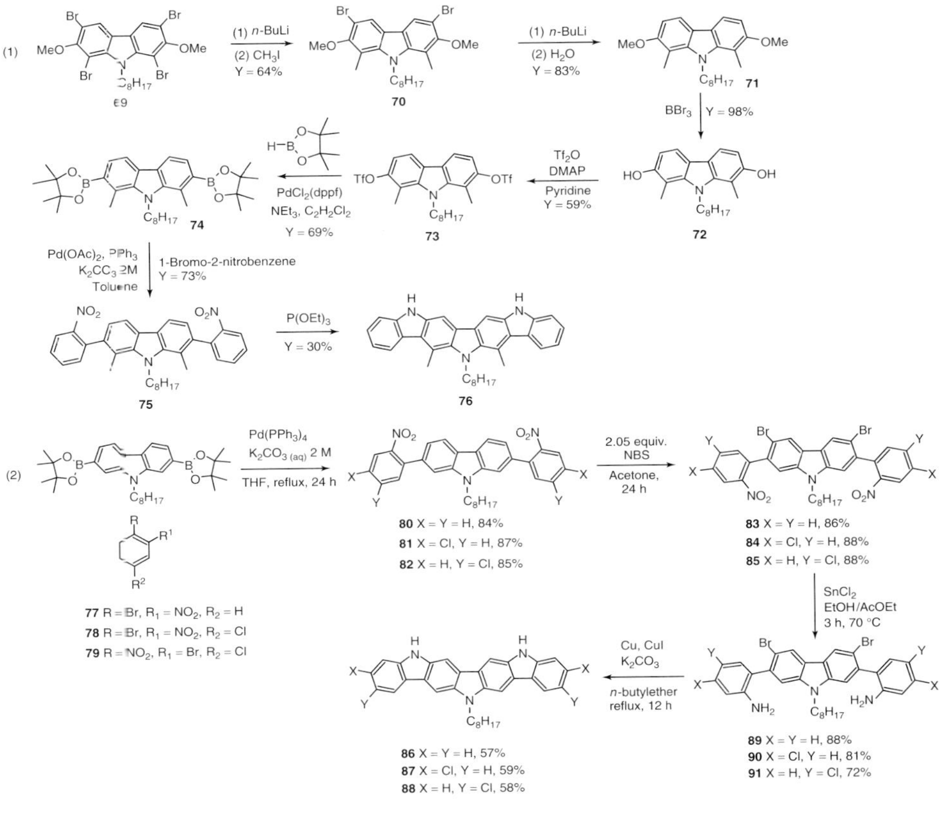

Scheme 6.18 Two different pathways to synthesize DIC units.

the final step is not regioselective, the 1- and 8-positions of the carbazole have to be protected by methyl groups. The other downside of this synthetic pathway is the complexity of functionalizing both ends of the DIC with halogens or any reactive groups.

Afterward, Leclerc *et al.* found that it was not necessary to use the Cadogan ring closure as the final step [138]. They were able to develop a synthetic pathway (Scheme 6.18, pathway 2), with the Ullmann reaction being the key ring-closure step. This synthesis takes advantage of the high reactivity of the 3- and 6-positions of the carbazole by adding the two bromine atoms (**83–85**) that are necessary to perform the Ullmann reaction (**86–88**). FETs were tested from these molecules and hole mobilities of 10^{-3} cm^2 (V s)$^{-1}$ were obtained. However, the clear advantage of this methodology, besides the overall yield, is that it is now possible to obtain quite efficiently 3,10-dichloro DICs. This opens the way for the synthesis of new polymers based on this rigid rod-like unit. In 2006, Leclerc *et al.* first reported the synthesis of DIC-based polymers [125]. Two different monomers were obtained with two completely different conjugation pathways: the 3,10- and 2,11-dichloro DICs. These two polymers were obtained via a Ni(0)-catalyzed Yamamoto homocoupling with good yields (83 and 80%, respectively) from the soluble fraction in chloroform. Unfortunately, only low molecular weights were obtained (3.0–6.0 kDa), which suggested that the monomers were not soluble enough to allow efficient polymerization of such rigid unit. Even though the performances were disappointing, this study still provided some interesting information about the conjugation along the π-conjugated system. In fact, it was shown that when the DIC units are linked at the 2- and 11-positions, the conjugation takes place through the nitrogen atoms, and in contrast the conjugation rather takes a *p*-phenylene pathway where the nitrogen atoms act as side chains instead of participating in the conjugation.

6.4 Concluding Remarks

From all the above-mentioned examples, it is clear that research in the synthesis of carbazole-containing conjugated polymers is still well underway. These 3.0–6.0 kDa monomers show a great versatility that allows the synthesis of polymers using various synthetic procedures. For instance, a great amount of research on poly(2,7-carbazole) derivatives is focused on finding the best comonomers for FET and PC applications. The synthesis of longer π-conjugated systems such as IC- and DIC-based polymers still attracts much interest, but some issues have to be addressed before the performances can reach those of carbazole derivatives. This is especially true for the synthesis of polymers because of their low solubility, but for oligomers and small molecules the performances are quite similar [130]. These ladder-like molecules are even easier to study because it is much easier to grow single crystals, which help defining some structure–property relationships [129, 133]. However, the main goal is to obtain highly performing polymers, where the monomers are quickly and easily synthesized with high overall yields. With the latest performances in PCs and FETs obtained with poly(2,7-carbazole) derivatives, it seems likely that this is an avenue that will continue to be widely investigated over the next years. Finally, with the recent report of poly(1,8-carbazole)s [139], completely new classes of conjugated polymers may soon emerge. Clearly, many new and exciting results should appear in the next few years with polycarbazoles.

References

1. Shirakawa, H., Louis, E.J., MacDiarmid, A.G., Chiang, C.K., and Heeger, A.J. (1977) *J. Chem. Soc. Chem. Commun.*, 578.
2. Skotheim, T.A. and Reynolds, J.R. (2007) *Handbook of Conducting Polymers*, Taylor & Francis Group, Boca Raton.
3. Klauk, H. (2006) *Organic Electronics*, Wiley-VCH Verlag GmbH, Weinheim.
4. Hadziioannou, G. and Malliaras, G.G. (2007) *Semiconducting Polymers*, Wiley-VCH Verlag GmbH, Weinheim.
5. Yu, L., Chen, M., and Dalton, L.R. (1990) *Chem. Mater.*, **2**, 649.
6. Leclerc M. and Faid, K. (1997) *Adv. Mater.*, **9**, 1087.
7. McQuade, D.T., Pullen, A.E., and Swager T.M. (2000) *Chem. Rev.*, **100**, 2537.
8. Burroughes, J.H., Bradley, D.D.C., Brown, A.R., Marks, R.N., Mackay, K., Friend, R.H., Burns, P.L., and Holmes, A.B. (1990) *Nature*, **347**, 539.
9. Dimitrakopoulos, C.D. and Malenfant, P.R.L. (2002) *Adv. Mater.*, **14**, 99.
10. Katz, H.E. (2004) *Chem. Mater.*, **16**, 4748.
11. Brabec, C.J., Sariciftci, N.S., and Hummelen, J.C. (2001) *Adv. Funct. Mater.*, **11**, 15.
12. Hoppe, H. and Sariciftci, N.S. (2004) *J. Mater. Res.*, **19**, 1924.
13. Coakley, K.M. and McGehee, M.D. (2004) *Chem. Mater.*, **16**, 4533.
14. Roncali J. (1992) *Chem. Rev.*, **92**, 711.
15. Kim, D.Y., Cho, H.N., and Kim, C.Y. (2000) *Prog. Polym. Sci.*, **25**, 1089.
16. Scherf, U. and List, E.J.W. (2002) *Adv. Mater.*, **14**, 477.
17. Leclerc, M. (2001) *J. Polym. Sci. Part A: Polym. Chem.*, **39**, 2867.
18. Kang, E.T., Neoh, K.G., and Tan, K.L. (1998) *Prog. Polym. Sci.*, **23**, 277.
19. Huang, J. and Kaner, R.B. (2004) *Angew. Chem. Int. Ed.*, **43**, 5817.
20. MacDiarmid, A.G. (2002) *Synth. Met.*, **125**, 11.
21. Chiang, J.C. and MacDiarmid, A.G. (1986) *Synth. Met.*, **13**, 193.
22. MacDiarmid, A.G., Chiang, J.C., Richter, A.F., and Epstein, A.J. (1987) *Synth. Met.*, **18**, 285.
23. Morin, J.-F., Leclerc, M., Adès, D., and Siove, A. (2005) *Macromol. Rapid Commun.*, **26**, 761.
24. Morin, J.F. and Leclerc, M. (2001) *Macromolecules*, **34**, 4680.
25. Huang, J., Virji, S., Weiller, B.H., and Kaner, B.H. (2003) *J. Am. Chem. Soc.*, **125**, 314.
26. Bailey, J.K., Pozarnsky, G.A., and Mecartney, M.L. (1992) *J. Mater. Chem.*, **7**, 2530.
27. Grazulevicius, J.V., Strohriegl, P., Pielichowski, J., and Pielichowski, K. (2003) *Prog. Polym. Sci.*, **28**, 1297.
28. Huang, J. and Kraner, R.B. (2007)in *Handbook of Conducting Polymers* (eds T.A. Skotheim and J.R. Reynolds), CRC Press, Boca Raton, p. 7.
29. Tang, X., Yu, J., Li, L., Wen, W., and Jiang, Y. (2008) *Chin. J. Chem. Phys.*, **21**, 510.
30. Xia, H., Li, M., Lu, D., Zhang, C.B., Xie, W.J., Liu, X.D., Yang, B., and Ma, Y.G. (2007) *Adv. Funct. Mater.*, **17**, 1757.
31. Ikeda, N. and Miyasaka, T. (2005) *Chem. Commun.*, 1886.
32. Lav, T.X., Tran-Van, F., Vidal, F., Péralta, S., Chevrot, C., Teyssié, D., Grazulevicius, J.V., Getautis, V., Derbal, H., and Nunzi, J.M. (2008) *Thin Solid Films*, **516**, 7223.
33. Dridi, C., Barlier, V., Chaabane, H., Davenas, J., and Ouada, H.B. (2008) *Nanotechnology*, **19**, 375201.

34. Barlier, V., Bounor-Legaré, V., Boiteux, G., Davenas, J., Slazak, A., Rybak, A., and Jung, J. (2009) *Synth. Met.*, **159**, 508.
35. Boudreault, P.-L.T., Blouin, N., and Leclerc, M. (2008) *Adv. Polym. Sci.*, **212**, 99.
36. Zotti, G., Zecchin, S., Schiavon, G., and Groenendaal, L. (2000) *Chem. Mater.*, **12**, 2996.
37. Schottland, P., Zong, K., Gaupp, C.L., Thompson, B.C., Thomas, C.A., Giurgiu, I., Hickman, R., Abboud, K.A., and Reynolds, J.R. (2000) *Macromolecules*, **33**, 7051.
38. Pearson, J.H. and Stolka, M. (1981) *Polymer Monographs*, Vol. 6 , Gordon and Breach, New York.
39. Narmann, H. and Strohriegl, P. (1992) *Handbook of Polymer Synthesis*, M. Dekker, New York.
40. Borsenberger, P.M. and Weiss, D. (1998) *Organic Photoreceptors for Xerography*, M. Dekker, New York.
41. Tang, X., Yu, J., Li, L., Zhang, L., and Jiang, Y. (2009) *Displays*, **30**, 123.
42. Wagner, J., Pielichowski, J., Hinsch, A., Pielichowski, K., Bogdal, D., Pajda, M., Kurek, S.S., and Burczyk, A. (2004) *Synth. Met.*, **146**, 159.
43. Beaupré, S., Breton, A.-C., Dumas, J., and Leclerc, M. (2009) *Chem. Mater.*, **21**, 1504.
44. Tsai, H.-C., Yu, I.-C., Chang, P.-H., Yu, D.-C., and Hsiue, G.-H. (2007) *Macromol. Rapid Commun.*, **28**, 334.
45. Jung, S.-H., Pisula, W., Rouhanipour, A., Räder, H.J., Jacob, J., and Müllen, K. (2006) *Angew. Chem. Int. Ed.*, **45**, 4685.
46. Ambrose, J.F. and Nelson, R.F. (1967) *J. Electrochem. Soc. Electrochem. Sci.*, **115**, 1159.
47. Ambrose, J.F., Carpenter, L.L., and Nelson, R.F. (1975) *J. Electrochem. Soc. Electrochem. Sci. Techn.*, **122**, 876.
48. Siove, A., David, R., Adès, D., Roux, C., and Leclerc, M. (1995) *J. Chim. Phys.*, **92**, 787.
49. Siove, A. and Adès, D. (2004) *Polymer*, **45**, 4045.
50. Gaupp, C.L. and Reynolds, J.R. (2003) *Macromolecules*, **36**, 6305.
51. Sotzing, G.A., Reddinger, J.L., Katritzky, A.R., Soloducho, J., Musgrave, R., and Reynolds, J R. (1997) *Chem. Mater.*, **9**, 1578.
52. Wellinghoff, S.T., Deng, Z., Reed, J.F., and Racchini, J. (1984) *Polym. Prepr.*, **25**, 238.
53. Siove, A., Adès, D., Chevrot, C., and Froyer, G. (1989) *Makromol. Chem.*, **190**, 1361.
54. Siove, A., Adès, D., Ngbilo, E. and Chevrot, C. (1990) *Synth. Met.*, **38**, 331.
55. Siove, A., Aboulkassim, A., Faïd, K., and Adès, D. (1995) *Polym. Int.*, **37**, 171.
56. Geissler, U., Hallensleben, M.L., Rienecker, A., and Rohde, N. (1997) *Polym. Adv. Technol.*, **8**, 87.
57. Iraqi, A. and Wataru, I. (2004) *J. Polym. Sci.: Polym. Chem. Ed.*, **42**, 6041.
58. Zhang, Z.B., Fujiki, M., Tang, H.-Z., Motonaga, M., and Torimistu, K. (2002) *Macromolecules*, **35**, 1988.
59. Romero, D.B., Schaer, M., Leclerc, M., Adès, D., Siove, A., and Zuppiroli, L. (1996) *Synth. Met.*, **80**, 271.
60. Adès, D., Boucard, V., Cloutet, E., Siove, A., Olivero, C., Castex, M.C., and Pichler, G. (2000) *J. Appl. Phys.*, **87**, 7290.
61. Zugang, L. and Nazarè, H. (2000) *Synth. Met.*, **111-112**, 47.
62. Kido, J., Hongawa, K., Okuyama, K., and Nagai, K. (1994) *Appl. Phys. Lett.*, **64**, 815.
63. Niu, Y.H., Liu, M.S., Ka, J.W., Bardeker, J., Zin, M.T., Schofield, R., Chi, Y., and Jen, A.K.Y. (2007) *Adv. Mater.*, **19**, 300.
64. Paik, K.L., Baek, N.S., Kim, H.K., Lee, J.-H., and Lee, Y. (2002) *Macromolecules*, **35**, 6782.
65. Smith, P.A.S. and Brown, B.B. (1951) *J. Am. Chem. Soc.*, **73**, 2435.
66. Smith, P.A.S. and Brown, B.B. (1951) *J. Am. Chem. Soc.*, **73**, 2438.
67. Leditschke, H. (1953) *Chem. Ber.*, **86** 522.
68. Yamato, T., Hideshima, C., Suehiro, K., Tashiro, M., Prakash, G.K.S., and Olah, G.A. (1991) *J. Org. Chem.*, **56**, 6248.
69. Cadogan, J.I.G., Cameron-Wood, M., Mackie, R.K., and Searle, R.J G. (1965) *J. Chem. Soc.*, 4831.

70. Cadogan, J.I.G. (1969) *Synthesis*, 11.
71. Freeman, A.W., Urvoy, M., and Criswell, M.E. (2005) *J. Org. Chem.*, **70**, 5014.
72. Appukkuttan, P., Van der Eycken, E., and Dehaen, W. (2005) *Synlett*, **2005**, 127.
73. Zotti, G., Schiavon, G., Zecchin, S., Morin, J.-F., and Leclerc, M. (2002) *Macromolecules*, **35**, 2122.
74. Dierschke, F., Grimsdale, A., and Müllen, K. (2003) *Synthesis*, 2470.
75. Morin, J.F., Drolet, N., Tao, Y., and Leclerc, M. (2004) *Chem. Mater.*, **16**, 4619.
76. Morin, J.-F. and Leclerc, M. (2002) *Macromolecules*, **35**, 8413.
77. Blouin, N., Michaud, A., and Leclerc, M. (2007) *Adv. Mater.*, **19**, 2295.
78. Blouin, N., Michaud, A., Gendron, D., Wakim, S., Blair, E., Neagu-Plesu, R., Belletete, M., Durocher, G., Tao, Y., and Leclerc, M. (2008) *J. Am. Chem. Soc.*, **130**, 732.
79. Bouchard, J., Wakim, S., and Leclerc, M. (2004) *J. Org. Chem.*, **69**, 5705.
80. Levesque, I., Gao, X., Klug, D.D., Tse, J.S., Ratcliffe, C.I., and Leclerc, M. (2005) *React. Funct. Polym.*, **65**, 23.
81. Levesque, I., Bertrand, P.O., Blouin, N., Leclerc, M., Zecchin, S., Zotti, G., Ratcliffe, C.I., Klug, D.D., Gao, X., Gao, F., and Tse, J.S. (2007) *Chem. Mater.*, **19**, 2128.
82. Nozaki, K., Takahashi, K., Nakano, K., Hiyama, T., Tang, H.-Z., Fujiki, M., Yamaguchi, S., and Tamao, K. (2003) *Angew. Chem. Int. Ed.*, **42**, 2051.
83. Kuwahara, A., Nakano, K., and Nozaki, K. (2005) *J. Org. Chem.*, **70**, 413.
84. Kitawaki, T., Hayashia, Y., Uenoa, A., and Chida, N. (2006) *Tetrahedron*, **62**, 6792.
85. Yamamoto, T. (2003) *Synlett*, 425.
86. Schlüter, A.D. (2001) *J. Polym. Sci. Part A: Polym. Chem.*, **39**, 1533.
87. Bao, Z., Chan, W.K., and Yu, L. (1995) *J. Am. Chem. Soc.*, **117**, 12426.
88. Blouin, N. and Leclerc, M. (2008) *Acc. Chem. Res.*, **41**, 1110.
89. Morin, J.-F., Beaupré, S., Leclerc, M., Lévesque, I., and D'Iorio, M. (2002) *Appl. Phys. Lett.*, **80**, 341.
90. Iraqi, A. and Wataru, I. (2004) *Chem. Mater.*, **16**, 442.
91. Shen, J.Y., Yang, X.L., Huang, T.H., Lin, J.T., Ke, T.H., Chen, L.Y., Wu, C.C., and Yeh, M.C.P. (2007) *Adv. Funct. Mater.*, **17**, 983.
92. Michinobu, T., Kumazawa, H., and Shigehara, K. (2007) *Chem. Lett.*, **36**, 620.
93. Michinobu, T., Kumazawa, H., Otsuki, E., Usui, H., and Shigehara, K. (2009) *J. Polym. Sci. Part A: Polym. Chem.*, **47**, 3880.
94. Kobayashi, N. and Kijima, M. (2007) *Appl. Phys. Lett.*, **91**, 081113.
95. Koguchi, R., Kobayashi, N., Shinnai, T., Oikawa, K., Tsuchiya, K., and Kijima, M. (2008) *Macromol. Chem. Phys.*, **209**, 439.
96. Kobayashi, N., Koguchi, R., and Kijima, M. (2006) *Macromolecules*, **39**, 9102.
97. Du, J., Xu, E., Zhong, H., Yu, F., Liu, C., Wu, H., Zeng, D., Ren, S., Sun, J., Liu, Y., Cao, A., and Fang, Q. (2008) *J. Polym. Sci. Part A: Polym. Chem.*, **46**, 1376.
98. Tang, W., Lin, T., Ke, L., and Chen, Z.-k. (2008) *J. Polym. Sci. Part A: Polym. Chem.*, **46**, 7725.
99. Qiu, S., Liu, L., Wang, B., Shen, F., Zhang, W., Li, M., and Ma, Y. (2005) *Macromolecules*, **38**, 6782.
100. Pan, X., Liu, S., Chan, H.S.O., and Ng, S.-C. (2005) *Macromolecules*, **38**, 7629.
101. Zhang, K., Tao, Y., Yang, C., You, H., Zou, Y., Qin, J., and Ma, D. (2008) *Chem. Mater.*, **20**, 7324.
102. Reitzenstein, D. and Lambert, C. (2009) *Macromolecules*, **42**, 773.
103. Yamamoto, N., Ito, R., Geerts, Y., and Nomura, K. (2009) *Macromolecules*, **42**, 5104.
104. Koguchi, R., Kobayashi, N., and Kijima, M. (2009) *Macromolecules*, **42**, 5946.
105. Wakim, S., Blouin, N., Gingras, E., Tao, Y., and Leclerc, M. (2007) *Macromol. Rapid Commun.*, **28**, 1798.
106. Iraqi, A., Simmance, T.G., Yi, H., Stevenson, M., and Lidzey, D.G. (2006) *Chem. Mater.*, **18**, 5789.
107. Yi, H., Iraqi, A., Stevenson, M., Elliott, C.J., and Lidzey, D.G. (2007) *Macromol. Rapid Commun.*, **28**, 1155.

108. Bellows, D., Gingras, E., Aly, S.M., Abd-El-Aziz, A.S., Leclerc, M., and Harvey, P.D. (2008) *Inorg. Chem.*, **47**, 11720.
109. Morin, J.-F., Boudreault, P.-L., and Leclerc, M. (2002) *Macromol. Rapid Commun.*, **23**, 1032.
110. Fu, Y. and Bo, Z. (2005) *Macromol. Rapid Commun.*, **26**, 1704.
111. Ajayaghosh, A. (2003) *Chem. Soc. Rev.*, **32**, 181.
112. Roncali, J. (2007) *Macromol. Rapid Commun.*, **28**, 1761.
113. Krebs, F.C., Nyberg, R.B., and Jorgensen, M. (2004) *Chem. Mater.*, **16**, 1313.
114. Belletête, M., Durocher, G., Hamel, S., Côté, M., Wakim, S., and Leclerc, M. (2005) *J. Chem. Phys.*, **122**, 104303.
115. Cho, S., Seo, J.H., Park, S.H., Kim, D.-Y., Beaupré, S., Leclerc, M., Lee, K., and Heeger, A.J. (2010) *Adv. Mater.*, Doi:10.1002/adma. 200903420.
116. Pisula, W., Mishra, A.K., Li, J., Baumgarten, M., and Müllen, K. (2008) in *Organic Photovoltaics: Materials, Device Physics and Manufacturing Technologies* (eds C.Brabec, V. Dyakonov, and U. Scherf), Wiley-VCH Verlag GmbH, Weinheim, p. 93.
117. Li, J., Dierschke, F., Wu, J., Grimsdale, A.C., and Mullen, K. (2006) *J. Mater. Chem.*, **16**, 96.
118. Ooi, Z.E., Tam, T.L., Shin, R.Y.C., Chen, Z.K., Kietzke, T., Sellinger, A., Baumgarten, M., Mullen, K., and de Mello, J.C. (2008) *J. Mater. Chem.*, **18**, 4619.
119. Leclerc, N., Michaud, A., Sirois, K., Morin, J.F., and Leclerc, M. (2006) *Adv. Funct. Mater.*, **16**, 1694.
120. Peet, J., Kim, J.Y., Coates, N.E., Ma, W.L., Moses, D., Heeger, A.J., and Bazan, G.C. (2007) *Nat. Mater.*, **6**, 497.
121. Zhang, F., Mammo, W., Andersson, L.M., Admassie, S., Andersson, M.R., and Inganäs, O. (2006) *Adv. Mater.*, **18**, 2169.
122. Park, S.H., Roy, A., Beaupre, S., Cho, S., Coates, N., Moon, J.S., Moses, D., Leclerc, M., Lee, K., and Heeger, A.J. (2009) *Nat. Photonics*, **3**, 297.
123. Wakim, S., Aïch, B.-R., Tao, Y., and Leclerc, M. (2008) *Polym. Rev.*, **48**, 432.
124. Wakim, S., Bouchard, J., Simard, M., Drolet, N., Tao, Y., and Leclerc, M. (2004) *Chem. Mater.*, **16**, 4386.
125. Blouin, N., Michaud, A., Wakim, S., Boudreault, P.L.T., Leclerc, M., Vercelli, B., Zecchin, S., and Zotti, G. (2006) *Macromol. Chem. Phys.*, **207**, 166.
126. Robinson, B. (1963) *J. Chem. Soc.*, 3097.
127. Li, Y., Wu, Y., Gardner, S., and Ong, B.S. (2005) *Adv. Mater.*, **17**, 849.
128. Wu, Y., Li, Y., Gardner, S., and Ong, B.S. (2005) *J. Am. Chem. Soc.* **127**, 614.
129. Boudreault, P.L.T., Wakim, S., Blouin, N., Simard, M., Tessier, C., Tao, Y., and Leclerc, M. (2007) *J. Am. Chem. Soc.*, **129**, 9125.
130. Boudreault, P.-L.T., Wakim, S., Tang, M.L., Tao, Y., Bao, Z., and Leclerc, M. (2009) *J. Mater. Chem.*, **19**, 2921.
131. Blouin, N., Leclerc, M., Vercelli, B., Zecchin, S., and Zotti, G. (2006) *Macromol. Chem. Phys.*, **207**, 175.
132. Li, Y., Wu, Y., and Ong, B.S. (2006) *Macromolecules*, **39**, 6521.
133. Belletete, M., Blouin, N., Boudreault, P.-L.T., Leclerc, M., and Durocher, G. (2006) *J. Phys. Chem. A*, **110**, 13696.
134. Belletete, M., Boudreault, P.-L.T., Durocher, G., and Leclerc, M. (2007) *THEOCHEM*, **824**, 15.
135. Drolet, N., Morin, J.F., Leclerc, N., Wakim, S., Tao, Y., and Leclerc, M. (2005) *Adv. Func. Mater.*, **15**, 1671.
136. Lu, J., Liang, F., Drolet, N., Ding, J., Tao, Y., and Movileanu, R. (2008) *Chem. Commun.*, 5315.
137. Tsai, J.-H., Chueh, C.-C., Lai, M.-H., Wang, C.-F., Chen, W.-C., Ko, B.-T., and Ting, C. (2009) *Macromolecules*, **42**, 1897.
138. Wakim, S., Bouchard, J., Blouin, N., Michaud, A., and Leclerc, M. (2004) *Org. Lett.*, **6**, 3413.
139. Tsuyoshi, M., Haruka, O., and Kiyotaka, S. (2008) *Macromol. Rapid Commun.*, **29**, 111.

7
Phenylene-Based Ladder Polymers

Andrew C. Grimsdale and Klaus Müllen

7.1
Introduction

Ladder polymers are double-stranded polymers in which each monomer unit is attached to its neighbors by two bonds [1]. Whereas a single-stranded linear polymer such as poly(*para*-phenylene) (PPP, **1a**, Figure 7.1) is often thought of as an essentially one-dimensional wirelike entity, a ladder polymer such as ladder-type poly(*para*-phenylene) (LPPP, **2**) has a more two-dimensional ribbon-like geometry, and so constitutes a structure intermediate between linear polymers and a truly two-dimensional structure such as graphene. Much of the interest in ladder polymers stems from the potentially beneficial effects their rigid structure has upon their properties, in particular their sharp absorption and emission bands, small Stokes shifts, and high photoluminescence (PL) quantum efficiencies, which makes them attractive materials for potential use in organic electronic devices such as light-emitting devices (LEDs) [2], organic lasers [3], field-effect transistors (FETs) [4], and solar cells [5]. In the case of ladder-type polyphenylenes a particular advantage of the ladder structure is that it enhances the conjugation along the backbone by reducing torsion between adjacent phenylene rings and so reduces the bandgap; as a result LPPPs are blue or blue-green emitters while substituted PPPs **1b** emit in the violet as steric interactions between the solubilizing side chains induce marked chain twisting [2]. For a synthetic chemist, the preparation of ladder polymers also presents a fascinating intellectual challenge.

Ladder polymers can be prepared in two ways: (i) by iterative multicenter condensation or addition (e.g., Diels–Alder cycloaddition) reactions and (ii) by polymer-analogous conversion of suitably functionalized single-stranded precursors [1]. A major feature of the second method is that the polymer-analogous reactions must proceed quantitatively to avoid formation of defects in the final polymer. Though ribbon-like polyacenes can be prepared by polycycloaddition methods, linear ladder-type PPPs are only accessible through the conversion of single-stranded PPPs. If methane or ethene bridges are used to connect benzene rings, the phenylene backbone is forced to be coplanar, but use of ethane or longer alkane bridges allows some torsion between adjacent phenylene rings (Figure 7.2).

Design and Synthesis of Conjugated Polymers. Edited by Mario Leclerc and Jean-François Morin

ISBN: 978-3-527-32474-3

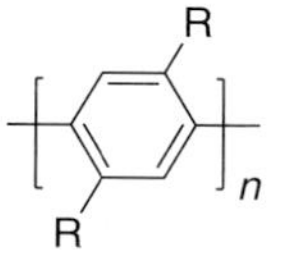

1a R = H
1b R = alkyl, alkoxy

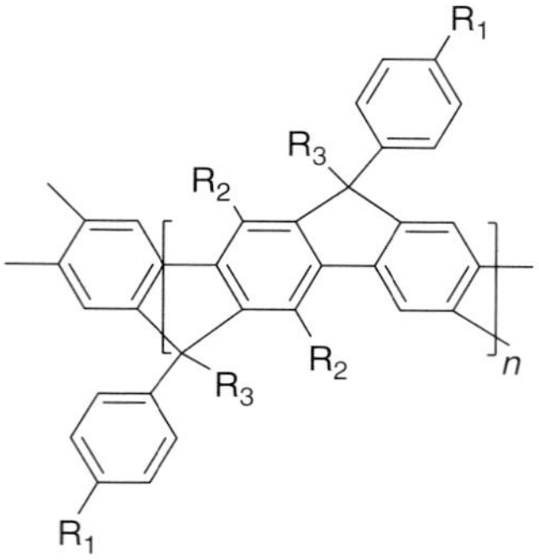

2 R_1,R_2 = alkyl, R_3 = H, alkyl

Figure 7.1 Single-stranded and ladder-type PPPs.

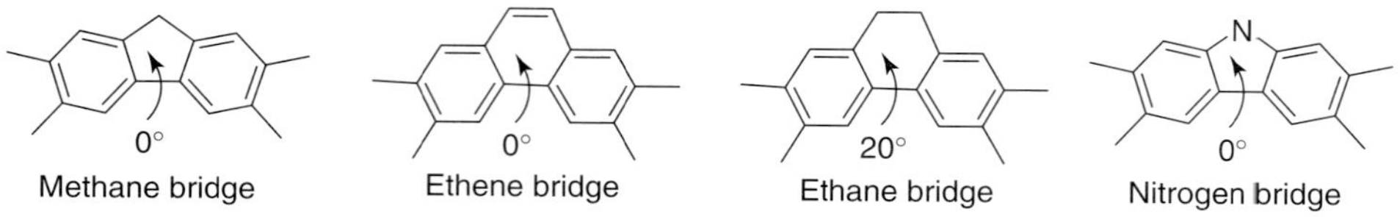

0° Methane bridge
0° Ethene bridge
20° Ethane bridge
0° Nitrogen bridge

Figure 7.2 Bridges and associated torsion angles in ladder polymers.

For an ethane bridge, the torsion angle is predicted to be about 20° [6]. Longer bridges would produce even larger torsion angles, thus removing one of the main advantages of the ladder structure; therefore, such bridges have, to date, not been used in ladder-type polyphenylenes. In place of a single-carbon bridge a nitrogen bridge can also be utilized and here also the torsion angle is 0°. The effects of replacing carbon with nitrogen are that only one instead of two substituents can be attached to the bridgehead, which affects the solubility, and that the electronic properties are altered (oxidation potential lowered) by the incorporation of the more electronegative element. As will be shown below, one of the carbon atoms in an ethene bridge has also been replaced by a nitrogen atom, but bridges containing more than one nitrogen atom have not yet been reported. Thus, some degree of control over the optical properties can be achieved by varying the type of bridges used, as the more coplanar the polymer the greater the expected degree of conjugation, and thus the longer the wavelengths of the absorption and emission maxima. As will be shown, there have even been examples of ladder polymers containing more than one type of bridge. The methane, nitrogen, and ethene bridges also impart greater rigidity to the structure and thus reduce the Stokes shift between the absorption and emission maxima.

A major problem in the synthesis of ladder-type polyphenylenes is the difficulty in avoiding defects due to incomplete ring-forming reactions. As a result there has been much interest in so-called "stepladder" polyphenylenes in which only some of the phenylene units are doubly linked – that is, there is a mixture of single and double linkages along the polymer backbone. Two types of stepladder polymers

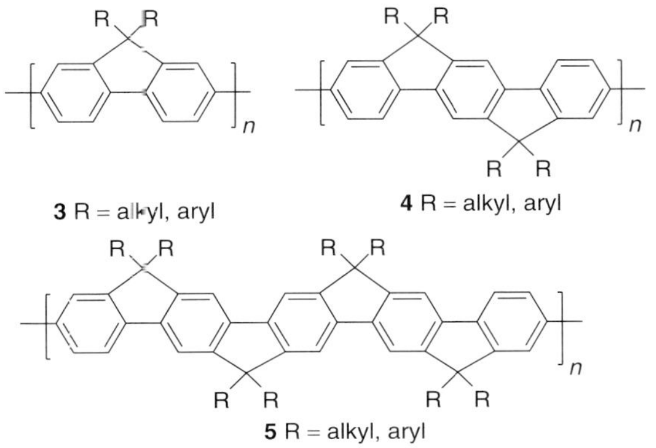

3 R = alkyl, aryl

4 R = alkyl, aryl

5 R = alkyl, aryl

Figure 7.3 Regular stepladder polyphenylenes.

exist: random stepladder polymers in which the single linkages occur at irregular intervals along the polymer chain, and regular stepladder polymers in which the polymer consists of a series of linked ladder-type oligophenylenes of equal size. Examples of these latter type of polymers are polyfluorenes (Figure 7.3, **3**) [7], polyindenofluorenes (**4**) [8], and poly(ladder-type pentaphenylenes) (**5**) [9], in which the single linkages occur after every second, third, and fifth phenylene unit respectively. The advantages of these polymers are that defects due to incomplete ring-closing reactions arise only during monomer synthesis where they can easily be removed by standard chemical purification techniques. A detailed discussion of the synthesis and properties of these stepladder polymers, which have been reviewed elsewhere [10], lies outside the scope of this chapter; so, we will merely note here that their optical properties are intermediate between those of PPPs **1b** and LPPPs **2**.

7.2 LPPPs with Single-Atom Bridges

The most widely studied class of LPPPs has been polymers such as **2a**–**c** with methane bridges. These are made [1, 11–16] (Scheme 7.1) from a poly(diacylphenylene-*co*-phenylene) precursor copolymer **6**, obtained by an AA-BB-type Suzuki polycondensation [17]. The key step is the polymer-analogous Friedel–Crafts ring-closing reaction on a polyalcohol made by reduction of **6** (**7a**), or by addition of a lithium reagent to **6** (**7b** and **c**). The ring closure of **7a** to **2a** proceeded smoothly upon addition of boron trifluoride (BF_3) in dichloromethane. NMR spectroscopy and mass spectral analysis suggested complete ring closure, indicating the presence of less than 1% of defects due to incomplete reaction. Polymers with number-average molecular weights (M_n) of up to 50 000 g mol^{-1} (50 kDa) corresponding to 150 phenylene units have been obtained, though values of 10–30 kDa are more typical. Since this synthesis was performed there have been considerable

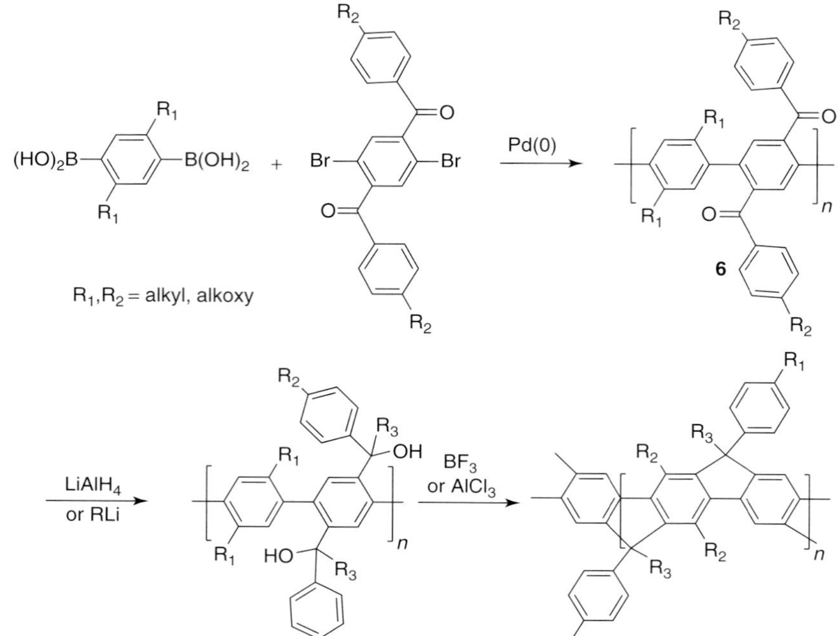

Scheme 7.1 Synthesis of methane-bridged LPPPs.

advances in the field of Suzuki polycondensation, which if applied to this system could produce significantly higher molar masses. For example, a group at Dow have developed highly efficient routes for preparing the structurally closely related polydialkylfluorenes with $M_n > 100\,000\,g\,mol^{-1}$ from an AA-BB-type Suzuki coupling after a less than 24-hour reaction time [18–23]. Microwave-assisted Suzuki polymerization [24] is another modern technique that could well lead to improved yields and molecular weight, but to date the best yields (61%) and molar masses ($M_n = 12\,600\,g\,mol^{-1}$) reported for **6** made by this method have been lower than those from the conventional method [25]. In view of the considerable savings in time, energy, and solvent offered by the microwave method, this method is well worth further investigation. Since the yields and molar masses obtained from other Suzuki reactions have been as good or better than those using conventional heating, it seems likely that with further optimization, molar masses of **6** comparable to the best obtained from the conventional method could be obtained. Ring closure of **7b** to Me-LPPP (**2b**) also proceeds smoothly with use of boron trifluoride [13], but to obtain complete ring closure of the phenyl-substituted polymer **7c** to give Ph-LPPP **2c**, aluminum trichloride must be used [1]. While complete ring closure is indicated by mass spectral and NMR analysis, characterization of **2a** by X-ray and neutron scattering [26], and by dynamic light-scattering experiments [27] indicate this polymer has a wormlike structure with a persistence length of only 6.5 nm, rather than the rigid ribbon-like structure expected for a fully ladder-type material. This suggests that there are considerable numbers of defects in the structure, so that polymer **2a** is best represented as a series of rigid ribbon-like segments linked by single bonds. The measured persistence length of 6.5 nm for **2a** is actually shorter than the values obtained for regular stepladder polymers such as polyfluorenes **3** (7 nm) or poly(ladder-type pentaphenylene)s **5** (25 nm) [28]. This picture is also supported by studies of the optical properties of the ladder polymers.

The parent LPPP **2a** with hydrogens on the bridgeheads displays an absorption maximum at 440–450 nm, with the PL emission maximum at 450–460 nm. This small Stokes shift, combined with the very sharp absorption and emission edges, and the good vibrational resolution in the absorption and emission spectra, is characteristic of a very rigid polymer [29]. This rigidity also results in LPPPs having very high fluorescence quantum efficiencies (up to 90% in solution for **2b**) as nonradiative decay pathways are seriously reduced [30]. The emission from the substituted LPPPs **2b** and **2c** is slightly redshifted compared to **2a** with emission maxima at 461 nm. Single-molecule spectroscopy studies on **2b** have provided new understanding of the nature of this and other conjugated polymers [31]. Whereas a molecule of low molar mass **2b** (M_w = 25 kDa, corresponding to about 62 phenylene units) produced only a single emission peak at 459 nm, which matched the emission maximum observed in ensemble measurements, a larger molecule (M_w = 67 kDa, about 165 phenylene units) produced up to five emission peaks at wavelengths between 450 and 461 nm. By performing a large number of single-molecule experiments it was established that there was a linear correlation between the average number of chromophores and the chain length. This indicates that the polymer **2b** consists of a series of linked chromophores, each of which can emit separately. This is consistent with the model described above, derived from light-scattering and other studies, of **2a** as a series of ladder-type segments linked together by single bonds. The size of these chromophores has been determined by comparing their emission with that of well-defined oligomers. Extrapolation from the absorption maxima of bridged oligomers (Figure 7.4) ranging from 3 rings (**8**) to 7 rings (**9**) has previously been used to estimate the effective conjugation length (ECL) for absorption in LPPPs as being about 11–12 benzene rings [6]. When a ladder-type undecamer **10** was made, its absorption and emission were found to be slightly blueshifted compared with Me-LPPP (**2b**) suggesting a larger value for the ECL [31]. Single-molecule spectroscopy studies on **10** demonstrated that the PL emission from the oligomer at 451 nm matches that from the smallest chromophores on the polymer, so that the emission maxima for the larger chromophores must correspond to longer segments of up to about 16 benzene rings, with the emission maximum at 459 nm seen in the bulk sample corresponding to a chromophore of about 15 phenylenes. The ECL for emission in LPPPs is thus around 15 benzene rings, which is similar to the values estimated for the structurally related stepladder polymers, polyfluorenes **3** (12 rings) [32–34] and polyindenofluorenes **4** (15–18 rings) [35]. The ECL for absorption and emission

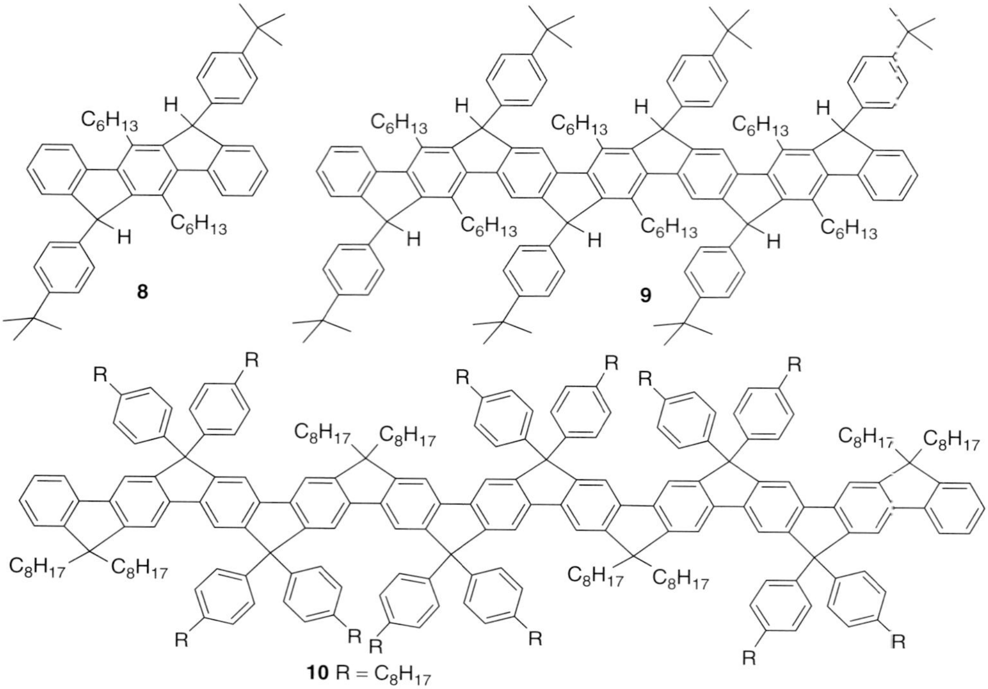

Figure 7.4 Ladder-type oligophenylenes.

of **2b** are also clearly very similar, as would be expected for a rigid ladder-type structure in which the ground state and excited state geometries are constrained to be very similar.

The emission maximum from **2a** is in the blue, making this a potentially promising material for blue LEDs. Unfortunately, the solid-state emission is dominated by a broad, featureless band of around 600 nm in the yellow part of the spectrum [36–39]. The relative intensities of blue and yellow emission bands from **2a** are heavily dependent upon the method used to prepare the films, the solvent used, and the film thickness. Annealing films of **2a** leads to the disappearance of the blue band. As a result, while blue electroluminescence (EL) has been seen from LEDs using **2a**, it is very unstable and yellow EL is more commonly observed [38, 40]. This emission was at first attributed to emission from aggregates, but more recent work has suggested it arises from a chemical defect [41], most probably a ketone similar to the fluorenone defects that have been detected in polyfluorenes [42]. By contrast, the emission from Me-LPPP **2b** is a stable blue-green color in both solution and thin films, with a solid-state fluorescence quantum efficiency of 60% [1, 13]. A long wavelength emission band at 560 nm is seen from films of **2b**, but it is much less intense than the yellow emission band from **2a** and does not increase in intensity upon annealing, so that it is attributed to aggregates [43]. This stability in the emission spectrum is to be expected if the defect in **2a** arises from oxidation of the bridgehead, as work on polyfluorenes has shown that materials with hydrogens at the bridgeheads are much more susceptible to oxidation [42], which can be prevented by ensuring complete alkylation of the bridgeheads [44]. Blue-green-emitting LEDs have been constructed using Me-LPPP with efficiencies of up to 4% [45–49]. In addition, Me-LPPP has been shown to be a very promising material for use in organic solid-state lasers with optically pumped lasing having been observed in both waveguide and distributed feedback configurations [50–54]. The phenyl-substituted polymer Ph-LPPP **2c** shows emission similar to Me-LPPP with emission maxima at 460 and 490 nm, but with an additional long wavelength band with well-resolved maxima at 600 and 650 nm [55]. This proved to be emission from triplet states (phosphorescence) arising from palladium-containing centers (the polymer contains about 80 ppm of palladium) bound into the polymer chain. These are considered to arise from interaction between phenyl lithium and the residues from the palladium catalyst used to make the precursor polymer **6**. A chiral LPPP has been prepared by using a resolved cyclophane bisboronic acid as one component in the initial Suzuki polycondensation [56–58]. Such a chiral polymer is a candidate for obtaining circularly polarized EL.

Random copolymers of LPPP with single-stranded PPPs [38] – the so-called random "stepladder" polymers have been found to show stabler blue EL than LPPP **2a** [59]. This is now considered to be due to reduced exciton migration to defect sites as a result of the twisting of the backbone due to steric interactions between the ladder and single-stranded PPP units. This torsion also causes the copolymers to display luminescence, which is blueshifted compared to the homopolymer **2a**. Blueshifted emission compared to LPPPs is also seen for regular stepladder polymers, for example, polyfluorenes **3**, in which the breaks between the double-stranded sections occur at regular intervals [60]. As the proportion of double-stranded linkages decreases the emission redshifts so that polymers based on ladder-type pentaphenylenes **5** exhibit pure blue emission with spectra very similar to those of LPPPs [9]. As with LPPPs, instability in the emission color due to formation of emissive defects has been a problem for regular stepladder polyphenylenes [42, 61]. This can be suppressed by using aryl substituents at the bridgeheads [62–65] or by ensuring their complete alkylation [66]. Defects due to incomplete ring closure are not a problem in these materials as the ring-closing reactions are being performed on small molecules where the ring-closed and unclosed compounds are readily separable.

A way to reduce defects through incomplete ring closure in LPPPs is to replace the benzene-bisboronic acid in Scheme 7.1 with a fluorenebisboronate. LPPPs **11a** and **b** have been formed in this way (Scheme 7.2), which show blue-green EL ($\lambda_{max} = 460$ nm) with the emission color even from **11a** reported to be more stable than for **2a** [67].

The molar masses of the polymers obtained by this method were clearly capable of improvement ($M_n = 16–22$ kDa), but the route does offer a number of possible advantages. First, the number of bonds formed in the ring-closing step is significantly reduced so the amount of defects produced should be much lower, especially if the recently developed method for producing defect-free dialkylfluorenes [44]

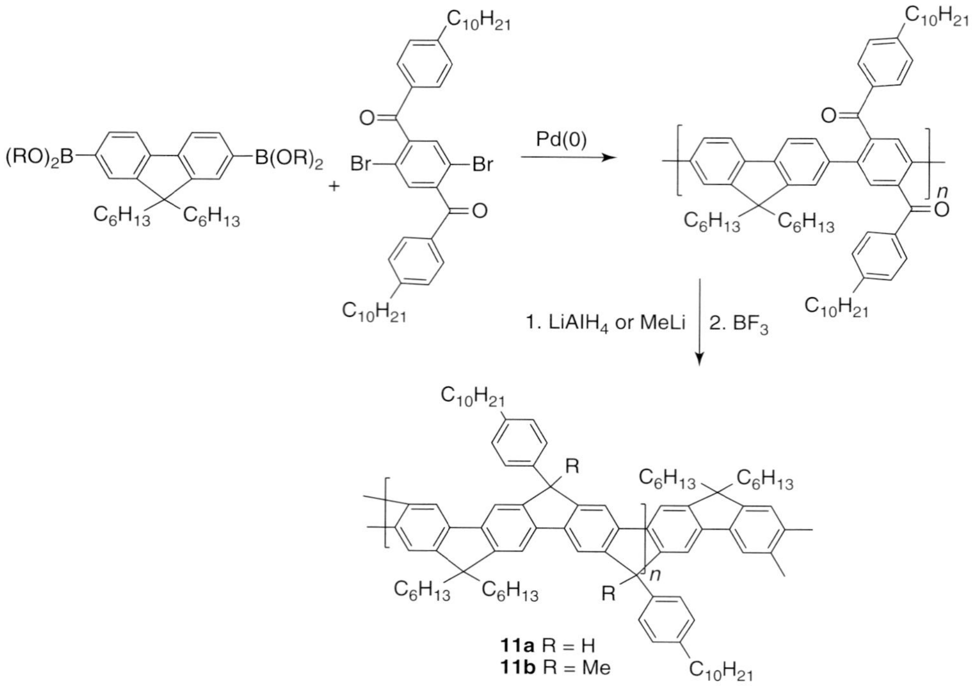

Scheme 7.2 Synthesis of relatively low-defect LPPPs using a fluorenebisboronate.

is used to make the boronate. Second, since fluorenes with electroactive groups such as hole-transporting triarylamines [63] or electron-transporting oxadiazoles [68] at the bridgehead have been prepared, this route could be used to introduce such groups into LPPPs. As the polyfluorenes bearing these groups show much improved device performance in LEDs due to superior charge injection [63, 68, 69], it can be reasonably expected that the performance of LPPPs would be enhanced by this means. Random copolymers of LPPP with oxadiazoles in the main chain have previously been reported to show improved EL efficiency than the LPPP homopolymer **2a** [70]. Incorporation of larger ladder-type oligophenylene units, for example, indenofluorenes, in LPPP synthesis could in principle reduce the amount of defects even further, but the longer and more complicated monomer synthesis involved would outweigh the likely benefits.

If instead of a 1,4-dibenzoylbenzene, the 1,3-diketone **12** is used in the Suzuki reaction with the 1,4-benzenediboronic acid, a ladder polymer **13** with alternating *meta*- and *para*-phenylene linkages is obtained (Figure 7.5) [71]. As would be expected, the *meta*-linkages disrupt the conjugation so the absorption is hypsochromically shifted to $\lambda_{max} = 389$ nm cf. 450 nm for **2a**. If the ketone **12** is reacted with a 1,3-benzene diboronic acid the all-*meta*-linked ladder polymer **14** is obtained, whose absorption maximum is even more blueshifted to 356 nm [30].

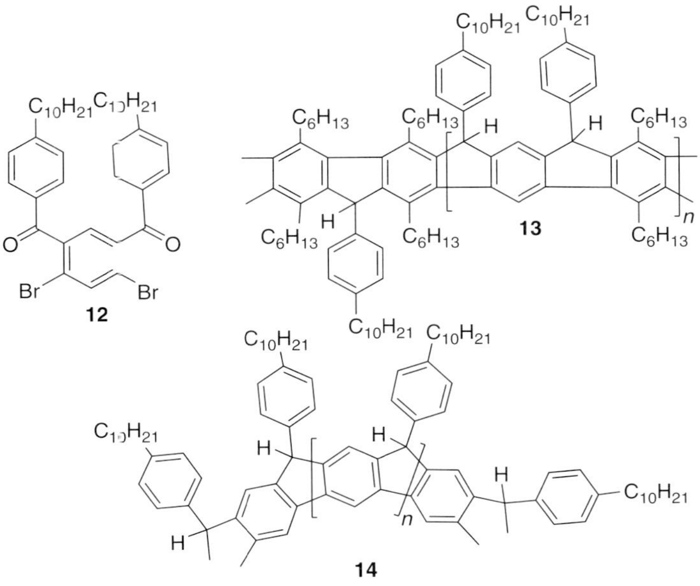

Figure 7.5 *meta*-Linked ladder polyphenylenes and the diketone used to make them.

Another aromatic molecule that has been introduced into ladder-type polyphenylenes is naphthalene. Ladder polymers **15–17** (Figure 7.6) incorporating naphthalene have been made by using a naphthalene-2,6-diboronate **18** or the naphthalene-1,5-diboronates **19** and **20** in microwave-assisted Suzuki polycondensations with the 1,4-diketone used to make Me-LPPP **2b** followed by treatment with methyl lithium and ring closure [24, 25, 72]. An interesting feature of the structure of polymer **16** is that it contains only six-membered rings. The molar masses from the microwave-assisted reactions (12–30 kDa) were higher than from conventionally heated reactions (4–11 kDa). The absorption (λ_{max} = 419 nm) and emission spectra (λ_{max} = 458 nm) of the 2,6-naphthalene-based polymer **15** are slightly blueshifted compared to Me-LPPP (λ_{max} = 426 nm and 464 nm for absorption and emission respectively). By contrast, the 1,5-naphthalene-based polymers **13** and **17** display markedly redshifted absorption (λ_{max} = 438 and 449 nm respectively) and emission (λ_{max} = 485 and 508 nm respectively). Lasing has been observed from polymer **15**.

Binaphthyls are well-known chiral units and a regular chiral stepladder polymer has been prepared [73] by using a chiral binaphthyl diboronate in place of a phenyldiboronate in the route shown in Scheme 7.1. The molar mass using the microwave-assisted Suzuki method was around 13 kDa, and the emission maxima were at 427 and 458 nm – as with the stepladder LPPPs mentioned above the breaks in the ladder produce a blue shift in absorption and emission. This polymer should produce circularly polarized emission and may also have applications in nonlinear optics.

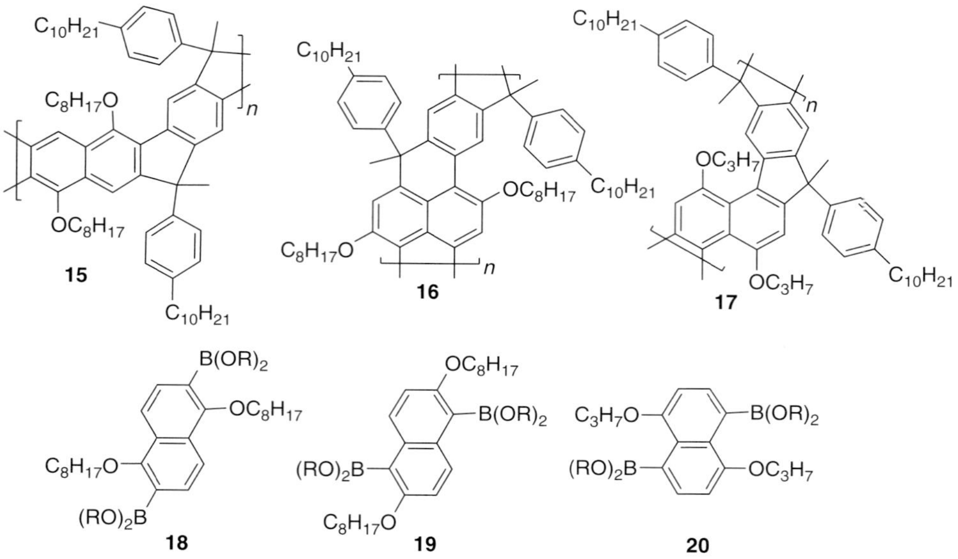

Figure 7.6 Naphthalene-based ladder polymers and the naphthalenediboronates used to make them.

Heteroarylenes can also be incorporated into ladder-type polymers. The polymer **21** (Figure 7.7) containing alternating phenylene and thiophene units was prepared by a modification of the standard LPPP synthesis (Scheme 7.1) using a Stille coupling of a thiophene bisstannylate with a dibromophenylene diketone to make the precursor polymer **22**, followed by reduction and ring closure [74]. This route had to be used because the Suzuki coupling with a thiophene bisboronate did not produce any polymer, due to the instability of the boronate under the reaction conditions. The molecular weight of the polymer was around 15 kDa corresponding to about 45 aromatic and heteroaromatic rings. The highly electron-rich thiophene

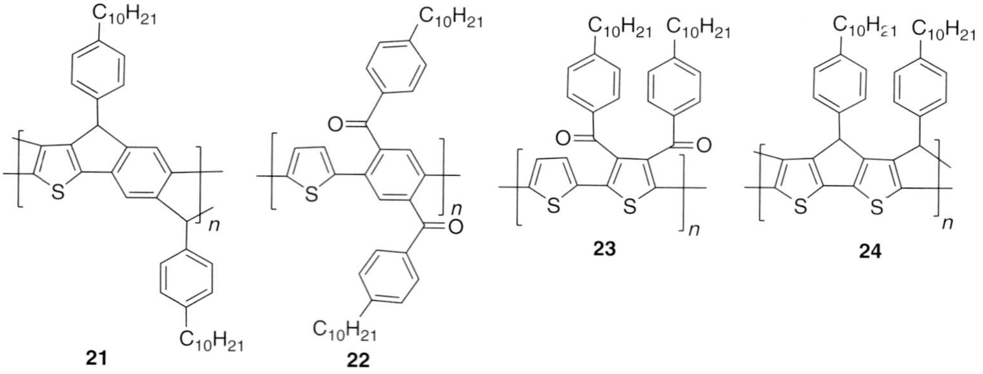

Figure 7.7 Thiophene-containing ladder polymers and their polyketone precursors

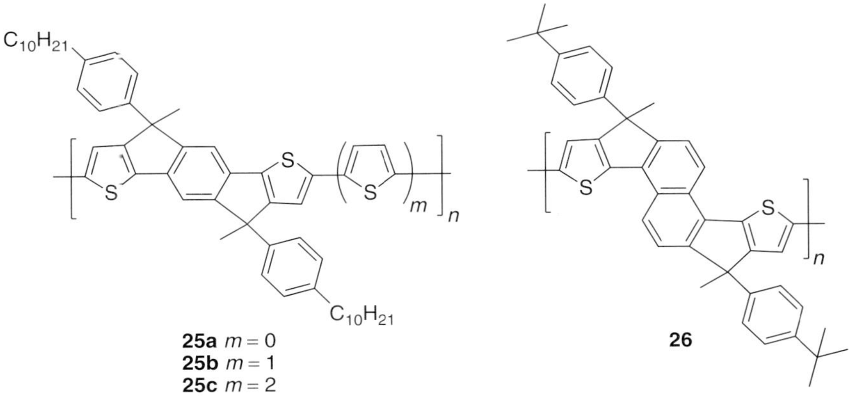

Figure 7.8 Stepladder thiophene-phenylene/naphthalene copolymers.

rings induce a marked redshift in the absorption and emission spectra so that polymer **21** produces orange PL (λ_{max} = 536 nm). Attempted ring closure of the all-thienyl polyketone **23** to give **24** was unsuccessful with only partially ring-closed materials being obtained. This can be attributed to the high inherent steric strain in the fully ring-closed polymer due to the structure containing only five-membered rings.

Like Suzuki couplings, Stille polycouplings can be promoted by the use of microwaves [25]. Microwave-assisted Stille couplings of oligothienyl bisstannylates with benzene or naphthalene diketones have been used to make stepladder polymers **25** and **26** (Figure 7.8) [75]. Their number-averaged molar masses are between 12 and 16 kDa. For the phenylene copolymers **25** the emission maximum redshifts as the number of thiophene units increases from **25a** (λ_{max} = 563 nm) to **25c** (λ_{max} = 581 nm). Replacement of the phenylene in **25a** with a naphthalene in **26** produces a blueshift in emission (λ_{max} = 535 nm).

Apart from single-carbon bridges both nitrogen and sulfur single-atom bridges have been incorporated into ladder-type polyphenylenes. A fully nitrogen-bridged polymer **27** (Figure 7.9) is of considerable interest as a synthetic target as it can be thought of as a ladder-type polyaniline, and so might reasonably be expected to have a very small bandgap and might even become conductive like polyaniline if doped. To date, the longest oligomer with all nitrogen bridges is a pentaphenylene **28**

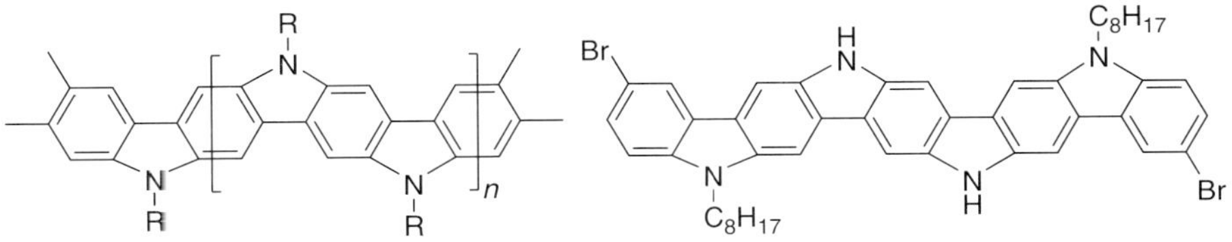

Figure 7.9 Ladder-type polyaniline and tetraaniline.

(or ladder-type tetraaniline), which has a bandgap of 2.42 eV [76]. Obtaining high yields in nitrogen bridge–forming reactions is one of the main synthetic problems that would need to be overcome in any proposed synthesis of **27**. To date, no method for forming such bridges is known whose yield is high enough to enable the bridge formation to be performed as a polymer-analogous reaction.

Ladder polymers containing a mixture of nitrogen and methane bridges have been prepared by incorporating carbazole units into LPPPs. Thus, replacing the fluorene bisboronate in the synthesis shown in Scheme 7.2 with a carbazole 2,7-bisboronate leads to the polymer **29** (Figure 7.10) [77]. Coupling a carbazole bisboronate with a 3,6-diacyl-2,7-dibromocarbazole [78] produces the poly(2,7-carbazole) **30a**, which can then be converted into the ladder-type polycarbazole **31a** [79]. The number-averaged molecular masses (M_n) for **30a** and **31a** by gel permeation chromatography (GPC) analysis were 17 and 13.5 kDa respectively. The drop in molar mass upon ring closure probably reflects the change in hydrodynamic volume due to the polymer's shape changing from a coil to a ribbon-like structure. The ladder polymer **31a** displays marked aggregation in solution resulting in broad peaks in the NMR absorption and the PL emission spectra, which made it difficult to confirm the structure. Considerable dilution was required before the sharp emission peaks, small Stokes' shift, and other characteristic features of the emission spectrum of a ladder polymer could be observed. This aggregation is not seen for LPPPs as in those polymers there are bulky aryl groups on both edges of the ribbon-like polymer, whereas in **31a**, such groups are found on only one edge thus permitting the chains to pack closely. Attaching a longer alkyl chain as in

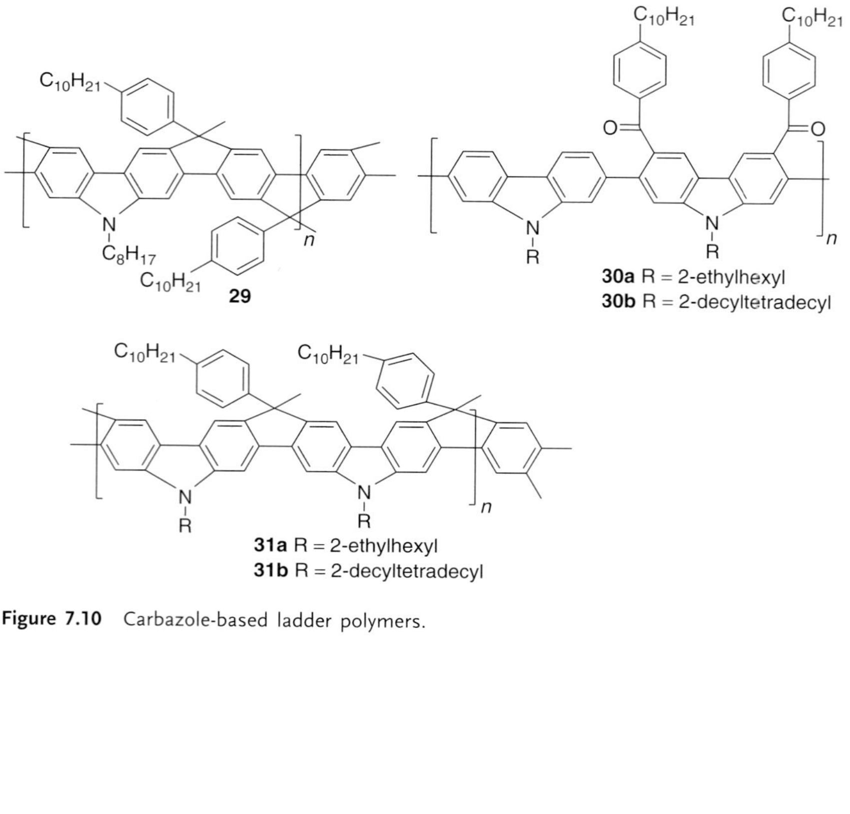

Figure 7.10 Carbazole-based ladder polymers.

31b improved the solubility of the ladder polymer and increased the M_n slightly to 15 kDa. The ladder polymers **29** and **31** possess higher highest occupied molecular orbital (HOMO) energy levels due to the electron donating effects of the nitrogens, reducing their bandgaps compared to LPPPs; as a result they exhibit blue-green emission (λ_{max} = about 470 nm). Polycarbazoles have shown promising results as donors in bulk heterojunction photovoltaic devices with perylene diimide as acceptors [30], and it was anticipated that the ladder polymer **31b** might display even better performance given its smaller bandgap, but the results were very disappointing. From the current versus voltage curves it appears that there is a problem with charge transport. Regular stepladder polymers with a mixture of carbon and nitrogen bridges show better results but the highest efficiency under solar illumination is still under 2% [10].

The final type of single-atom bridge that has been used in ladder-type polyphenylenes is a sulfur bridge. Sulfur bridges have the disadvantage that they cannot bear solubilizing groups, so an all-sulfur-bridged polymer would be insoluble but soluble polymers with a mixture of sulfur and other bridges can be made. Whereas nitrogen bridge–forming reactions are generally not suitable for use as polymerization steps, polydiaryl sulfides can be formed by nucleophilic attack of aryl disulfides upon dihaloarenes. The polysulfide **32** has been prepared in such a manner with a high molar mass (M_n = 150 kDa) and then converted to the soluble ladder-type polymers **33** as shown in Scheme 7.3 [81]. These can be oxidized to form conjugated polycations with very small bandgaps.

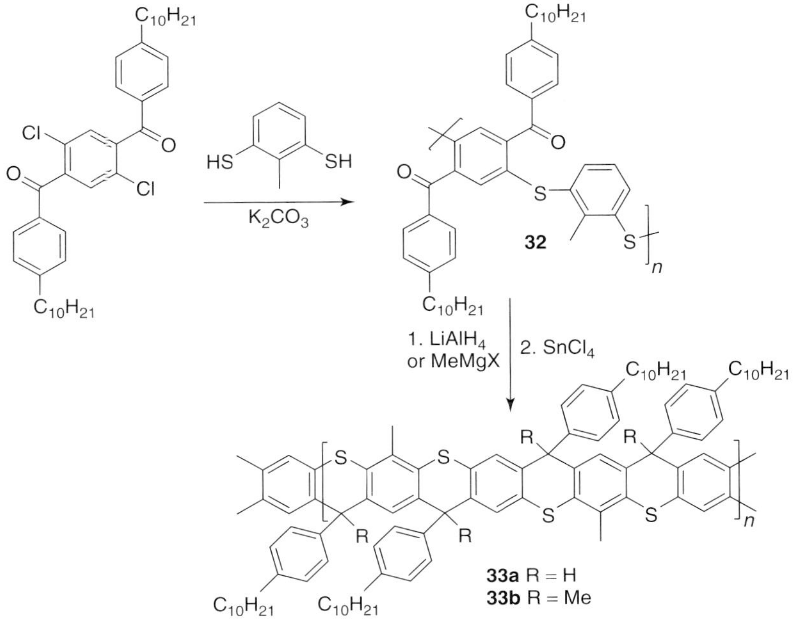

$C_{10}H_{21}$ Cl O Cl O $C_{10}H_{21}$ HS SH K_2CO_3 $C_{10}H_{21}$ O O S **32** S n $C_{10}H_{21}$ 1. $LiAlH_4$ or MeMgX 2. $SnCl_4$ $C_{10}H_{21}$ $C_{10}H_{21}$ R R S S S S R R n $C_{10}H_{21}$ $C_{10}H_{21}$ **33a** R = H **33b** R = Me

Scheme 7.3 Synthesis of ladder polymers containing sulfur bridges.

Ladder polymers with a mixture of nitrogen and sulfur bridges might possess interesting optoelectronic properties. To date no such materials have been reported but they should be fairly readily accessible, for example, from phenothiazine. Manipulation of the electronic properties of sulfur-bridged polymers by oxidation of the bridges to the *S*-oxide or *S,S*-dioxide, as has been used to affect the properties of oligomers containing thiophenes [82], is another potentially interesting area of research that is yet to be explored. Other elements, for example, silicon or phosphorus, have been used to bridge biphenyls [10], but to date have yet to appear in ladder polymers. The synthetic difficulties involved in making silicon bridges will probably outweigh any possible benefits they might offer (e.g., oxidative stability), but the recent interest in phosphorus-based oligoarenes as n-type semiconductors [83, 84] suggests that incorporating such units into ladder polymers might well be worth investigating.

7.3
LPPPs with Two-Atom Bridges

Ladder-type PPPs have also been made with two-carbon bridges. Samarium(II) iodide has been used to convert the poly(diacylphenylene) **34a** into a ladder polymer **35** with dihydroxyethane bridges (Scheme 7.4) [85]. The polymer is obtained with a molar mass (M_n) of 53 kDa corresponding to a degree of polymerization of

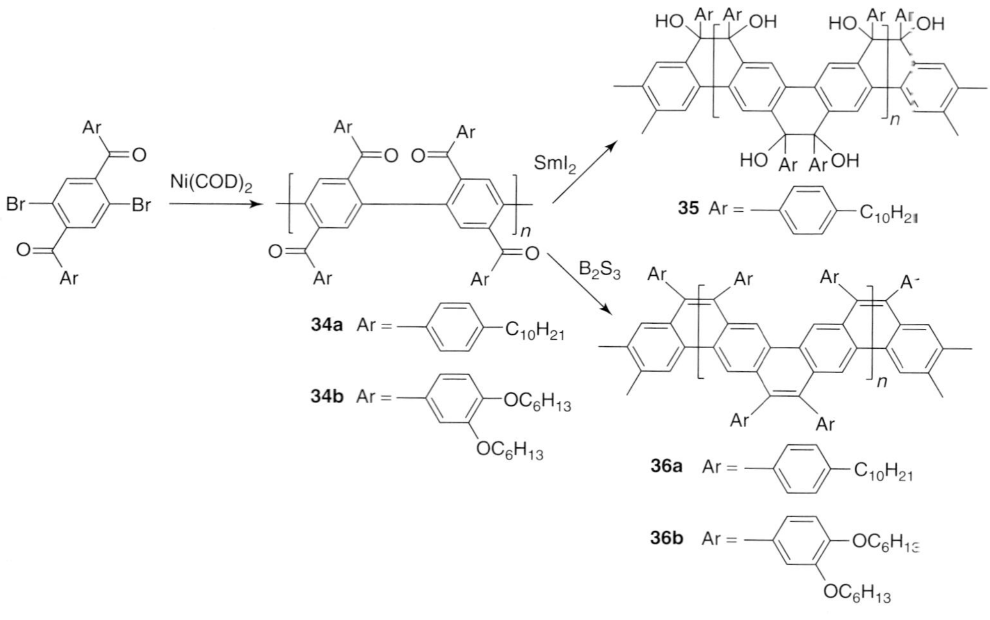

Ar Ar O Br Br O Ar $Ni(COD)_2$ Ar O O Ar O Ar Ar n SmI_2 B_2S_3 HO Ar Ar OH HO Ar Ar OH n HO Ar Ar OH **35** Ar = $C_{10}H_{21}$ **34a** Ar = $C_{10}H_{21}$ **34b** Ar = OC_6H_{13} OC_6H_{13} Ar Ar Ar Ar n Ar Ar **36a** Ar = $C_{10}H_{21}$ **36b** Ar = OC_6H_{13} OC_6H_{13}

Scheme 7.4 Synthesis of ladder-type PPPs with ethane and ethene bridges.

over 95 benzene rings. This material displays blue-green PL in both solution ($\lambda_{max} = 459$ nm) and the solid state ($\lambda_{max} = 482$ nm). This redshift in going from solution to thin films indicates **35** is a less rigid polymer than the methane-bridged polymers **2**, which show no such effect, but the small Stokes' shift indicates only weak geometrical changes in going from the ground to the excited state. No long wavelength emission band is seen in the PL spectrum of **35**. Treatment of **34a** and **b** with boron sulfide produces acene-like polymers **36a** and **b** with ethene bridges through a McMurry-like coupling of intermediate thioketones [86, 87]. These soluble polymers can be prepared with quite high molar masses ($M_n = 25$ kDa). They exhibit blue-green luminescence ($\lambda_{max} = 478$ and 484 nm) with some long wavelength emission in the solid state, which is attributed to aggregates. Their EL efficiency is very low (<0.1%).

Another acene-like polymer with ethene bridges **37** has been made from the precursor polymer **38** (Scheme 7.5) by acid-catalyzed ring closure [88]. There is no report on the emission from this material, but the absorption edge was reported to be at 478 nm, suggesting it should be a blue-green or green emitter. A major issue in this synthesis is whether the ring-forming reaction is complete. The UV–Vis spectrum of the product contains features of unknown origin suggesting that the ring closure is incomplete.

The McMurry-like coupling of the carbazole-based polyketone **39** (Figure 7.11, $M_n = 21$ kDa) with boron sulfide to produce the polymer **40** does not go to completion as shown by the presence of a residual C=O stretch in the IR absorption spectrum [79]. This appears to be due to the lower solubility of the ring-closed polymer causing it to fall out of solution before the reaction attains completion. Replacing the ethylhexyl substituents on the carbazole units with a longer group as in **31b** might enhance its solubility enough to overcome this. Unlike the methane-bridged ladder-type polycarbazoles **31** discussed above no aggregation is observed in solution for this material, presumably because the alternating

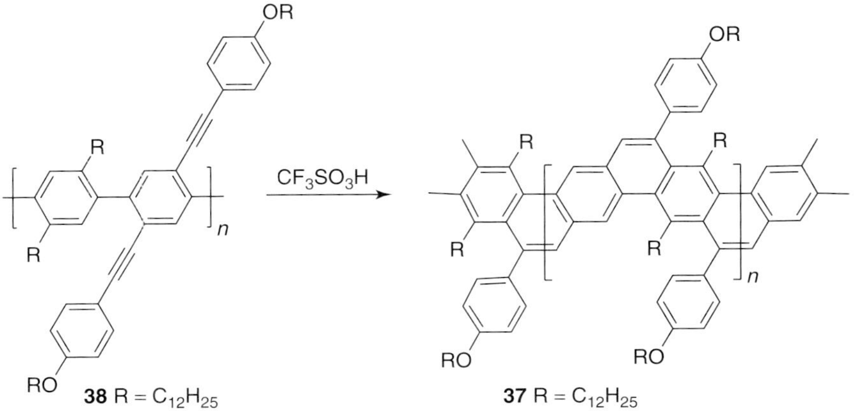

38 R = $C_{12}H_{25}$ $\xrightarrow{CF_3SO_3H}$ **37** R = $C_{12}H_{25}$

Scheme 7.5 Formation of an acene-like polymer through acid-mediated ring closure.

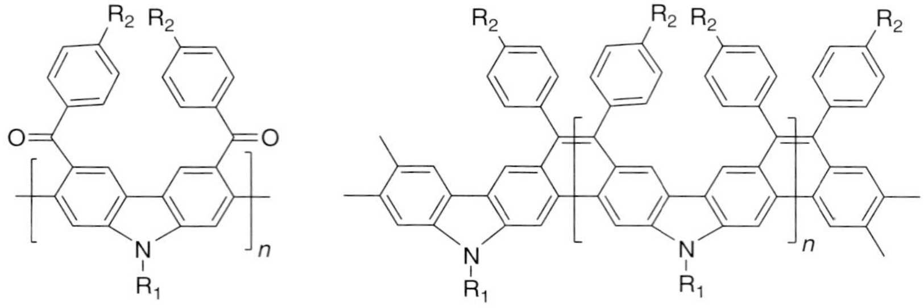

39 R_1 = 2-ethylhexyl, $R_2 = C_{10}H_{21}$

40 R_1 = 2-ethylhexyl, $R_2 = C_{10}H_{21}$

Figure 7.11 Ladder-type polycarbazole with ethene bridges.

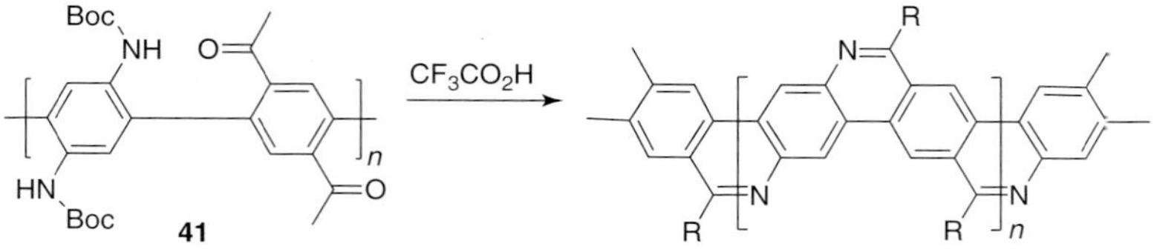

41 —CF_3CO_2H→ **42** R = alkyl or alkylphenyl

Scheme 7.6 Synthesis of a ladder-type PPPs with imine bridges.

five- and six-membered rings along its backbone would induce it to form a helical structure.

Two-atom bridges with a nitrogen atom can be prepared by formation of imines in a polymer-analogous reaction. Thus acid-promoted deprotection of the boc-protected polyamines **41** produces the aza-substituted ladder polymers **42** (Scheme 7.6) [89, 90]. These materials are only soluble in strongly protic solvents, and their luminescent properties, if any, have not been reported. The absorption maxima in protic solvents are around 400 nm, with secondary bands between 510 and 550 nm, while in the solid state, maxima between 460 and 490 nm were seen, indicating that protonation has a major effect on the conjugation length. As with polymer **37**, the absorption spectra of **42** contain unexplained features suggesting incomplete or nonregiospecific ring closure may be occurring.

While the synthetic routes to these polymers with two-atom bridges are interesting and they have some fascinating molecular topologies, they have as yet shown no signs of matching the performance of their brethren with single-atom bridges.

7.4 Conclusion

Efficient synthetic routes have been demonstrated for the preparation of phenylene-based ladder polymers and copolymers with a variety of single- and double-atom bridges linking the phenylene units. These materials have shown some very interesting optoelectronic properties, with excellent performances being obtained from organic electronic devices using them. While problems remain in obtaining structurally perfect materials, and the molar masses of some materials could be further improved; ladder-type polyphenylenes have already demonstrated that they are not just objects of scientific curiosity, but have the potential to be the active components in commercially viable products. There seems little reason to doubt that ladder-type polymers based on phenylene rings will continue to fascinate synthetic chemists, photophysicists, and materials scientists for some time to come.

References

1. Scherf, U. (1999) *J. Mater. Chem.*, **9**, 1853–1864.
2. Kraft, A., Grimsdale, A.C., and Holmes, A.B. (1998) *Angew. Chem. Int. Ed. Engl.*, **37**, 403–428.
3. McGehee, M.D. and Heeger, A.J. (2000) *Adv. Mater.*, **12**, 1655–1668.
4. Dimitrakopoulos, C.D. and Malenfant, P.R.L. (2002) *Adv. Mater.*, **14**, 99–117.
5. Brabec, C.J., Sariciftci, N.S., and Hummelen, J.C. (2001) *Adv. Funct. Mater.*, **11**, 15–26.
6. Grimme, J., Kreyenschmidt, M., Uckert, F., Müllen, K., and Scherf, U. (1995) *Adv. Mater.*, **7**, 292–295.
7. Neher, D. (2001) *Macromol. Rapid Commun.*, **22**, 1365–1385.
8. Setayesh, S., Marsitzky, D., and Müllen, K. (2000) *Macromolecules*, **33**, 2016–2020.
9. Jacob, J., Sax, S., Piok, T., List, E.J.W., Grimsdale, A.C., and Müllen, K. (2004) *J. Am. Chem. Soc.*, **126**, 6987–6995.
10. Grimsdale, A.C. and Müllen, K. (2007) *Macromol. Rapid Commun.*, **28**, 1676–1702.
11. Scherf, U. and Müllen, K. (1997) *ACS Symp. Ser.*, **672**, 358–380.
12. Scherf, U. and Müllen, K. (1995) *Adv. Polym. Sci.*, **123**, 1–40.
13. Scherf, U., Bohnen, A., and Müllen, K. (1992) *Makromol. Chem.*, **193**, 1127–1133.
14. Scherf, U. and Müllen, K. (1992) *Macromolecules*, **25**, 3546–3548.
15. Scherf, U. and Müllen, K. (1991) *Makromol. Chem., Rapid Commun.*, **12**, 489–497
16. Scherf, U. (1999) *Top. Curr. Chem.*, **201**, 163–222.
17. Schlüter, A.D. (2001) *J. Polym. Sci., Part A: Polym. Chem.*, **39**, 1533–1556.
18. Woo, E.P., Inbasekaran, M., Shiang, W., and Roof, G.R. (1997) PCT Intl. Pat. Appl., WO97/05184.
19. Inbasekaran, M., Wu, W., and Woo, E.P. (1998) US Patent 5777070.
20. Bernius, M.T., Inbasekaran, M., Woo, E.P., Wu, W., and Wujkowski, L. (1999) *Proc. SPIE*, **3797**, 129–137.
21. Bernius, M., Inbasekaran, M., Woo, E., Wu, W., and Wujkowski, L. (2000) *J. Mater. Sci. Mater. Electron.*, **11**, 111–116.
22. Bernius, M., Inbasekaran, M., O'Brien, J., and Wu, W. (2000) *Adv. Mater.*, **12**, 1737–1750.
23. Inbasekaran, M., Woo, E., Bernius, M., and Wujkowski, L. (2000) *Synth. Met.*, **111–112**, 397–401.
24. Galbrecht, F., Bünnagel, T.W., Scherf, U., and Farrell, T. (2007) *Macromol. Rapid Commun.*, **28**, 387–394.
25. Nehls, B.S., Asawapirom, U., Füldner, S., Preis, E., Farrell, T., and Scherf, U. (2004) *Adv. Funct. Mater.*, **14**, 352–356.
26. Hickl, P., Ballauff, M., Scherf, U., Müllen, K., and Lindner, P. (1997) *Macromolecules*, **30**, 273–279.
27. Petekidis, G., Fytas, G., Scherf, U., Müllen, K., and Fleischer, G. (1999) *J. Polym. Sci., Part A: Polym. Chem.*, **37**, 2211–2220.
28. Somma, E., Loppinet, B., Fytas, G., Setayesh, S., Jacob, J., Grimsdale, A.C.,

and Müllen, K. (2004) *Colloid Polym. Sci.*, **282**, 867–873.
29. Graupner, W., Grem, G., Meghdadi, F., Paar, C., Leising, G., Scherf, U., Müllen, K., Fischer, W., and Stelzer, F. (1994) *Mol. Cryst. Liq. Cryst.*, **256**, 549–554.
30. Stampfl, J., Graupner, W., Leising, G., and Scherf, U. (1995) *J. Lumin.*, **63**, 117–123.
31. Schindler, F., Jacob, J., Grimsdale, A., Scherf, U., Müllen, K., Lupton, J.M., and Feldmann, J. (2005) *Angew. Chem. Intl. Ed.*, **44**, 1520–1525.
32. Klaerner, G. and Miller, R.D. (1998) *Macromolecules*, **31**, 2007–2009.
33. Miller, R.D., Klaerner, G., Fuhrer, T., Kreyenschmidt, M., Kwak, J., Lee, V., Chen, W.-D., and Scott, J.C. (1999) *Nonlinear Opt.*, **20**, 269–295.
34. Lee, S.H. and Tsutsui, T. (2000) *Thin Solid Films*, **363**, 76–80.
35. Setayesh, S., Marsitzky, D., and Müllen, K. (2000) *Macromolecules*, **33**, 2016–2020.
36. Leising, G., Grem, G., Leditzky, G., and Scherf, U. (1993) *Proc. SPIE*, **1910**, 70–77.
37. Grem, G. and Leising, G. (1993) *Synth. Met.*, **55–57**, 4105–4110.
38. Hüber, J., Müllen, K., Saalbeck, J., Schenk, H., Scherf, U., Stehlin, T., and Stern, R. (1994) *Acta Polym.*, **45**, 244–247.
39. Grem, G., Martin, V., Meghdadi, F., Paar, C., Stampfl, J., Sturm, J., Tasch, S., and Leising, G. (1995) *Synth. Met.*, **71**, 2193–2194.
40. Grüner, J., Wittmann, H.F., Hamer, P.J., Friend, R.H., Huber, J., Scherf, U., Müllen, K., Moratti, S.C., and Holmes, A.B. (1994) *Synth. Met.*, **67**, 181–185.
41. Lupton, J.M. (2002) *Chem. Phys. Lett.*, **365**, 366–368.
42. Scherf, U. and List, E.W.J. (2002) *Adv. Mater.*, **14**, 477–487.
43. Haugeneder, A., Lemmer, U., and Scherf, U. (2002) *Chem. Phys. Lett.*, **351**, 354–358.
44. Cho, S.Y., Grimsdale, A.C., Jones, D.J., Watkins, S.E., and Holmes, A.B. (2007) *J. Am. Chem. Soc.*, **130**, 11910–11911.
45. Leising, G., Köpping-Grem, G., Meghdadi, F., Niko, A., Tasch, S., Fischer, W., Pu, L., Wagaman, M.W., Grubbs, R.H., Althouel, L., Froyer, G., Scherf, U., and Huber, J. (1995) *Proc. SPIE*, **2528**, 307–314.
46. Leising, G., Ekström, O., Graupner, W., Meghdadi, F., Moser, M., Kranzelbinder, G., Jost, T., Tasch, S., Winkler, B., Athouel, L., Froyer, G., Scherf, U., Müllen, K., Lanzani, G., Nisoli, M., and DeSilvestri, S. (1996) *Proc. SPIE*, **2852**, 189–200.
47. Leising, G., Tasch, S., Meghdadi, F., Athouel, L., Froyer, G., and Scherf, U. (1996) *Synth. Met.*, **81**, 185–189.
48. Tasch, S., Niko, A., Leising, G., and Scherf, U. (1996) *Appl. Phys. Lett.*, **68**, 1090–1092.
49. Tasch, S., Niko, A., Leising, G., and Scherf, U. (1996) *Mat. Res. Soc. Symp. Proc.*, **413**, 71–76.
50. Leising, G., Tasch, S., Brandstätter, C., Graupner, W., Hampel, S., List, E.W.J., Meghdadi, F., Zenz, C., Schlichting, P., Rohr, U., Geerts, Y., Scherf, U., and Müllen, K. (1997) *Synth. Met.*, **91**, 41–47.
51. Leising, G., List, E.W.J., Zenz, C., Tasch, S., Brandstätter, C., Graupner, W., Markart, P., Meghdadi, F., Kranzelbinder, G., Niko, A., Resel, R., Zojer, E. Schlichting, P., Rohr, U., Geerts, Y., Scherf, U., Müllen, K., Smith, R., and Gin, D. (1998) *Proc. SPIE*, **3476**, 76–87.
52. Stagira, S., Zavelani-Rossi, M., Nisoli, M., DeSilvestri, S., Lanzani, G., Zenz, C., Mataloni, P., and Leising, G. (1998) *Appl. Phys. Lett.*, **73**, 2860–2862.
53. Kallinger, C., Hilmer, C., Haugeneder, A., Perner, M., Spirkl, W., Lemmer, U., Feldmann, J., Scherf, U., Müllen, K., Gombert, A., and Wittwer, V. (1998) *Adv. Mater.*, **10**, 920–923.
54. Riechel, S., Kallinger, C., Lemmer, U., Feldmann, J., Gombert, A., Wittwer, V., and Scherf, U. (2000) *Appl. Phys. Lett.*, **77**, 2310–2312.
55. Lupton, J.M., Pogantsch, A., Piok, T., List, E.W.J., Patil, S., and Scherf, U. (2002) *Phys. Rev. Lett.*, **89**, 7401–7404.
56. Fiesel, R., Huber, J., and Scherf, U. (1998) *Enantiomer*, **3**, 383–389.

57. Fiesel, R., Huber, J., Apel, V., Enkelmann, V., Hentsche, R., Scherf, U., and Cabrera, K. (1997) *Macromol. Chem. Phys.*, **198**, 2623–2650.
58. Fiesel, R., Huber, J., and Scherf, U. (1996) *Angew. Chem., Intl. Ed. Engl.*, **35**, 2111–2113.
59. Grüner, J., Hamer, P.J., Friend, R.H., Huber, H.-J., Scherf, U., and Holmes, A.B. (1994) *Adv. Mater.*, **6**, 748–752.
60. Grimsdale, A.C. and Müllen, K. (2006) *Adv. Polym. Sci.*, **199**, 1–82.
61. Grimsdale, A.C., Leclère, P., Lazzaroni, R., MacKenzie, J.D., Murphy, C., Setayesh, S., Silva, C., Friend, R.H., and Müllen, K. (2002) *Adv. Funct. Mater.*, **12**, 729–733.
62. Setayesh, S., Grimsdale, A.C., Weil, T., Enkelmann, V., Müllen, K., Meghdadi, F., List, E.J.W., and Leising, G. (2001) *J. Am. Chem. Soc.*, **123**, 946–953.
63. Ego, C., Grimsdale, A.C., Uckert, F., Yu, G., Srdanov, G., and Müllen, K. (2002) *Adv. Mater.*, **14**, 809–811.
64. Jacob, J., Zhang, J., Grimsdale, A.C., Müllen, K., Gaal, M., and List, E.J.W. (2003) *Macromolecules*, **36**, 8240–8245.
65. Jacob, J., Grimsdale, A.C., Müllen, K., Sax, S., Gaal, M., and List, E.J.W. (2005) *Macromolecules*, **38**, 9933–9938.
66. Cho, S.Y., Grimsdale, A.C., Jones, D.J., Watkins, S.E., and Holmes, A.B. (2007) *J. Am. Chem. Soc.*, **129**, 11910–11911.
67. Qiu, S., Lu, P., Liu, X., Lu, F.S., Liu, L., Ma, Y., and Shen, J. (2003) *Macromolecules*, **36**, 9823–9829.
68. Wu, F.-I., Reddy, S., Shu, C.-F., Liu, M.S., and Jen, A.K.-Y. (2003) *Chem. Mater.*, **15**, 269–274.
69. Shu, C.-F., Dodda, R., Wu, F.-I., Liu, M.S., and Jen, A.K.-Y. (2003) *Macromolecules*, **36**, 6698–6703.
70. Grüner, J., Friend, R.H., Huber, J., and Scherf, U. (1996) *Chem. Phys. Lett.*, **251**, 204–210.
71. Scherf, U. and Müllen, K. (1992) *Polymer*, **33**, 2443–2446.
72. Nehls, B.S., Füldner, S., Preis, E., Farrell, T., and Scherf, U. (2005) *Macromolecules*, **38**, 687–694.
73. Nehls, B.S., Galbrecht, F., Brauer, D.J., Lehmann, C.W., Scherf, U., and Farrell, T. (2006) *J. Polym. Sci., Part A: Polym. Chem.*, **44**, 5533–5545.
74. Forster, M., Annan, K.O., and Scherf, U. (1999) *Macromolecules*, **32**, 3159–3162.
75. Bünnagel, T.W., Nehls, B.S., Galbrecht, F., Schottler, K., Kudla, C.J., Volk, M., Pina, J., Seixas de Melo, J., Burrows, H.D., and Scherf, U. (2008) *J. Polym. Sci., Part A: Polym. Chem.*, **46**, 7342–7353.
76. Wakim, S. and Leclerc, M. (2005) *Synlett*, 1223–1234.
77. Patil, S.A., Scherf, U., and Kadashchuk, A. (2003) *Adv. Funct. Mater.*, **13**, 609–614.
78. Dierschke, F., Grimsdale, A.C., and Müllen, K. (2003) *Synthesis*, **35**, 2470–2472.
79. Dierschke, F., Grimsdale, A.C., and Müllen, K. (2004) *Macromol. Chem. Phys.*, **205**, 1147–1153.
80. Li, J., Dierschke, F., Wu, J., Grimsdale, A.C., and Müllen, K. (2006) *J. Mater. Chem.*, **16**, 96–100.
81. Freund, T., Scherf, U., and Müllen, K. (1994) *Angew. Chem. Int. Ed. Engl.*, **33**, 2424–2426.
82. Melucci, M., Frere, P., Allain, M., Levillain, E., Barbarella, G., and Roncali, J. (2007) *Tetrahedron*, **63**, 9774–9783.
83. Baumgartner, T. and Reau, R. (2006) *Chem. Rev.*, **106**, 4681–4727.
84. Dienes, Y., Eggenstein, M., Karpati, T., Sutherland, T.C., Nyulaszi, L., and Baumgartner, T. (2008) *Chem. Eur. J.*, **14**, 9878–9889.
85. Forster, M. and Scherf, U. (2000) *Macromol. Rapid Commun.*, **21**, 810–813.
86. Chmil, K. and Scherf, U. (1993) *Makromol. Chem., Rapid Commun.*, **14**, 217–222.
87. Chmil, K. and Scherf, U. (1997) *Acta Polym.*, **48**, 208–211.
88. Goldfinger, M.B. and Swager, T.M. (1994) *J. Am. Chem. Soc.*, **116**, 7895–7896.
89. Tour, J.M. and Lambda, J.J.S. (1993) *J. Am. Chem. Soc.*, **115**, 4935–4936.
90. Lambda, J.J.S. and Tour, J.M. (1994) *J. Am. Chem. Soc.*, **116**, 11723–11736.

8
Silole-Containing Conjugated Polymers

Junwu Chen and Yong Cao

8.1
Introduction

Conjugated polymers have drawn broad attention owing to their important applications in polymeric light-emitting diodes (PLEDs) [1, 2], photovoltaic cells (PVCs) [3], field-effect transistors (FETs) [4], and their chemosensing abilities of various targets [5]. Solution processing of the conjugated polymers, such as spin coating and printing, is the attractive advantage in the fabrications of large-area optoelectronic devices. Conjugated polymers are the best candidates that can be used to fabricate flexible optoelectronic devices [6]. A notable feature of conjugated polymers also lies in the enormous versatility of synthetic methodology, which affords wide scope to construct new polymers with improved properties.

Siloles [7–10] or silacyclopentadienes are a group of five-membered silacyclics that possess $\sigma^*-\pi^*$ conjugation arising from the interaction between the σ^* orbital of two exocyclic σ-bonds on the silicon atom and the π^* orbital of the butadiene moiety [11]. The calculated LUMO level of a silole ring is lower than that of other heterocyclopentadienes, such as pyrrole, furan, and thiophene [10, 11]. The unique aromaticity and the low-lying LUMO level endue siloles with intriguing optoelectronic properties. Siloles **1–4**, as shown in Scheme 8.1, are the typical building blocks to construct various silole-containing polymers (SCPs).

Up to six substituents can be attached to the simple silole ring of **1**, which allows considerable room for tuning the optoelectronic properties. 2,3,4,5-Tetraarylsiloles, normally possessing noncoplanar geometry, are the widely studied silole small molecules [7, 12–23]. The distances between silole cores of any two adjacent 2,3,4,5-tetraarylsilole molecules, even in the solid state, are far from the normal $\pi-\pi$ interaction distance (about 3–4 Å), as found in the crystal structures [13]. Two unusual photophysical properties, aggregation-induced emission (AIE) [13, 24] and blueshift of photoluminescence (PL) emission in the crystal state compared to that in amorphous solids [25], have been found for 2,3,4,5-tetraphenylsiloles.

Design and Synthesis of Conjugated Polymers. Edited by Mario Leclerc and Jean-François Morin

ISBN: 978-3-527-32474-3

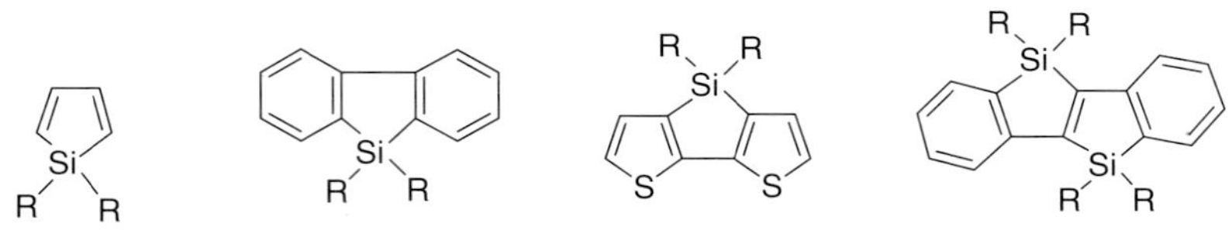

Scheme 8.1 Typical siloles for construction of conjugated polymers.

Owing to the AIE characteristics, 2,3,4,5-tetraphenylsiloles can show extremely high PL quantum yields (up to 100%), even in the crystalline form [25, 26]. Thereby, 2,3,4,5-tetraphenylsiloles are excellent emitters in the fabrication of electroluminescence (EL) devices [26, 27] showing external quantum efficiency (η_{EL}) up to 8%, close to the theoretical limit for a singlet emitter [25, 27]. Some 2,3,4,5-tetraarylsiloles with 2,5-electron deficient heterocycles exhibit high electron mobility and have been utilized as the electron-transporting layers for EL devices [26, 28]. Recently, chemosensing behaviors with 2,3,4,5-tetraphenylsiloles as the fluorescent chemosensor for explosives and organic solvent vapors, have also been reported [29, 30]. Dibenzosilole or 9-silafluorene **2** [31], dithienosilole or silicon-bridged bithiophene **3** [32], and bis-silicon-bridged stilbene **4** [33], are the typical examples of siloles fused with other aromatic rings. The enlarged skeletons of the siloles are coplanar, showing normal PL properties of typical organic emitters.

The incorporation of siloles in polymers is of interest and importance in chemistry and materials science [34]. Some optoelectronic properties, impossible for silole small molecules, may be realized with SCPs. The first synthesis of SCPs was reported by Tamao *et al.* in 1992 [35]. Since then, more research groups have been attracted to this field. Different types of SCPs, such as main chain-type π-conjugated SCPs catenated through aromatic carbon of a silole, main chain-type σ-conjugated SCPs catenated through silicon atom of a silole, SCPs with silole pendants, and hyperbranched or dendritic SCPs, have been synthesized [34].

In this chapter, three types of conjugated SCPs, π-conjugated SCPs, σ-conjugated SCPs, and SCPs with mixed $\sigma-\pi$-conjugations, are described. The widely variable chemical structures and the versatile functionalities, such as the tunable bandgaps from 4.0 to 1.45 eV, variable fluorescent emissions from UV to blue, green, and red (RGB) lights, AIE, fluorescent chemosensors for 2,4,6-trinitrotoluene (TNT)-type explosives, efficient EL emissions for RGB colors and white light with simultaneous RGB emission, phosphorescent hosts with high triplet energy level, efficient bulk-heterojunction solar cells, stable FETs with high hole mobility in air, and attenuation of strong laser power are reviewed and discussed. The notable features of conjugated SCPs indicate that they form an important group of polymeric semiconductors.

8.2
π-Conjugated Silole-Containing Polymers

8.2.1
Simple Silole Ring-Based Polymers

Poly(2,5-silole)s are the simplest π-conjugated polysiloles. However, the synthesis of the polymer was not so successful in the beginning, despite a few attempts [36–38]. The first proposal, in 1989, for poly(2,5-silole), was thereafter revised by the authors [36, 37]. Some 2,5-difunctionalsiloles such as 5-stannyl- and 5-boryl-2-halosilole had been demonstrated as not suitable for the preparation of the poly(2,5-silole)s, possibly due to the steric hindrance from the 3,4-substituents and the low reactivity of the 2,5-functional groups [38]. The first successful synthesis of the poly(2,5-silole) **7** was reported in 1999 with the introduction of a well-designed 2,5-difunctionalsilole monomer **6** (Scheme 8.2) [39]. Although the Yamamoto coupling reactions of 2,5-diiodosilole **5** using Ni(0) catalysts failed, compound **5** could be readily converted to the 2-iodo-5-zincated silole **6** via monolithiation with *n*-butyllithium and further addition of $ZnCl_2$(tmen) (tmen = *N,N,N′,N′*-tetramethylethylenediamine), at low temperatures. The $PdCl_2(PPh_3)_2$ catalyzed cross-coupling reaction of **6** under reflux afforded **7** as a red powder in 29% yield. The M_w and M_n of **7** were 4560 and 3440, respectively, giving a degree of polymerization (DP) of 13. The absorption peak (λ_{max}(abs)) of **7** in 2-MeTHF was at 482 nm at 293 K, but λ_{max}(abs) shifted to longer wavelengths of 542 nm at a lower temperature of 153 K. This may be due to the change of effective conjugation length by the conformational change of the main chain. The bandgap of **7** was not given in the report. However, a bandgap of the polymer, if calculated with the absorption edge (650 nm), is 1.9 eV, which was very close to 1.88 eV that was theoretically predicted by Hong *et al.* [40]. The polymer showed no fluorescence at room temperature.

Tremendous efforts have been devoted to the synthesis of copolymers with the simple silole ring. Silole–thiophene copolymers were the early examples [35]. The synthesis routes for the monomers are shown in Scheme 8.3. Compounds **8** and **9** were prepared by Ni-catalyzed intramolecular cyclization of thiophene or bithiophene-containing 1,6-diynes with hydrodisilane in moderate yields [35, 41]. The *N*-bromosuccinimide (NBS) bromination of **8** afforded the 2,5-thienylsilole dibromide **10**. The distannylated compound **11** was obtained by lithiation of **8** with *n*-BuLi and then treatment with $n\text{-}Bu_3SnCl$. Similarly, after lithiation

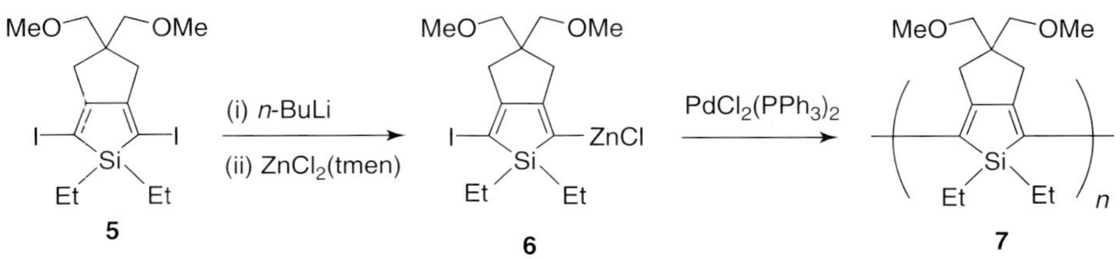

Scheme 8.2 Preparation of **6** and **7**.

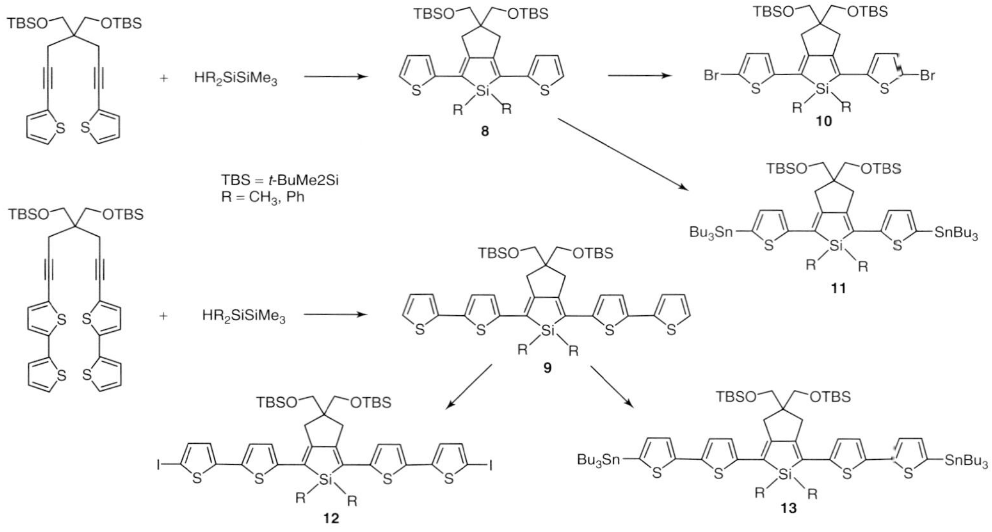

TBS = t-BuMe2Si
R = CH_3, Ph

Scheme 8.3 Preparation of **8–13**.

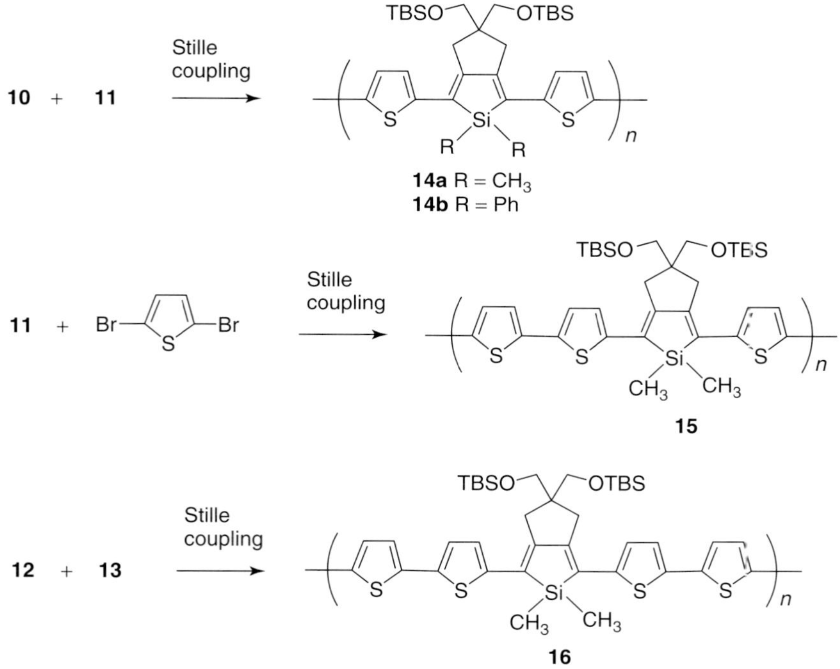

14a R = CH_3
14b R = Ph

Scheme 8.4 Preparation of **14–16**.

of **9**, compounds **12** and **13** were afforded by the treatments with iodine and n-Bu_3SnCl, respectively. Alternating silole–bithiophene, silole–terthiophene, and silole–quaterthiophene copolymers, named as **14**, **15**, and **16** respectively, were prepared by Stille coupling reactions (Scheme 8.4) [40, 41]. The M_w of the copolymers ranged from 12 600 to 69 400 with $M_w/M_n \leq 2.19$. Copolymers **15** and **16** showed comparable absorption peaks at ~547 nm while both **14a** and **14b** copolymers showed two bathochromic shifted absorption peaks (576 and 618 nm for **14a**; 594 and 615 nm for **14b**). Electrical conductivities of the copolymers doped with iodine were measured by the use of four-probe technique. Among the copolymers, the highest conductivity of 2.4 S cm^{-1} was achieved for **16**.

Yamaguchi *et al.* reported silole–thiophene alternating copolymer **17**, prepared by Suzuki coupling reaction of 2,5-diboronic acid of silole and 2,5-thienylsilole dibromide (Scheme 8.5) [42]. The M_w and M_n of **17** were 48 700 and 18 400, respectively. Copolymer **17** showed a broad absorption spectrum with λ_{max}(abs) at 648 nm in chloroform. The calculated bandgap from the absorption edge was only 1.55 eV, a very small value so far reported for the synthesized SCPs. The bandgap also coincides well with the theoretical value predicted by Hong *et al.* [40].

Reynolds and coworkers synthesized 2,5-bis[2-(3,4-ethylenedioxy)thienyl]silole (BEDOT-silole, **19**) by the Stille coupling reaction of 2,5-dibromosilole **18** with 2-stanyl-EDOT (Scheme 8.6) [43]. The targeted silole-BEDOT alternating copolymer **20** was prepared by electrochemical oxidative polymerization, by utilizing repeated potential scan mode. The spectroelectrochemical spectra of **20** indicated that the copolymer showed very narrow bandgap of 1.3–1.4 eV.

Reductive intramolecular cyclization by treatments of bis(phenylethynyl)silanes with LiNaph readily gave 2,5-dilithiosilole that was readily transformed to

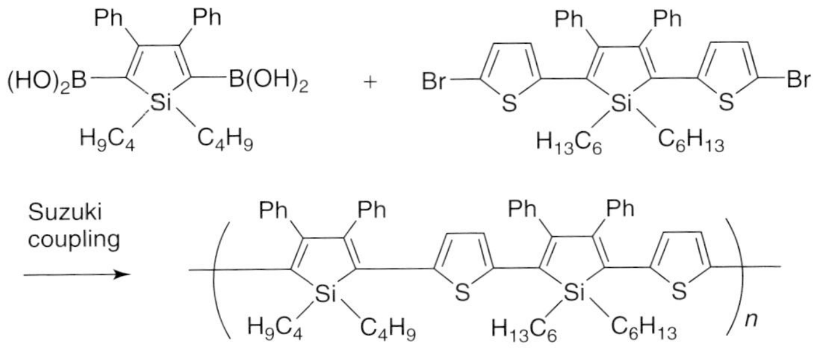

Scheme 8.5 Preparation of **17**.

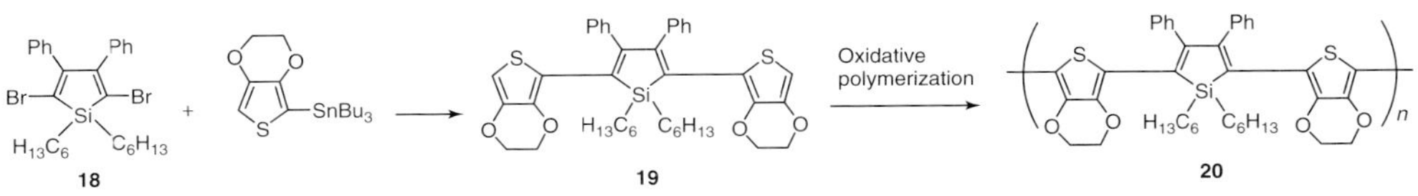

Scheme 8.6 Preparation of **19** and **20**.

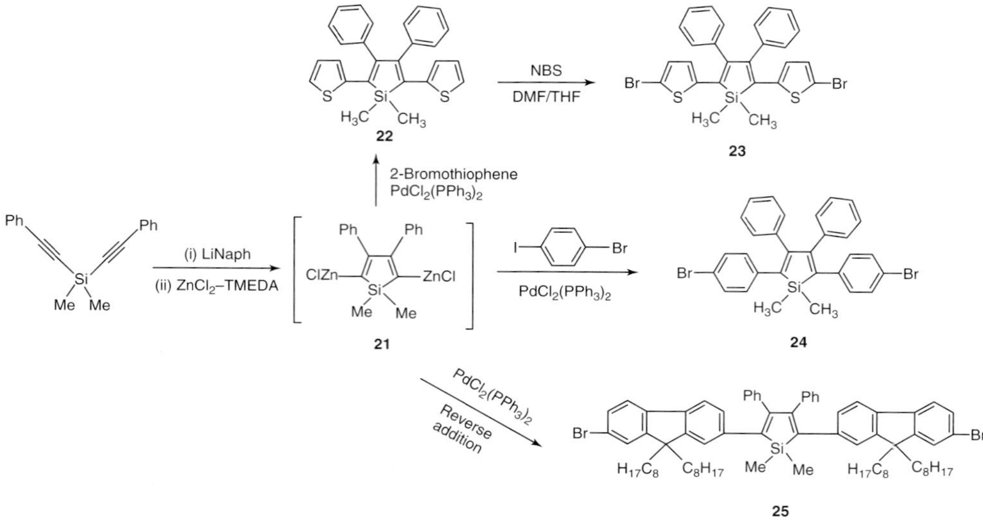

Scheme 8.7 Preparation of **21**–**25**.

2,5-dizincated siloles **21** (Scheme 8.7) [12, 38]. The intermediate **21** was a powerful reagent to prepare 2,5-difunctional silole monomers. 2,5-Bis(2'-thienyl)silole **22** was obtained by the Pd-catalyzed cross-coupling of **21** with 2-bromothiophene, which was transformed to 2,5-bis(5'-bromo-2'-thienyl)silole **23** by NBS bromination [44]. We have described the one-pot synthesis methods for some 2,5-difunctional silole monomers that are not suitable to prepare by conventional brominations of their precursors [45, 46]. The selective cross-coupling reaction of **21** with 4-bromoiodobenzene afforded the targeted silole monomer **24** in a high yield of 74% [45]. A reverse addition process was introduced in the one-pot synthesis of a big monomer **25**, in which the solution of **21** was dropwise added to the mixture of excess 2,7-dibromo-9,9-dioctylfluorene and the Pd catalyst [46]. The reverse addition method and the excessive amount of 2,7-dibromo-9,9-dioctylfluorene can protect the generated **25** and greatly hinder the formation of undesired oligomers from the coupling reaction between **21** and **25**.

In the past decade, polyfluorenes have emerged as emitting materials suitable for use in PLEDs because of their highly efficient PL and EL emissions, their thermal and oxidative stability, and their good solubility [47, 48]. Copolymerizations of fluorene and silole monomers would afford novel copolymers as promising candidates for applications in optoelectronic devices. Some typical 2,7-difunctional fluorene monomers are shown in Scheme 8.8.

Random and alternating copolymers **30** derived from fluorene and 2,5-dithienylsilole were prepared by Suzuki coupling reactions (Scheme 8.9) [44]. The copolymers possessed moderate molecular weights with M_w up to 44 400. The absorption spectra of the copolymers displayed the narrow bandgap nature of the silole unit. In particular, only a silole-dominant broad absorption band

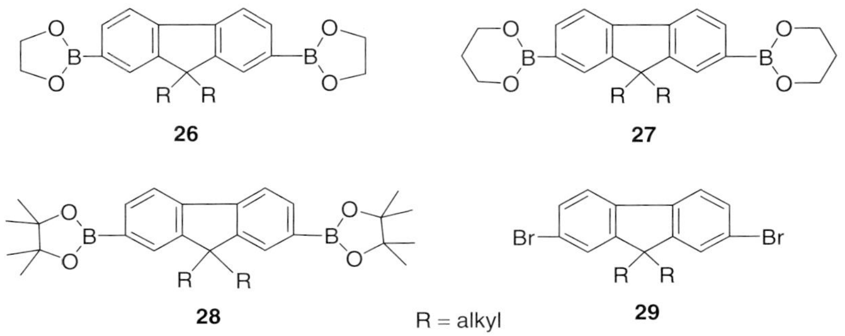

Scheme 8.8 Typical examples of 2,7-difunctional fluorene monomers.

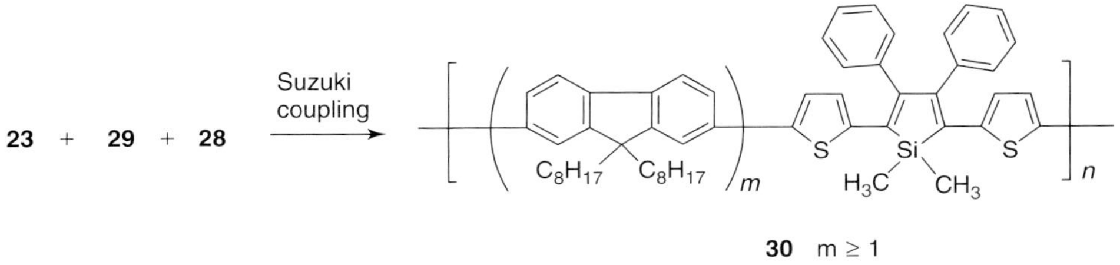

Scheme 8.9 Preparation of **30**.

with λ_{max}(abs) at 520 nm and the edge at ~600 nm was found for the film of the alternating copolymer. The HOMO and LUMO levels of the alternating copolymer were −5.71 and −3.60 eV, respectively. EL devices with **30** as the emissive layer emitted red light with emission peak (λ_{max}(EL)) up to 638 nm. The η_{ELmax} of the devices reached 0.89%. In a bulk-heterojunction PVC with methanofullerene [6,6]-phenyl C61-butyric acid methyl ester (PCBM) as an electron acceptor, the alternating copolymer showed impressive performance as the electron donor. In a device configuration of ITO/PEDOT/active layer/Ba/Al, the dark current density–bias curve showed a small leakage current, suggesting a continuous, pinhole-free active layer in the device. Under illumination of an AM1.5 solar simulator at 100 mW cm^{-2}, a high short-circuit current (J_{sc}) of 8.67 mA cm^{-2}, an open-circuit voltage (V_{oc}) of 0.65 V, and a fill factor (FF) of 35.8% were achieved. The calculated power conversion efficiency (PCE) was 2.01%. A top-contact FET device, with gold as the source and drain electrodes, ITO as the gate electrode, and polyacrylonitrile as the organic insulator on the gate, was fabricated so as to evaluate the carrier transport property of the alternating copolymer. On the basis of the transfer characteristics of the transistor, an FET hole mobility of 4.5×10^{-5} cm^2 $(V\ s)^{-1}$ was obtained.

We also reported the synthesis and properties of random and alternating copolymers **31** derived from fluorene and 2,3,4,5-tetraphenylsilole (Scheme 8.10) [45]. Though 2,3,4,5-tetraphenylsiloles are almost not fluorescent in solutions with relative quantum efficiency (Φ_{PL}) as low as ~0.1% due to rotational quenching effect by

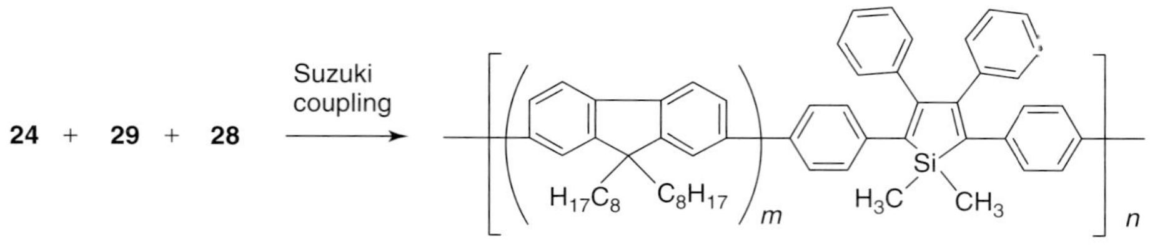

24 + 29 + 28
Suzuki
coupling
H17C8
C8H17
m
Si
H3C
CH3
n

31 m ≥ 1

Scheme 8.10 Peparation of **31**.

the peripheral phenyl groups [13], copolymers **31** could show much strong solution PL. For the alternating copolymer, its solution displayed silole-contributed green emission with Φ_{PL} of 13%, demonstrating that the phenyl rotations in solution were largely restricted when the siloles were incorporated in the rigid main chain. The films of **31** were highly fluorescent, with absolute quantum yields up to 34%. The EL spectra of **31** showed the exclusive green emissions (λ_{max}(EL) ~ 528 nm) from the silole units, which were almost not changed with the copolymer compositions. An η_{ELmax} of 1.51% was achieved with **31** as the emissive layer in EL devices.

White EL from a single polymer that can display simultaneous RGB emission is a promising candidate for applications in full-color display with a color filter and a backlight for liquid crystal display. Three-color white EL had been reported for a simple silole-containing polyfluorene **32** (Scheme 8.11) [49]. Incorporating small amounts of a green-emissive 2,5-diphenylsilole and a red-emissive 2,5-dithienylsilole in the blue-emissive polyfluorene backbone, efficient and stable white-light EL from a single polymer with a simultaneous RGB emission could be realized. The Commision International de L'Eclairage (CIE) coordinates ($x = 0.33$, $y = 0.36$) of the white-light EL spectra were very close to ($x = 0.33$, $y = 0.33$) for pure white light. The relative intensities for the three RGB peaks, at 450, 505, and 574 nm, were 0.94, 1, and 0.97, respectively, showing a balanced simultaneous RGB emission. The EL device displayed a maximum luminous efficiency of 2.03 cd A^{-1} for a brightness of 344 cd m^{-2} and a luminous efficiency of 1.86 cd A^{-1} for a more practical brightness of 2703 cd m^{-2}.

Well-defined alternating copolymer **33** possessed a repeating unit comprising of terfluorene and silole ring, but copolymers **34** were random (Scheme 8.12) [46]. These copolymers had binary chemical structures, in which the simple silole ring was incorporated in the polyfluorene main chain. A copolymer with very high M_w of 509 000 and M_w/M_n of 3 was realized during the polymerizations of **34**,

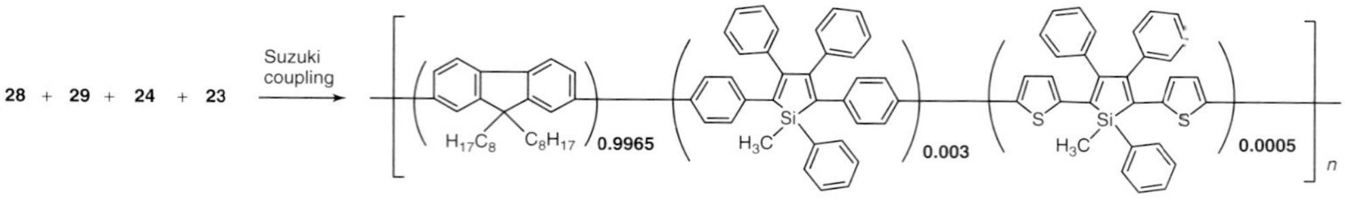

28 + 29 + 24 + 23
Suzuki
coupling
H17C8
C8H17
0.9965
Si
H3C
0.003
S
S
Si
H3C
0.0005
n

32

Scheme 8.11 Preparation of **32**.

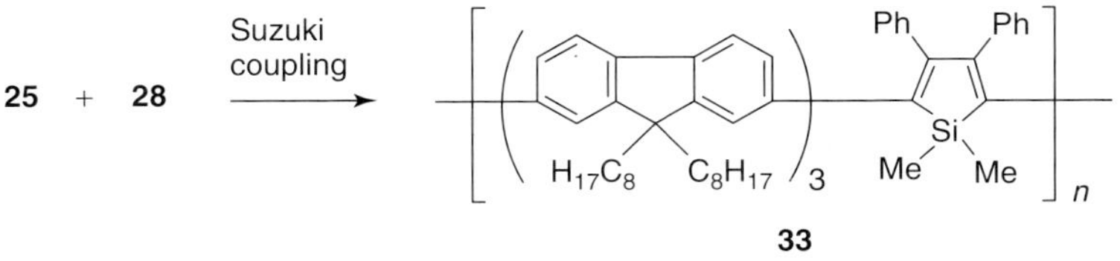

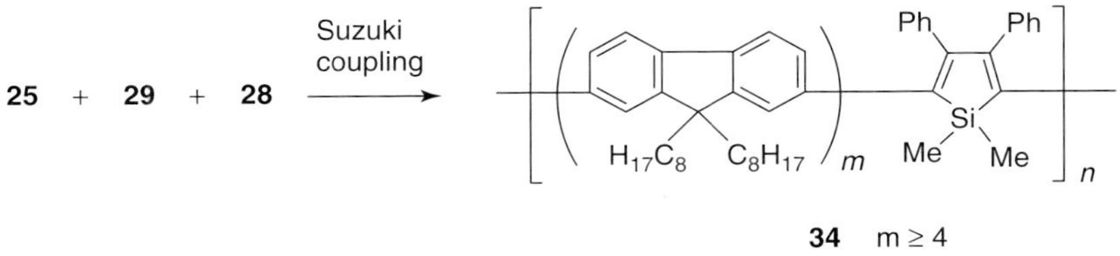

Scheme 8.12 Preparation of **33** and **34**.

demonstrating that monomer **25**, a fluorenyl end-capped dibromide, was highly reactive. Similar to **31**, films of copolymers **33** and **34** were highly fluorescent, with absolute quantum yields up to 83%. The EL emission peaks of **33** and **34** were between 539 and 546 nm and η_{ELmax} of 0.47% was achieved with neat film of **33** as the emissive layer. A largely improved η_{ELmax} of 1.99% was obtained when a blend film of **33** and poly(9,9-dioctyl-2,7-fluorene) (PF8) was utilized as the emissive layer.

Random fluorene-2,5-diphenylsilole copolymers **35** were prepared by Ni(0)-catalyzed Yamamoto reactions of the fluorene and silole dibromides with molar feed ratio of 90 : 10 (Scheme 8.13) [50]. The films of the copolymers showed PL peaks (λ_{max}(PL)) at ~550 nm. The organic light-emitting electrochemical cells (OLECs) based on blends of **35**/ionic liquid were evaluated.

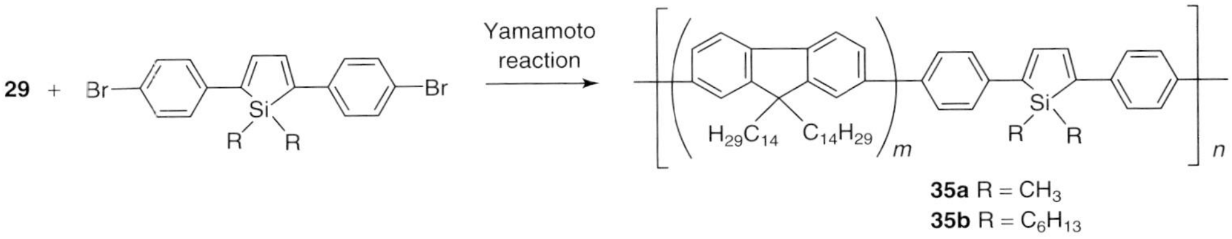

Scheme 8.13 Preparation of **35**.

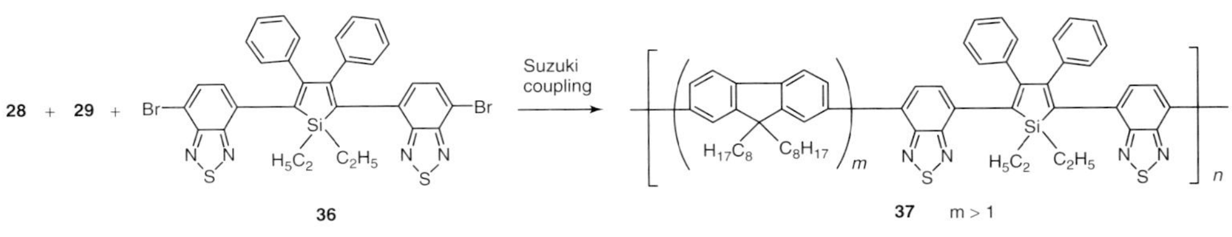

Scheme 8.14 Preparation of **37**.

Dibromides **36** of bis(2,1,3-benzothiadiazolyl)silole (Scheme 8.14) were prepared in a similar procedure as described in the preparation of **25** [51]. Random copolymers **37** were prepared by palladium(0)-catalyzed Suzuki coupling reaction. The copolymers possessed moderate molecular weights with M_w of 54 400–83 500 and M_w/M_n from 1.5 to 2.4. The optical bandgaps of the copolymers varied from 2.79 to 2.13 eV, depending on the silole content. Under excitation at 390 nm, **37** emitted orange lights with absolute quantum yields between 22 and 34%. The copolymers displayed red-light emissions with λ_{max}(EL) of 601 nm. With its neat film as the emissive layer, the EL device showed η_{ELmax} of 0.51%, but the device performance could be improved to η_{ELmax} of 1.37% when using a **37**/PF8 blend film as the emissive layer.

We prepared random and alternating silole-containing poly(3,6-carbazole) **38** by Suzuki coupling reactions of 3,6-carbazole diboronic ester with 2,5-bis(*p*-bromophenyl)silole **24**, in the presence or absence of 3,6-dibromocarbazole (Scheme 8.15) [52]. The M_w of the copolymers ranged from 11 100 to 16 700 with $M_w/M_n \leq 1.7$. From the absorption spectrum of the alternating copolymer, the absorption peak for the silole units was 395 nm. The films of **38** showed optical bandgaps between 2.53 and 2.79 eV. At the onset of the oxidation waves, it was observed from cyclic voltammetry (CV) measurements that the HOMO levels of **38** decreased from −5.15 to −5.34 eV along with the increase of silole contents in the copolymers. The excitation energy transfer from carbazole segments to the silole units was obvious from the PL spectra of **38** in solutions and films. The absolute PL quantum yields of films of **38** were between 16 and 43%. Since the HOMO levels of **38** met good hole injections from ITO anode (work function ~−4.7 eV), single-layer PLEDs with a simple configuration of ITO/**38**/Ba/Al were fabricated. An η_{EL} of 0.77% at a practical brightness of 333 cd m^{-2} could be achieved from the EL devices, which is much better than those of the reported single-layer devices based on conjugated carbazole copolymers containing comonomers such as pyridine, quinoline, oxadiazole, and phenylenevinylene. On the basis of the transfer

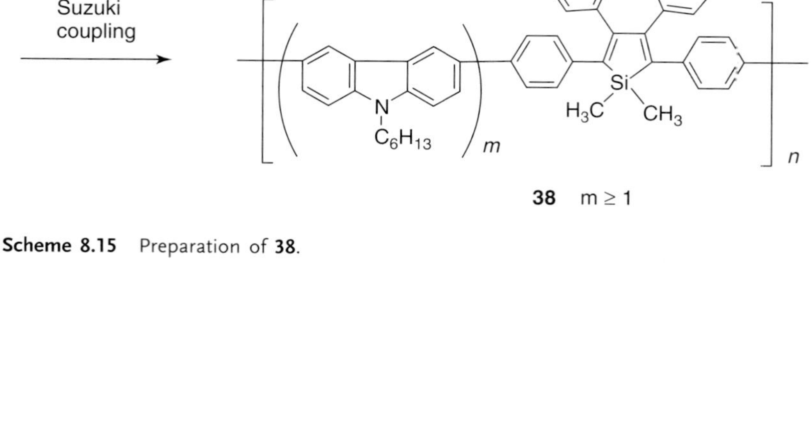

Scheme 8.15 Preparation of **38**.

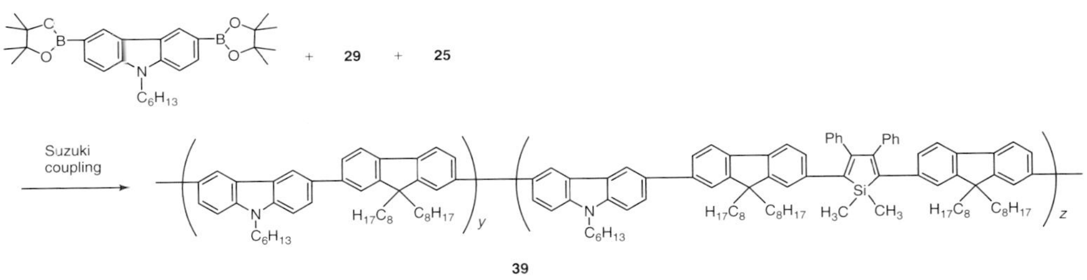

Scheme 8.16 Preparation of **39**.

characteristics of **38** in FET devices, hole mobility up to 9.3 × 10^{-6} cm^2 $(V\ s)^{-1}$ were obtained.

3,6-Carbazole-2,7-fluorene-silole random copolymers **39** (Scheme 8.16) with M_w up to 52 100 were synthesized by Suzuki coupling reactions [53]. Under excitation, the films of the copolymers show silole-dominated green emissions because of PL excitation energy transfer, with high absolute PL quantum yields up to 86%. The copolymers possessed HOMO levels of around 5.36 eV, mainly from the contribution of the 3,6-carbazole. Since the HOMO level of 1,3,5-tris(*N*-phenylbenzimidizol-2-yl)benzene (TPBI) is −6.2 eV [54], it is expected that effective hole blocking may be achieved in a device configuration of ITO/PEDOT/**39**/TPBI/Ba/Al. The hole blocking by the TPBI layer can significantly improve the EL efficiency, a high η_{ELmax} of 3.03% and maximum luminous efficiency of 7.59 cd A^{-1} can be achieved from the device configuration, in comparison to η_{ELmax} of 0.48% and efficiency of 1.20 cd A^{-1} for an EL device without the TPBI layer.

Poly(2,5-silole-ethynylene) **40** was prepared by the Stille coupling reaction of 2,5-dibromosilole **18** with bis(tributylstannyl)acetylene in a high yield of 93% (Scheme 8.17) [55]. The M_w and M_w/M_n of the polymer were 6520 and 3.6, respectively. Polymer **40** showed λ_{max}(abs) at 604 and 632 nm for its solution and film, respectively. The absorptions were largely bathochromic shifted relative to that of poly(thiophene-ethynylene) (λ_{max}(abs) = 438 nm). The absorption edge of **40** was at 740 nm, showing a bandgap of 1.67 eV. Some 2,5-diethynylsilole-based polymers **41** as shown in Scheme 8.18, were synthesized similarly by two research groups [55, 56]. It should be noted that **41a** and **b** were insoluble polymeric materials, possibly due to too strong intermolecular forces, while the other three polymers **41c–e** were soluble. The values of M_w of **41c** and **d** could reach

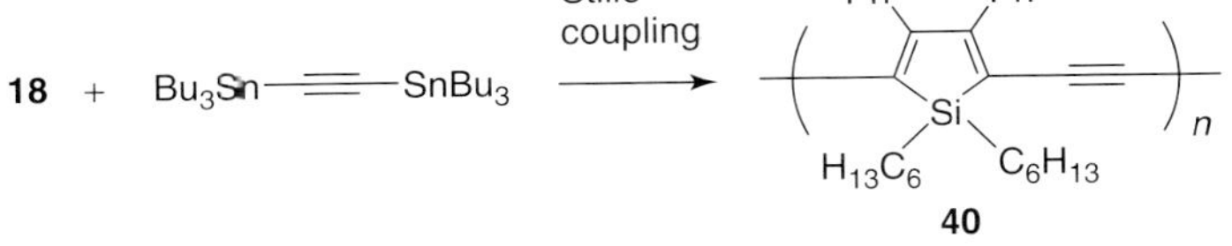

Scheme 8.17 Preparation of **40**.

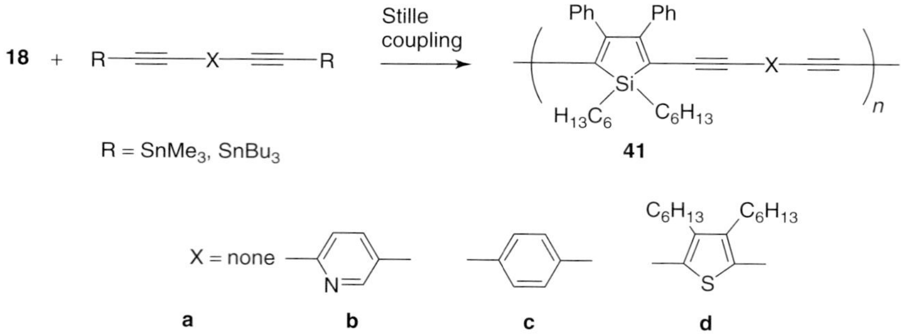

18 + R—≡—X—≡—R → (Stille coupling) 41

$R = SnMe_3, SnBu_3$

X = none (a), b, c, d

Scheme 8.18 Preparation of **41**.

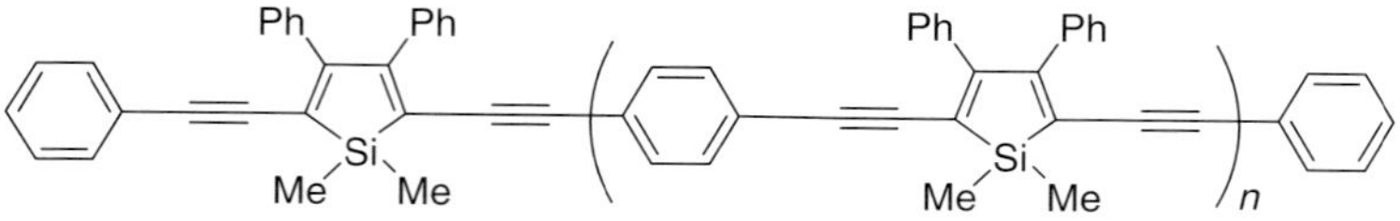

42 $n = 0, 1, 2, 3, 4$

Scheme 8.19 2,5-Diethynylsilole-based π-conjugated oligomers **42**.

~63 000 with M_w/M_n of ~5–6. The M_w of **41e** was several thousands. Polymers **41c** and **d** showed λ_{max}(abs) at ~500 and 576 nm in solutions, respectively. The iodine doped **40** and **41c** showed low conductivities (10^{-5}–10^{-2} S cm^{-1}). The syntheses of a group of 2,5-diethynylsilole-based π-conjugated oligomers **42**, with a similar repeating unit to **41c**, had been reported by Pagenkopf and coworkers (Scheme 8.19) [57]. The absorption and PL properties of **42** were also described.

Tang and coworkers synthesized three substituted polyacetylenes **45** carrying 1,2,3,4,5-pentaphenylsilole pendants (Scheme 8.20) [58]. The reaction of 1-chlorosilole **43** with ethynyl-Grignard reagent afforded 1-ethynylsilole **44a**. Monomers **44b** and **c** were obtained from reactions of **43** with 10-undecyn-1-ol and 11-phenyl-10-undecyn-1-ol, respectively. Metathesis catalysts $NbCl_5$– and WCl_6–Ph_4Sn worked well in the polymerizations of the acetylene monomers. The values of M_w of **45a–c** were 68 800, 11 500, and 33 400, respectively, with $M_w/M_n \leq 3.5$. The glass temperatures (T_g) of **45b** and **c**, measured by differential scanning calorimetry (DSC), were 56 and 82 °C. The higher T_g of **55c** should be due to its more rigid disubstituted polyacetylene backbone. For the DSC scanning in a temperature range of 30–270 °C, no T_g could be detected for **45a**, suggesting that the polymer has a very high T_g, due to the extremely rigid repeating units. The decomposition temperatures for the three polymers were all higher than 350 °C. Polymers **45b** and **c** showed comparable bandgaps with absorption edges at ~490 nm. The absorption edge of polymer **45a**, with direct silole linkage on the polyacetylene backbone, was at longer wavelength of ~550 nm, due to larger conjugation length. Under excitations, polymers **45c** emitted faint light in chloroform

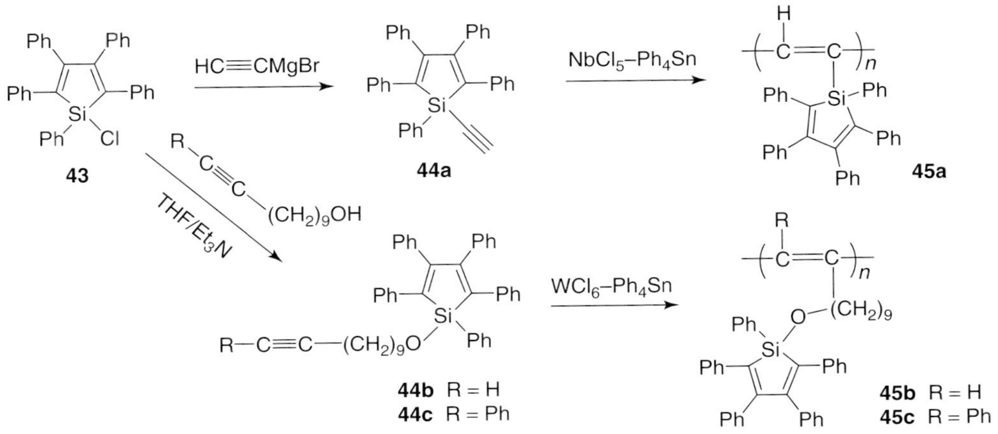

Scheme 8.20 Preparation of **44** and **45**.

solution ($\Phi_{PL} = 0.2\%$), but both its film and nanoaggregates showed bright green emissions at 512 nm. The Φ_{PL} for nanoaggregates formed in the 90% methanol mixture is 9.25%, which was ~46 times to that of the chloroform solution. Similar behavior was also found for **45b**, with ~20 times of Φ_{PL} changing. So far, polymers **45b** and **c** are the rare cases that a polymer can show the AIE property, already found from many small molecules [59, 60]. While **45a** did not show the AIE-active property, the long flexible spacer of nonanyloxy group decoupled the silole pendants of **45b** and **c** from the rigid polyacetylene backbone and enabled the silole groups to pack during aggregate formations. The mechanism for the AIE behaviors was revealed by the observations of cooling-enhanced emissions of the polymer solutions, from which a model of restricted intramolecular rotations of the peripheral phenyls on the silole ring was proposed for the first time [13, 58]. An η_{ELmax} of 0.55% was achieved with **45c** as the emissive polymer in an EL device.

The 1-chlorosilole **43** was also transformed to a silole-containing diphenylacetylene monomer **46** (Scheme 8.21) [61]. Though the monomer contained two ethynyl groups, it was found that only the ethynyl group of the diphenylacetylene moiety was polymerized by the metathesis catalyst, affording **47** with M_w of several thousands [61]. No T_g could be detected for **47** during the DSC scanning from room temperature to 300 °C, owing to the rigidity of the polydiphenylacetylene backbone. The polymer showed a decomposition temperature of 399 °C. The solution of **47** at room temperature only showed the PL emission of the polydiphenylacetylene backbone. No enhanced silole emission was found for the aggregates of **47**, indicating that silole clusters were difficult to form during the aggregation of the polymer chains, and that the rigid main chain and the short alkoxy spacer hindered the siloles from packing together and the large free volumes, normally possessed by films of polydiphenylacetylenes, allowed the peripheral benzenes on the silole ring to undergo rotation-induced nonradiative decay under the excitations. Upon

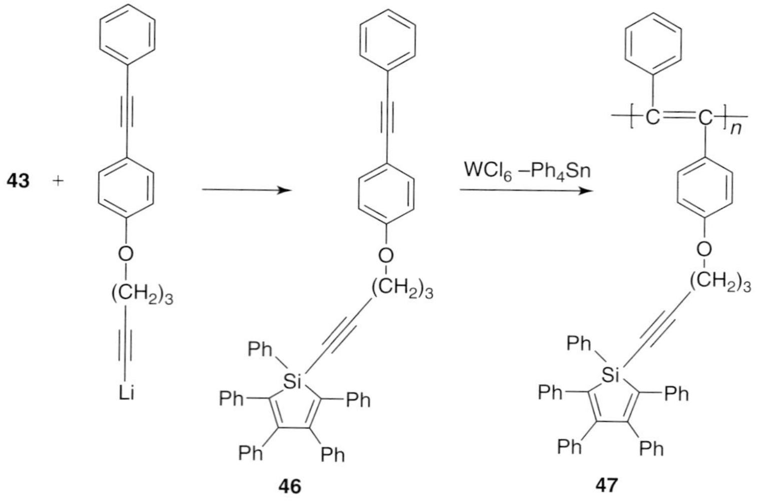

Scheme 8.21 Preparation of **46** and **47**.

cooling of the polymer solution to low temperature, strong silole emission was found, demonstrating that the benzene rotations were readily restricted.

8.2.2
Dibenzosilole-Based Polymers

The preparations of 2,7- or 3,6-difunctional dibenzosiloles are crucial to the synthesis of π-conjugated poly(dibenzosilole)s. Large advances in the preparations of the functional monomers have been demonstrated since 2005, which promote the development of π-conjugated poly(dibenzosilole)s.

In a report by Holmes and coworkers [62], as shown in Scheme 8.22, 2,2'-diiodo-4,4'-dibromobiphenyl was synthesized with 2,5-dibromonitrobenzene as the starting material, which was an important intermediate for 2,7-dibromodibenzosilole **48** and 2,7-diboronic ester **49**, the two 2,7-difunctional dibenzosilole monomers. The higher reactivity of the two iodine atom at 2,2'-positions of 2,2'-diiodo-4,4'-dibromobiphenyl than the two bromine atoms at 4,4'-positions is of importance in the synthesis of **48**, which was carried out by lithiation with *t*-butyllithium and cyclization with dichlorosilane. Monomer **49** was formed by lithiation and boronation reactions. 2,7-Dibromo-3,5-dimethoxyl-dibenzosilole **50** was prepared according to a similar procedures for **48** (Scheme 8.22) [63].

2,2'-Dibromo-5,5'-dichlorobiphenyl and 2,2'-Diiodo-5,5'-dibromobiphenyl were the important compounds for the syntheses of 3,6-difunctional dibenzosilole monomers (Scheme 8.23). Different synthetic routes had been introduced to prepare the two intermediates by us and Holmes's group, from which some 3,6-difunctional monomers **51–53** were obtained according to similar procedures as in the cases of **48** and **49** [64–66].

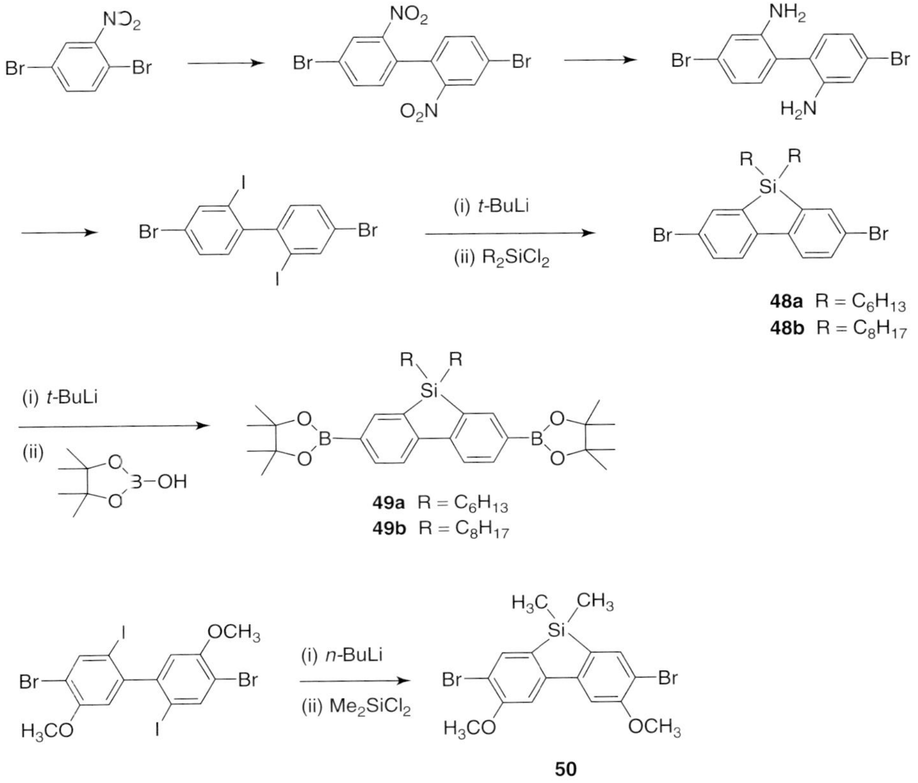

NO_2
Br Br

NO_2
Br Br
O_2N

NH_2
Br Br
H_2N

I
Br Br
I

(i) *t*-BuLi
(ii) R_2SiCl_2

R R
Si
Br Br

48a $R = C_6H_{13}$
48b $R = C_8H_{17}$

(i) *t*-BuLi
(ii) O B–OH O

R R
Si
O B O   O B O

49a $R = C_6H_{13}$
49b $R = C_8H_{17}$

I OCH_3
Br Br
H_3CO I

(i) *n*-BuLi
(ii) Me_2SiCl_2

H_3C CH_3
Si
Br Br
H_3CO OCH_3

50

Scheme 8.22 Preparation of **48**–**50**.

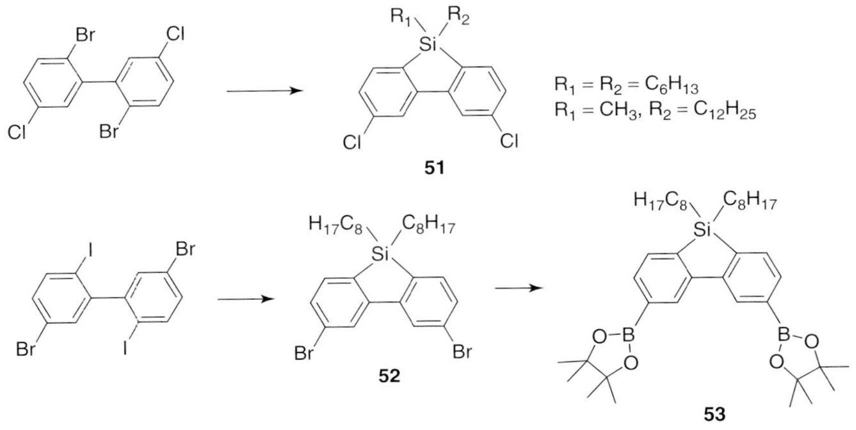

Br Cl
Cl Br

R_1 R_2
Si
Cl Cl

51

$R_1 = R_2 = C_6H_{13}$
$R_1 = CH_3$, $R_2 = C_{12}H_{25}$

I Br
Br I

$H_{17}C_8$ C_8H_{17}
Si
Br Br

52

$H_{17}C_8$ C_8H_{17}
Si
O–B O   B–O O

53

Scheme 8.23 Preparation of **51**–**53**.

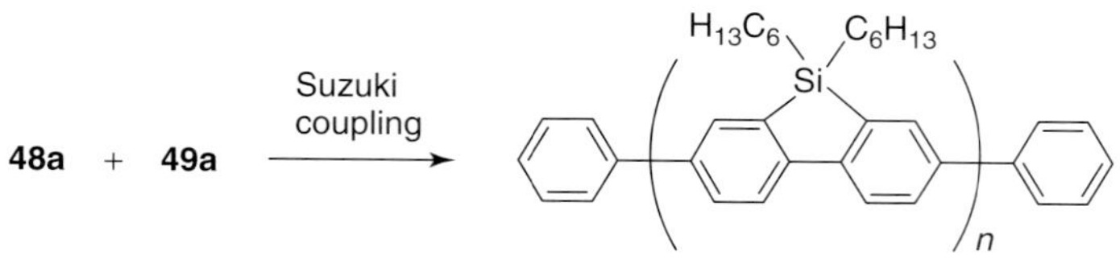

48a + 49a — Suzuki coupling → 54 ($H_{13}C_6$, C_6H_{13}, Si, n)

Scheme 8.24 Preparation of **54**.

The Pd(OAc)$_2$/tricyclohexylphosphine catalyzed Suzuki coupling reaction of **48a** and **49a** readily afforded the phenyl group end-capped poly(2,7-dibenzosilole) **54** in a high yield of 93% (Scheme 8.24) [62]. The M_w and M_n of **54** were 220 000 and 31 000, respectively, giving DP of ~90. The λ_{max}(abs) of a thin film of **54** was at 390 nm, which was comparable to that of PF8 [67]. The optical bandgaps of **54** and PF8 (2.93 eV) were identical. The HOMO and LUMO levels of **54**, determined by CV, were −5.77 and −2.18 eV, respectively. The T_g of **54** was 149 °C, much higher than that of PF8. Under an excitation at 325 nm, the polymer film emitted a highly efficient blue light (CIE coordinates $x = 0.15$, $y = 0.11$, absolute PL quantum yield = 62%) with a λ_{max}(PL) at 425 nm and two vibronic sidebands at 449 and 482 nm. Notably, the blue light emission was quite stable even after annealing at 250 °C for 16 hours, indicating the much better spectral stability of a poly(2,7-dibenzosilole) than that of PF8. Polymer **54** was electroluminescent and showed a blue emission with maxima at 431 and 451 nm in an EL device.

Mo *et al.* prepared two poly(3,6-dibenzosilole)s **55** via a nickel-catalyzed Yamamoto reaction (Scheme 8.25) [64]. The M_w and M_n of **55a** with two hexyl substituents at silicon atom were 16 000 and 10 000, respectively. This polymer

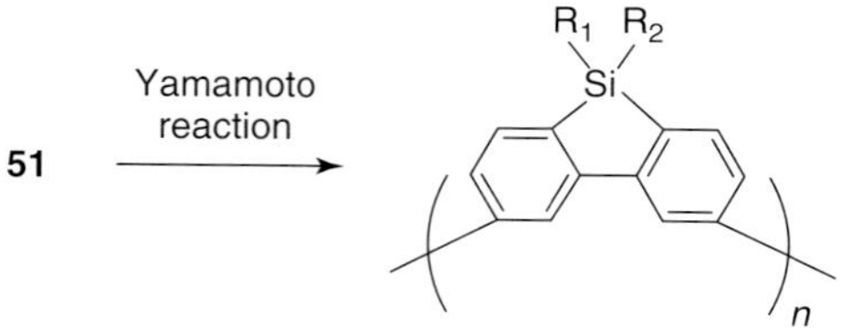

51 — Yamamoto reaction → 55 (R_1, R_2, Si, n)

55a $R_1 = R_2 = C_6H_{13}$
55b $R_1 = CH_3$, $R_2 = C_{12}H_{25}$

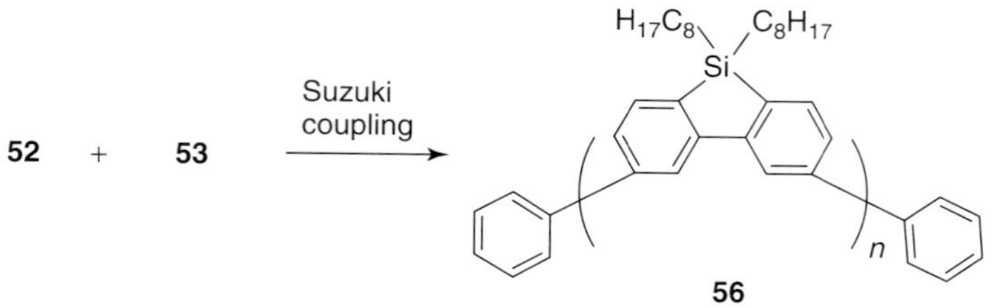

52 + 53 — Suzuki coupling → 56 ($H_{17}C_8$, C_8H_{17}, Si, n)

Scheme 8.25 Preparation of **55**.

displayed a T_g of 83 °C, slightly higher than the 75 °C for PF8. The two polymers exhibited identical λ_{max}(abs) at 283 nm and identical absorption edges at ~310 nm in solutions as well as in thin films. The absorption peaks and edges of **55** are significantly blueshifted compared with those of **54** and PF8, demonstrating that the conjugation lengths of a poly(3,6-dibenzosilole) is shorter than that of a poly(2,7-dibenzosilole), probably owing to the less coplanar linkages of the 3,6-dibenzosilole units. The bandgaps of the two poly(3,6-dibenzosilole)s were 4.0 eV, which is the widest bandgap so far reported for conjugated polymers. The results also suggest that poly(3,6-dibenzosilole)s possess high triplet energy levels. The oxidation onsets of **55** were at 1.7 V in CV scans, giving the estimated HOMO levels of −6.1 eV. Though reduction potentials were not recorded in the CV scans, LUMO levels of −2.1 eV of **55** were obtained with their HOMO levels and the optical bandgaps. Under excitation at 325 nm, polymers **55** emitted UV lights with λ_{max}(PL) of 355–360 nm in solutions as well as in films. We have utilized **55a** as the polymeric hosts for the fabrications of blue polymer phosphorescent light-emitting diodes [68]. High η_{EL} of 4.8% and luminance efficiency of 7.2 cd A^{-1} at 644 cd m^{-2} had been achieved in blue phosphorescence devices (λ_{max}(EL) = 462 nm, CIE coordinates $x = 0.15$, $y = 0.26$). The performance of the devices is much better than that of reported blue phosphorescent devices with poly(*N*-vinylcabarzole) (PVK) as the host.

Holmes and coworkers prepared a poly(3,6-dibenzosilole) **56** via a Suzuki coupling reaction in a yield of 93% (Scheme 8.25) [66]. The M_w and M_n of **56** were 23 000 and 11 000, respectively. The optical bandgap of the polymer was 3.5 eV, somewhat lower than those of **55**. CV measurements of **55** gave the LUMO level of −2.15 eV, but the oxidation wave was not recorded. With the optical bandgap and the LUMO level of **56**, an HOMO level of −5.65 eV was obtained. The PL spectra of films of **56** at room temperature and 77 K were compared. It was found that the PL spectrum at 77 K was from its triplet emission. At the onset of the triplet emission, a triplet energy of 2.55 eV was obtained, which was significantly higher than that of PF8 (2.1 eV). With the polymer as the host for a green phosphorescent dopant, exclusively green EL emission was observed, implying that the energy transfer from the host to the dopant was complete.

A fluorene-2,7-dibenzosilole alternating copolymer **57** was prepared by the Suzuki coupling reaction in a yield of 94% (Scheme 8.26) [62]. The molecular weights of the copolymer were extremely high (M_w = 425 000, M_n = 109 000, DP ~ 160).

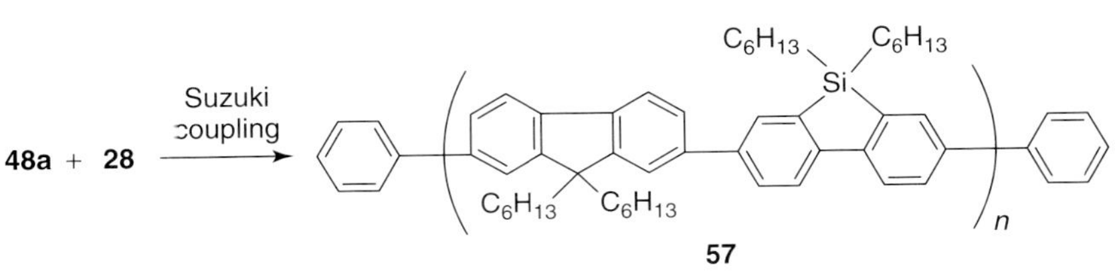

48a + **28** —Suzuki coupling→ **57**

C_6H_{13} C_6H_{13} Si C_6H_{13} C_6H_{13} *n*

Scheme 8.26 Preparation of **57**.

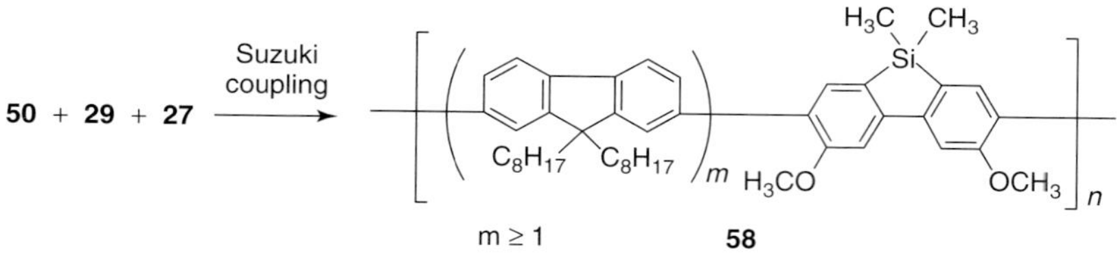

50 + 29 + 27 → (Suzuki coupling) 58 ($m \geq 1$)

Scheme 8.27 Preparation of **58**.

The absorption and fluorescent properties of **57** were quite similar to those of poly(2,7-dibenzosilole) **54**.

Random and alternating fluorene-2,7-dibenzosilole copolymers **58** were reported by Huang and coworkers (Scheme 8.27) [63]. Three feed ratios of fluorene to the silole (90 : 10, 80 : 20, and 75 : 25) were chosen for the preparations of the random copolymers. The values of M_w for the random copolymers ranged from 20 500 to 40 400, with M_w/M_n between 1.4 and 1.8. The M_w of 7100 and M_n of 5800 for the alternating copolymer were significantly lower than those of **57**, possibly due to the steric hindrance from the methoxy groups during the polymerization. Copolymers **58** showed polyfluorene-segment-dominant absorptions and PL emissions in solutions, but the absorption and PL spectra of the alternating copolymer showed dibenzosilole-contributed blueshifts. Slight redshifts of the PL spectra were found for films of **58**. The absolute PL quantum yields of **58** ranged from 51.8 to 59.9%, close to that of PF8. Green emission bands were found when **58** were heated in air, showing thermal stability comparable to that of PF8.

Wang *et al.* incorporated 3,6-dibenzosilole in the main chain of the polyfluorene [65]. As shown in Scheme 8.28, random and alternating copolymers **59**, with molar ratios of fluorenes to the 3,6-dibenzosilole including 50 : 50, 70 : 30, 90 : 10, and 95 : 5, were synthesized by Suzuki coupling reactions. The M_w of **59** ranged from 50 000 to 94 000 with $M_w/M_n \leq 2.5$. With higher silole contents, the copolymers possessed lowered HOMO levels and showed blueshifted absorptions and PL emissions in solutions as well as films. The films of **59** showed intense PL emissions, with absolute PL quantum yields ranging from 38 to 84%. Utilizing the copolymers as the emissive layer in EL devices, highly efficient pure blue emissions with CIE coordinates of ($x = 0.16$, $y = 0.07$), η_{EL} of 3.34%, and luminance efficiency of 2.02 cd A^{-1} at 326 cd m^{-2} were achieved from the copolymer with 90% fluorene

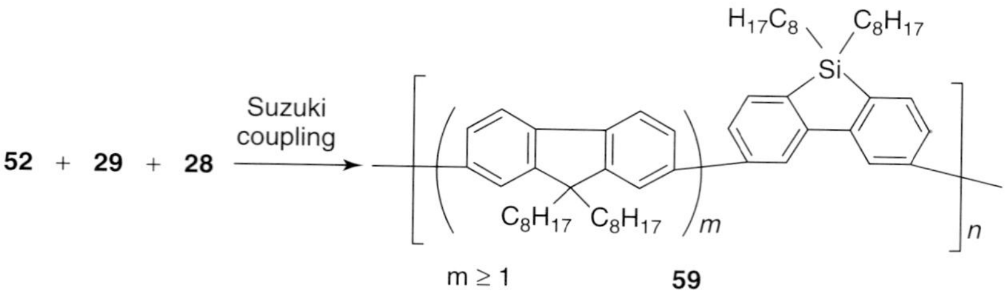

52 + 29 + 28 → (Suzuki coupling) 59 ($m \geq 1$)

Scheme 8.28 Preparation of **59**.

content. The blue color matched the NTSC blue standard ($x = 0.14$, $y = 0.08$) quite well. The EL spectral stability of the devices was quite good even under operation at elevated temperatures. It is well known that good color purity, spectral stability, and high device efficiency remains a great challenge for blue-emitting polyfluorenes. The color purity and color stability achieved with **59** demonstrated that fluorene-3,6-dibenzosilole copolymers are outstanding candidates for blue emitter.

A random copolymer **60** was prepared with the molar ratio of 9 : 1 for 2,7-dibenzosiloles to 3,6-dibenzosilole (Scheme 8.29) [69]. The copolymer showed blue emission with λ_{max}(PL) at ~425 nm, with absolute PL quantum yield of 83%. No undesired long-wavelength green emission was observed in the PL spectrum after thermal annealing of the copolymer film at 200 °C for 8 hours. Copolymer **60** exhibited high performance of blue EL with η_{EL} of 1.95%, a luminous efficiency of 1.69 cd A^{-1} and a maximal brightness of 6000 cd m^{-2} with the CIE coordinates of ($x = 0.162$, $y = 0.084$). Thus the incorporation of the 3,6-silafluorene unit into the poly(2,7-silafluorene) main chain could not only improve the color purity of the EL devices but also enhanced its device efficiency.

Random copolymers **61** and **62** derived from 2,7-dibenzosilole and 2,1,3-benzothiadiazole derivatives were prepared with a molar ratio of 9 : 1 for 2,7-dibenzosilole to 2,1,3-benzothiadiazole (Scheme 8.30) [70]. Copolymer **61** showed a green emission with λ_{max}(PL) of 530 nm while **62** emitted a red light with λ_{max}(PL) of 629 mm due to the efficient intramolecular energy transfer from dibenzosilole unit to benzothiadiazole (**61**) and dithienylbenzothiadiazole (**62**), respectively. Both copolymers possessed comparable absolute PL quantum yield

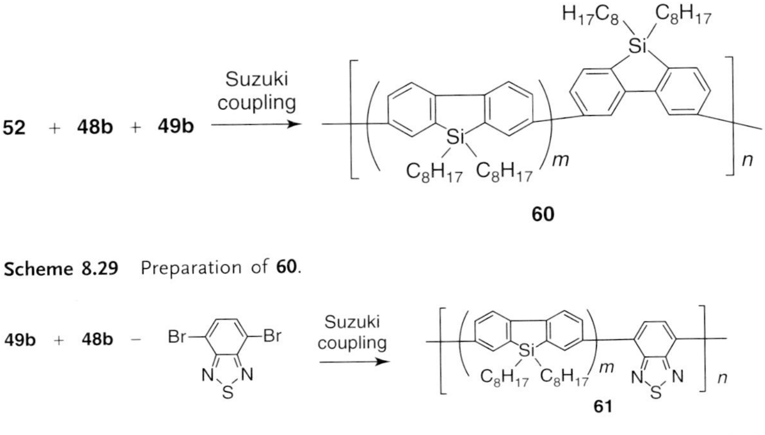

Scheme 8.29 Preparation of **60**.

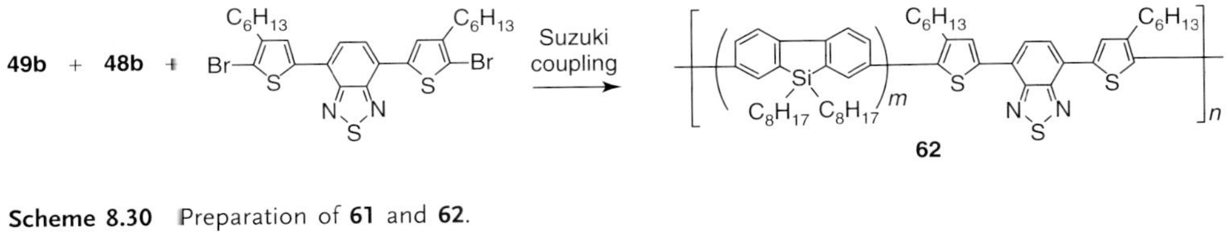

Scheme 8.30 Preparation of **61** and **62**.

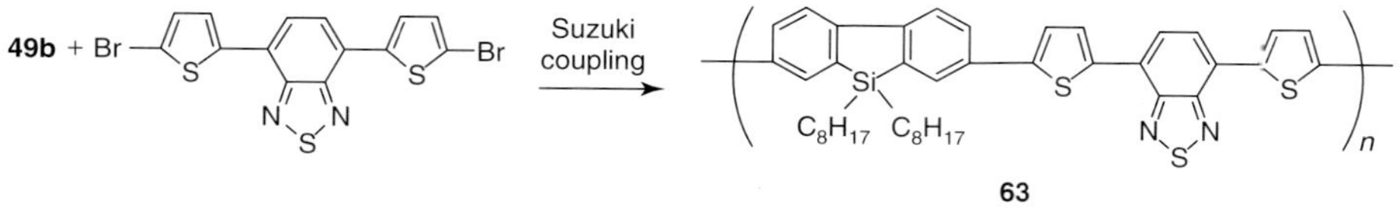

Scheme 8.31 Preparation of **63**.

of ~52%. Copolymer **61** was an excellent green EL polymer, and a maximum η_{EL} of 3.81% was realized in EL devices. With the copolymer **62** as the emissive layer, good device performance with a maximum η_{EL} of 2.89% was achieved for a saturated red EL (λ_{max}(EL) = 643 nm).

Leclerc and coworkers prepared alternating copolymer **63** derived from 2,7-dibenzosilole and 4,7-dithienyl-2,1,3-benzothiadiazole (Scheme 8.31) [71]. The M_n and M_w of **63** were 15 000 and 20 000, respectively. The 4,7-dithienyl-2,1,3-benzothiadiazole-contributed λ_{max}(abs) was at 539 nm for a polymer solution or 576 nm for a thin film. The optical bandgap of **63** was 1.85 eV, calculated with the onset wavelength (669 nm) of the film absorption spectrum. The electrochemical HOMO and LUMO levels of **63** were −5.7 and −3.81 eV respectively, giving an electrochemical bandgap of 1.89 eV that matched the optical bandgap. A photovoltaic device, with a configuration of ITO/PEDOT/(**63** : PCBM = 1 : 4)/Al, displayed J_{sc} of 2.80 mA cm^{-2}, V_{oc} of 0.97 V, and fill factor (FF) of 55% under illumination at 90 mW cm^{-2}, giving a PCE of 1.6%. An independent work published almost at same time by our group reported **63** with much higher molecular weights (M_n = 79 000 and M_w = 330 000) and significantly enhanced photovoltaic performance [72]. PVC devices with similar device configuration but with a different component ratio (**63** : PCBM = 1 : 2)showed much higher J_{sc} of 9.50 mA cm^{-2}, V_{oc} of 0.90 V, and FF of 50.7% under illumination of AM 1.5 solar simulator at 80 mW cm^{-2}, corresponding to a high PCE of 5.4%. The FET hole mobility for **63**, measured under ambient conditions without any encapsulation, reached $\sim 1 \times 10^{-3}$ cm^2 V^{-1} s^{-1}.

Usta *et al.* prepared dibenzosilole–thiophene copolymers **64** with $y = 1$ and 2, by Suzuki coupling reactions (Scheme 8.32) [73]. Copolymers **64** displayed high T_g over 200 °C and good thermal stability with decomposition temperatures of over 400 °C. The bandgap of 2.5 eV for **64a** was higher than the 2.3 eV for **64b**. FET devices of **64** with a top-contact configuration were fabricated to evaluate their FET

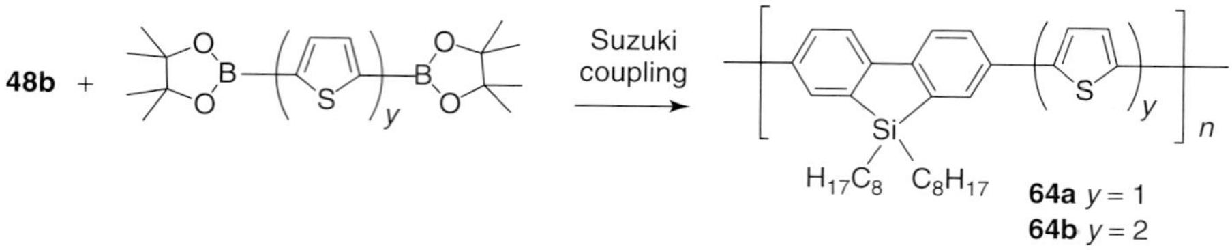

Scheme 8.32 Preparation of **64**.

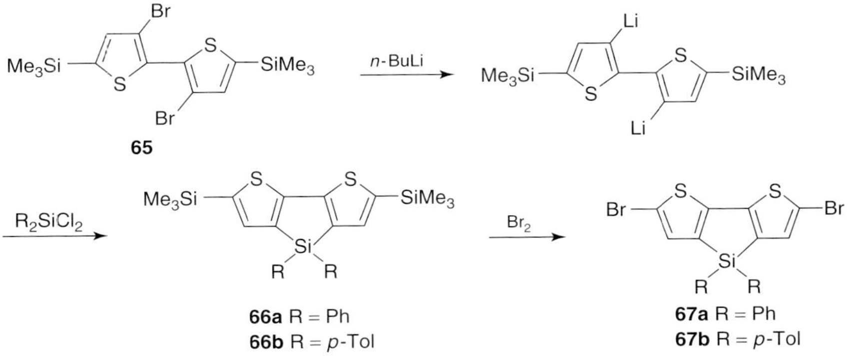

Scheme 8.33 Preparation of **66** and **67**.

hole mobility. Copolymer **64b** showed higher hole mobility of 6×10^{-3} cm^2 (V s)$^{-1}$ when compared with 6×10^{-5} cm^2 (V s)$^{-1}$ for **64a**.

8.2.3 Dithienosilole-Based Polymers

Ohshita *et al.* reported the synthesis of dithienosiloles and the synthetic method was also effective to prepare some dithienosilole monomers [32]. As shown in Scheme 8.33, with 3,3'-dibromo-5,5'-bis(trimethylsilyl)-2,2'-bithiophene **65** as the starting material, dithienosiloles **66** were formed via the lithiation and cyclization reactions. The trimethyl groups of **66** were replaced with bromine atoms by the treatment with bromine, readily affording **67** in good yields.

Usta *et al.* described another synthetic route for dithienosilole monomers [73]. As shown in Scheme 8.34, the lithiation of 3,3'-dibromo-2,2'-bithiophene and the cyclization that followed readily afforded dithienosilole **68**. Further bromination of **68** with NBS gave dibromodithienosilole **69**. Further transformations afforded bis(trimethylstannyl) compound **70**, an important silole monomer.

Anodic oxidation polymerization was used for the preparation of a poly (2,6-dithienosilole) **71**, a dithienosilole-based homopolymer (Scheme 8.35) [74]. By applying constant potential scan to the electrolyte solution of **66b**, rather than repeated potential scans, the possible decomposition in terms of liberation of the organosilicon moieties could be largely suppressed. The molecular weights (M_w) of **71**, prepared under different conditions, ranged from several thousands to 17 800, with M_w/M_n between 1.1 and 2.6. The varied absorption features indicated

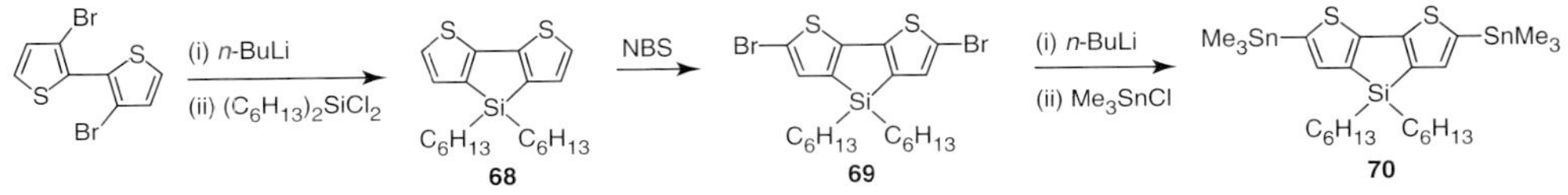

Scheme 8.34 Preparation of **68**–**70**.

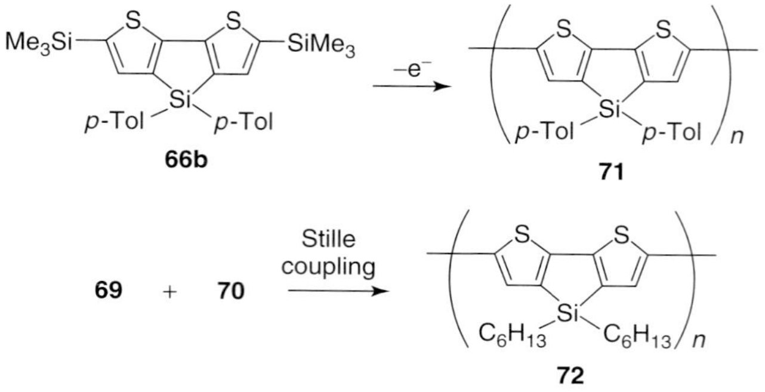

Scheme 8.35 Preparation of **71** and **72**.

that **71** should possess varied polymer structures under the different oxidation runs. Typically, **71** showed λ_{max}(abs) at ~487 nm. The polymer displayed red-light emission with λ_{max}(PL) at ~656 nm. A single-layer EL device also emitted a red light, with a low brightness even at a high voltage. Poly(2,6-dithienosilole) **72** was prepared by Stille coupling reaction (Scheme 8.35) with M_w of 65 200 and wide molecular distribution ($M_w/M_n = 8$) [75]. The λ_{max}(abs) values for solution and film of **72** were comparable (~500 nm). The copolymer showed poor photovoltaic performance, with PCE of 0.02 and 0.05% for as-fabricated device and annealed device, respectively.

In a short report by Barton and coworkers, the synthesis and some properties of dithienosilole–thiophene alternating copolymer **73a** were described (Scheme 8.36) [55]. Copolymer **73a** showed λ_{max}(abs) at 556 and 582 nm in THF and in film respectively. Excited at 510 nm, the copolymer emitted a red light. When the copolymer was doped with iodine, a high electrical conductivity of 400 S cm^{-1} was achieved. This value is the highest among SCPs and is also close to that of well-defined poly(3-alkylthiophene) [41]. Usta *et al.* prepared two dithienosilole–thiophene copolymers **73** with $y = 1$ and 2 (Scheme 8.36) [73]. The M_w of **73** ranged from 30 000 to 41 000 with $M_w/M_n \leq 3$. The unaligned films of copolymers **73a** and **b** showed remarkably high FET hole mobility of 0.02 and 0.06 cm^2 $(V\ s)^{-1}$, respectively, as measured in air. Furthermore, the FET devices possessed high current on/off ratios of ~10^5–10^6. The FET device with **73b** also displayed impressive stabilities under repeated on/off cycles up to 2000 cycles. No obvious

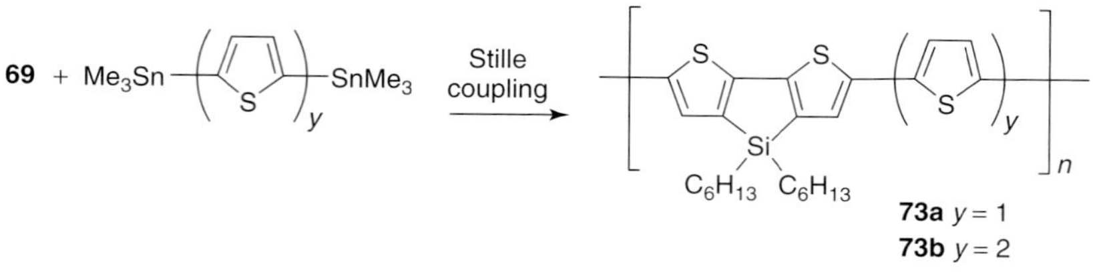

Scheme 8.36 Preparation of **73**.

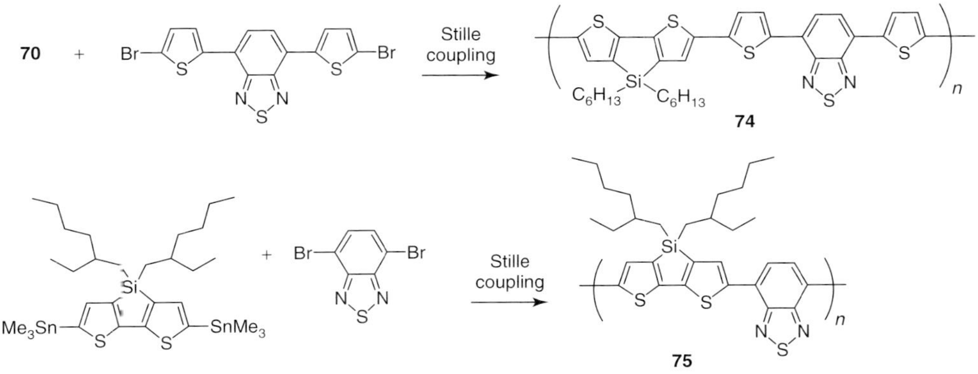

Scheme 8.37 Preparation of **74** and **13**.

deterioration was also found for the FET device with **73a** after 6000 cycles. The results show the impressive performances of FETs with SCPs as the active layer.

Liao *et al.* prepared alternating copolymer **74** derived from dithienosilole and 4,7-dithienyl-2,1,3-benzothiadiazole (Scheme 8.37), with M_w of 29 200 and $M_w/M_n = 5.2$ [75]. The copolymer showed λ_{max}(abs) at 563 nm for a solution and 593 nm for a thin film. The optical bandgap of **74** was 1.4 eV. The electrochemical HOMO and LUMO levels of **74** were −5.13 and −3.23 eV respectively. A photovoltaic device with **74** : PCBM = 1 : 1 as the active layer displayed PCE of 0.07 and 0.18% for as-fabricated device and annealed device, respectively. Compared with **74**, the dithienosilole-2,1,3-benzothiadiazole alternating copolymer **75** reported by Hou *et al.* [76] had a simpler repeating unit. The M_n of **75** was 18 000, with very narrow molecular weight distribution ($M_w/M_n = 1.2$). The optical bandgap of **75** was 1.45 eV. The electrochemical HOMO and LUMO levels of **75** were −5.05 and −3.27 eV respectively, giving an electrochemical bandgap of 1.78 eV. The copolymer showed FET hole mobility of 3×10^{-3} cm^2 (V s)$^{-1}$. With an active layer of **75** : $PC_{70}BM$ = 1 : 1, photovoltaic devices showed J_{sc} of 12.7 mA cm^{-2}, V_{oc} of 0.68 V, and FF of 55% under illumination of AM 1.5G solar simulator at 100 mW cm^{-2}, giving an outstanding PCE of 5.1%. The device showed a broad spectral response range covering 350–800 nm.

Jen and coworkers synthesized two random copolymers **76** based on fluorene and dithienosilole (Scheme 8.38) [77]. The 2,6-dibromodithienosilole **67a** was fed in low quantities (silole/fluorene ratios ≤ 0.05/0.95). The M_w of the two polymers ranged

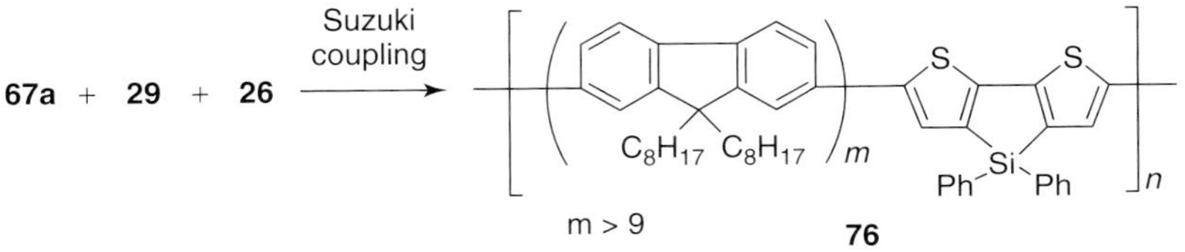

Scheme 8.38 Preparation of **76**.

from 76 000 to 146 000, with $M_w/M_n \leq 2.8$. Under excitation, strong Forster energy transfer from fluorene segments to the silole units was found and the PL spectra of films of **76** only showed the silole emissions at 500 and 527 nm. EL devices with a simple configuration of ITO/**76**/Ca emitted green light with η_{ELmax} up to 0.3%. Inserting a PEDOT layer as hole-injection and -transporting layer, η_{ELmax} was raised to 0.41%. With a cross-linked hole-injection and -transporting layer to replace the PEDOT, η_{ELmax} was boosted to 1.64% and the maximum brightness of 25 900 cd m^{-2} was reached.

8.2.4
Poly(bis-silicon-bridged stilbene)

Yamaguchi and coworkers reported a new synthetic route for bis-silicon-bridged stilbene **77** by intramolecular reductive cyclization of bis(*o*-silyphenyl)acetylene with lithium naphalenide (LiNaph) followed by quenching with iodine (Scheme 8.39) [78]. After lithiations of 3,8-dimethoxy compounds **77a** and **b** with *sec*-BuLi, 2,7-diiodide compounds **78a** and **b** were obtained by treatment with iodine. Also, the treatment with *i*-PrOB(pin) (pin = pinacolato) gave the 2,7-diboronic ester **78c**. In the preparation of **79**, it was found that organozinc reagent Li^t-Bu$_2$ZnTMP worked well to transform **77c** into 2,7-dizincated intermediate and the following treatment with iodine afforded the targeted 2,8-diiodide. The 2,8-diboronic ester **80** was synthesized by traditional lithiation of **79** and then treatment with *i*-PrOB(pin).

Poly(bis-silicon-bridged stilbene)s **81** were prepared by Suzuki coupling reactions of the diiodides and diboronic esters (Scheme 8.40) [78]. The values of M_n for **81a–c** were 7000, 23 000, and 8000, respectively, with $M_w/M_n \leq 2.4$. Though solutions of the three polymers showed comparable λ_{max}(abs) at ~420 nm, obvious difference of the methoxy- or fluoro-substitution effect was found from their PL emissions of the solutions. The λ_{max}(PL) of **81a** was at 476 nm, with Φ_{PL} of 36% while that of **81c** with fluorine substitutions, was blueshifted to 454 nm, with increased Φ_{PL} of 63%. Polymer **81b**, derived from the methoxy- and fluoro-monomers, showed λ_{max}(PL) at 467 and Φ_{PL} of 51%, which lie between those of **81a** and **c**.

Bis-silicon-bridged stilbene-bithiophene alternating copolymer **82** was prepared by the Stille coupling reaction (Scheme 8.41) [78]. The M_n of the copolymer was 12 000, with M_w/M_n of 2.6. The THF solution of copolymer **82** showed λ_{max}(abs) at 494 nm, with a redshift of 74 nm relative to the poly(bis-silicon-bridged stilbene) **81a**. The solution displayed a green emission with λ_{max}(PL) at 547 nm. The Φ_{PL} of the PL emission was only 0.025%, significantly lower than 36% for **81a**, indicating that the incorporation of the bithiophene greatly opens the nonradiative pathway. Bis-silicon-bridged stilbene-1,4-diethynylbenzene alternating copolymer **83** was prepared by Sonogashira coupling reaction (Scheme 8.41) [78]. The M_n of the copolymer was 37 500, with M_w/M_n of 2.6. THF solution of copolymer **83** showed λ_{max}(abs) at 482 nm, with a redshift of 62 nm relative to the poly(bis-silicon-bridged stilbene) **81a**. The solution displayed a green emission with λ_{max}(PL) at 506 nm and Φ_{PL} of 15%.

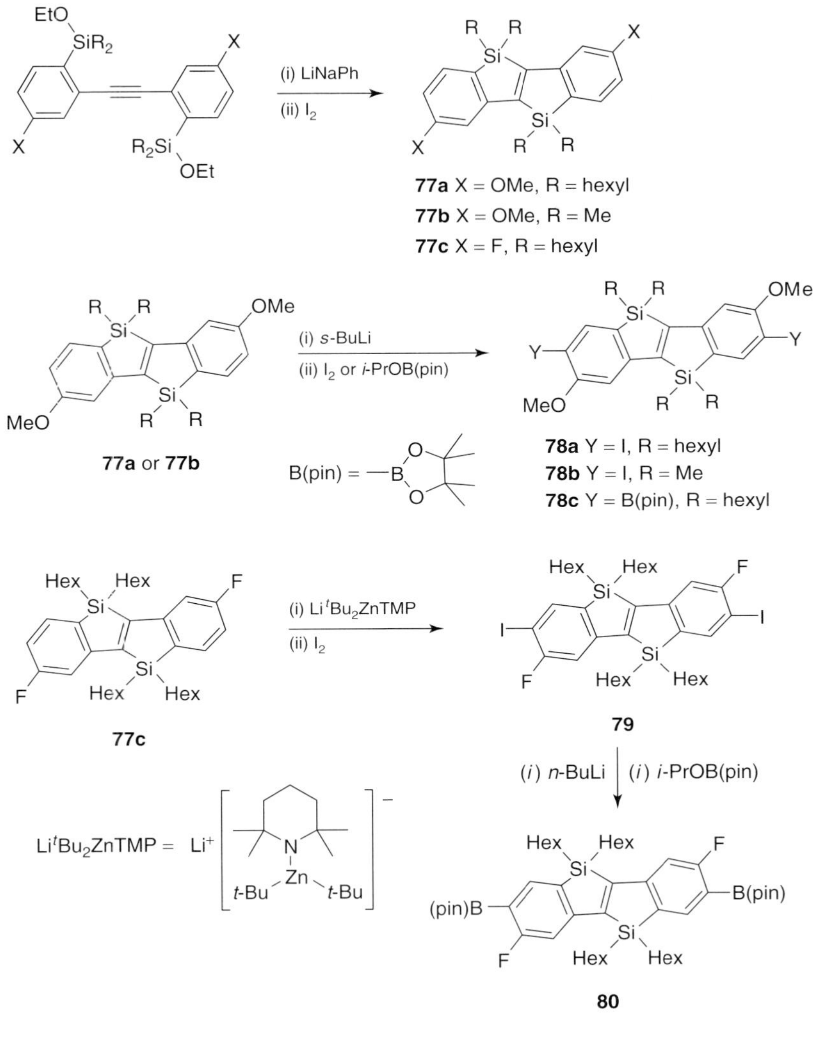

EtO
SiR2
X
X
R2Si
OEt
(i) LiNaPh
(ii) I2
R R
Si
X
Si
X
R R
77a X = OMe, R = hexyl
77b X = OMe, R = Me
77c X = F, R = hexyl
R R
Si
OMe
Si
MeO
R R
77a or 77b
(i) s-BuLi
(ii) I2 or i-PrOB(pin)
B(pin) =
O
B
O
R R
Si
OMe
Y
Y
Si
MeO
R R
78a Y = I, R = hexyl
78b Y = I, R = Me
78c Y = B(pin), R = hexyl
Hex Hex
Si
F
Si
F
Hex Hex
77c
(i) Li^tBu2ZnTMP
(ii) I2
Hex Hex
Si
F
I
I
Si
F
Hex Hex
79
(i) n-BuLi
(i) i-PrOB(pin)
Li^tBu2ZnTMP = Li+
N
Zn
t-Bu
t-Bu
Hex Hex
Si
F
B(pin)
(pin)B
Si
F
Hex Hex
80

Scheme 8.39 Preparation of **77**–**80**.

8.3
σ-Conjugated Silole-Containing Polymers

8.3.1
Poly(1,1-silole)s

Poly(1,1-silole)s, SCPs catenated through the ring silicon atom, can be regarded as a new class of polysilanes [79]. Polysilanes normally possess wide bandgaps

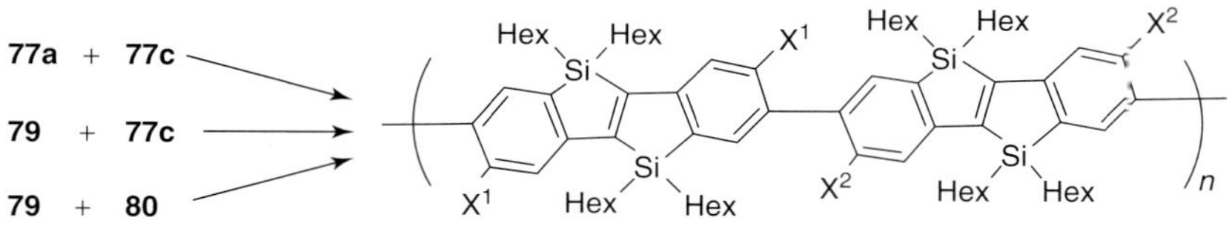

77a + **77c** → ; **79** + **77c** → ; **79** + **80** →

81a $X^1 = X^2 =$ OMe
81b $X^1 =$ OMe, $X^2 =$ F
81c $X^1 = X^2 =$ F

Scheme 8.40 Preparation of **81**.

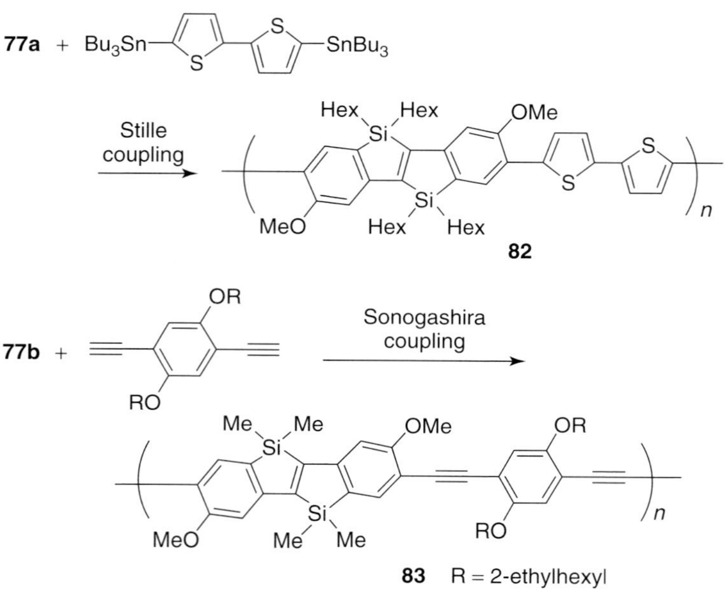

77a + Bu_3Sn–(bithiophene)–$SnBu_3$ → Stille coupling → **82**

77b + (diethynyl dialkoxybenzene) → Sonogashira coupling → **83** R = 2-ethylhexyl

Scheme 8.41 Preparation of **82** and **13**.

due to the σ-conjugation nature of the Si–Si main chain. Compared with the traditional polysilanes, poly(1,1-silole)s are expected to show narrowed bandgaps due to the extension of conjugation length through the conjugated silole ring. Poly(1,1-silole)s can be synthesized by Wurtz-type polycondensation and dehydrogenation polymerization.

The most straightforward route to poly(1,1-silole)s is the Wurtz-type polycondensation of 1,1-dichlorosilole with alkaline metals (Scheme 8.42). Generally, 1,1-dichlorosilole was allowed to react with 2 equiv. of an alkaline metal. Methanol or EtMgBr had been utilized to quench reactive terminal groups [80, 81].

As shown in Scheme 8.43, the reaction of diphenylacetylene with lithium readily gave 1,4-dilithio-1,2,3,4-tetraphenyl-1,3-butadiene **84**, an important intermediate to

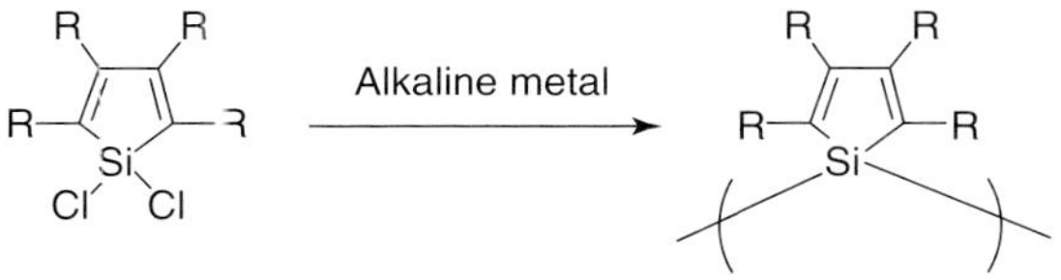

Scheme 8.42 Wurtz-type polycondensation.

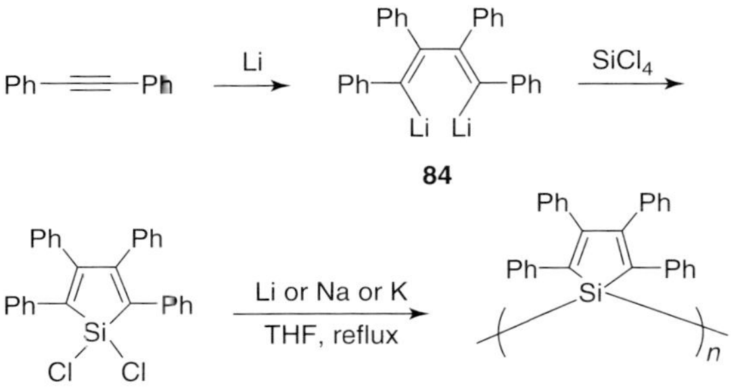

Scheme 8.43 Preparation of **84**–**86**.

prepare many 2,3,4,5-tetraphenylsiloles [13]. The 1,1-dichlorosilole **85** was formed via the cyclization reaction of **84** with silicon tetrachloride. In a report by West and coworkers, lithium, sodium, and potassium were utilized in the preparations of poly(1,1-silole) **86** [80]. The reactions afforded **86** in comparable isolated yields of ~33%. The molecular weights ($M_w \approx 5500$, $M_w/M_n \approx 1.1$) of **86** indicated that its DP was about 15. Polymer **86** showed λ_{max}(abs) at ~370 nm and λ_{max}(PL) at 520 nm in solution. The electroluminescent property of **86** was characterized. In a simple device configuration of ITO/**86**/Mg : Ag, the device emitted a green light (λ_{max}(EL) = 520 nm) with η_{EL} of 0.03% at a current density of 0.3 mA cm^{-2}.

Sohn *et al.* found that PL intensities of the toluene solution of **86** could be quenched by the additions of tiny amounts of TNT, 2,4,6-trinitrophenol (picric acid), 2,4-dinitrotoluene (DNT), and nitrobenzene, demonstrating that poly(1,1-silole)s are potential chemosensors for explosives [82]. TNT could also be detected utilizing the polymer film. In an air stream containing 4 ppb TNT, 8.2% decrease of PL intensity was found from the film. PL quenching was also detected when the film contacted a 50 ppb TNT–water solution.

1,1-Dichlorosilole can also be prepared via transformation of 1,1-diaminosilole [83]. As shown in Scheme 8.44, the 1,1-diaminosilole **87** was synthesized via intramolecular reductive cyclization of a bis(arylethynyl)silane in the presence of 4 amounts of LiNaph and the next trapping with dimethyl sulfate. The treatment of **87** with dry HCl gas at a low temperature readily afforded the 1,1-dichlorosilole. The polymerization was carried out in THF at −20 °C, giving poly(1,1-silole) **88** in a yield of 61% [81]. The M_w and M_w/M_n of **88** were 7200 and 1.14, respectively, giving DP of about 20. Polymer **88** displayed an absorption shoulder at ~320 nm and λ_{max}(PL) at 460 nm in a chloroform solution, all blueshifted relative to those of **86**. The Φ_{PL}

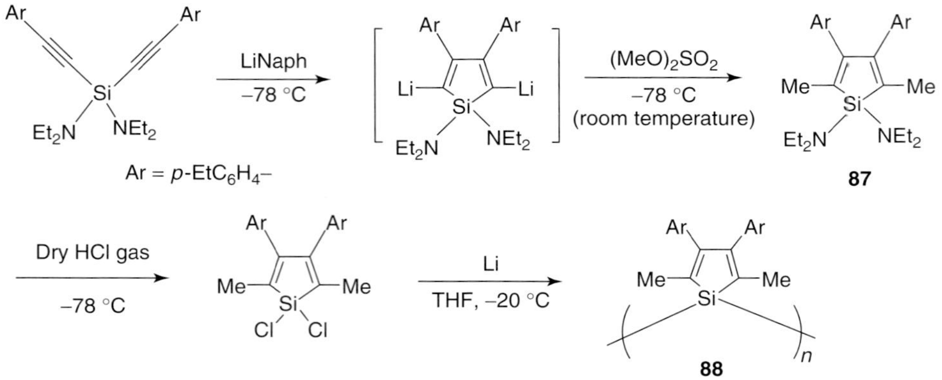

Scheme 8.44 Preparation of **87** and **88**.

of **88** in a solution is rather low (0.03%). This might be attributed to the phenyl rotational effect and the 2,5-dimethyl substitution effect. In the polymerization, a tiny amount (6%) of cyclic hexamer, an end-capped silole dendrimer, was also obtained.

In the report by Tamao and coworkers, an alternative route to **88** was also established (Scheme 8.45) [81]. The coupling reaction of the 1,1-dilithiosilole with chlorine-terminated tersilole in THF at $-78\,^{\circ}C$ gave **88** in 48% yield, although the molecular weight was relatively lower ($M_w \sim 3000$, $M_n \sim 2400$, DP ~ 8). Polymer **88** could be completely degraded in the presence of excess lithium [81].

Beside the preparation of poly(1,1-silole)s by Wurtz-type polycondensation, dehydrogenation polymerization is also an effective method. As shown in Scheme 8.46, Tanaka and coworkers utilized $Pd(cod)_2$ as the dehydrogenation catalyst, giving the poly(9,9-dibenzosilole) **89** in nearly 100% yield [84]. The product contained 40% toluene-insoluble fraction. Gel permeation chromatography (GPC) profiles of the toluene-soluble fraction displayed a bimodal distribution ($M_w = 8100$, $M_n = 3200$, 47%; $M_w = 3000$, $M_n = 2200$, 51%).

In a recent report by Toal *et al.*, dehydrogenation polymerization was also utilized to prepare 2,3,4,5-tetraphenyl-type poly(1,1-silole) **90** (Scheme 8.47), a polymer with comparable chemical structures to **86** [85]. Three catalysts, H_2PtCl_6, $Rh(PPh_3)_3Cl$,

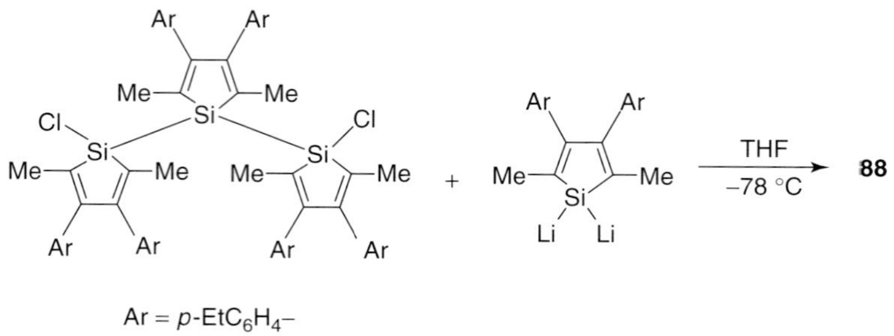

Scheme 8.45 An alternative route for the preparation of **88**.

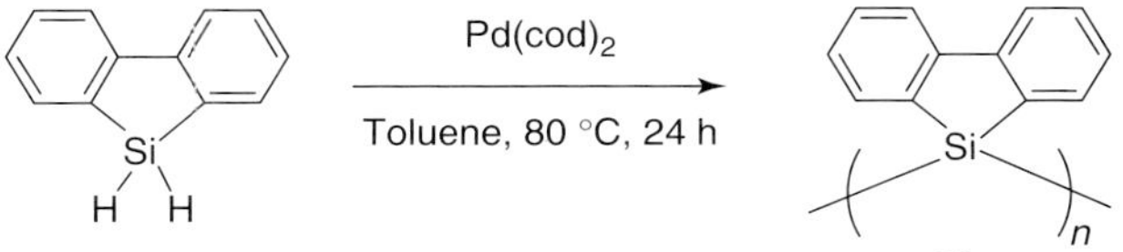

Pd(cod)$_2$

Toluene, 80 °C, 24 h

89

Scheme 8.46 Preparation of **89**.

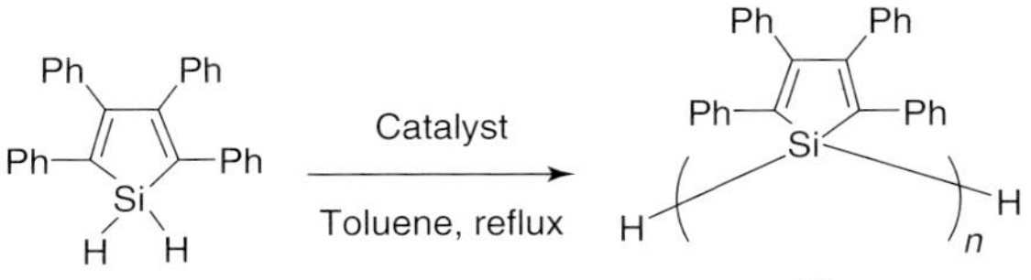

Catalyst

Toluene, reflux

90

Scheme 8.47 Preparation of **90**.

and Pd(PPh$_3$)$_4$, were introduced. Among the three catalysts, H_2PtCl_6 gave **90** in the highest yield of 80%. The molecular weights (M_w) of **90** were on an order of 10^3, showing trimodal distributions.

8.3.2 Copolymers of 1,1-Silole

Sanji *et al.* prepared a silole-incorporated polysilane **93** via ring-opening polymerization of tetraphenyl silole spirooctamethyl cyclopentasilane **92** (Scheme 8.48) [86]. The butyllithium-catalyzed ring-opening polymerization of **92** was carried out smoothly in THF at −40 °C, giving **93** in 41% yield. The M_n and M_w/M_n of **93** were 17 000 and 1.3, respectively, higher than aforementioned poly(1,1-silole)s. Polymer **93** displayed a λ_{max}(abs) at 320 nm and an absorption shoulder at 360 nm, assignable to the polysilane skeleton and the silole ring, respectively. The λ_{max}(PL)

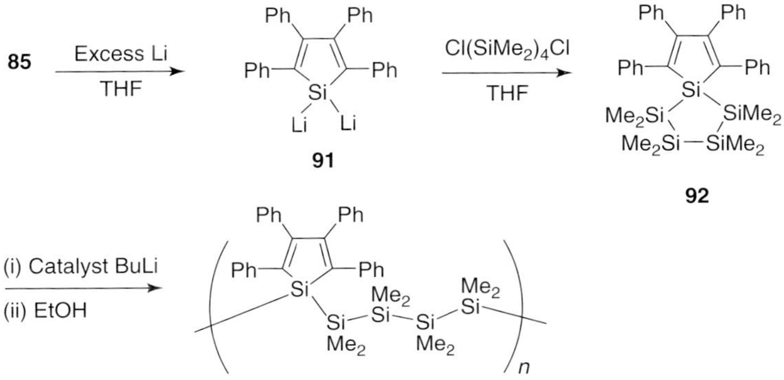

85 → Excess Li, THF → **91** → Cl(SiMe$_2$)$_4$Cl, THF → **92**

(i) Catalyst BuLi

(ii) EtOH

93

Scheme 8.48 Preparation of **91**–**93**.

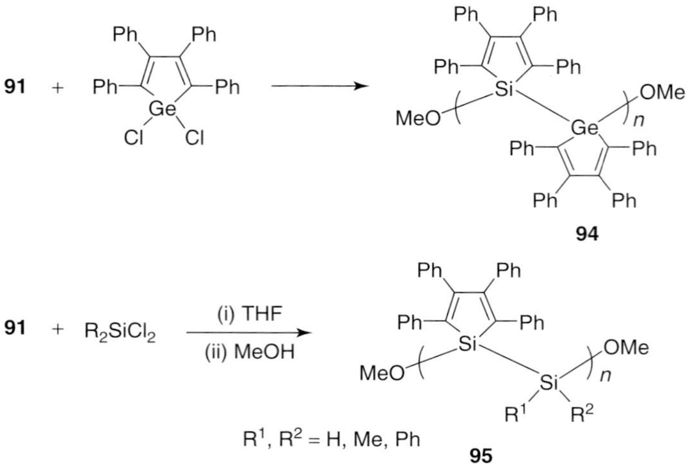

Scheme 8.49 Preparation of **94** and **95**.

of **93** was at 520 nm, coincidently identical to that of **86**. The Φ_{PL} of the copolymer was 4.1%. EL devices were fabricated but showed very low efficiency. The time of flight (TOF) hole mobility of **93** was ~10^{-4} cm^2 $(V\ s)^{-1}$ based on the dispersive TOF signal.

In a report by Sohn *et al.*, a 1-1-silole-1,1-germole alternating copolymer **94** was synthesized by Wutz-type polycondensation (Scheme 8.49) of 1,1-dilithiosilole **91** with 1,1-dichlorogermole [87]. The molecular weights of **94** (M_w = 5500, M_n = 5000) were comparable to those of the poly(1,1-silole) **86**. The reaction of **91** with dichlorosilanes afforded the silole–silane alternating copolymers **95**. The values of M_w of **95** ranged from 4400 to 5000. The $\pi-\pi^*$ transitions of **94** and **95** were comparable at ~367 nm, and the two polymers showed blue-greenish emissions (488–510 nm) in solutions. The chemosensor applications of the two polymers to a series of nitroaromatic explosives (TNT, picric acid, etc.) were explored, from which an electron-transfer mechanism was proposed.

8.4
Silole-Containing Polymers with $\sigma-\pi$-Mixed Conjugation

$\sigma-\pi$-Conjugated 2,5-diethynylsilole-organosilicon alternating copolymer **96** were synthesized by the Sonogashira coupling reactions of 2,5-dibromosilole and organosilicon diynes (Scheme 8.50) [88, 89]. The M_w of **96** ranged from 8000 to 21 000 with $M_w/M_n \leq 2.5$. The copolymers showed comparable λ_{max}(abs) between 408 and 415 nm. Copolymers **96b–e** showed decomposition temperatures over 400 °C but **96a** showed a lower decomposition temperature of 354 °C. The $FeCl_3$ vapor doped **96a–c** showed low conductivities (10^{-4}–10^{-3} S cm^{-1}). The Fe(0)-coordination to the silole ligand of **97** were investigated. When **96d** was heated at 150 °C in the presence of excess $Fe(CO)_5$ in benzene, only an insoluble

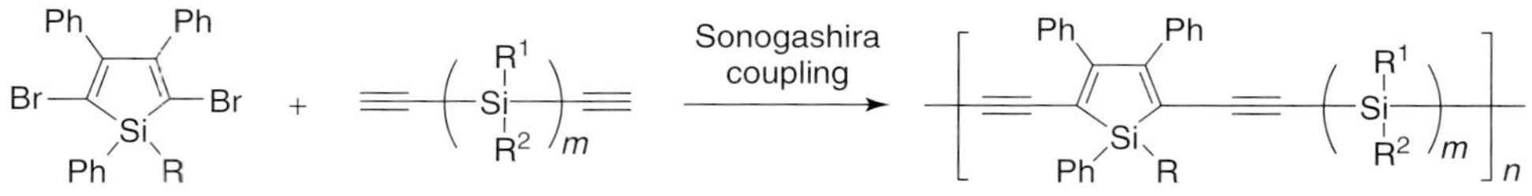

96a $m = 2$, R = Ph, R^1 = Ph, R^2 = Me
96b $m = 2$, R = Ph, R^1 = hex, R^2 = Me
96c $m = 1$, R = Ph, $R^1 = R^2$ = *n*-Bu
96d $m = 2$, R = Me, $R^1 = R^2$ = Et
96e $m = 1$, R = Me, $R^1 = R^2$ = *n*-Bu

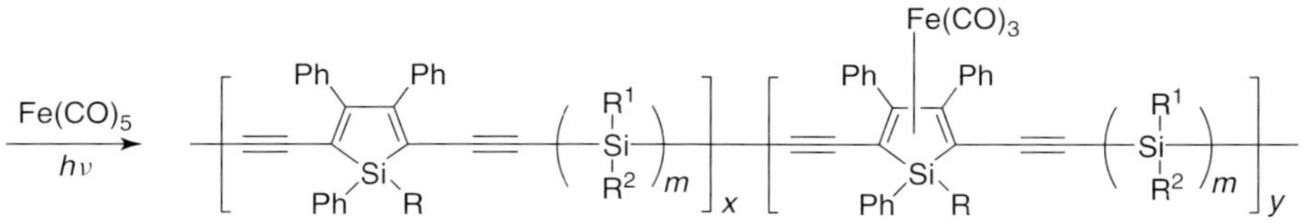

97d $m = 2$, R = Me, $R^1 = R^2$ = Et
97e $m = 1$, R = Me, $R^1 = R^2$ = *n*-Bu

Scheme 8.50 Preparation of **96–97**.

product could be obtained. Instead of heating, with photochemical irradiation for 2 hours by a high-pressure mercury lamp, benzene-soluble $Fe(CO)_3$-coordinated **97d** was afforded in a yield of 79%. The ^{1}H-NMR signals revealed that 90% silole rings on the main chain had been coordinated. Similarly, polymer **97e** was prepared but the polymer showed lower solubility to perform a successful NMR test. Elemental analysis indicated that the silole units were wholly coordinated by $Fe(CO)_3$. It should be noted that the coordination process decreased the molecular weights of the resulting products. Copolymers **97d** and **e** showed decomposition temperatures of 214 and 220 °C, respectively, much lower than those of their precursors **96d** and **e**; this can be attributed to the liberation of $Fe(CO)_3$ during the heating process [89]. UV absorption spectra of **97d** and **e** (λ_{max}(abs) ~405 nm) showed that the $Fe(CO)_3$-coordination slightly decreased conjugation lengths of the polymers. For polymers **96a–c**, however, no coordination could be realized, possibly because of steric hindrance from the 1,1-diphenyl groups on the silole units. The utilization of **96d** and **e** as a hole-transporting layer in organic light emitting diodes (OLED) were also investigated.

2,5-Diethynylsilole-based platinayne polymer **98** was prepared by Wong *et al.* (Scheme 8.51) [90]. The M_w of **98** reached 17 530 with M_w/M_n of 1.77. Polymer **98** showed a T_g of 228 °C and a decomposition temperature of 348 °C. The absorption spectrum of **98** showed a main peak at 370 nm with a tail extending to ~600 nm in solution, which gave a bandgap of 2.1 eV. The solution of **98** at room temperature emitted a silole-dominant light peaked at 537 nm and no triplet emission from the platinayne was found.

Toyoda *et al.* synthesized 3,4-diethynylsilole–organosilicon alternating copolymer **99** by anionic ring-opening polymerization of cyclic silole derivatives, in yields of 75–87% (Scheme 8.52) [91, 92]. The M_w of the copolymers ranged from 9400

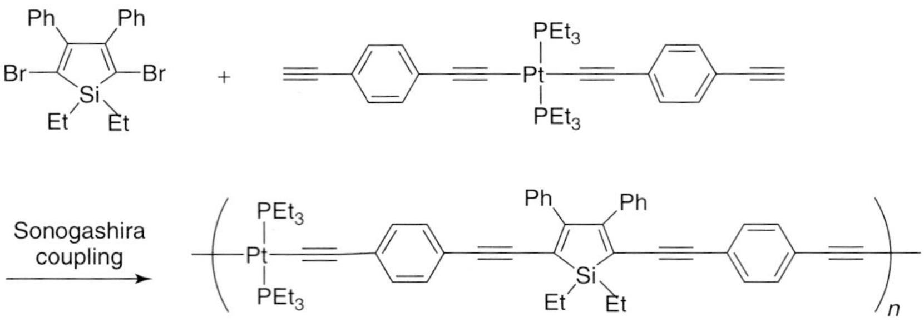

Scheme 8.51 Preparation of **98**.

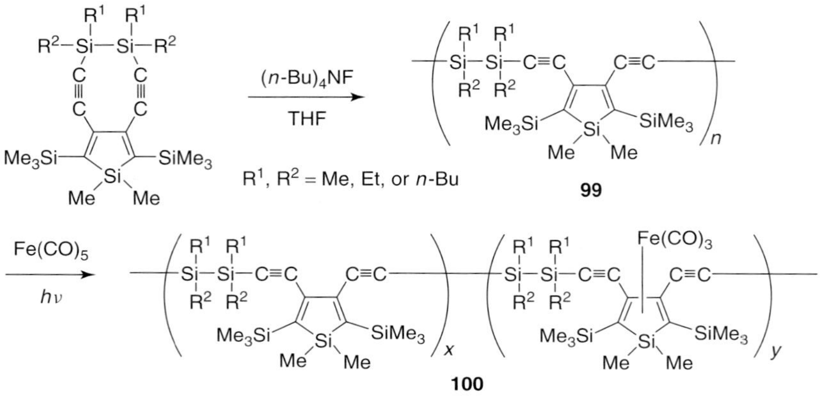

Scheme 8.52 Preparation of **99** and **100**.

to 47 900 with $M_w/M_n \leq 1.76$. Copolymers **99** showed decomposition temperatures of ~300 °C. The $Fe(CO)_3$-coordinated **100** were prepared by photochemical irradiations [92]. Similar to the cases of **97d** and **e**, the molecular weights and decomposition temperatures of copolymers **100** were all lower than those of the precursors **99**. The absorption spectra of **99** showed λ_{max}(abs) at 250 nm with edges at 380 nm in solution. The λ_{max}(abs) of **100** all blueshifted to 209 nm but the absorption edges redshifted to longer wavelengths of 420 nm. The $FeCl_3$ vapor doped **99** showed very low conductivities of 10^{-8} S cm^{-1} but $FeCl_3$ or iodine vapor doped **100** showed higher conductivities of 10^{-5} S cm^{-1}.

Corriu *et al.* reported the preparations of silole-incorporated polycarbosilane **102** via Sonogashira cross-coupling reactions of a 1,1-diethynylsilole **101** with a series of (hetero)aromatic dihalides in good yields (Scheme 8.53) [93]. The building blocks of the acromatic cycles, Z, included substituted benzenes, biphenyl, anthracene, fluorene, pyridine, bipyridine, thiophene, flourenone, and so on. The values of M_w of **102** varied largely from 1610 to 10 190, depending on the dihalides. Under UV irradiation, **102** (Z = 1,4-phenylene) could react with iron carbonyl complex $Fe(CO)_5$

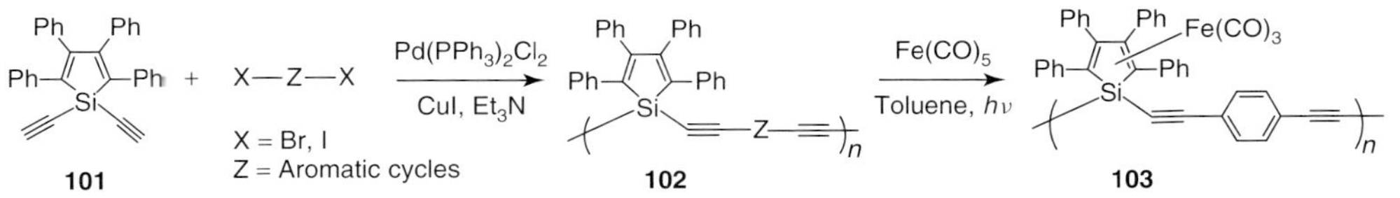

Scheme 8.53 Preparation of **102** and **103**.

under irradiation, giving $Fe(CO)_3$-coordinated **103** containing ~30% silole–iron carbonyl complexes.

Constructions of hyperbranched polysiloles **104** were carried out by polycyclotrimerizations of 1,1-diethynylsilole **101** in the presence or absence of 1-octyne, catalyzed by $TaCl_5$–Ph_4Sn (Scheme 8.54) [94]. The cyclotrimerization of three ethynyl groups readily formed phenyl connections between silole rings. Even without 1-octyne, **104** was still soluble owing to the large free volumes of hyperbranched polymers. The M_w of **104** ranged between 3530 and 5820, with $M_w/M_n \leq 1.7$. Polysiloles **104** showed decomposition temperatures over 343 °C. Under excitations, the solutions of the polysiloles **104** emitted faint blue-greenish light at ~500 nm, with comparable Φ_{PL} of 1%. It should be noted the Φ_{PL} was 10 times higher than that of a tetraphenylsilole ($\Phi_{PL} \approx 0.1\%$) in a solution, demonstrating that the structural rigidity effect of the polyarylene-like **104** can restrict the phenyl rotation-induced nonradiative decays of siloles. The observed cooling-enhanced

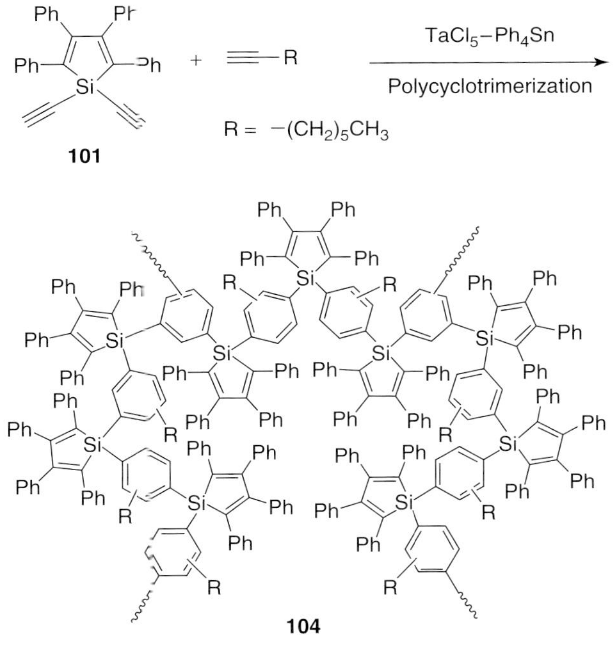

Scheme 8.54 Preparation of **104**.

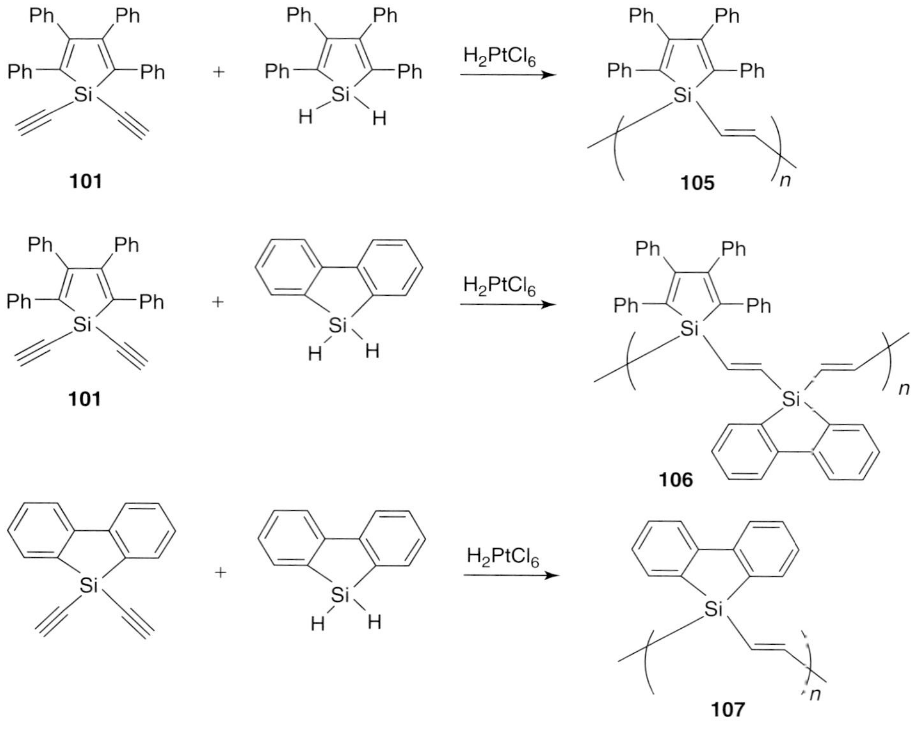

Scheme 8.55 Preparation of **105**–**107**.

emissions of the solutions of **104** did support the PL emission model for a substituted silole [13]. The polysiloles **104** were nonlinear optically active and strongly attenuated the optical power of intense laser pulses. The dichloromethane solution of **104** (0.7 mg ml^{-1}) started to limit the optical power at a low threshold of 185 mJ cm^{-2} and could suppress the optical signals to a great extent (81% power cutoff).

Recently, Sanchez *et al.* synthesized poly(1,1-silole-vinylene)s **105**–**107** via H_2PtCl_6-catalytic hydrosilylation reactions (Scheme 8.55) [95]. It was found that the polymers maintained regioregular *trans*-vinylene Si–C backbone. The M_w of **105**–**107** ranged between 4000 and 4500, with $M_w/M_n \leq 1.37$. The poly(tetraphenylsilole-vinylene) **105** emitted blue-green lights with λ_{max}(PL) at 493 nm in solution and 490 nm for solid film. The poly(dibenzosilole-vinylene) **107** emitted UV lights with λ_{max}(PL) at 362 nm in solution and 376 nm for solid film, which were comparable to that of poly(3,6-dibenzosilole) **55**. The light emission of **106** showed λ_{max}(PL) at 492 nm in solution and 486 nm for solid film, quite similar to the case of **105**. The Φ_{PL} values of the polymers in solution were between 1 and 4.3%. The fluorescent detections of the polymers to several nitro-based explosives were investigated.

As shown in Scheme 8.56, Tian and coworkers reported the synthetic route toward the spirodibenzosilole monomer **109** [96]. The lithiation of 2,2'-dibromobiphenyl

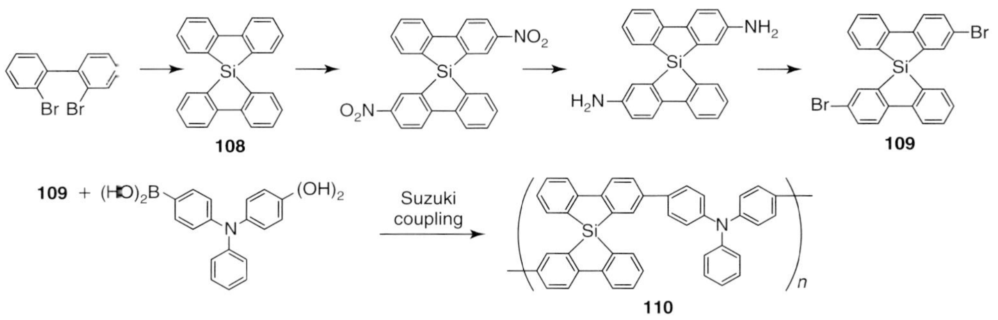

Scheme 8.56 Preparation of **108**–**110**.

and then cyclization with silicon tetrachloride afforded the spirodibenzosilole **108**. It should be noted attempts to prepare **109** by direct bromination of **108** with bromine were unsuccessful. Spirodibenzosilole–triphenylamine alternating copolymer **110** was synthesized by the Suzuki coupling reaction of **109** with triphenylamine diboronic acid. The M_w and M_n of the copolymer were 9577 and 7014, respectively. Copolymer **110** showed a T_g of 158.2 °C. The chloroform solution of **110** displayed λ_{max}(abs) at 342 nm and blue PL emission at 405 nm. But PL spectrum of the polymer in film became rather broad and weak, possibly owing to excimer formation. Owing to the contribution of the triphenylamine units, **110** possessed a high-lying HOMO of −5.47 eV. The EL property of the polymer was evaluated and an η_{ELmax} of 0.03% was obtained.

Ohshita *et al.* synthesized 2,6-dithienyldithienosilole–organosilicon alternating copolymer **111** by $NiCl_2$(dppe) catalyzed dehalogenative coupling reaction of **67** with diGrignard reagent (Scheme 8.57) [97, 98]. Copolymer **111** may be regarded as organosilicon-inserted **73b**, showing $\sigma-\pi$-conjugation on the main chain. The absorption spectra of **111** showed main peaks at ~440 nm in solution, with ~100-nm blueshift relative to that of **73b**, due to the obstruction of π-conjugation by the organosilicon units. The electron- and hole-transporting properties and electrical conductivity of **111** were also evaluated.

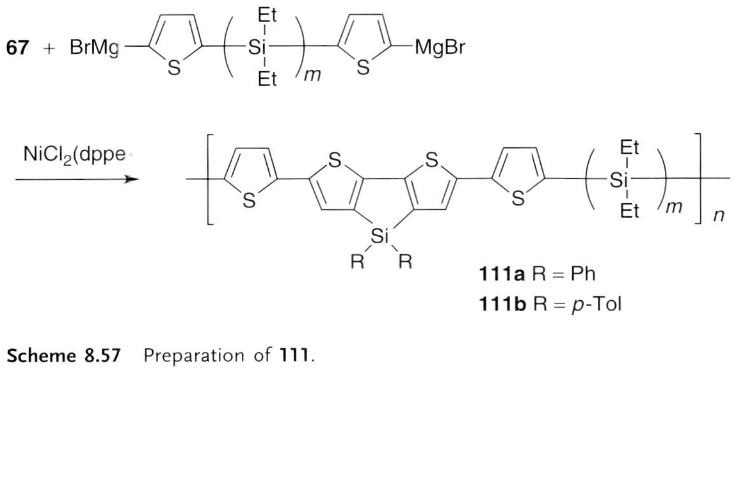

Scheme 8.57 Preparation of **111**.

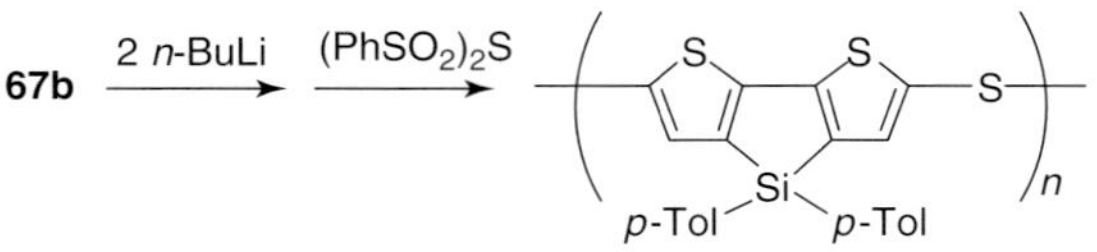

Scheme 8.58 Preparation of **112**.

As shown in Scheme 8.58, poly(2,6-dithienosilole sulfide) **112** was synthesized [99]. The solution and film of **112** showed comparable λ_{max}(abs) at ~385 nm and λ_{max}(PL) at ~488 nm. Compared with anodic oxidation-polymerized **71** (λ_{max}(abs) at ~487 nm and λ_{max}(PL) at ~656 nm), the sulfur isolation in **112** largely decreased the main-chain conjugation.

8.5 Summary

In this chapter, the versatility of synthetic chemistry and the optoelectronic properties of simple silole ring-, dibenzosilole-, dithienosilole-, and bis-silicon-bridged stilbene-containing conjugated polymers have been extensively described. The backbones of the SCPs possess three types of conjugations: π-, σ-, and $\sigma-\pi$-mixed conjugation. The diverse chemical structures of siloles and the widely tunable bandgaps from 4.0 to 1.45 eV of SCPs imply that the SCPs would exhibit many functionalities. With regard to fluorescence, widely variable fluorescent colors from UV light to RGB lights have been realized. Fluorescent chemosensing of TNT-type explosives, AIE, and attenuation of strong laser power reveal the attracting features of SCPs. Highly efficient EL including blue, green, red, and white lights, phosphorescent hosts with high triplet energy level, very efficient solar cells with PCE of over 5%, and stable FETs with high hole mobility in air, are the important applications of SCPs in optoelectronic film devices. These results indicate that SCPs are an important type of polymeric semiconductors and may play a significant role in organic electronics. The development of new SCP structures and new synthesis methods will continue to provide new and improved optoelectronic properties.

Acknowledgments

This work was partially supported by National Natural Science Foundation of China (#50773023, #50433030, and #U0634003) and Ministry of Science and Technology (#2009CB623600).

References

1. Burroughes, J.H., Bradley, D.D.C., Brown, A.R., Marks, R.N., Mackay, K., Friend, R.H., Burns, P.L., and Holmes, A.B. (1990) *Nature (London)*, **347**, 539.

2. Kraft, A., Grimsdale, A.C., and Holmes, A.B. (1998) *Angew. Chem., Int. Ed. Engl.*, **37**, 402
3. Coakley, K.M. and McGehee, M. (2004) *Chem. Mater.*, **16**, 4533.
4. Bao, Z., Dodabalapur, A., and Lovinger, A.J. (1996) *Appl. Phys. Lett.*, **69**, 4108.
5. Thomas, S.W.III, Joly, G.D., and Swager, T.M. (2007) *Chem. Rev.*, **107**, 1339.
6. Gustafsson, G., Cao, Y., Treacy, G.M., Klavetter, F., Colaneri, N., and Heeger, A.J. (1992) *Nature (London)*, **357**, 477.
7. Dubac, J., Laporterie, A., and Manuel, G. (1990) *Chem. Rev.*, **90**, 215.
8. Colomer, E., Corriu, R.J.P., and Lehureux, M. (1990) *Chem. Rev.*, **90**, 265.
9. Wrackmeyer, B. (1995) *Coord. Chem. Rev.*, **145**, 125.
10. Yamaguchi, S. and Tamao, K. (1998) *J. Chem. Soc., Dalton Trans.*, 3693.
11. Yamaguchi, S. and Tamao, K. (1996) *Bull. Chem. Soc. Jpn.*, **69**, 2327.
12. Yamaguchi, S., Endo, T., Uchida, M., Izumizawa, T., Furukawa, K., and Tamao, K. (2000) *Chem. Eur. J.*, **6**, 1683.
13. Chen, J., Law, C.C.W., Lam, J.W.Y., Dong, Y., Lo, S.M.F., Williams, I.D., Zhu, D., and Tang, B.Z. (2003) *Chem. Mater.*, **15**, 1535.
14. Chen, H. Chen, J., Qiu, C., Tang, B.Z., Wong, H., and Kwok, H.S. (2004) *IEEE J. Select. Topics Quantum Electron.*, **10**, 10.
15. Boydston, A.J., Yin, Y., and Pagenkopf, B.L. (2004) *J. Am. Chem. Soc.*, **126**, 3724.
16. Tracy, H.J., Mullin, J.L., Klooster, W.T., Martin, J.A., Haug, J., Wallace, S., Rudloe, I. and Watts, K. (2005) *Inorg. Chem.*, **44**, 2003.
17. Yu, G., Yin, S., Liu, Y., Chen, J., Xu, X., Sun, X., Ma, D., Zhan, X., Peng, Q., Shuai, Z., Tang, B., Zhu, D., Fang, W., and Luo, Y. (2005) *J. Am. Chem. Soc.*, **127**, 6335.
18. Mi, B., Dong, Y., Li, Z., Lam, J.W.Y., Haussler, M., Sung, H.H.Y., Kwok, H.S., Dong, Y., Williams, I.D., Liu, Y., Luo, Y., Shuai, Z., Zhu, D., and Tang, B.Z. (2005) *Chem. Commun.*, 3583.
19. Zhan, X., Risko, C., Amy, F., Chan, C., Zhao, W., Barlow, S., Kahn, A., Bredas, J.L., and Marder, S.R. (2005) *J. Am. Chem. Soc.*, **127**, 9021.
20. Lee, J., Yuan, Y.Y., Kang, Y., Jia, W.L., Liu, Z.H., and Wang, S. (2006) *Adv. Funct. Mater.*, **16**, 681.
21. Geramita, K., McBee, J., Shen, Y., Radu, N., and Tilley, T.D. (2006) *Chem. Mater.*, **18**, 3261.
22. Yin, S., Peng, Q., Shuai, Z., Fang, W., Wang, Y.H., and Luo, Y. (2006) *Phys. Rev. B*, **73**, 205409.
23. Mullin, J.L., Tracy, H.J., Ford, J.R., Keenan, S.R., and Fridman, F. (2007) *J. Inorg. Organomet. Polym. Mater.*, **17**, 201.
24. Luo, J., Xie, Z., Lam, J.W.Y., Cheng, L., Chen, H., Qiu, C., Kwok, H.S., Zhan, X., Liu, Y., Zhu, D., and Tang, B.Z. (2001) *Chem. Commun.*, 1740.
25. Chen, J., Xu, B., Yang, K., Cao, Y., Sung, H.H.Y., Williams, I.D., and Tang, B.Z. (2005) *J. Phys. Chem. B*, **109**, 17086.
26. Murata, H., Kafafi, Z.H., and Uchida, M. (2002) *Appl. Phys. Lett.*, **80**, 189.
27. Chen, H.Y., Lam, J.W.Y., Luo, J.D., Ho, Y.L., Tang, B.Z., Zhu, D.B., Wong, M., and Kwok, H.S. (2002) *Appl. Phys. Lett.*, **81**, 574.
28. Tamao, K., Uchida, M., Izumizawa, T., Furukawa, K., and Yamaguchi, S. (1996) *J. Am. Chem. Soc.*, **118**, 11974.
29. Toal, S.J., Jones, K.A., Magde, D., and Trogler, W.C. (2005) *J. Am. Chem. Soc.*, **127**, 11661.
30. Chen, J. and Cao, Y. (2006) *Sens. Actuators. B*, **114**, 65.
31. Gilman, H. and Gorsich, R.D. (1958) *J. Am. Chem. Soc.*, **80**, 1883.
32. Ohshita, J., Nodono, M., Kai, H., Watanabe, T., Kunai, A., Komaguchi, K., Shiotani, M., Adachi, A., Okita, K., Harima, Y., Yamashita, K., and Ishikawa, M. (1999) *Organometallics*, **18**, 1453.
33. Yamaguchi, S., Xu, C., and Tamao, K. (2003) *J. Am. Chem. Soc.*, **125**, 13662.
34. Chen, J. and Cao, Y. (2007) *Macromol. Rapid Commun.*, **28**, 1714.
35. Tamao, K., Yamaguchi, S., Shiozaki, M., Nakagawa, Y., and Ito, Y. (1992) *J. Am. Chem. Soc.*, **114**, 5867.

36. Shinar, J., Ijadi-Maghsoodi, S., Ni, Q.X., Pang, Y., and Barton, T.J. (1989) *Synth. Met.*, **28**, C593.
37. Barton, T.J., Ijadi-Maghsoodi, S., and Pang, Y. (1991) *Macromolecules*, **24**, 1257.
38. Tamao, K., Yamaguchi, S., and Shiro, M. (1994) *J. Am. Chem. Soc.*, **116**, 11715.
39. Yamaguchi, S., Jin, R.Z., Itami, Y., Goto, T., and Tamao, K. (1999) *J. Am. Chem. Soc.*, **121**, 10420.
40. Hong, S.Y. and Marynick, D.S. (1995) *Macromolecules*, **28**, 4991.
41. Tamao, K., Yamaguchi, S., Ito, Y., Matsuzaki, Y., Yamabe, T., Fukushima, M., and Mori, S. (1995) *Macromolecules*, **28**, 8668.
42. Yamaguchi, S., Goto, T., and Tamao, K. (2000) *Angew. Chem. Int. Ed.*, **39**, 1695.
43. Lee, Y., Sadki, S., Tsuie, B., and Reynolds, J.R. (2001) *Chem. Mater.*, **13**, 2234.
44. Wang, F., Luo, J., Yang, K., Chen, J., Huang, F., and Cao, Y. (2005) *Macromolecules*, **38**, 2253.
45. Wang, F., Luo, J., Chen, J., Huang, F., and Cao, Y. (2005) *Polymer*, **46**, 8422.
46. Liu, Z., Wang, L., Chen, J., Wang, F., Ouyang, X.Y., and Cao, Y. (2007) *J. Polym. Sci. Part A: Polym. Chem.*, **45**, 756.
47. Leclerc, M. (2001) *J. Polym. Sci. Part A: Polym. Chem.*, **39**, 2867.
48. Scherf, U. and List, E.J.W. (2002) *Adv. Mater.*, **14**, 477.
49. Wang, F., Wang, L., Chen, J., and Cao, Y. (2007) *Macromol. Rapid Commun.*, **28**, 2012.
50. Habrard, F., Ouisse, T., Stephan, O., Aubouy, L., Gerbier, Ph., Hirsch, L., Huby, N., and Van der Lee, A. (2006) *Synth. Met.*, **156**, 1262.
51. Liu, Y., Chen, Z., Chen, J., Wang, F., and Cao, Y. (2007) *Polym. Bull.*, **59**, 31.
52. Wang, Y., Hou, L., Yang, K., Chen, J., Wang, F., and Cao, Y. (2005) *Macromol. Chem. Phys.*, **206**, 2190.
53. Liu, Z., Zou, J., Chen, J., Huang, L., Peng, J., and Cao, J.Y. (2008) *Polymer*, **49**, 1604.
54. Thomas, K.R.J., Lin, J.T., Tao, Y.T., and Chuen, C.H. (2004) *Chem. Mater.*, **16**, 5437.
55. Chen, W., Ijadi-Maghsoodi, S., and Barton, T.J. (1997) *Polym. Prepr. Am. Chem. Soc., Div. Polym. Chem.*, **38**, 189.
56. Yamaguchi, S., Iimura, K., and Tamao, K. (1998) *Chem. Lett.*, 89.
57. Boydston, A.J., Yin, Y., and Pagenkopf, B.L. (2004) *J. Am. Chem. Soc.*, **126**, 10350.
58. Chen, J., Xie, Z., Lam, J.W.Y., Law, C.C.W., and Tang, B.Z. (2003) *Macromolecules*, **36**, 1108.
59. Zeng, Q., Li, Z., Dong, Y., Di, C.A., Qin, A., Hong, Y., Ji, L., Zhu, Z., Jim, C.K.W., Yu, G., Li, Q., Li, Z., Liu, Y., Qin, J., and Tang, B.Z. (2007) *Chem. Commun.*, 70.
60. Xie, Z., Yang, B., Xie, W., Liu, L., Shen, F., Wang, H., Yang, X., Wang, Z., Li, Y., Hanif, M., Yang, G., Ye, L., and Ma, Y. (2006) *J. Phys. Chem. B*, **110**, 20993.
61. Chen, J., Kwok, H.S., and Tang, B.Z. (2006) *J. Polym. Sci. Part A: Polym. Chem.*, **44**, 2487.
62. Chan, K.L., McKiernan, M.J., Towns, C.R., and Holmes, A.B. (2005) *J. Am. Chem. Soc.*, **127**, 7662.
63. Chen, R.F., Fan, Q.L., Liu, S.J., Zhua, R., Pu, K.Y., and Huang, W. (2006) *Synth. Met.*, **156**, 1161.
64. Mo, Y.Q., Tian, R.Y., Shi, W., and Cao, Y. (2005) *Chem. Commun.*, 4925.
65. Wang, E.G., Li, C., Mo, Y., Zhang, Y., Ma, G., Shi, W., Peng, J.B., Yang, W., and Cao, Y. (2006) *J. Mater. Chem.*, **16**, 4133.
66. Chan, K.L., Watkins, S.E., Mak, C.S.K., McKiernan, M.J., Towns, C.R., Pascu, S.I., and Holmes, A.B. (2005) *Chem. Commun.*, 5766.
67. Grice, A.W., Bradley, D.D.C., Bernius, M.T., Inbasekaran, M., Wu, W.W., and Woo, E.P. (1998) *Appl. Phys. Lett.*, **73**, 629.
68. Zhang, X., Jiang, C., Mo, Y., Xu, Y., Shi, H., and Cao, Y. (2006) *Appl. Phys. Lett.*, **88**, 629.
69. Wang, E., Li, C., Peng, J., and Cao, Y. (2007) *J. Polym. Sci. Part A: Polym. Chem.*, **45**, 4941.
70. Wang, E., Li, C., Zhuang, C.W., Peng, J., and Cao, Y. (2008) *J. Mater. Chem.*, **18**, 797.

71. Boudreault, P.L.T., Michaud, A., and Leclerc, M. (2007) *Macromol. Rapid Commun.*, **28**, 2176.
72. Wang, E., Wang, L., Luo, C., Zhuang, W., Peng, J., and Cao, Y. (2008) *Appl. Phys. Lett.*, **92**, 033307.
73. Usta, H., Lu, G., Facchetti, A., and Marks, T.J. (2006) *J. Am. Chem. Soc.*, **128**, 9034.
74. Ohshita, J., Hamamoto, D., Kimura, K., and Kunai, A. (2005) *J. Organomet. Chem.*, **690**, 3027.
75. Liao, L., Dai, L., Smith, A., Durstock, M., Lu, J., Ding, J., and Tao, Y. (2007) *Macromolecules*, **40**, 9406.
76. Hou, J., Chen, H.Y., Zhang, S., Li, G., and Yang, Y. (2008) *J. Am. Chem. Soc.*, **130**, 16144.
77. Liu, M.S., Luo, J., and Jen, A.K.Y. (2003) *Chem. Mater.*, **15**, 3496.
78. Xu, C., Yamada, H., Wakamiya, A., Yamaguchi, S., and Tamao, K. (2004) *Macromolecules*, **37**, 8978.
79. Tamao, K. and Yamaguchi, S. (2000) *J. Organomet. Chem.*, **611**, 5.
80. Sohn, H., Huddleston, R., Powell, D.R., West, R., Oka, K., and Yonghua, X. (1999) *J. Am. Chem. Soc.*, **121**, 2935.
81. Yamaguchi, S., Jin, R.Z., and Tamao, K. (1999) *J. Am. Chem. Soc.*, **121**, 2937.
82. Sohn, H., Calhoun, R.M., Sailor, M.J., and Trogler, W.C. (2001) *Angew. Chem. Int. Ed*, **40**, 2104.
83. Yamaguchi, S., Jin, R.Z., and Tamao, K. (1997) *Organometallics*, **16**, 2230.
84. Chauhan, B.P.S., Shimizu, T., and Tanaka, M. (1997) *Chem. Lett.*, 785.
85. Toal, S.J., Sohn, H., Zakarov, L.N., Kassel, W.S., Golen, J.A., Rheingold, A.L., and Trogler, W.C. (2005) *Organometallics*, **24**, 3081.
86. Sanji, T., Sakai, T., Kabuto, C., and Sakurai, H. (1998) *J. Am. Chem. Soc.*, **120**, 4552.
87. Sohn, H., Sailor, M.J., Magde, D., and Trogler, W.C. (2003) *J. Am. Chem. Soc.*, **125**, 3821.
88. Ohshita, J., Mimura, N., Arase, H., Nodono, M., Kunai, A., Komaguchi, K., and Shiotani, M. (1998) *Macromolecules*, **31**, 7985.
89. Ohshita, J., Arase, H., Sumida, T., Mimura, N., Yoshimoto, K., Tada, Y., Kunugi, Y., Harima, Y., and Kunai, A. (2005) *Inorg. Chim. Acta*, **358**, 4156.
90. Wong, W.Y., Wong, C.K., Poon, S.Y., Lee, A.W.M., Mo, T., and Wei, X. (2005) *Macromol. Rapid Commun.*, **26**, 376.
91. Ohshita, J., Hamaguchi, T., Toyoda, E., Kunai, A., Komaguchi, K., Shiotani, M., Ishikawa, M., and Naka, A. (1999) *Organometallics*, **18**, 1717.
92. Toyoda, E., Kunai, A., and Ishikawa, M. (1995) *Organometallics*, **14**, 1089.
93. Corriu, R.J.P., Douglas, W.E., and Yang, Z.X. (1993) *J. Organomet. Chem.*, **456**, 35.
94. Chen, J., Peng, H., Law, C.C.W., Dong, Y., Lam, J.W.Y., Williams, I.D., and Tang, B.Z. (2003) *Macromolecules*, **36**, 4319.
95. Sanchez, J.C., DiPasquale, A.G., Rheingold, A.L., and Trogler, W.C. (2007) *Chem. Mater.*, **19**, 6459.
96. Xiao, H., Leng, B., and Tian, H. (2005) *Polymer*, **46**, 5707.
97. Ohshita, J., Nodono, M., Watanabe, T., Ueno, Y., Kunai, A., Harima, Y., Yamashita, K., and Ishikawa, M. (1998) *J. Organomet. Chem.*, **553**, 487.
98. Ohshita, J., Nodono, M., Takata, A., Kai, H., Adachi, A., Sakamaki, K., Okita, K., and Kunai, A. (2000) *Macromol. Chem. Phys.*, **201**, 851.
99. Lee, K.H., Ohshita, J., and Kunai, A. (2004) *Organometallics*, **23**, 5481.

9 Polyfluorenes

Qiang Zhao, Shujuan Liu, and Wei Huang

9.1 Introduction

The rapid development of organic electronics has demonstrated that π-conjugated polymers are very promising materials for the fabrication of low-cost, yet high-quality novel electronic devices, such as polymer light-emitting devices (PLEDs) (ultrathin, flexible, full-color displays) [1–4], field-effect transistors (FETs) [5–7], photovoltaic cells (PCs) [8–12], and memory devices [13], owing to their easy processability and semiconducting properties. Current widely studied conjugated polymers include poly(phenylvinylene) (PPV), polythiophene (PT), poly(phenylene ethynylene)s (PPEs), poly(carbazole)s, poly(*para*-phenylene)s (PPPs), and polyfluorenes (PFs) [14–17]. So far, PFs have emerged as the most attractive blue-emitting materials because of their high efficiency, charge-transport properties, excellent solubility and film-forming ability, good chemical and thermal stability, and tunable properties through chemical modifications and copolymerization [18–28]. Therefore, they are widely used as active materials in optoelectronic devices, especially in light-emitting layers in PLEDs [29].

9.2 Chemical Structures of Polyfluorenes

The monomer unit of PFs consists of rigid planar biphenyl units, bridged by a carbon atom at carbon-9 position (Scheme 9.1), ensuring a high degree of conjugation. The fluorene repeat unit of PFs can be functionalized at the carbon-9 atom with different side groups, such as aryl or alkyl groups. The alkyl side groups can improve the solubility of PFs in common organic solvents, ensuring processability of high-quality thin films by spin casting and can modify the interchain interactions in films. The structures of side groups also have influence on the morphology of PFs. PFs with different morphology show distinct photophysical properties, which will be discussed in Section 9.4. Some functional groups, such as carbazole (Cz) and oxadiazole, also can be introduced into the side groups at the

Design and Synthesis of Conjugated Polymers. Edited by Mario Leclerc and Jean-François Morin

ISBN: 978-3-527-32474-3

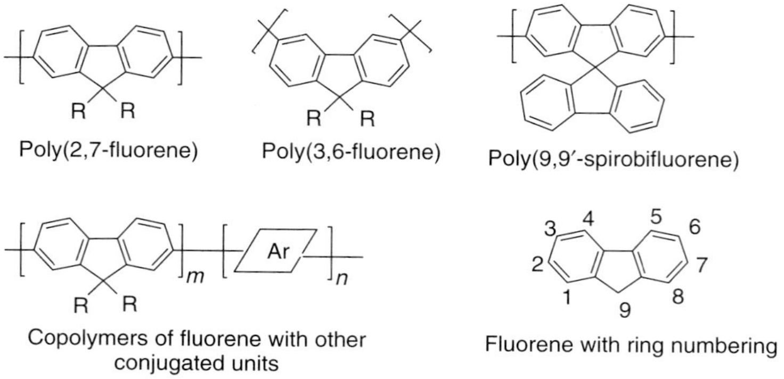

Poly(2,7-fluorene) Poly(3,6-fluorene) Poly(9,9′-spirobifluorene)

Copolymers of fluorene with other conjugated units

Fluorene with ring numbering

Scheme 9.1 Typical chemical structures of polyfluorenes and fluorene with ring numbering.

carbon-9 position of fluorene. PFs can also be functionalized by copolymerization with other conjugated units (Ar). The conjugated linkages are typically bonded at 2 and 7, or at 3 and 6 positions of fluorene units. Some typical chemical structures of PFs are shown in Scheme 9.1.

9.3 Synthesis of Polyfluorenes

9.3.1 Synthesis of Monomers

One major reason for the extensive investigation of PF derivatives is the ready accessibility of monomers. Dialkylfluorene (**2**) can be easily synthesized in high yield from fluorene (**1**) and alkyl bromide in the presence of base under inert atmosphere (Scheme 9.2). Then it can be converted to monomer 2,7-dibromo-9,9-dialkylfluorene (**3**) efficiently by bromination reaction. **3** can also be made in a high yield by direct alkylation of 2,7-dibromofluorene in the presence of base under inert atmosphere. Next, trimethyl borate (or triisopropyl borate) is added into the suspension of the lithium salt formed by reaction of **3** with *n*-butyllithium in dry THF to give **4**, which can be efficiently converted to monomer 9,9-dialkylfluorene-2,7-bis(trimethylene boronate) (**5**) by refluxing with 1,3-propanediol in toluene. The monomer **4** can also be obtained through the Grignard reagent approach. 2,7-Dibromo-9,9-dialkylfluorene first reacts with magnesium turnings in dry THF to form a Grignard reagent. Then the Grignard reagent solution is slowly dropped into a stirred solution of trimethyl borate (or triisopropyl borate) in dry THF at −78 °C and then slowly warmed to room temperature to obtain **4**.

The synthesis of 3,6-dibromo-9,9-dialkylfluorene (**10**) is not as straightforward as that of 2,7-dibromo-9,9-dialkylfluorene. The 3 and 6 positions in fluorene cannot be directly brominated by standard electrophilic aromatic substitution.

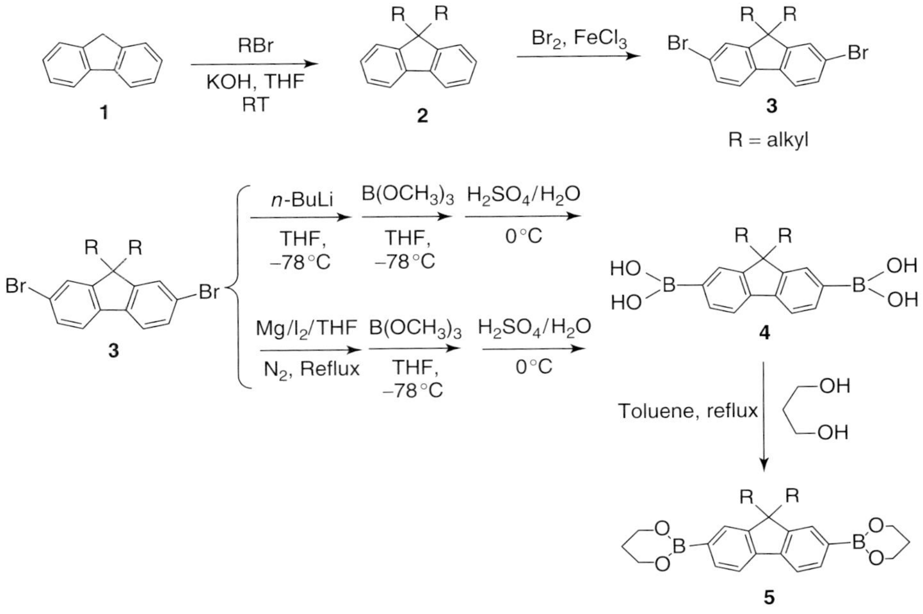

Scheme 9.2 Synthesis of monomer 2,7-dibromo-9,9-dialkylfluorene (**3**) and 9,9-dialkylfluorene-2,7-bis(trimethylene boronate) (**5**).

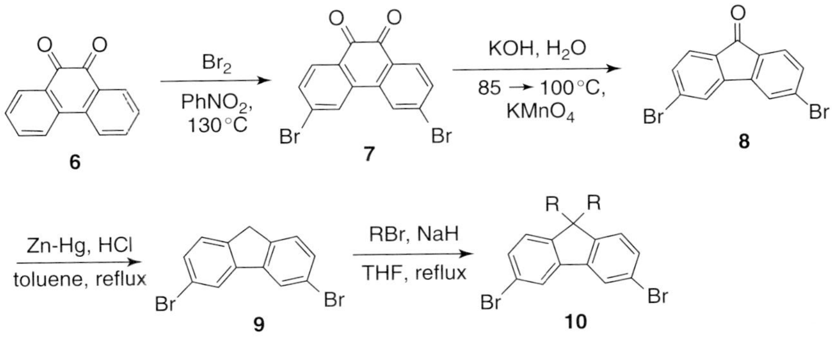

Scheme 9.3 Synthesis of monomer 3,6-dibromo-9,9′-dialkylfluorene.

The synthesis of **10** is shown in Scheme 9.3 [30]. First, the bromination of 9,10-phenanthrenequinone (**6**) affords 3,6-dibromophenanthrenequinone (**7**). Next, 3,6-dibromofluorenone (**8**) is obtained by exploiting benzilic acid rearrangement and decarboxylation. Then the reduction of **8** using Zn–Hg catalyst gives the important intermediate 3,6-dibromofluorene (**9**), which can be easily converted to monomer **10** by alkylation reaction.

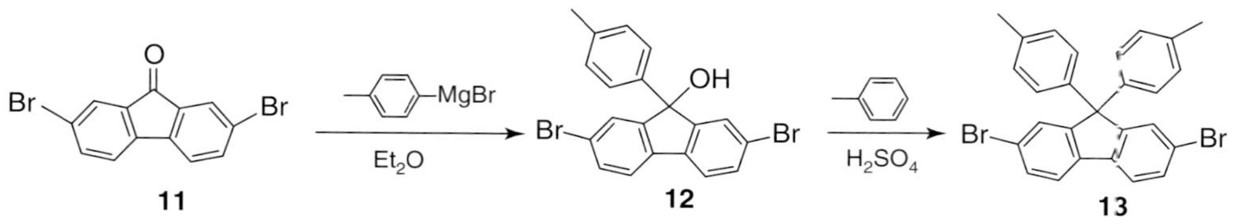

Scheme 9.4 Synthesis of fluorene monomer with aryl substituents at the C-9 position.

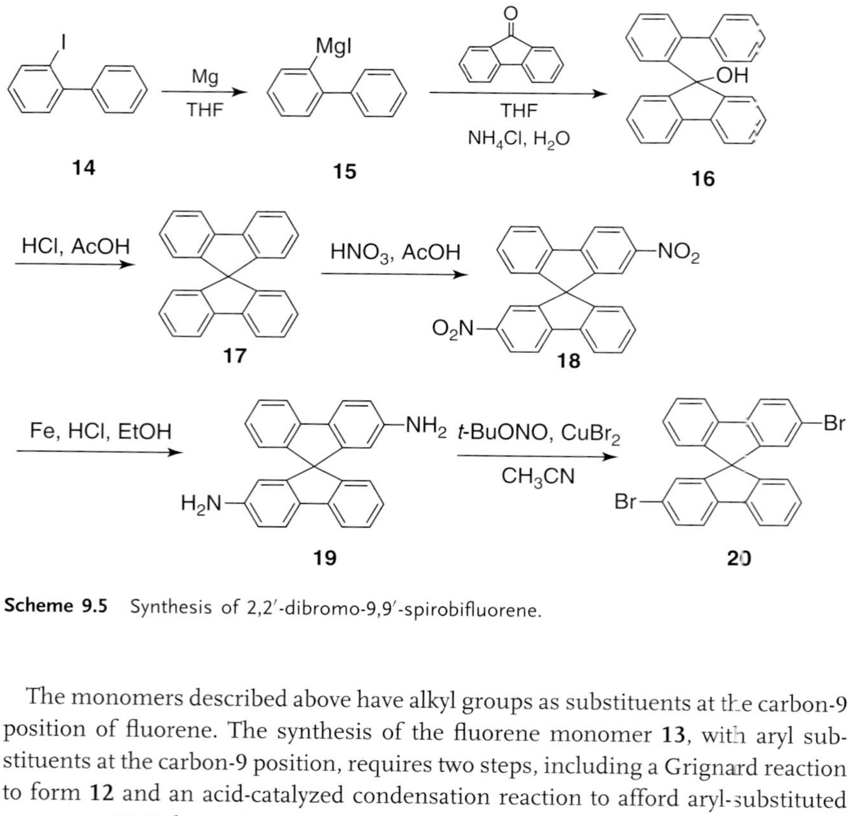

Scheme 9.5 Synthesis of 2,2′-dibromo-9,9′-spirobifluorene.

The monomers described above have alkyl groups as substituents at the carbon-9 position of fluorene. The synthesis of the fluorene monomer **13**, with aryl substituents at the carbon-9 position, requires two steps, including a Grignard reaction to form **12** and an acid-catalyzed condensation reaction to afford aryl-substituted monomer **13** (Scheme 9.4) [31].

9,9′-Spirobifluorene (**17**) can be prepared by the Clarkson and Gomberg method [32]. The addition of the Grignard reagent from 2-iodobiphenyl to 9-fluorenone affords the 9-(biphenyl-2-yl)-9-fluorenol (**16**) (Scheme 9.5). Next, a catalytic amount of hydrochloric acid is added to a boiling solution of the carbinol in acetic acid, leading to the formation of 9,9′-spirobifluorene by an instantaneous ring-closure reaction. For the synthesis of dibromo monomer of **20**, direct bromination methods lead to a mixture of mono-, di-, and polybromides, which are difficult to separate. Hence, another synthetic pathway is adopted as shown in Scheme 9.5.

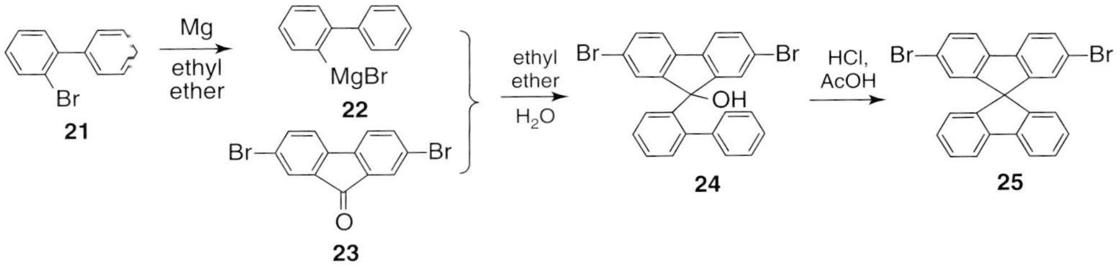

Scheme 9.6 Synthesis of 2,7-dibromo-9,9′-spirobifluorene.

First, the nitration of **17** with nitric acid in an acetic acid medium gives the dinitro derivative **18**, which subsequently, affords the spirodiamine (**19**) by reduction. Substitutive deamination of the diamine **19** with *tert*-butylnitrite and copper(II) bromide in acetonitrile produces dibromo monomer 2,2′-dibromo-9,9′-spirobifluorene (**20**) [33].

The monomer 2,7-dibromo-9,9′-spirobifluorene (**25**) is synthesized from 2,7-dibromo-9-fluorenone (**23**) by reacting with a Grignard reagent of 2-bromobiphenyl (**22**), followed by acid treatment, as shown in Scheme 9.6.

9.3.2 Synthesis of Polyfluorenes

Three kinds of carbon–carbon cross-coupling reactions are utilized to synthesize PFs from fluorene monomers, including oxidative coupling and transition-metal mediated polycondensation reactions (Scheme 9.7).

PFs can be synthesized by oxidative coupling of monomer **2** with iron(III) chloride as catalyst according to Scholl reaction [34]. Yoshino and coworkers first reported the synthesis of poly(9,9-dihexyl-2,7-fluorene) using this method [35, 36]. But this method will lead to high level of defects and low molecular weight with some degree of branching and nonconjugated linkages through positions other than 2 and 7. This method has been supplanted by transition-metal mediated polycondensation reactions.

Transition-metal mediated polycondensation reactions include nickel-catalyzed Yamamoto coupling and palladium(0)-catalyzed Suzuki coupling reactions (Scheme 9.7). Yamamoto coupling starts from dibromo-monomers and can be efficiently used for the synthesis of PFs. However, the scope of this method is restricted to the preparation of homopolymers and statistical copolymers. Suzuki cross-coupling route developed by a group at Dow can obtain high-molecular-weight PFs [37, 38]. The Suzuki coupling reaction starts from precursor materials bearing halides and boronic acids or boronic acid esters in the presence of palladium(0) catalyst, such as tetrakis(triphenylphosphine)palladium(0). This method can be utilized to synthesize homopolymers and copolymers. Compared with the Yamamoto coupling reaction, the Suzuki coupling reaction is more complex, but gives higher molecular weight.

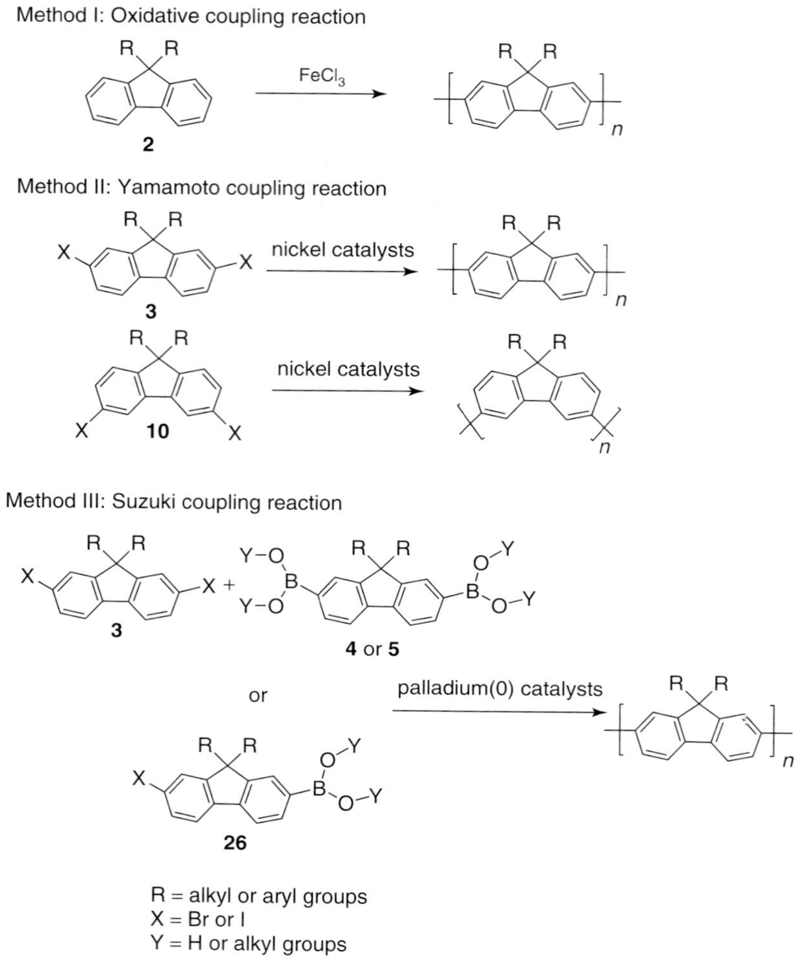

Method I: Oxidative coupling reaction

R R

$FeCl_3$

2

n

Method II: Yamamoto coupling reaction

X X

nickel catalysts

3

nickel catalysts

10

Method III: Suzuki coupling reaction

Y–O B O–Y

3

4 or 5

or

palladium(0) catalysts

26

R = alkyl or aryl groups
X = Br or I
Y = H or alkyl groups

Scheme 9.7 Three methods for the synthesis of polyfluorenes.

9.4 Basic Properties of Polyfluorenes

9.4.1 Phase Behavior

The phase behavior of PFs is crucial for their photophysical properties. There are three distinguishable classes of conformational isomers termed the *α-phase* (disordered and twisted structure, glassy phase), *β-phase* (a flatter conformation), and *γ-phase* (with an emission energy and a backbone conformation in between that of the α- and β-phases) for PFs [39]. A well-studied example is the β-phase, originally reported by Bradley *et al.* for poly(9,9-di-*n*-octylfluorene) (PF8) [40–42]. For this phase, the polymer chains approach a planar zigzag conformation resulting in an extended conjugation length [43–51], and exhibits the largest intrachain correlation and conjugation lengths of all the phases [42]. PFs might exhibit β-phase upon thermal treatment of polymer film. The β-phase might also be induced by spin coating from a poor solvent [41], or from a high-boiling-point solvent [52]. It might also be formed by drop casting followed by slow solvent evaporation [50], or by exposing a pristine film to solvent vapors or low-temperature thermal cycling [41]. These treatments cause local stresses in the material, which force some of the molecules to adopt a distinct planar, extended conformation, often described in terms of a 2_1 helical conformation with torsion angle of about 160° between neighboring monomers [39, 41].

The phase behaviors of PFs can be influenced strongly by the structure of the alkyl side chains at C-9 position of fluorene. The formation of β-phase can be suppressed for the branched alkyl chain-substituted PFs (such as 2-ethylhexyl, 3,7-dimethyloctyl, or 3,7,11-trimethyldodecyl side chains) [53]. Monkman *et al.* investigated the influence of *n*-alkyl-chain length of 6, 7, 8, 9, and 10 carbon atoms (PF6, PF7, PF8, PF9, and PF10) on β-phase formation in PFs. They found that no β-phase is formed in the PF6, and PF8 has an optimal side chain length for the formation of this phase, which are attributed to a balance between two factors. The first is the dimer/aggregate formation efficiency, which is poorer for longer (more disordered) alkyl-chain lengths, and the second is the van der Waals bond energy available to overcome the steric repulsion and planarize the conjugated backbone, which is insufficient in the PF6 with a shorter alkyl chain [54].

9.4.2 Basic Photophysical Properties

PFs display rich and attractive optical properties. PFs with conjugated linkages bonded at 2 and 7 positions (poly(2,7-fluorene) in Scheme 9.1) and those with conjugated linkages bonded at 3 and 6 positions (poly(3,6-fluorene) in Scheme 9.1) of fluorene units exhibit different optical properties. The absorption of poly(2,7-fluorene) consists of a strong featureless $\pi-\pi^*$ transition that peaks at about 3.2 eV (380 nm). The unstructured absorption is characteristic of many π-conjugated polymers and arises in part from the distribution of the conjugation lengths. It causes a distribution of energies that leads to a broadened absorption and obscures the vibronic structure of any particular segment [55]. In contrast, the emission of the poly(2,7-fluorene) in solution typically shows a vibronically well-resolved structure with peaks at about 420, 448, and 472 nm, assigned to the 0–0, 0–1, and 0–2 intrachain singlet transitions, with the 0–0 transition being the most intense. Poly(2,7-fluorene)s show intense sky-blue light under excitation by UV light with high photoluminescence (PL) quantum efficiency (Φ_f) of more than 50% [56, 57] both in solution and in the solid state [58]. However, poly(3,6-fluorene)s show different optical properties from their 2,7-counterparts discussed above. The absorption spectrum of poly(3,6-fluorene) in

THF solution features a sharp absorption peak at 259 nm with a vibronic shoulder at 321 nm. The optical bandgap (E_g) is estimated to be about 3.65 eV from the onset of the absorption peak (340 nm) in the film spectrum, which is evidently a hypsochromic shift compared to 380 nm ($E_g = 2.95$ eV) for poly(2,7-fluorene). In addition, the absorption spectrum of the poly(3,6-fluorene) thin film is almost identical to that in solution without any redshift, which is typically observed for planar-conjugated polymers. This seems to indicate that the 3,6-fluorene linkage might have a much less planar conformation than poly(2,7-fluorene)s due to the reduced interchain interaction. Poly(3,6-fluorene)s show an emission maximum of 344 nm with a vibronic shoulder at 358 nm in THF solution, which is also significantly a hypsochromic shift compared to poly(2,7-fluorene)s [30].

As mentioned above, the β-phase can be formed in linear alkyl chain-substituted PFs, such as for PF8. And this phase exhibits the largest intrachain correlation and conjugation lengths among all phases [42]. Compared with the glassy phase, the β-phase shows a distinct redshift of absorption and emission peaks with a remarkably well-resolved vibronic progression both in absorption and emission due to the chain geometry with an extended π-conjugation and increased electronic delocalization. In contrast, the glassy phase of PF only reveals a well-resolved vibronic progression in the emission spectrum. The addition of nonsolvent into the solution of PF8 can lead to the agglomeration of polymer chains accompanied by the occurrence of new, redshifted absorption peaks at about 2.85, 3.04, and 3.22 eV, which indicates the formation of the β-phase [21]. The relative intensity of the new β-phase-related bands increases with increasing nonsolvent content, and simultaneously the intensity of the broad unstructured PF absorption decreases. The PL spectrum of PF8 also significantly changes with increasing nonsolvent content (increasing β-phase content). New bands at lower energies (2.81, 2.65, and 2.49 eV) arise and the emission bands of "isolated" (molecular-dissolved) PF chains diminished at higher nonsolvent contents [21]. The increased electronic delocalization in the β-phase leads to the redshifted absorption and emission, together with picosecond Förster energy transfer from the less-ordered PF8 [43, 44, 59]. Thus only a small fraction of β-phase can completely dominate the emission characteristics as a result of ultrafast migration of excitons to the lower-energy β-phase segments [44, 46, 51, 52]. Therefore, the β-phase can act as a self-doped system where a fraction of chains with lower energy gap are dispersed in a host matrix of polymer chains with larger energy gap.

In addition to the redshift of absorption and emission peaks, the π-electron delocalization in the β-phase can improve photophysical stability in terms of both increased lifetime and reduced spectral meander [60]. However, the PL quantum yield would be reduced owing to the larger population of polaron and triplet excitons in the β-phase than in other phase morphologies [47, 52, 61, 62].

PFs with branched alkyl-chain substituents typically display a solid-state emission spectrum that is nearly identical to the spectrum in dilute solution, with only a very slight bathochromic shift of about 100 meV is observed. The absorption band in the solid state is slightly broadened as a result of small local variations of the π-overlap because of some conformational disorder [63].

9.5
Polyfluorene-Based Blue-Emitting Materials

PFs are a class of popular blue-emitting materials. The applications of PFs in light-emitting diodes (LEDs), however, has been hampered by the appearance of long-wavelength green emission, which leads to the problems of color purity (whitish or greenish-blue rather than pure blue) and color stability. The green emission has been attributed to either intermolecular interaction, which leads to the formation of aggregates and/or interchain excimers, or to the presence of emissive keto defect sites that arise as a result of thermo- or electrooxidative degradation of the PF backbone [64–66]. As a result, the expected blue emission from PFs becomes an undesired blue–green color in LED applications. Several approaches have been adopted to reduce the formation of aggregation or keto defect in PFs, including the introduction of bulky side chains, using cross-linked structures, improving the oxidative stability of pendant groups or chain ends [67], and limiting chain mobility by blending it with a high-glass-transition temperature (T_g) polymer [68].

9.5.1
Polyfluorenes Modified by Aryl Groups at C-9 Position

9.5.1.1 Spiro-Functionalized Polyfluorenes

Spiro-annulated molecules use the spiro-bridge to connect two conjugated moieties. The tetrahedral bonding atom at the center of a spiro-annulated molecule maintains a 90° angle between the connected conjugated moieties via a σ-bonded network [69]. This structural feature minimizes the close packing of spiro-annulated molecules in the solid state. And spiro-annulated fluorene derivatives exhibit high T_g, good thermal stability, and no long-wavelength emission [70].

Poly(9,9′-spirobifluorene)s show poor solubility in common organic solvents [71]. Flexible alkyl chains introduced into spirofluorene units serve to improve the solubility of poly(9,9′-spirobifluorene)s, such as polymer **P1** synthesized by the Suzuki coupling reaction through an AB-type monomer route (Scheme 9.8) [72]. This polymer exhibits good luminescent stability and no green emission was observed after the films were annealed at 200 °C for 3 hours in air.

Huang *et al.* synthesized soluble spiro-functionalized polymer **P2** by copolymerization of spirobifluorene monomer **25** and monomer **5** through the Suzuki coupling reaction, as shown in Scheme 9.8 [24]. The T_g of **P2** was 105 °C, which is much higher than that of PF8 (75 °C), indicating that the incorporation of spirobifluorene into the polymer chain serves to increase the T_g. More importantly, **P2** exhibits narrower and more stable PL spectrum compared with conventional PFs. The pure blue emission is attributed to the amorphous structure and weak interchain interaction of the polymer in the solid state as a result of the steric hindrance of the spiro structure. The higher T_g is favorable for improving spectral stability. Using this kind of polymer as light-emitting layer, pure blue electroluminescence (EL)

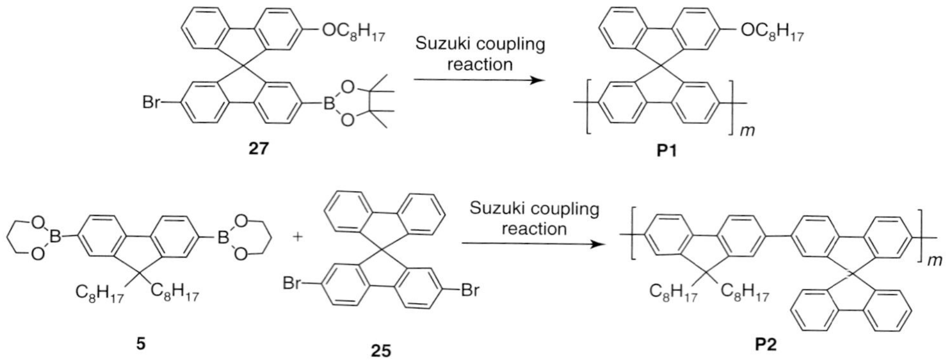

Scheme 9.8 Synthesis of polymer **P1** and **P2**.

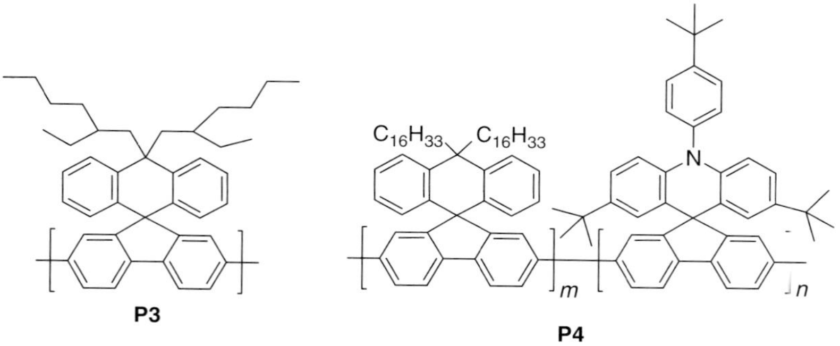

Scheme 9.9 Chemical structures of polymers containing the spiroanthracenefluorene units (**P3** and **P4**).

was realized. The results indicate that spiro-structure modification is a promising approach to solving the problems of color impurity and poor spectral stability of PFs as blue EL materials.

Kim *et al.* synthesized soluble homopolymer **P3** containing a spiro anthracene fluorene unit by Yamamoto coupling reactions (Scheme 9.9) [73]. The remote carbon-10 position of the spiroanthracenefluorene unit provides facile substitution of alkyl groups. No clear phase transition including T_g was observed before the decomposition. The polymer showed a deep-blue emission with a high spectral stability against heat treatment, UV irradiation, and high current passage due to the bulky spiroanthracenefluorene structure. Furthermore, they synthesized copolymer **P4** containing spiroanthracenefluorene unit and spiro-triphenylamine (TPA) structure [74]. This polymer showed thermal decomposition temperature (T_d) from 385 to 404 °C. Both higher T_d and T_g were observed for the polymer with high TPA contents.

Realizing a three-dimensionally spiro-annulated structure is also an effective way to obtain highly stable blue emitters. For **P5–P8** (Scheme 9.10), these rigid

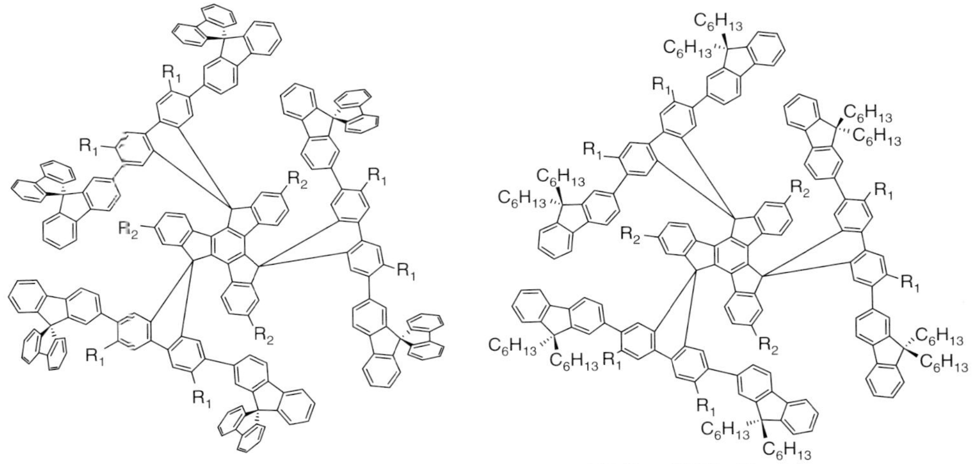

P5 $R_1 = OCH_3$, $R_2 = H$
P6 $R_1 = H$, $R_2 = OCH_3$

P7 $R_1 = OCH_3$, $R_2 = H$
P8 $R_1 = H$, $R_2 = OCH_3$

Scheme 9.10 Chemical structures of polymers with three-dimensionally spiro-annulated structure (**P5–P8**).

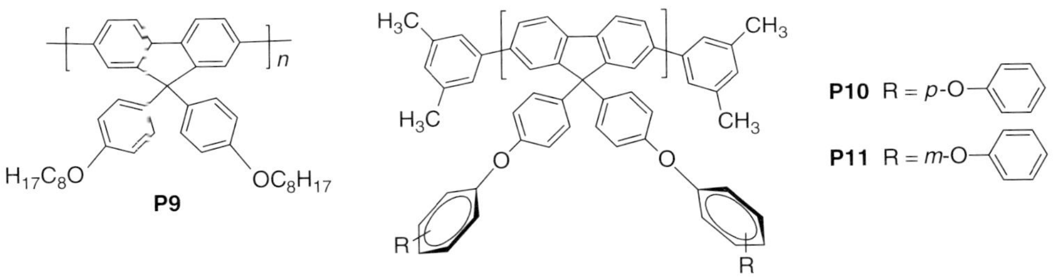

P9

P10 R = *p*-O–Ph

P11 R = *m*-O–Ph

Scheme 9.11 Chemical structures of polymers with alkoxyphenyl or aromatic ether groups at C-9 position (**P9–P11**).

three-dimensional spirotruxene-type materials can form excellent amorphous states to prevent the formation of the long-wavelength emission induced by excimers (aggregates) or ketone defects [75]. The PL spectra of these materials exhibit excellent thermal stability upon air annealing. OLEDs fabrication reveals that these materials exhibit both pure blue EL performance and good color integrity at different operating voltages.

9.5.1.2 Aryl Group-Modified Polyfluorenes at C-9 Position

Introducing aryl groups into C-9 position of fluorene will improve the emission color purity and stability. **P9** with alkoxyphenyl group side chains emits pure and stable blue light, suggesting that incorporation of the alkoxyphenyl group at the 9-position of fluorene is an effective way to suppress aggregation (Scheme 9.11) [76]. **P10** and

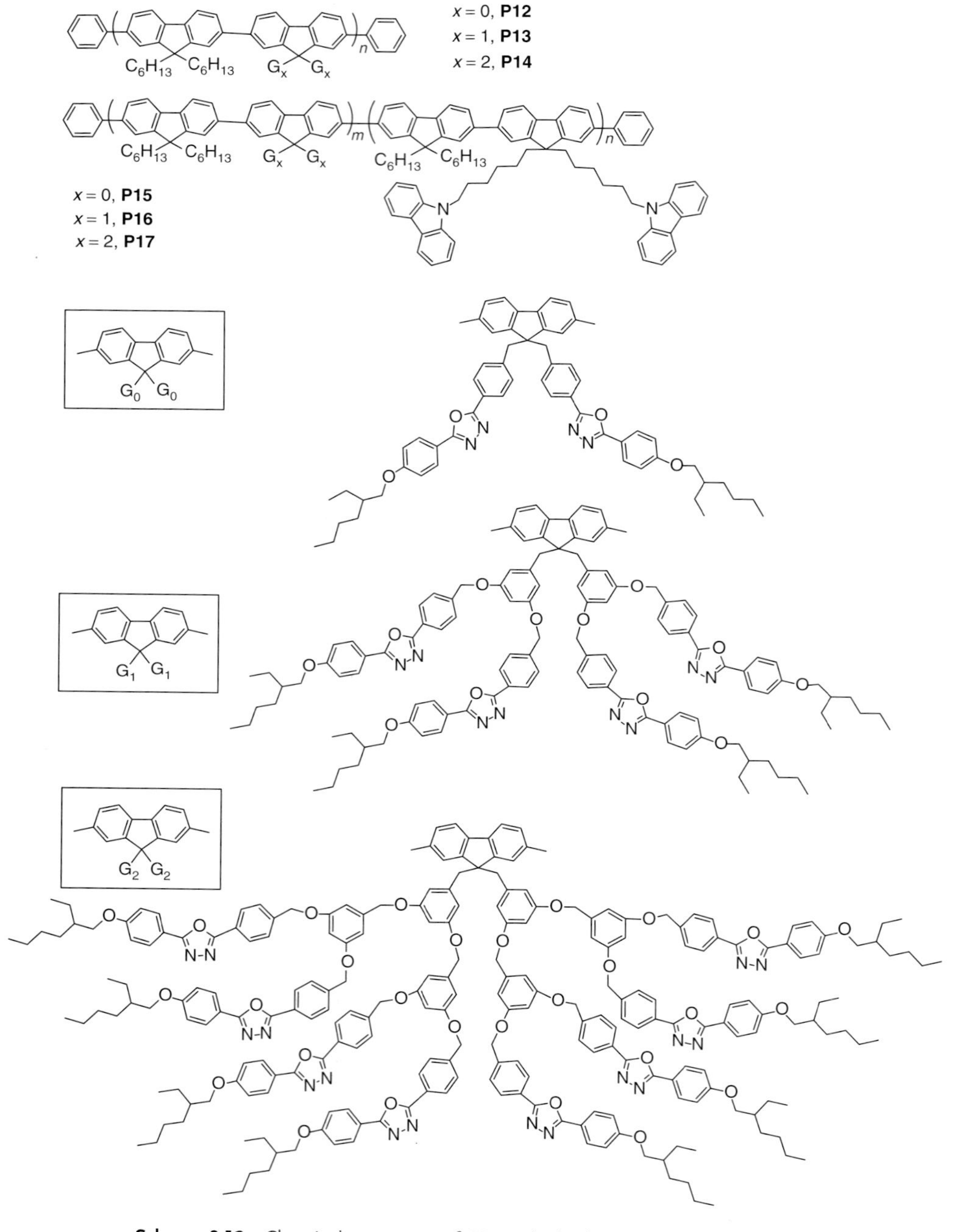

$x = 0$, **P12**
$x = 1$, **P13**
$x = 2$, **P14**

$x = 0$, **P15**
$x = 1$, **P16**
$x = 2$, **P17**

Scheme 9.12 Chemical structures of PFs with dendritic side chains at C-9 position.

P11 containing covalently linked aromatic ether (AE) moieties were synthesized via microwave (MW) Yamamoto coupling reactions [77]. Thermogravimetric analysis (TGA) and thermal oxidative degradation studies confirm that the incorporation of AE units at the C-9 position substantially improves the thermal, oxidative, and color stability of these new materials. The two polymers were annealed in N_2 atmosphere at 200 °C for 72 hours and at 150 °C in ambient atmosphere for 1 hour showing no evidence of green emission.

Lin *et al.* synthesized a series of polymers (**P12–P17** in Scheme 9.12) containing different generations of poly(benzyl ether) dendritic wedges with oxadiazole peripheral functional groups, including copolymers bearing Cz pendant groups [78]. The dendritic side chain polymers possess excellent solubility in common organic solvents and good thermal stability with T_d at more than 370 °C. The G1- and G2-substituted polymers may narrow down the emission spectra by reducing the emission spectral tail (extending to longer wavelength regions) to induce pure blue PL emission. The improvement in the quality of the emission spectra is attributed to less molecular packing caused by the steric hindrance of bulky dendritic side chains.

Introducing polyhedral oligomeric silsesquinoxane (POSS) into PFs can significantly reduce the formation of aggregation or keto defect in the PFs [79, 80]. As in the POSS-tethered PF **P18** (Scheme 9.13), they have well-defined architectures [81]. Both T_d and T_g increase as the amount of POSS is increased, presumably because the tethered POSS enhances the thermal stability and retards the polymer chain mobility. This particular molecular architecture increases the quantum yield of PFs significantly by reducing the degree of interchain aggregation and results in a purer and stronger blue light by preventing the formation of keto defects.

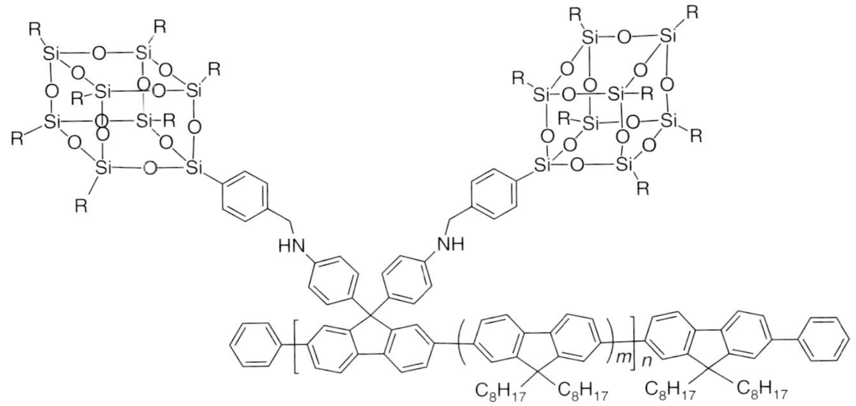

P18

Scheme 9.13 Chemical structures of PFs with POSS at side chains.

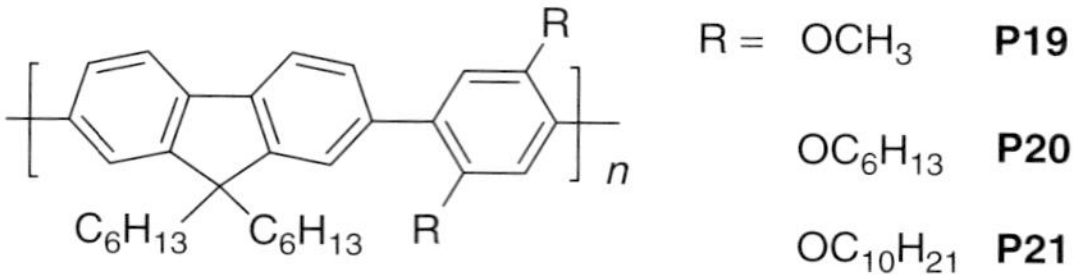

R = OCH_3 **P19**

OC_6H_{13} **P20**

$OC_{10}H_{21}$ **P21**

Scheme 9.14 Chemical structures of **P19–P21**.

9.5.2
Fluorene-Based Copolymers with Other Conjugated Units

Backbone structural modification for PFs by alternatively inserting substituted aryl units provides another opportunity to improve the quality of the emission spectra of fluorene-based blue-light-emitting polymers and to suppress the excimer formation in the polymers.

Polymer **P19–P21** (Scheme 9.14) with alkoxy chains as substituents show much better thermally spectral stability than poly(9,9-dihexyl-fluorene-2,7-diyl) and poly[(9,9-dihexyl-fluorene-2,7-diyl)-*alt-co*-(9,9′-spirobifluorene-2,7-diyl)] and longer side chains on the phenylene rings can enhance the spectral stability [26]. Interestingly, **P20** and **P21** easily form thermotropic liquid crystalline states. In a study of a series of poly(9,9-dialkylfluorenes) with different length of the linear alkyl chains, it is reported that the polymer with the greatest liquid crystalline order produces the greatest excited-state interchain communication or excimer emission [82]. However, the same tendency was not observed for **P19–P21**. On the contrary, **P19**, a polymer without crystallization, shows the strongest tendency in forming excimer, while **P20** and **P21**, which easily form crystalline states, exhibit better thermal spectral stability than **P19**. This might be due to the poor planar configuration in these polymers and the enhancement of longer side chains for separating backbones, even at the cost of forming crystals. **P21** emits deep-blue light without longer wavelength components in the emission spectrum. The absolute PL efficiency of polymer in neat films was found to be about 40%. The maximum external quantum efficiency of 0.60% was obtained for the EL devices based on this polymer [83, 84].

Huang and Lai *et al.* synthesized a series of copolymers **P20** and **P22–P29** (Scheme 9.15) composed of alternating 9,9-dihexylfluorene with different aryl comonomers and investigated the tuning of electronic properties [25]. The number average molecular weights (M_n) of these polymers range from ~7300 to ~60 000 with the polydispersity index (PDI) of 1.4–2.0. Efficient blue-light emission, good solubility, and thermal stability and relatively high T_g were observed for all polymers. By varying the aryl comonomers, the bandgaps of the polymers can be tuned from 2.81 to 3.35 eV, which provide an effective synthetic methodology to adjust the highest occupied molecular orbital (HOMO) and lowest unoccupied molecular orbital (LUMO) energy levels of the conjugated polymers. For the polymers with the same main-chain structure of polyfluorene-*co-alt*-phenylene, attachment of electron-donating alkoxy groups on phenylene ring (**P20**) gives rise to a spectral redshift, corresponding to slightly decreased HOMO and increased LUMO energy

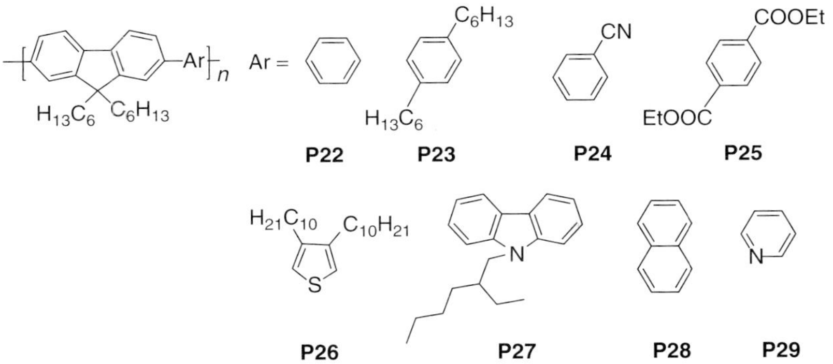

P22 P23 P24 P25

P26 P27 P28 P29

Scheme 9.15 Chemical structures of **P22–P29**.

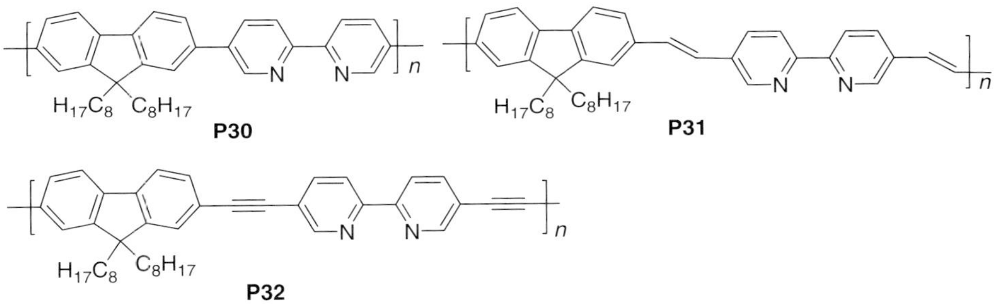

P30 P31 P32

Scheme 9.16 Chemical structures of polymers **P30–P32**.

levels, while attachment of electron-withdrawing ester groups (**P25**) leads to an obvious blueshift in the absorption spectrum with a decrement in both the HOMO and LUMO energy levels as compared to that of the unsubstituted polymer **P22**. As for **P27**, in comparison with the homopolymer, the introduction of Cz comonomer has caused an obvious spectral blueshift. A decrement in both the HOMO and LUMO energy levels has been observed for **P28** in which naphthalene was chosen as the comonomer. For **P29**, however, although there is no obvious difference in its absorption and emission spectra as compared with those of the homopolymer, both the HOMO and LUMO energy levels were reduced greatly.

Polymer **P30–32** (Scheme 9.16) composed of 9,9-dioctylfluorene and 2,2′-bipyridine (bpy) units, which are alternatively linked by the C–C single bond (**P30**), vinylene bond (**P31**), or ethynylene bond (**P32**), have been synthesized via the Suzuki reaction, the Wittig–Horner reaction, and the Heck reaction, respectively [85]. **P30** exhibits a much larger Stokes shift compared with other two polymers, indicative of higher extended and rigid backbone conformations in the polymers linked by the vinylene and ethynylene bonds. Owing to the presence of bpy units, all three polymers are responsive to a wide variety of transition-metal ions by an absorption spectral redshift and fluorescence quenching. **P30** gives the highest sensitivity due to the highest backbone flexibility among three polymers.

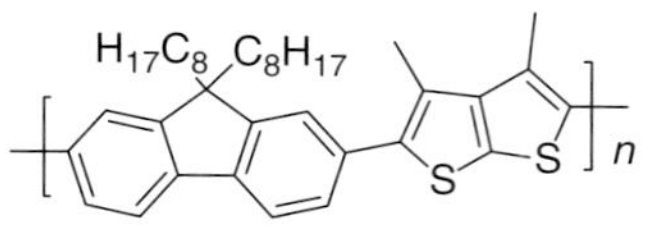

P33

Scheme 9.17 Chemical structures of fluorene thienothiophene copolymer **P33**.

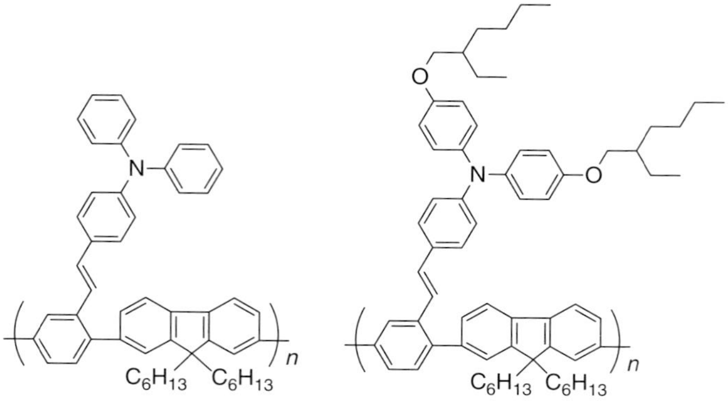

P34 P35

Scheme 9.18 Chemical structures of fluorene-based copolymers containing triphenylamine in the side chains.

Campbel *et al.* synthesized a fluorene thienothiophene copolymer **P33** (Scheme 9.17) [86]. The *syn* arrangement of the thienothiophene units disrupts conjugation, leading to deep-blue PL and EL emission. The polymer also has liquid-crystal properties and the fluorene thienothiophene backbone leads to a high mobility. Hence, this polymer should be suitable for use in LEDs, light-emitting transistors, and photonic structures that require deep-blue polarized emission and anisotropic charge transport.

P34 and **P35** (Scheme 9.18) bear TPA moiety through a vinylene bridge [87]. The copolymers possess good thermal stability with T_d of 415–448 °C and relatively high T_g of about 91–159 °C. The absorption and PL spectra of the polymers in the solid state are similar to those in dilute chloroform solution. And the film PL of both polymers shows no longer wavelength emission, showing that the two-dimensional conjugated system can effectively suppress π-stacking/aggregation.

Poly[2,7-(9,9-dioctylfluorene)-alt-1,3-(5-carbazolphenylene)] (PFCz) emitted a UV blue light with an emission peak at about 396 nm due to a metaconjugated structure. It was selected as the host with a small amount (1%) of blue-light-emitting 4-N,N-diphenylaminostilbene (DPS) units incorporated as dopant (**P36–P41** in Scheme 9.19) [88]. The dopant is covalently connected to the host to realize molecular dispersion of the dopant and avoid phase separation. A small amount of electron-transporting oxadiazole units were also incorporated into the polymer main chain to improve the electron-transporting ability. The polymers show a pure

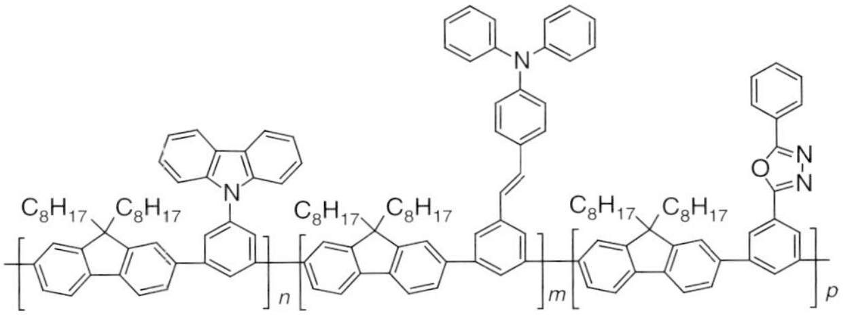

P36 $n:m:p = 100:0:0$
P37 $n:m:p = 98:2:0$
P38 $n:m:p = 94:2:4$
P39 $n:m:p = 88:2:10$
P40 $n:m:p = 78:2:20$
P41 $n:m:p = 0:100:0$

Scheme 9.19 Chemical structures of **P36–P41**.

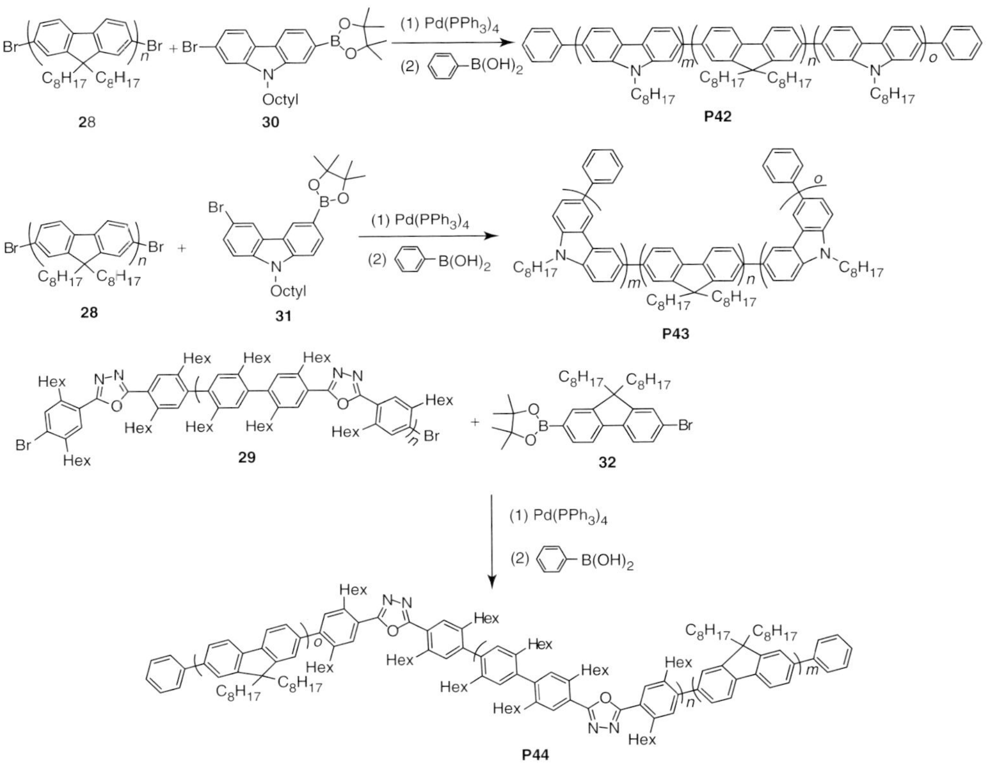

Scheme 9.20 Synthesis of triblock copolymer **P42–P44**.

blue-light emission even with only 1% of DPS units because of efficient energy transfer from the UV-blue-light emitting PFCz segments to the blue-light-emitting DPS units. PLEDs based on these copolymers show very stable pure blue-light EL emission with excellent device performance.

Bo *et al.* synthesized a series of conjugated triblock copolymer **P42–P44** containing hole-transporting polycarbazole segments, electron-transporting polyoxadiazole segments, and blue-light-emitting PF segments through a two-step Suzuki coupling reaction (Scheme 9.20) [89]. First dibromo-terminated polymer precursors

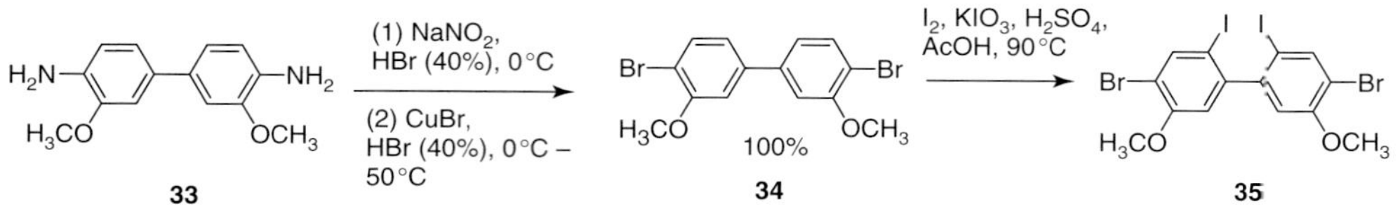

Scheme 9.21 Synthesis of 6,6′-diiodo-4,4′-dibromo-3,3′-dimethoxylbiphenyl (**35**).

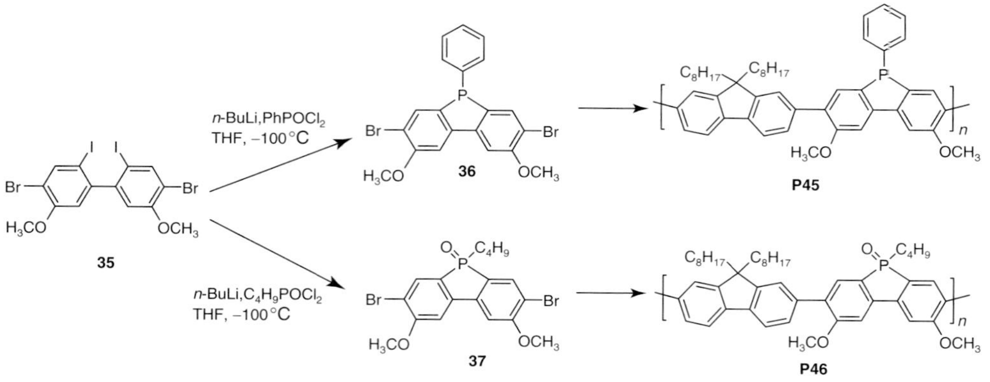

Scheme 9.22 Synthesis of copolymers of phosphafluorene and fluorene.

(polyfluorenes **28** and polyoxadiazole **29**) were synthesized as the central building blocks. Then, these precursors were further polymerized with AB-type monomers **30–32** respectively to achieve the target triblock copolymers. All the polymers exhibit good thermal stability and efficient through-bond energy transfer.

Polyheterofluorenes, in which the sp^3 hybridized carbons at C-9 position of fluorene in PFs are substituted by other atoms (such as phosphorus, silicon, sulfur, and boron), are receiving increasing attention because of their unique electronic structures that result from the particular interactions between the heteroatom and the π-conjugated polyphenylene-like framework. The synthesis of dibromo (or diiodo) substituted heterofluorene monomers is crucial for the preparation of polyheterofluorenes.

Recently, Huang *et al.* found a facile, highly efficient, and economical route to prepare the important intermediate 6,6′-diiodo-4,4′-dibromo-3,3′-dimethoxylbiphenyl (**35**) in two steps with very high yield from the cheap and readily available *o*-dianisidine (Scheme 9.21) [90]. A general synthetic strategy for the preparation of 2,7-dibrommo-9-heterofluorenes, such as phosphafluorene (**36** and **37** in Scheme 9.22) and silafluorene (**38** in Scheme 9.23), was developed successfully from this important intermediate. And the copolymers of fluorene with phosphafluorene or silafluorene were obtained.

The incorporation of phosphorus moieties into conjugated molecules offers considerable promise for the development of new functional materials with novel

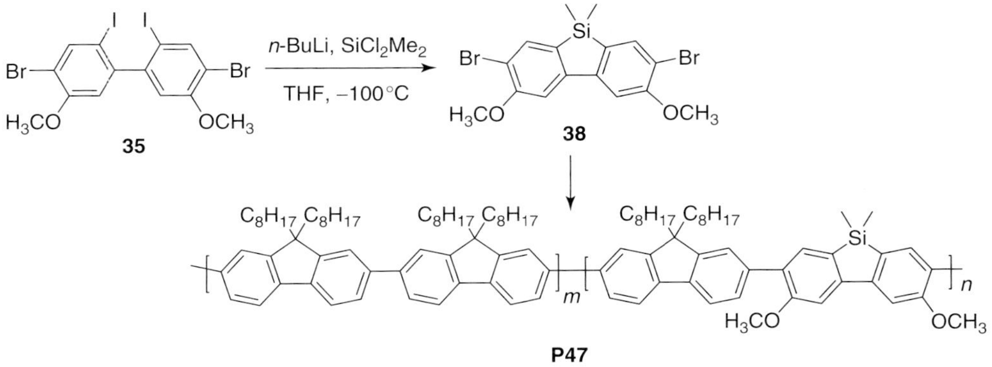

Scheme 9.23 Synthesis of copolymer of 2,7-silafluorene and 2,7-fluorene (**P47**).

properties because of the versatile reactivity and electronic properties of phosphorus centers. Huang *et al.* synthesized copolymers (**P45** and **P46**) of phosphafluorene and fluorene via Suzuki polymerization for the first time (Scheme 9.22) [91]. Both polymers exhibit good thermal stability with T_d higher than 403 °C. The optical bandgap of polymers in solid state decreases with the introduction of the phosphafluorene unit into PF8, suggesting that the phosphafluorene incorporation leads to increased conjugation length despite the spatial hindrance of the methoxyl substituents of phosphafluorene. In contrast to the polysilafluorenes, which emit only blue light like polyfluorene, phosphafluorene-containing copolymers show not only strong blue-light emitting but also single-layer white EL. The oxidized phosphorus atom on phosphafluorene of **P46** significantly changed the blue EL of PF into white.

Silafluorene is an excellent blue or deep-blue chromophore with high thermal stability and low-lying LUMO. In addition, silafluorene has a chemical structure remarkably similar to that of fluorene. So, the copolymers of silafluorene and fluorene are of interes. Huang *et al.* synthesized a series of blue-emitting, 2,7-silafluorene and 2,7-fluorene conjugated copolymers **P47** (Scheme 9.23) by Suzuki coupling reaction, confirming that silafluorene is an excellent building unit for optoelectronic materials [92]. By investigating the optical properties of **P47** in solid state, the introduction of silafluorene can result in considerably increased optical bandgap and blueshifted absorption and emission spectra through the spatial hindrance of methoxyl substituents, which is different from **P45** and **P46**. Another approach to obtain copolymers of silafluorene and fluorene was reported by Peng and Cao *et al.* They synthesized the monomer 3,6-dibromo-9,9-di-*n*-octylsilafluorene from 2,2′-dinitrobiphenyl as starting materials (Scheme 9.24). Copolymer **P48** based on 3,6-silafluorene and 2,7-fluorene was synthesized via the Suzuki coupling reaction and the bandgaps of this polymer increase with increasing 3,6-silafluorene content [93]. The incorporation of 3,6-silafluorene into the PF main chain not only suppressed the long-wavelength emission but significantly improved the efficiency and color purity of the **P48**-based devices.

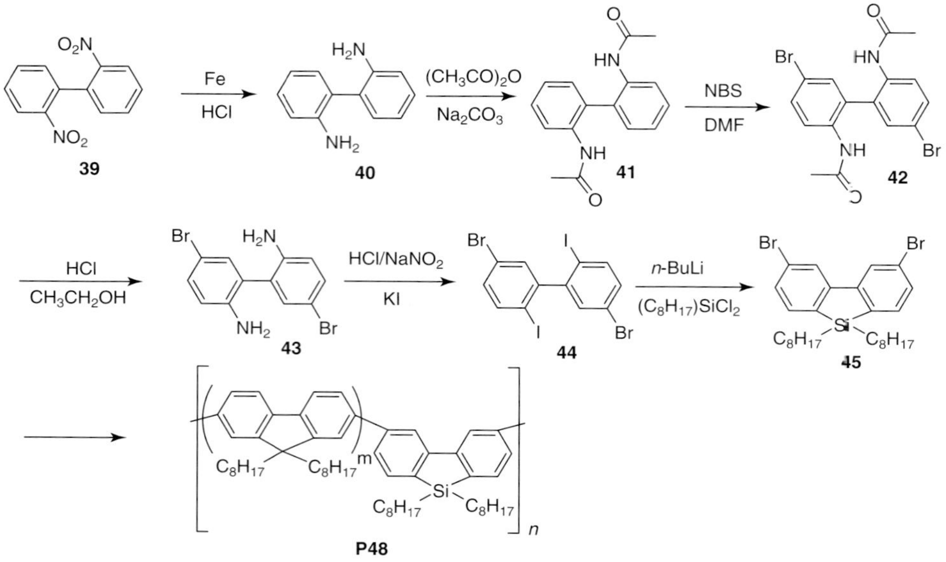

Scheme 9.24 Synthesis of copolymer of 3,6-silafluorene and 2,7-fluorene (**P48**).

Dibenzothiophene-S,S-dioxide (SO) is a electron-deficient moiety, which is topologically similar to the fluorene moiety but its LUMO energy level is lower than that of 9,9-dialkylfluorene. Incorporating SO (3,7-diyl or 2,8-diyl) into the poly[9,9-bis(4-(2-ethylhexyloxy) phenyl)fluorene-2,7-diyl] (PPF) backbone, blue-light-emitting **P49** and **P50** with stable spectra and high efficiency were realized (Scheme 9.25) [94]. The extent of intramolecular charge transfer(s) in **P49** is more significant than that in **P50**, indicating that the introduction of the meta-linkage SO unit into the polymer backbone can effectively disturb the effective conjugation length, which offsets the ICT effect. The EL spectra of all copolymers remain almost unchanged, and no green emission is observed even at high applied current density (more than 360 mA cm^{-2}) or high annealing temperature (~200 °C) due to the bulk alkoxyphenyl substituted at the C-9 position of the fluorene unit and the electron-withdrawing SO_2 group in the SO unit. **P51** with different contents of SO moieties (2–30 mol%) exhibits excellent thermal stabilities with T_d above 420 °C [95]. Electrochemical experiment shows that the introduction of SO moieties decreases both HOMO and LUMO energy levels and the LUMO energy level decreases more than the HOMO compared with PF homopolymer, which indicates the better electron-injection and -transport properties for **P51**. In addition, the absorption and PL spectra undergo progressive bathochromic shifts with increasing SO content. Dual excited states of local exciton (LE) and charge transfer (CT) character are observed, which depend on the local environment. The PL spectra change from well-structured LE emission (typical for PF8) in low polar toluene to a structureless redshifted CT emission in more polar chlorcform.

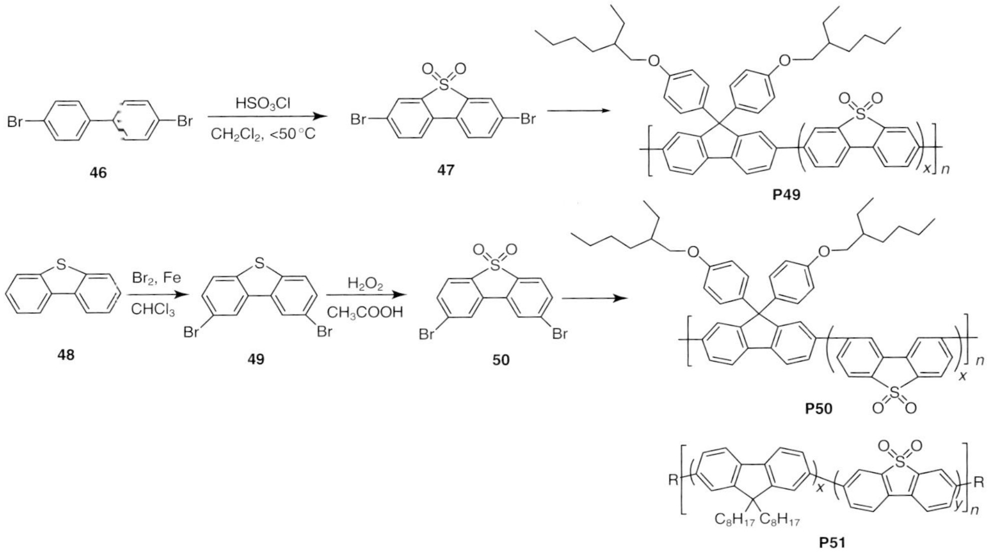

Scheme 9.25 Synthesis of copolymers based on dibenzothiophene-S,S-dioxide and fluorene.

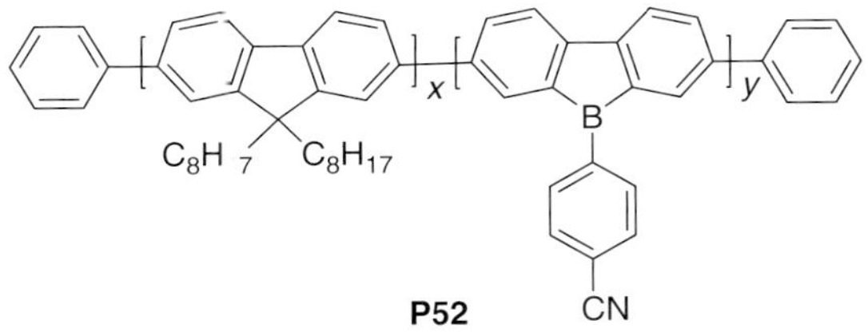

Scheme 9.26 Chemical structure of polyfluorene containing dibenzoborole units.

Bonifácio *et al.* introduced dibenzoborole units into PF and synthesized **P52** (Scheme 9.26) [96]. This polymer can respond to fluoride, cyanide, and iodide anions. The addition of fluoride and cyanide can lead to incomplete fluorescence quenching and characteristic isosbestic points, which is in accordance with a static-quenching mechanism. In the case of iodide, a nearly complete fluorescence quenching and the absence of an isosbestic point is observed, indicating a collisional-quenching mechanism.

9.5.3 Hyperbranched Polyfluorenes

The highly branched and globular features can depress the aggregation and excimer formation so as to make the materials form good-quality amorphous films, and consequently improve the light-emitting efficiencies as well. Hence, electroactive

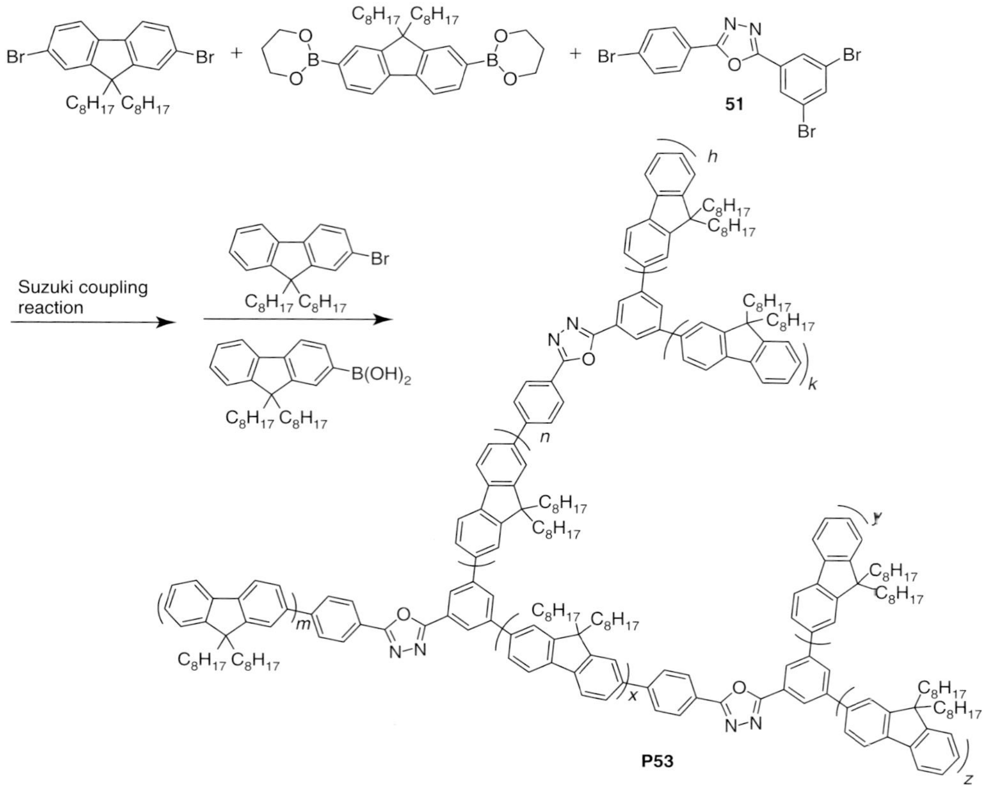

Scheme 9.27 Synthesis of hyperbranched fluorene-based polymer **P53**.

and light-emitting hyperbranched polymers are of current interest for developing efficient EL and other photonic devices.

The conventional methods to synthesize the hyperbranched polymers are the AB_2 and $A_2 + B_3$ approaches [97]. Huang *et al.* reported a novel "A2 + A′2 + B3" approach based on the Suzuki coupling reaction for the synthesis of hyperbranched fluorene-based polymer (**P53** in Scheme 9.27) [98]. To avoid the excimer formation and improve the charge transport-balance properties of PFs, oxadiazole was introduced into the hyperbranched polymer as a branching unit. The molar ratio was found to be a key factor in the solubility of hyperbranched products. The solubility was poor in common organic solvents for polymers with higher content of the branching unit. The resulting hyperbranched oxadiazole-containing PFs show stable blue-light emission even in the air at the elevated temperatures.

By the "A2 + A′2 + B3" approach, a series of hyperbranched triazine-containing polymer based on fluorene (**P54** in Scheme 9.28) were synthesized [99]. The resulting polymers show pure blue emission and distinct suppression of aggregation-induced green emission. By introducing triazine into PF, the

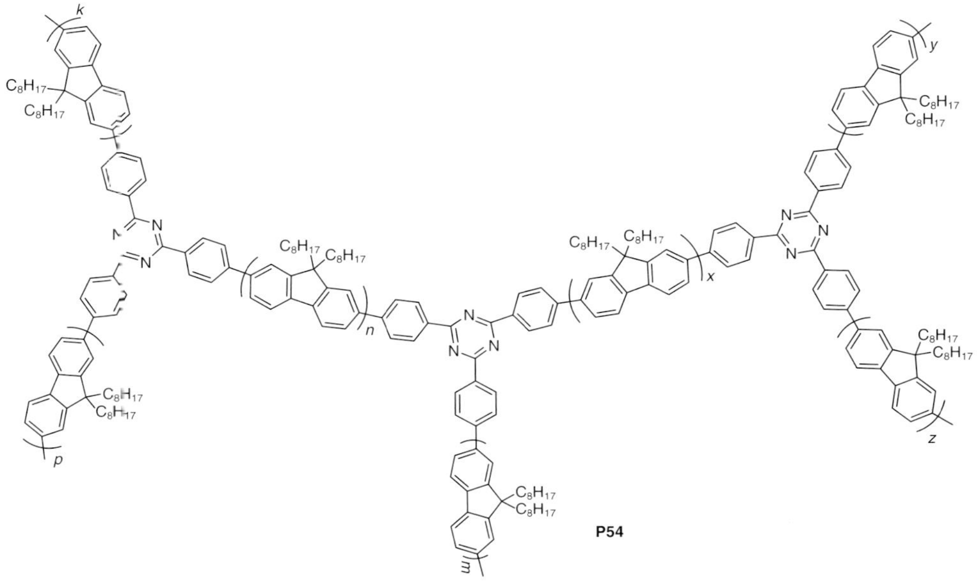

P54

Scheme 9.28 Chemical structure of hyperbranched triazine-containing polyfluorene.

electron-injection and -transport properties are improved. External quantum efficiency of 1.4% was achieved for the PLED based on these hyperbranched polymers. In addition, a series of hyperbranched polymers with a 3,4,5-triphenyl-1,2,4-triazole branching unit (**P55** in Scheme 9.29) were also synthesized through "A2 + A′2 + B3" approach based on the Suzuki condensation coupling [100]. The resulting polymers exhibit excellent thermal stability with T_d higher than 446 °C. Neither melting temperature nor obvious T_g is observed below 300 °C, suggesting that they are basically amorphous materials.

Liu and He *et al.* synthesized a hyperbranched tetrahedral **P56** with a δ-Si interrupted structure (Scheme 9.30) [101]. For its synthesis, the reaction was carried out in a dilute solution to minimize the formation of cross-linked insoluble polymer. The polymer showed high thermal stability and was not prone to self-aggregation and concentration quenching in the film state because its film emission spectrum is essentially the same as its solution one, with almost the same emission maximum, full width at half-maximum (fwhm) and comparable quantum efficiency. For **P57** containing hyperbranched Cz units (Scheme 9.30), the polymer films also exhibit very stable blue-light emission even annealing at 200 °C for 1 hour under nitrogen [102].

Photo-cross-linkable groups were introduced into the side chains at C-9 positions of fluorene, and then hyperbranched PFs of **P58** and **P59** (Scheme 9.31) were synthesized by the Suzuki coupling reaction [103, 104]. The polymers with higher content of the core exhibit more stable PL spectra even after annealing at high

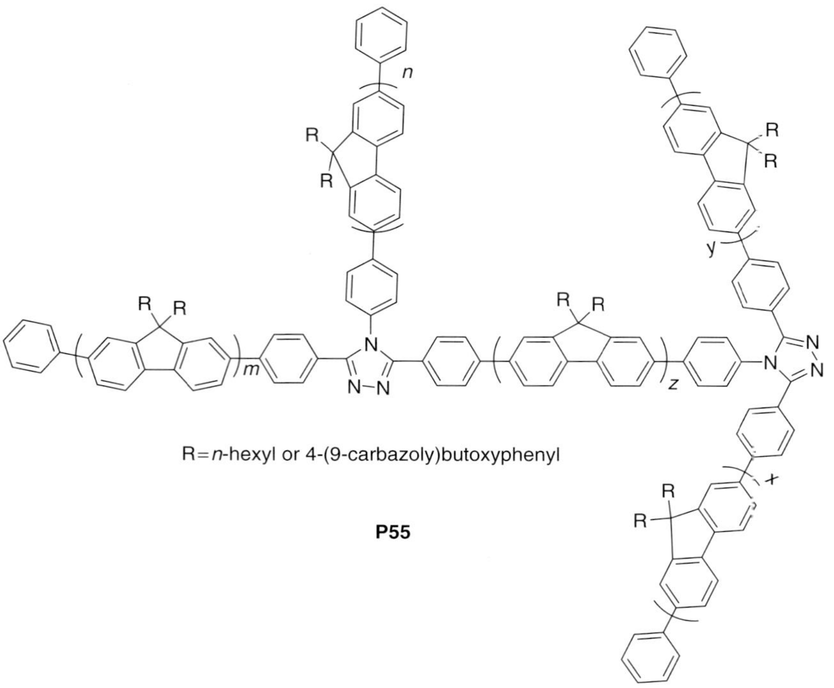

R=*n*-hexyl or 4-(9-carbazoly)butoxyphenyl

P55

Scheme 9.29 Chemical structure of hyperbranched polyfluorene **P55** with a 3,4,5-triphenyl-1,2,4-triazole branching unit.

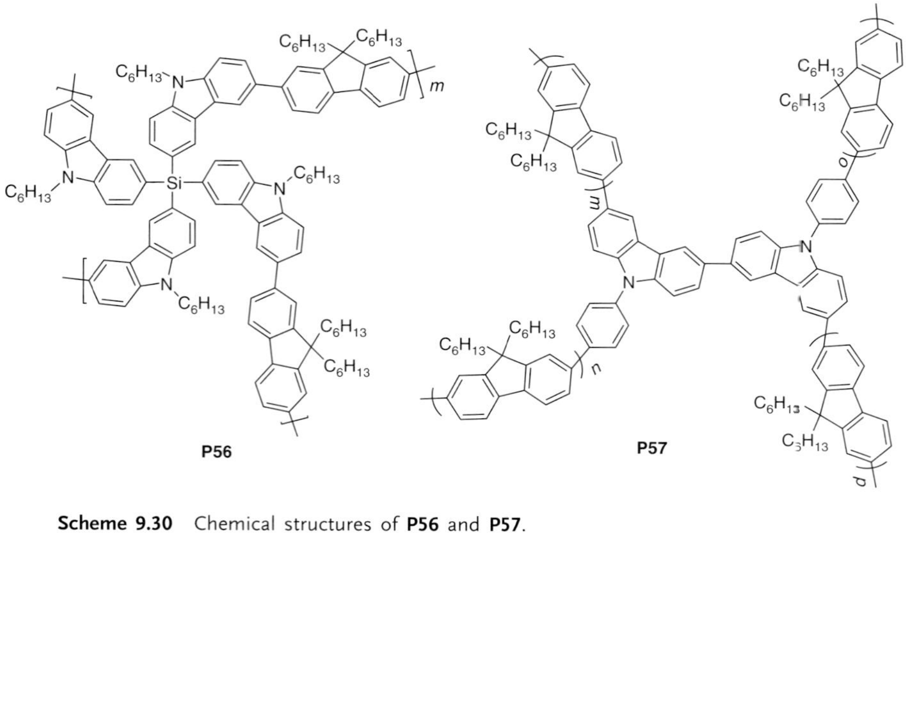

P56

P57

Scheme 9.30 Chemical structures of **P56** and **P57**.

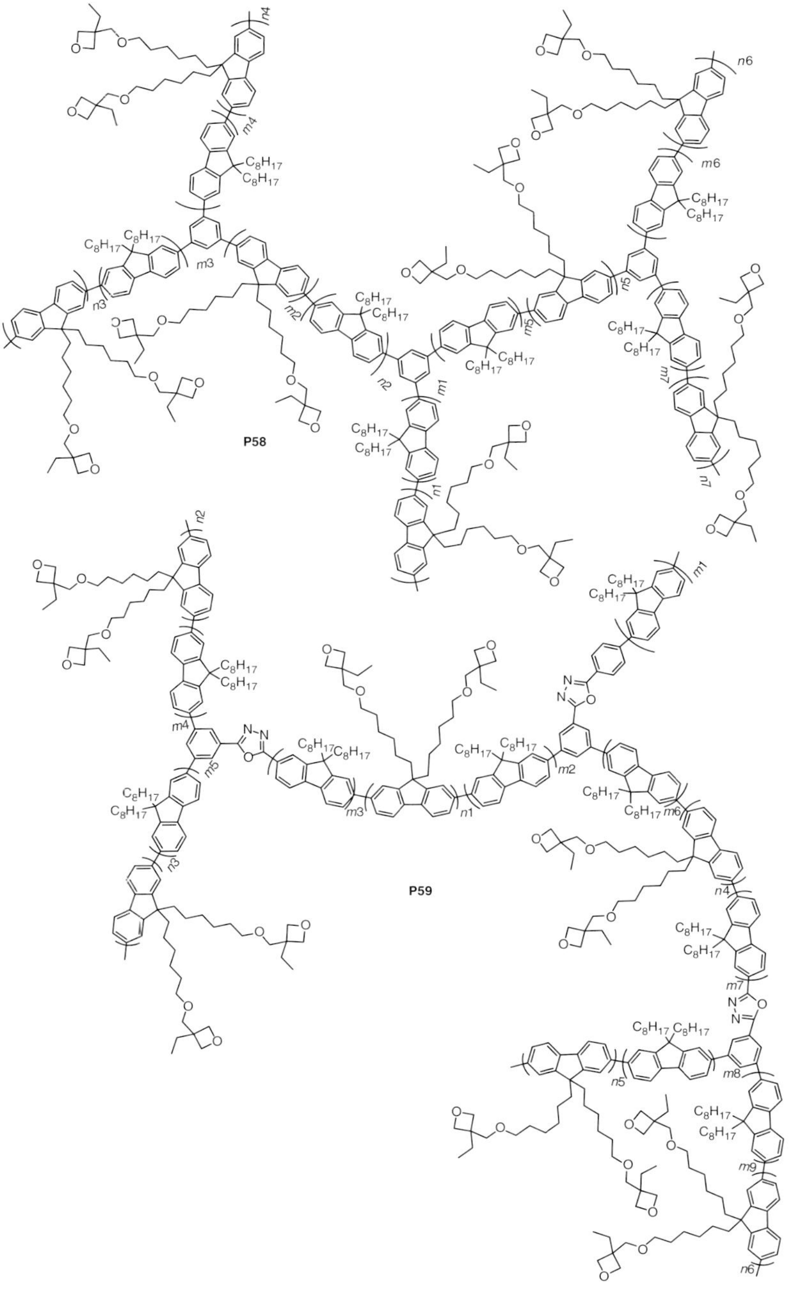

P58

P59

Scheme 9.31 Hyperbranched polyfluorenes containing photo-cross-linkable groups.

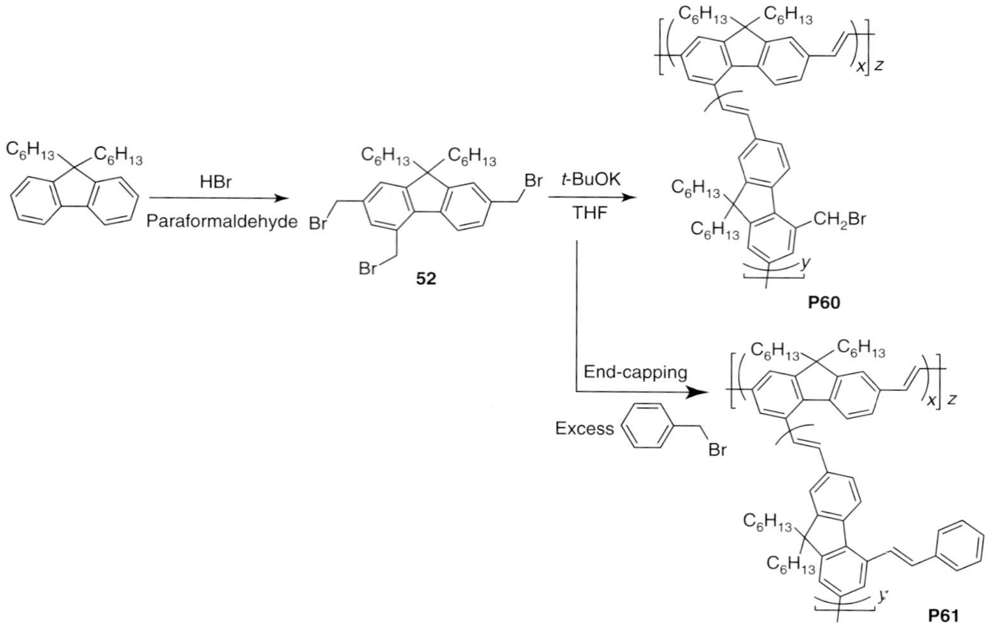

Scheme 9.32 Synthesis of **P60** and **P61**.

temperature. The resulting polymers can be cross-linked photochemically by the cross-linkable group and the photo-cross-link has little influence on their PL spectra.

Chen *et al.* employed a trifunctional derivative of fluorene [2,4,7-tris(bromomethyl)-9,9-dihexylfluorene] (**52**) as a branch unit to synthesize hyperbranched **P60** and **P61** by the Gilch reaction (Scheme 9.32) [105]. Polymerization of a trifunctional monomer easily results in fast gelation due to formation of three-dimensional networks. Therefore, the degree of conversion (polymerization time) should be strictly controlled to prevent undesirable gelation.

9.5.4
Star-Shaped Polyfluorenes

Luminescent monodisperse PFs with well-defined star-shaped structures as well as superior chemical purity are rather appealing and have attracted increasing interest. These starburst compounds are characterized by superior structural uniformity and chemical purity as well as highly branched and dendritic structures, which are capable of preventing close packing and spatial reorientation of the molecules, thus suppressing self-aggregation and favoring the formation of high-quality amorphous films. The molecular structure and the chemical and physical properties can be well controlled by judicious choice of synthetic methodology. High purity is one of the most important factors that govern device performance. In contrast to conjugated polymers, these starburst compounds can be purified by simple chemistry approaches such as column chromatography [106, 107]. These unique features render this kind of material rather promising in achieving high-performance optoelectronics.

The extensive applications of star-shaped PFs are often hampered by cumbersome and time-consuming preparation processes. Especially, performing the multiple transformations usually involved generally results in fairly low yield and remarkably difficult purification of the desired products from partially substituted by-products. Constructing these complex dendritic materials conveniently with high yield and purity is thus greatly challenging.

With the advance of dedicated MW equipment, microwave-assisted organic synthesis (MAOS) has attracted widespread use and acceptance as a valuable alternative to the use of conductive heating in dramatically reducing reaction times (from days and hours to minutes and seconds). Huang *et al.* developed a facile and powerful MW-enhanced multiple Suzuki coupling methodology for the preparation of highly luminescent star-shaped monodisperse PFs (**P62–P72** in Scheme 9.33) [108]. A series of highly luminescent six-arm monodisperse

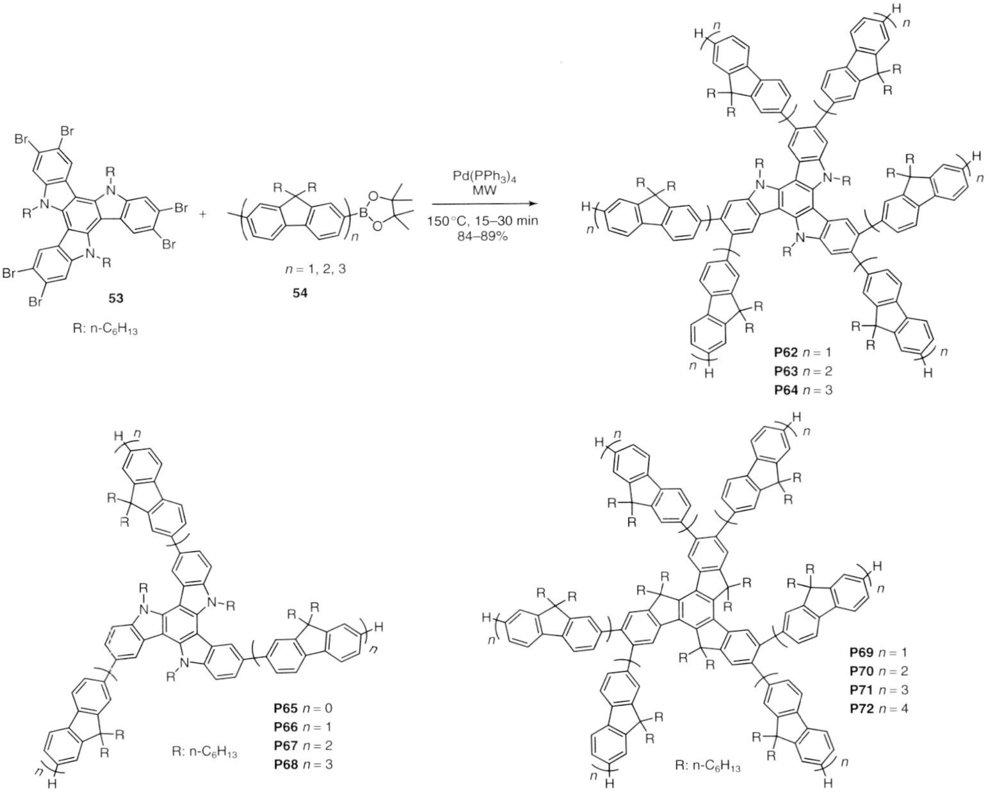

Scheme 9.33 Synthesis of star-shaped polyfluorenes by MW-enhanced methodology.

triazatruxene derivatives **P62–P64** containing 6–18 fluorene units can be synthesized in minutes with high yields (**P63** 86%; **P64** 84%) and purity by the MW-enhanced multiple-coupling methodology (Scheme 9.33). These materials can be purified by column chromatography. Using traditional methods to perform these multiple transformations results in quite low yields and presents serious challenges during the separation of the desired products from partially substituted by-products, especially tetra- and pentasubstituted by-products, which proved that MW-enhanced multiple Suzuki coupling methodology is a convenient and powerful way for the preparation of well-defined conjugated functional dendritic materials. The PL maximum peaks in films of **P63** (439 nm) and **P64** (437 nm) are quite close to those in solutions (both at 440 nm) but slightly blueshifted. This implies that the intermolecular interactions in the solid state are effectively suppressed and good amorphous states are formed in films, which might be attributed to the novel six-arm dendritic hindered structure. A single-layer EL device based on **P64** is fabricated with the configuration of ITO (indium tin oxide)/PEDOT:PSS poly(3,4-ethylenedioxythiophene):poly(styrenesulfonate)/**P64** (spin-cast, 130 nm)/Ba/Al. Highly efficient (2.07 cd A^{-1} and 2.0%) and pure-deep-blue EL with good color stability has been achieved [109].

Through the similar MW-enhanced coupling methodology, Huang *et al.* also prepared a series of three-arm triazatruxene derivatives **P65–P68** [110] and six-armed nanostar macromolecules **P69–P72** with a central truxene (Scheme 9.33) [111]. Excellent lasing characteristics (e.g., $E_{th}{}^{laser} \approx 0.4$ nJ/pulse (1.3 μJ cm^{-2}), slope efficiency $\approx$ 5.3% for $\lambda^{laser} = 437$ nm) are found for 1D-DFB (distributed feedback laser) lasers using **P71** as gain medium. The threshold represents one of the lowest reported for an organic semiconductor laser, irrespective of cavity design [48,112–115]. These results illustrate the great potential of starburst macromolecular materials for use in a variety of luminescence applications.

Skabara and Perrpichka synthesized **P73–P76** (Scheme 9.34) with a central truxene core and from monofluorene to quaterfluorene arms by attaching oligofluorene arms of corresponding length directly to the central truxene core [116]. These star-shaped macromolecules demonstrate good thermal stability (up to 400–420 °C) and improved glass transition temperatures with an increase in length of the oligofluorene arms (from $T_g = 63\,^{\circ}$C for T1 to 116 °C for T4). Efficient blue PL ($\lambda_{PL} = 398$–422 nm) in both solution ($\Phi_f = 70$–86%) and solid state ($\Phi_f = 43$–60%) was observed. In addition, these materials show electrochromic behavior, which reversibly change their color from colorless in the neutral state (~340–400 nm) to red or purple color in the oxidized state (~500–600 nm).

Star-shaped PFs introduced certain organic chromophores show attractive properties. **P77–P79** (Scheme 9.35) based on a pyrene core with four oligofluorene arms of different length have excellent thermal stability with T_d that range from 377 to 391 °C [117]. The single-layered device made of **P79** shows sky-blue emission similar to that of the film state PL with a maximum brightness of over 2700 cd m^{-2} and a maximum current efficiency of 1.75 cd A^{-1}. Ding *et al.* synthesized **P80–P83** (Scheme 9.35) bearing a 4,4′,4″-tris(carbazol-9-yl)-triphenylamine (TCTA) core and six oligofluorene arms [118]. All these oligomers have good film-forming

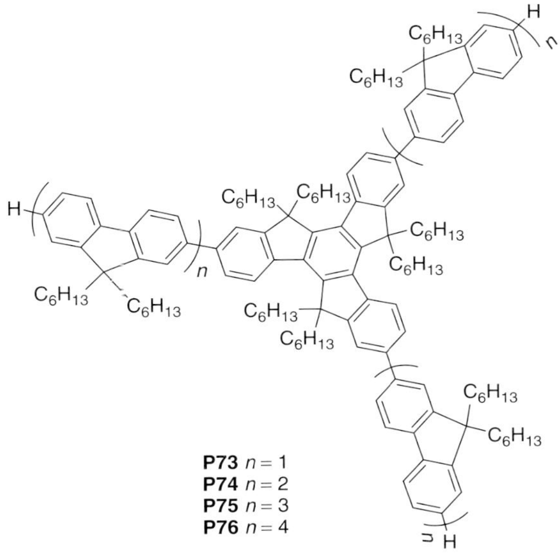

H
C_6H_{13}
C_6H_{13}
n
H
C_6H_{13}
C_6H_{13}
n
C_6H_{13}
C_6H_{13}
C_6H_{13}
C_6H_{13}
C_6H_{13}
C_6H_{13}
C_6H_{13}
C_6H_{13}
n
H

P73 $n = 1$
P74 $n = 2$
P75 $n = 3$
P76 $n = 4$

Scheme 9.34 Chemical structures of three-arm monodisperse derivatives with a central truxene core (**P73–P76**).

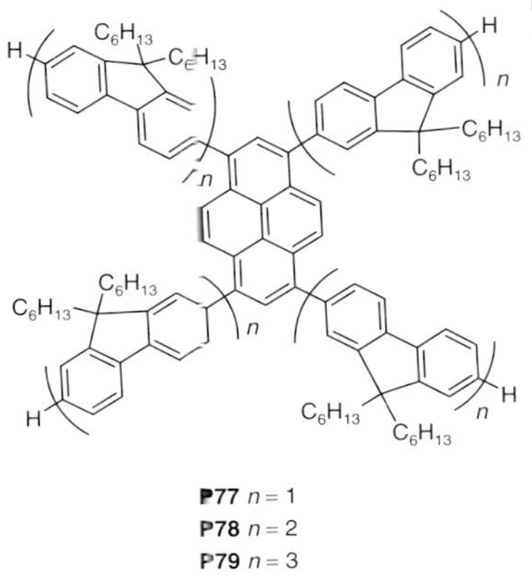

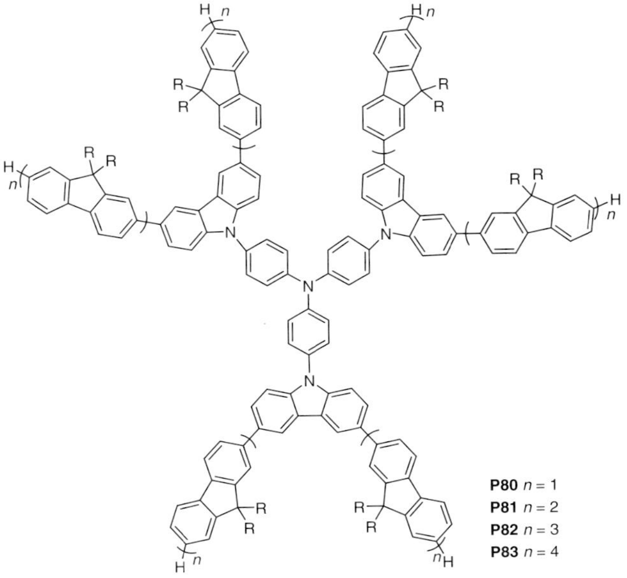

H
C_6H_{13}
C_6H_{13}
n
H
n
C_6H_{13}
C_6H_{13}
C_6H_{13}
C_6H_{13}
n
H
C_6H_{13}
C_6H_{13}
n
H

P77 $n = 1$
P78 $n = 2$
P79 $n = 3$

H n
R
R
H n
R
R
R
R
H
n
N
R
R
H
n
N
N
N
R
R
R
R
H n
n H

P80 $n = 1$
P81 $n = 2$
P82 $n = 3$
P83 $n = 4$

Scheme 9.35 Chemical structures of star-shaped PFs with pyrene (**P77–P79**) and TCTA core (**P80–P83**).

capabilities, and display bright, deep-blue fluorescence ($\lambda_{max} = 395-416$ nm) both in solution and in the solid state. EL devices are successfully fabricated using these materials as hole-transporting emitters, and emit deep-blue light. In addition, these large-energy-gap starburst oligomers are good host materials for red electrophosphorescence.

9.6
Emission Color Tuning of Polyfluorenes

The blue emission of PF itself restricts its application in the field of full-color PLED displays. One successful method for tuning the emission color covering the entire visible range was the copolymerization of fluorene with lower bandgap comonomers, including organic chromophores and metal complexes. In these copolymers, the PF family acts as energy donors, and the lower bandgap segment acts as energy acceptor. Energy transfer and trapping of excitons or electrons between the chromophoric segments in conjugated polymer chains are known to be fast and efficient, which shifts the wavelength of emission to a longer wavelength.

9.6.1
Polyfluorene Copolymers with Organic Chromophores

9.6.1.1 Green-Light-Emitting Polyfluorene Derivatives
Green emission is easily obtained by introducing thiophene or bithiophene moieties into the PF chains, which can be realized by the Yamamoto or Suzuki coupling reaction as well as the facile $FeCl_3$ oxidative polymerization (Scheme 9.36). **P84** was synthesized through the Yamamoto coupling reaction from the monomer of **55** (2,7-bis(5-bromo-4-hexylthienyl)-9,9-dihexyl-9H-fluorene) and its emission color was tuned from blue to green with an emission peak at 493 nm and a shoulder at 515 nm [119]. Green-light-emitting polymers **P85** and **P86** composed of bithiophene moieties were synthesized through the Suzuki coupling reaction (Scheme 9.36) [120]. The polymer **P85** with the head-to-head (HH) structural regularity can also be prepared with even higher molecular weights from **58** by the facile $FeCl_3$ oxidative polymerization. However, the synthesis of **P86** with the same $FeCl_3$ oxidation approach failed [121, 122]. The HH or tail-to-tail (TT) structural regioregularity between the two substituted thiophene rings does not affect the optical and electrochemical properties of the resulting polymers. However, the T_g and the environmental stability of the polymers depend on the structural regioregularity, and HH coupling configuration corresponds to a higher T_g and better stability.

In order to improve the luminescence properties of PF, Tsiang *et al.* introduced gold nanoparticles into the end of **P87** (Scheme 9.37) and obtained polymer **P88** [123]. Compared with **P87**, the threshold voltage of PLED devices based on **P88** is lowered and the efficiency is markedly improved (an order of magnitude increase). The light emitting from the **P88**-based device is very close to the standard green

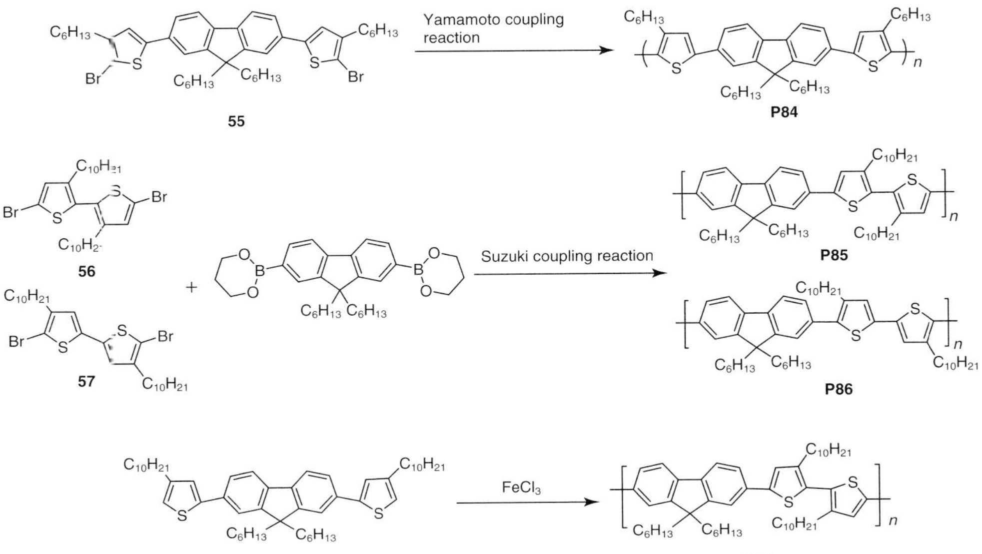

Scheme 9.36 Synthesis of copolymers based on fluorene and thiophene or bithiophene.

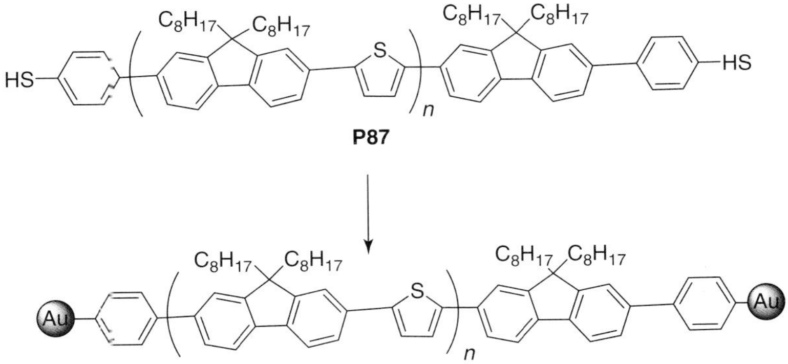

Scheme 9.37 Chemical structures of **P87** and **P88**.

demanded by the National Television System Committee (NTSC), making **P88** an excellent candidate for a green-light-emitting material.

Introducing 4,7-diphenyl-2,1,3-benzothiadiazole into the backbone of PF is also an effective way to obtain green-light-emitting polymers. **P89** (Scheme 9.38) containing 4,7-diphenyl-2,1,3-benzothiadiazole units was synthesized by Wang *et al.* The emission maximum at 521 nm and Commission Internationale de l'Eclairage (CIE) coordinates of (0.29, 0.63), which are very close to the (0.26, 0.65) of standard

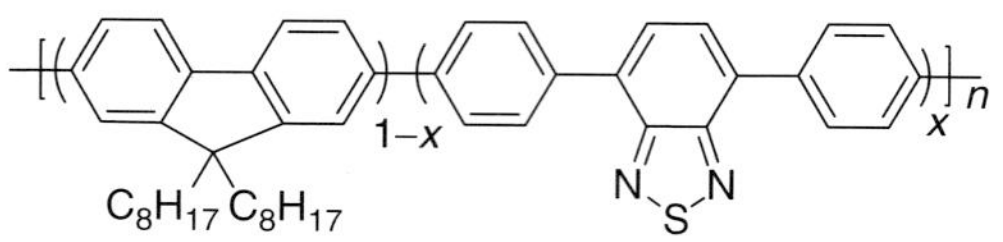

P89

Scheme 9.38 Chemical structure of **P89**.

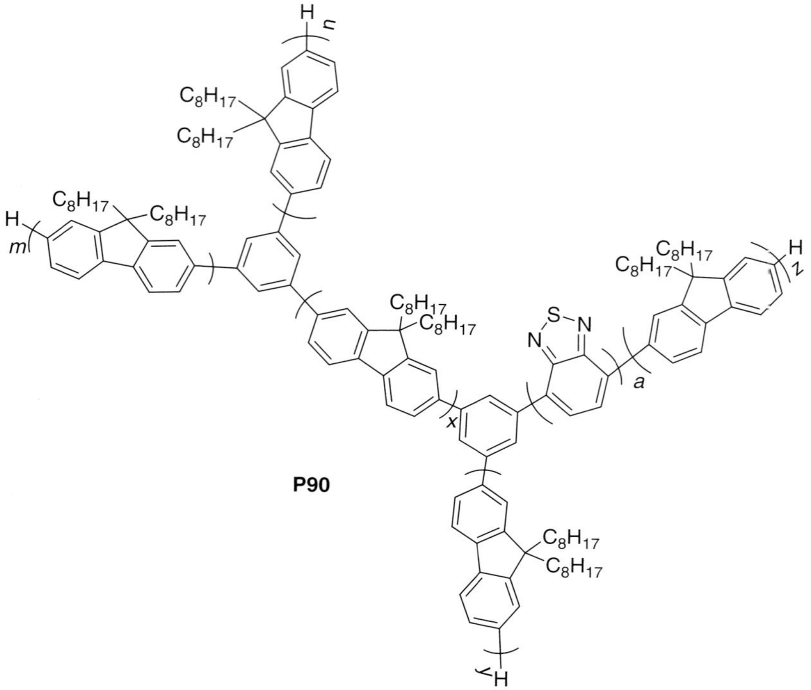

P90

Scheme 9.39 Chemical structure of hyperbranched polyfluorene derivative **P90** with green emission.

green emission demanded by the NTSC, was realized [124]. The single-layer devices (ITO/PEDOT/**P89**/Ca/Al) exhibit a luminous efficiency of 5.96 cd A^{-1} and a power efficiency of 2.21 lm W^{-1}.

Huang *et al.* introduced different contents of branching units into the poly(fluorene-*co*-benzothiazole) to construct a series of hyperbranched **P90** (Scheme 9.39) [125]. These polymers are thermally stable with T_d ranging from 280 to 376 °C, confirming the common advantages of hyperbranched structure. The comonomer 2,1,3-benzothiadiazole (BT) with narrow bandgap acts as a powerful exciton trap, which allows efficient energy transfer of the exciton from the fluorene segment to BT unit. As a result, the blue emission from the fluorene segment is completely quenched in concentrated solutions and in the solid state. Furthermore, green light from BT units was also observed.

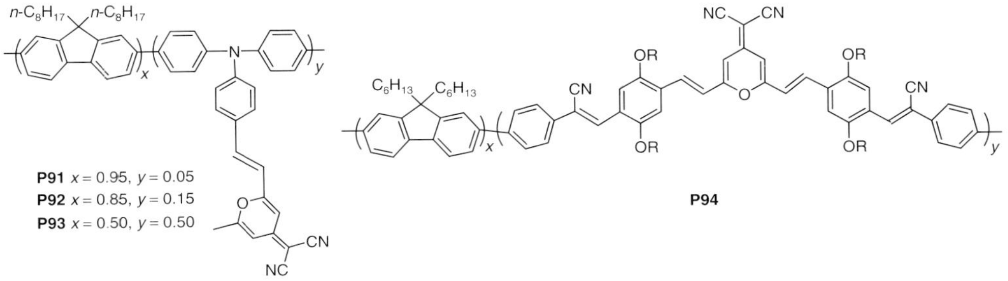

P91 $x = 0.95$, $y = 0.05$
P92 $x = 0.85$, $y = 0.15$
P93 $x = 0.50$, $y = 0.50$

P94

Scheme 9.40 Structures of red-light-emitting polyfluorene derivatives with DCM.

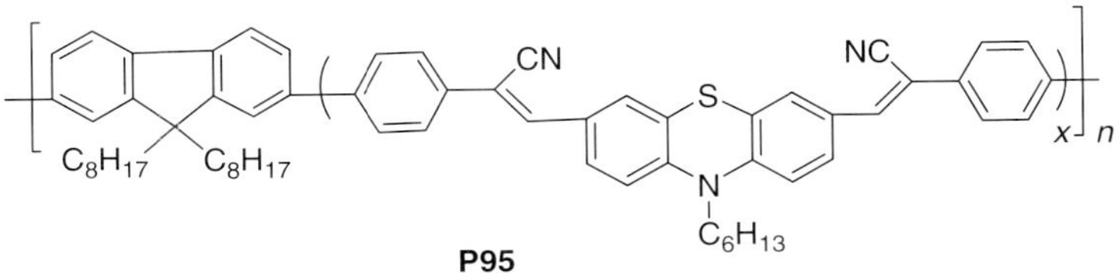

P95

Scheme 9.41 Structures of red-light-emitting polyfluorene derivative with PZB.

9.6.1.2 Red-Light-Emitting Polyfluorene Derivatives

4-Dicyanomethylene-6-methyl-4H-pyran (DCM) is a well-known red-emitting chromophore. Jin *et al.* introduced DCM into polymers **P91–P93** as pendants through the TPA (Scheme 9.40) [126]. In solid state, only a long-wavelength emission at 620 nm originating from the DCM moieties is observed. This is interpreted as an efficient energy transfer or exciton migration from the PF main chain to the low-band-gap trap sites DCM moieties. LED devices were fabricated and the EL spectra showed only red emissions as observed in the PL spectra of the polymer in solid state. Chen *et al.* synthesized **P94** with DCM moieties on the main chain (Scheme 9.40) [127]. To extend the conjugation length and enhance the electron affinity of the copolymers, they also introduced the *p*-phenylenevinylene segment with a cyano group on the vinylene unit into the polymer. The polymer exhibited moderately high T_g (>100 °C) and good thermal stability (T_d > 410 °C). In the film state, the emission maximum originated from DCM segments redshifted gradually from 560 to 596 nm with increasing DCM contents.

Similarly, the emission spectra of the PF derivative **P95** (Scheme 9.41) containing lower bandgap bis(2-phenyl-2-cyanovinyl)-10-hexylphenothiazine) (PZB) comonomer on the main chain are significantly redshifted as the fraction of PZB in the copolymer is increased [128]. The EL device based on the copolymer containing 50% PZB shows almost pure red emission. The maximum brightness and external quantum efficiency can reach 3440 cd m^{-2} and 0.45%.

Copolymer **P96** of a 3,4-diphenylmaleimide-thiophene-fluorene triad is an excellent orange–red light-emitting material (Scheme 9.42) [129]. It has a solution emission wavelength of around 598 nm with Φ_f of 37%. In the solid state, the fluorescent wavelength is only slightly redshifted to 601 nm, and the fluorescent

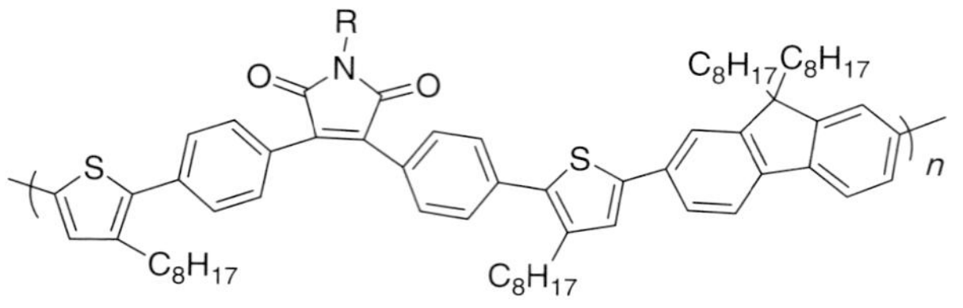

P96

Scheme 9.42 Chemical structures of red-light-emitting copolymer **P96**.

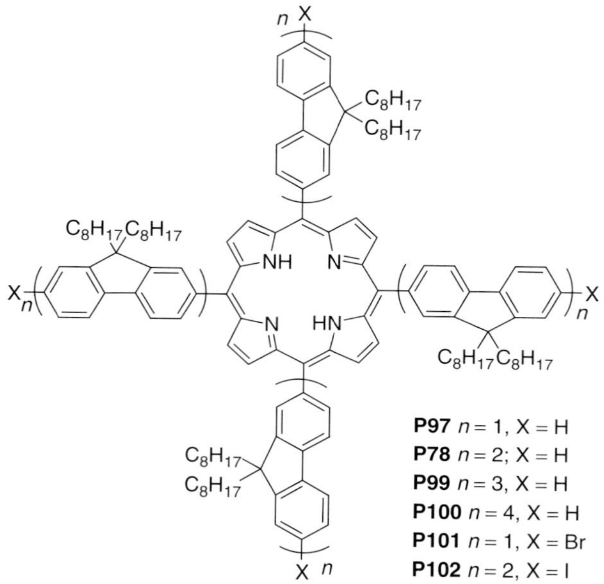

P97 $n = 1$, X = H
P78 $n = 2$; X = H
P99 $n = 3$, X = H
P100 $n = 4$, X = H
P101 $n = 1$, X = Br
P102 $n = 2$, X = I

Scheme 9.43 Star-shaped polyfluorene derivatives with porphyrin core.

intensity still remains high ($\Phi_f \sim 22\%$). Its PLEDs yielded bright orange–red EL with a high intensity of over 2000 cd m^{-2}, a luminous efficiency of 1.25 cd A^{-1} and an external quantum efficiency of 0.74%.

Bo *et al.* synthesized a series of well-defined star-shaped porphyrins **P97–P102** with four oligofluorene arms at their meso positions (Scheme 9.43) [130]. The oligofluorene arms not only conjugate with the porphyrin rings but also provide good solubility for the whole molecules. The side octyl chains on the fluorene units can efficiently suppress the aggregation of the porphyrin rings. The star-shaped oligomers show red emission due to efficient energy transfer from the oligofluorene arms to the porphyrin core. The efficiency of light absorption and energy transfer was intensified with increasing conjugated length of the oligofluorene arms.

Swager *et al.* synthesized red-light-emitting **P103** containing iptycene units (Scheme 9.44) [131]. The absorption and emission spectra of **P103** in solid state are

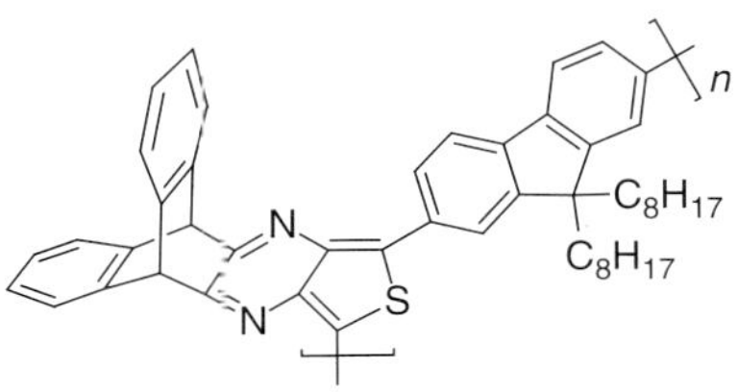

P103

Scheme 9.44 Chemical structure of polyfluorene derivative with iptycene unit.

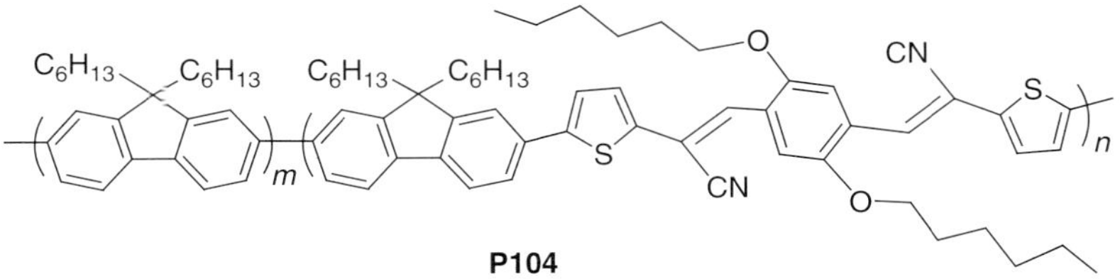

P104

Scheme 9.45 White-light-emitting polymer **P104**.

almost identical to its spectra in solution. Shoulder peaks and tails at the longer wavelength range were not observed. This indicates that the insulating effect of iptycene units prevents the aggregation of the polymer chains and the formation of excimers in the solid state.

9.6.1.3 White-Light-Emitting Polyfluorene Derivatives

Two strategies have been commonly adopted to develop white-light-emitting single polymers. One way is to incorporate an orange-emissive unit into a blue-light-emitting PF host. By decreasing the content of the orange-emissive unit, incomplete energy transfer from the PF host to the orange-emissive unit can be achieved, resulting in the emission of white light. Another method is to introduce low contents of both a green chromophore and a red chromophore into the blue-light-emitting PF host. The energy transfer from the blue host to the green and red components is incomplete and that from the green to the red component is negligible. Hence, white emission composed of the three primary colors can be realized.

Chen *et al.* synthesized **P104** containing very low content (0.1 or 0.025 mol%) of orange-emissive unit 2,5-dihexyloxy-1,4-bis(2-thienyl-2-cyanovinyl)benzene (Scheme 9.45) [132]. White-light emission is realized via the incomplete energy transfer between the fluorene moieties and 2,5-dihexyloxy-1,4-bis(2-thienyl-2-cyanovinyl)benzene. For the white-light-emitting EL devices based on polymers with 0.1 and 0.025 mol% of 2,5-dihexyloxy-1,4-bis(2-thienyl-2-cyanovinyl)benzene units, the maximal brightness was 5419 and 3011 cd m^{-2}, and the current efficiency was 1.92 and 1.98 cd A^{-1}, respectively.

Wang *et al.* designed four polymers **P105–P108** with two kinds of attachment of orange chromophores, 4-(4-alkyloxy-phenyl)-7-(4-diphenylamino-phenyl)-2,1,3-

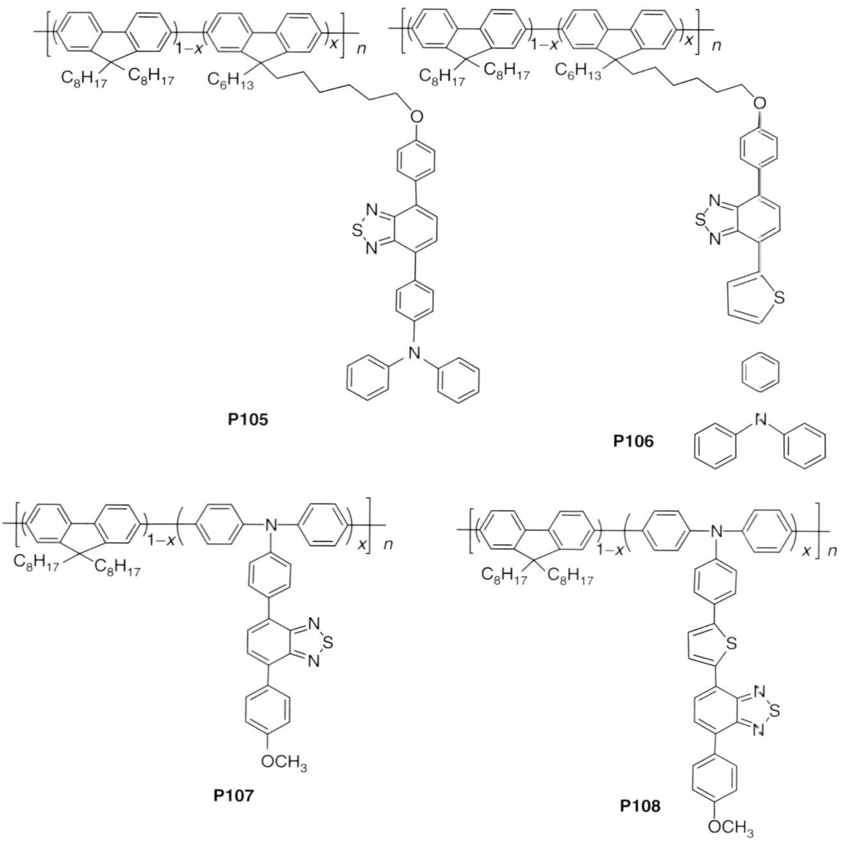

P105

P106

P107

P108

Scheme 9.46 White-light-emitting polymer **P105–P108**.

benzothiadiazole (MOB-BT-TPA) and 4-(4-alkyloxy-phenyl)-7-(5-(4-diphenylamino-phenyl)-thiophene-2-yl)-2,1,3-benzothiadiazole (MOB-BT-Th-TPA), into PFs (Scheme 9.46) [133]. By decreasing the content of the orange chromophores, incomplete energy transfer from blue host to orange dopant was achieved and simultaneous blue and orange emission for white EL was realized. The effects of the side- and main-chain attachments on the EL efficiencies of the resulting polymers were compared. The side chain–type single polymers are found to exhibit more efficient white EL than the main-chain-type single polymers. A single-layer device of the resulting side chain–type polymers exhibited a turn-on voltage of 3.5 V, luminous efficiency as high as 10.66 cd A^{-1}, and power efficiency up to 6.68 lm W^{-1}, with CIE coordinates of (0.30, 0.40). Using 4,7-(4-(diphenylamino)phenyl)-2, 1,3-benzothiadiazole (TPABT) as the orange-emissive core and fluorene segments as the blue-emissive arms, they also designed star-like polymers **P109** (Scheme 9.47) [134]. The star-shaped polymers were synthesized by a one-pot Suzuki coupling reaction with an A_4-type monomer **59** and an AB-type monomer

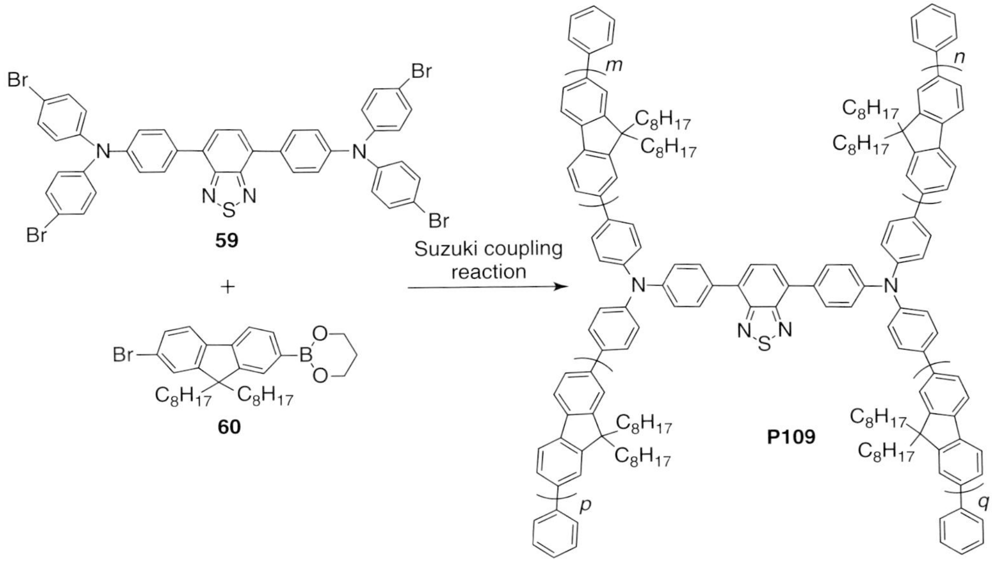

Scheme 9.47 Synthesis of star-like white-light-emitting polymer **P109**.

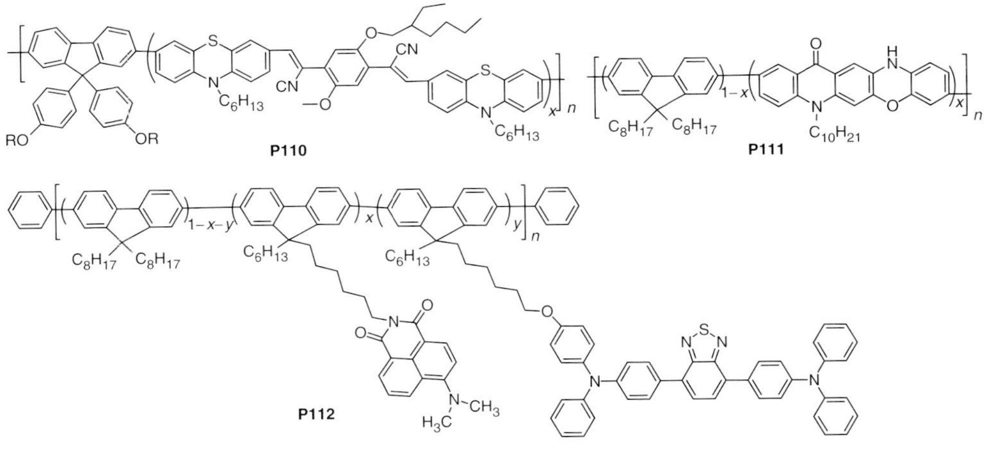

Scheme 9.48 Chemical structures of **P110–P112**.

60 (Scheme 9.45). The M_n of polymers ranges from 14 000 to 27 900 with the PDI ranging from 2.17 to 2.53. For the polymer with an orange core unit content of 0.03 mol%, white EL is observed with simultaneous blue emission from the fluorene segments and orange emission from the [4,7-bis(5-(4-(*N*-phenyl-*N*-(4- methylphenyl)amino)phenyl)-thienyl-2-)-2,1,3-benzothiadiazole] (TPABT) unit.

Using a similar design principle, other white-light-emitting polymers **P110–P112** were also obtained (Scheme 9.48) [135–137].

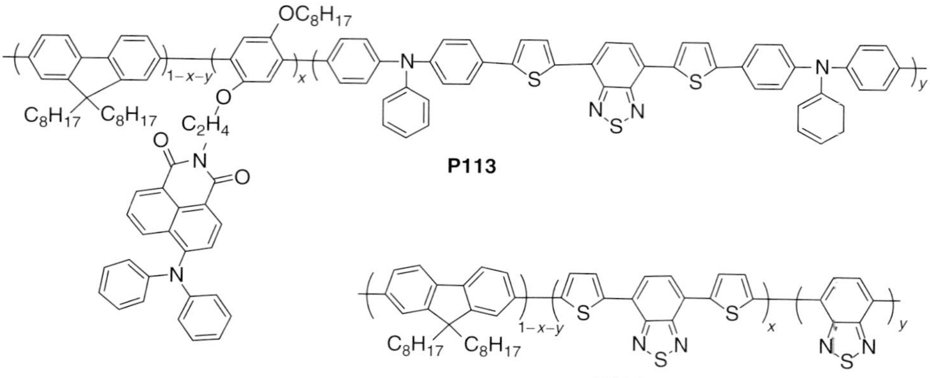

P113

P114

Scheme 9.49 White-light-emitting polymers with three emissive species.

By covalently attaching both a small amount of a green-light-emitting component of DPAN (4-diphenylamino-1,8-naphthalimide) to the side chain and a small amount of a red-light-emitting component of TPATBT to the main chain of blue PF host, a single polymer **P113** with three emissive species was designed and synthesized (Scheme 9.49) [138]. White EL with simultaneous blue (445 nm), green (515 nm), and red (624 nm) emission is successfully achieved from a single-layer device fabricated from the polymer. Peng and Cao *et al.* also realized white-light emission with three primary-color-emissive species from **P114** with a backbone consisting of fluorene, BT, and 4,7-bis(2-thienyl)-2,1,3-benzothiadiazole (DBT) as the blue-, green-, and red-light-emitting units, respectively (Scheme 9.49) [139]. The external quantum efficiency and luminance efficiency of white-light-emitting device based on this polymer reach 3.84% and 6.20 cd A^{-1}, respectively.

9.6.2
Polyfluorene Copolymers with Phosphorescent Heavy-Metal Complexes

Conjugated polymers comprising phosphorescent heavy-metal complexes have attracted increasing attention recently because phosphorescent heavy-metal complexes can increase the luminescence efficiency in LEDs by harvesting the large percentage of triplet excitons created upon electron–hole recombination [140–143]. PFs, as the most attractive blue-emitting materials, are usually selected as host materials for phosphorescent heavy-metal complexes due to their high efficiency and charge-transport properties. The energy can be transferred efficiently from the polymer main chain (as host materials) to the metal complexes (as guest materials) in the main or side chain. To date, various phosphorescent heavy-metal complexes, such as iridium(III) (Ir), platinum(II) (Pt), europium(III) (Eu), rhenium(I) (Re), and ruthenium(II) (Ru) complexes, are introduced into the main or side chain of PFs by Suzuki- or Yamamoto-type polymerization method of fluorene monomers and complex monomers containing active groups on ligands.

9.6.2.1 Polyfluorene Copolymers with Ir(III) Complexes

Ir(III) complexes can be introduced into the side or main chain of PFs.

Chen *et al.* first reported red-emitting PFs grafted with neutral iridium complexes (**P115**) on the side chain (Scheme 9.50) [144]. The incorporation of Cz can significantly increase the efficiency and lower the turn-on voltage. They demonstrated that both the energy transfer from the PF main chain and from an electroplex formed between main-chain fluorene and side chain Cz moieties to the red Ir(III) complex can significantly enhance the device performance. Subsequently, Yang *et al.* also reported a series of fluorene-*alt*-carbazole copolymers grafted with neutral cyclometalated Ir(III) complexes (**P116–P120**) [145]. The red-light-emitting devices based on these polymers show the highest luminous efficiency of 4.0 cd A^{-1} and a peak emission of 610 nm. PF with charged Ir(III) complexes in the side chain can be obtained by introducing N^N ligand into the side chain of polymer and then chelating with Ir(III) chloride-bridged dimmer. Yang and Peng *et al.* [146] synthesized a series of copolymers with charged Ir(III) complexes in the side chain (Scheme 9.50) (**P121**). Excellent device performances of maximum external quantum efficiency of 7.3% and luminous efficiency of 6.9 cd A^{-1} were obtained using this kind of polymer as emission materials.

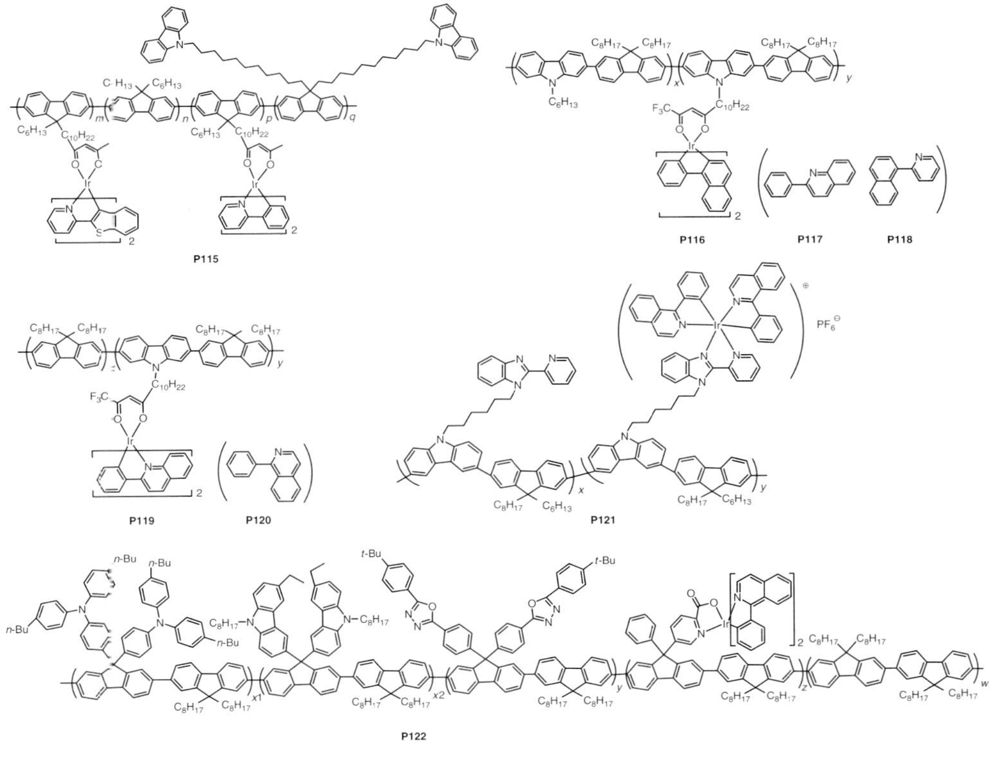

P115 P116 P117 P118 P119 P120 P121 P122

Scheme 9.50 Polyfluorenes with Ir complexes in the side chain.

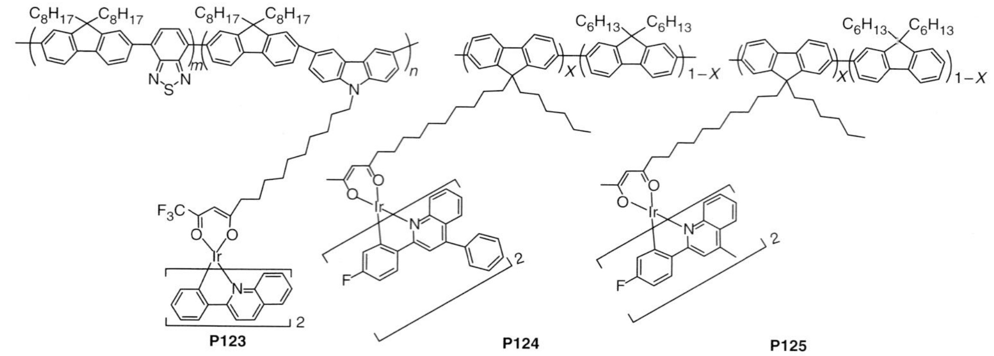

C_8H_{17} C_8H_{17} C_8H_{17} C_8H_{17} m n N S N N F_3C O O Ir N 2

P123

C_6H_{13} C_6H_{13} X 1−X O O Ir N F 2

P124

C_6H_{13} C_6H_{13} X 1−X O O Ir N F 2

P125

Scheme 9.51 Polymers with Ir(III) complexes in the side chain for white-light-emitting devices.

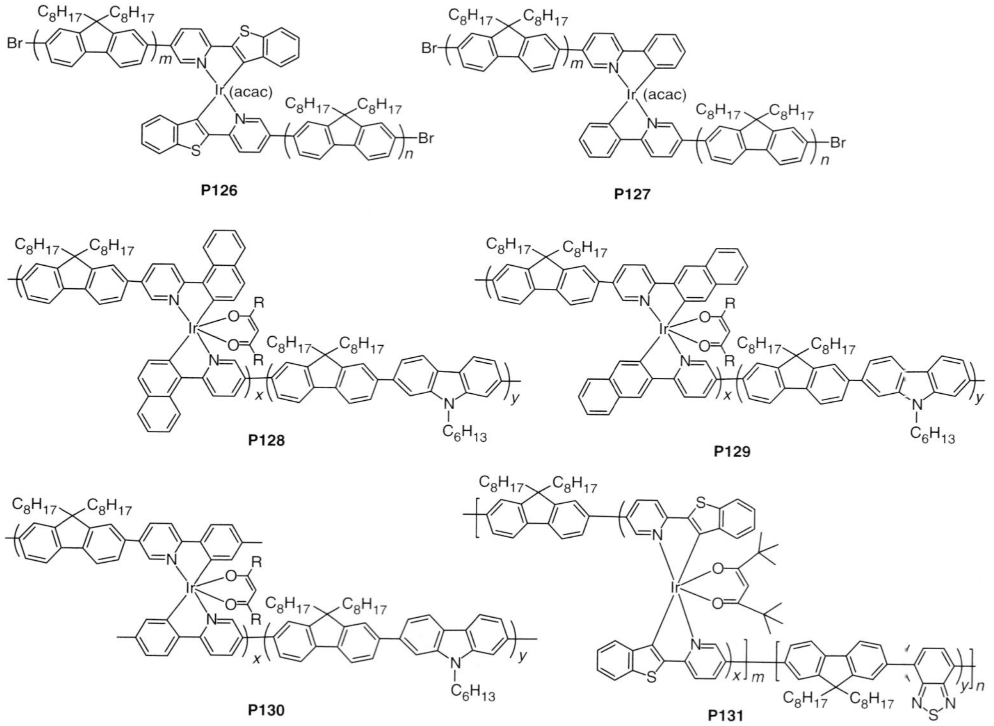

C_8H_{17} C_8H_{17} Br m N S Ir (acac) C_8H_{17} C_8H_{17} N S n Br

P126

C_8H_{17} C_8H_{17} Br m N Ir (acac) C_8H_{17} C_8H_{17} N n Br

P127

C_8H_{17} C_8H_{17} N Ir O O R R C_8H_{17} C_8H_{17} N x y N C_6H_{13}

P128

C_8H_{17} C_8H_{17} N Ir O O R R C_8H_{17} C_8H_{17} N x y N C_6H_{13}

P129

C_8H_{17} C_8H_{17} N Ir O O R R C_8H_{17} C_8H_{17} N x y N C_6H_{13}

P130

C_8H_{17} C_8H_{17} S N Ir O O N S x m C_8H_{17} C_8H_{17} y n N S N

P131

Scheme 9.52 Polymers with Ir(III) complexes incorporated into the main chain through C ^ N ligands.

Neher and Shu *et al.* synthesized red-emitting **P122** containing a red-emitting iridium complex and carrier-transporting units as the substitutes of the C-9 position of fluorene [147]. Saturated red-emitting PLED with peak luminance efficiency of 9.3 cd A^{-1} was realized.

White-light emission could be obtained from a single polymer **P123** with three individual emission species by introducing a small number of neutral Ir(III) complexes into the side chain of the polymer. In **P123**, fluorene is used as the blue-emissive component, BT and the Ir(III) complex act as green- and red-emissive chromophores, respectively. By changing the proportion of BT and Ir(III) complex in the polymer, the EL spectrum from a single polymer can be adjusted to achieve white-light emission. White-light-emitting polymers (polymer **P124** and **125**) (Scheme 9.51) [148] can also be obtained by introducing orange phosphorescent iridium complex into the PF. The white-light emissions from such polymers are stable in the white-light region, indicating that the approach of incorporating singlet and triplet species into the polymer backbone is promising for white-light PLEDs.

PFs with neutral Ir(III) complexes in the main chain can be prepared by introducing the ligands of complexes into the main chain of polymers [149, 150]. PFs with red or green neutral Ir(III) complexes in the main chain (polymer **P126** and **127**, Scheme 9.52) were first reported by Holmes *et al.* Subsequently, Cao *et al.* [151, 152] reported polymers **P128–P129** and the highly efficient saturated red-phosphorescent PLEDs were achieved on the basis of polymer **P128**.

Similar to **P123**, **P131** containing blue, green- and red-emissive components can realize white-light-emitting devices.

Another way to introduce neutral Ir(III) complexes into the main chain of PFs was realized by the Suzuki coupling reaction of fluorene segments and β-diketone ligand chelated with Ir(III) chloride-bridged dimmer (polymer **P132** and **P133** in Scheme 9.53) [153, 154]. Furthermore, using a similar method, white EL emission was realized by incorporating green-emitting fluorenone and red-emitting iridium complex into the PF main chain (**P134**) [155]. In addition, Ir(III) complexes can also be introduced into the end of the PF main chain via β-diketone ligand (**P135**).

Huang *et al.* synthesized a series of PF derivatives (**P136–P138**) with charged red-light-emitting Ir(III) complexes in the backbones (Scheme 9.54) [156, 157]. Saturated red-light emission can be realized by choosing appropriate ligands. Cz units were also introduced into the backbones of polymers. Almost complete energy transfer from the host fluorene segments to the guest Ir(III) complexes was achieved in solid state even at the low feed ratio of complexes. Intra- and interchain energy transfer mechanisms coexist in the energy-migration process of this host–guest system, and the intramolecular energy transfer might be a more efficient process. Chelating polymers show more efficient energy transfer than the corresponding blended system and exhibit good thermal stability, redox reversibility, and film formation as well. Preliminary results concerning PLEDs indicate that these materials offer promising opportunities in optoelectronic applications.

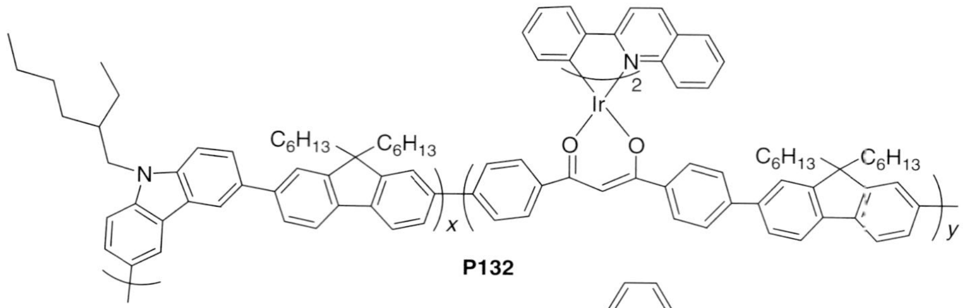

P132

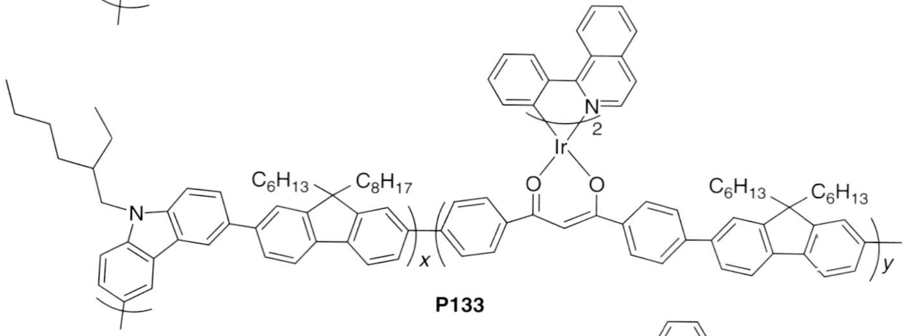

P133

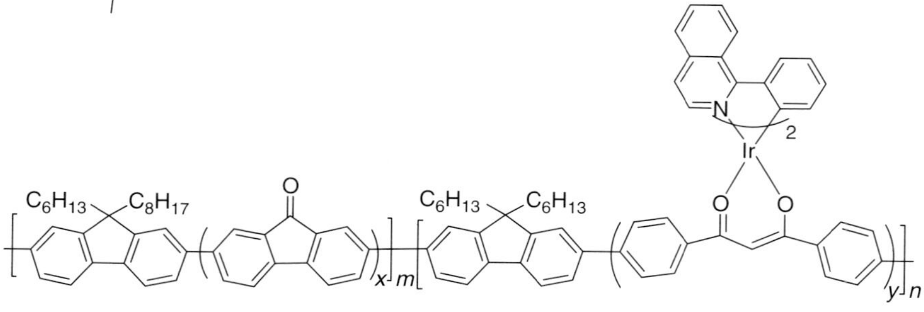

P134

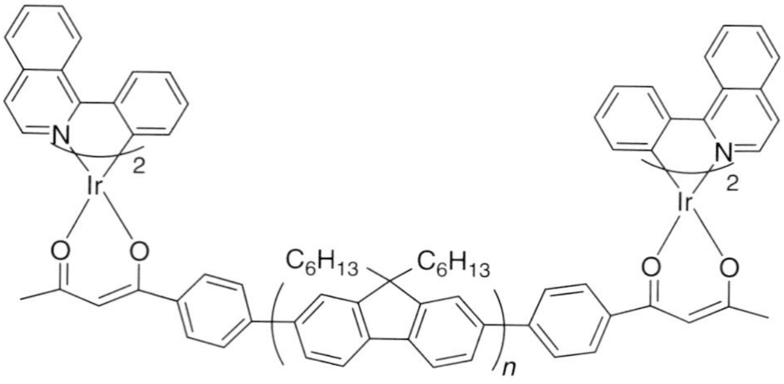

P135

Scheme 9.53 Polymers with Ir(III) complexes incorporated into the main chain through β-diketone ligands.

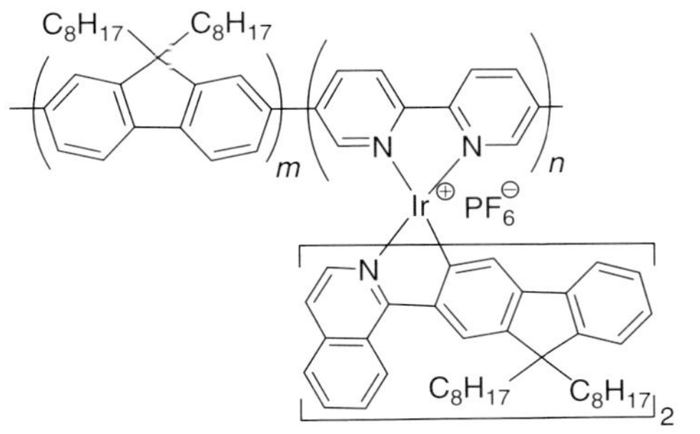

P136

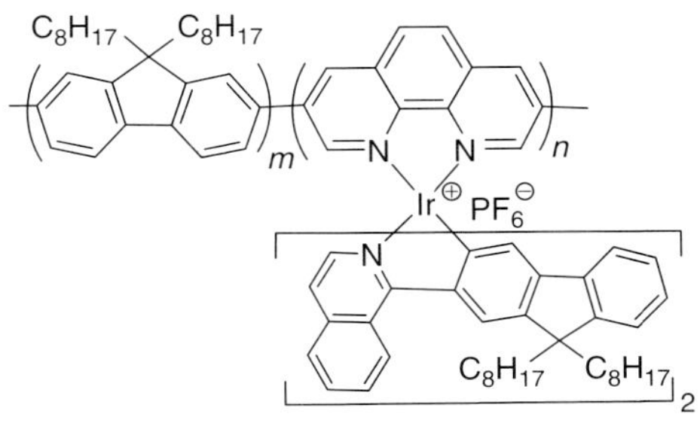

P137

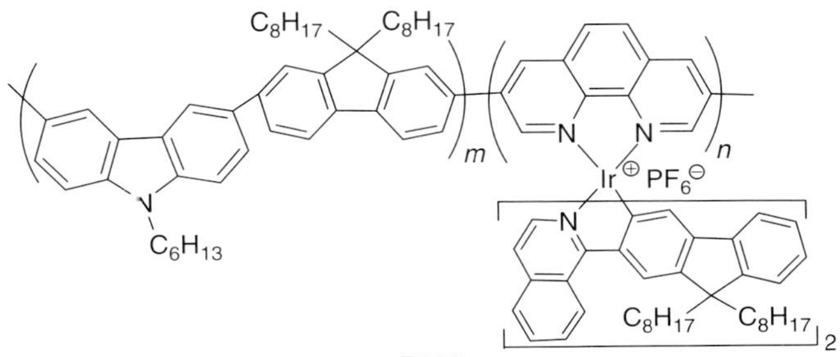

P138

Scheme 9.54 Polymers with charged Ir(III) complexes in the main chain.

9.6.2.2 Polyfluorene Copolymers with Eu(III) Complexes

Kang *et al.* [158–161] synthesized a series of conjugated copolymers (polymer **P139–P142** in Scheme 9.55) containing fluorene and Eu(III) complexes chelated to the main chains through a three-step process, involving Suzuki coupling copolymerization, hydrolysis of benzoate units, and postchelation. In the copolymer films casting from solutions, emission from the fluorene moieties could be suppressed, and the absorbed excitation energy was transferred effectively to the Eu(III) complexes. Subsequently, they first demonstrated a conjugated copolymer of 9,9-dialkylfluorene and Eu-complexed benzoate for write-once read-many-times memory application in a sandwich structure of Al/polymer/ITO [162]. In these active polymers, the fluorene moiety served as the electron donor, while the europium complex served as the electron acceptor. An electrical bistability phenomenon was observed on this device: low-conductivity state for the as-fabricated device and high-conductivity state after device transition by applying a voltage.

Huang *et al.* synthesized a main-chain-type Eu(III)-containing **P143** with photo-cross-linkable group [163]. The effective inter- and/or intrachain Förster energy transfer occurs from the fluorene-excited state to the Eu(III) complex. The photochemical cross-linking of polymer by the addition of a photoinitiator yields an insoluble network. Primary emission was attributed to the intrachain interaction of fluorene–bipyridine polymer, and the characteristic emission of Eu^{3+} ion became much weaker than that before cross-linking.

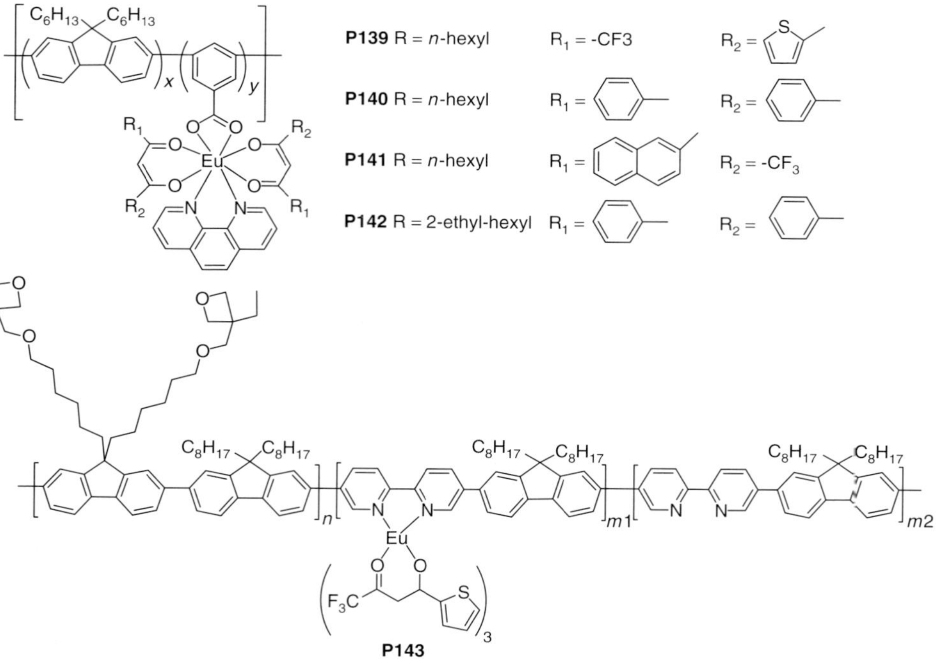

Scheme 9.55 Polyfluorene derivatives containing Eu(III) complexes.

9.6.2.3 Polyfluorene Copolymers with Pt(II), Re(I), Ru(II), and Os(II) Complexes

The Pt complex is one of the phosphorescent materials with high performance, which can be used in optoelectronic devices and sensors. **P145** containing Pt(II) complexes can be synthesized by direct metalation reaction of the metal-free polyfluorene-*co*-tetraphenylporphyrin copolymer **P144** with $PtCl_2$ (route 1 in Scheme 9.56), or by copolymerization of fluorene monomers and Pt complex monomer **61** containing active groups on ligands through the Suzuki coupling reaction (route 2 in Scheme 9.56) [164]. In comparison with the copolymers synthesized via a postpolymerization metalation route, copolymerization from Pt(II) complexes proved to be a more efficient synthetic route for high-efficiency electrophosphorescent polymers [165].

Owing to the longer lifetimes of triplet excitons in conjugated materials and the energy transfer process in the host–guest system, PFs containing phosphorescent metal complexes show increased sensitivity to trace analytes. Swager *et al.* [166] synthesized copolymer **P146** and **P147** (Scheme 9.57) containing fluorene and Pt(II) complex and sought to investigate the potential of phosphorescent conjugated polymers as chemosensing materials. The polymeric nature of this conjugated material gives a sensitivity improvement for dissolved oxygen quantification, demonstrating the potential use of phosphorescent conjugated polymers as chemosensing materials.

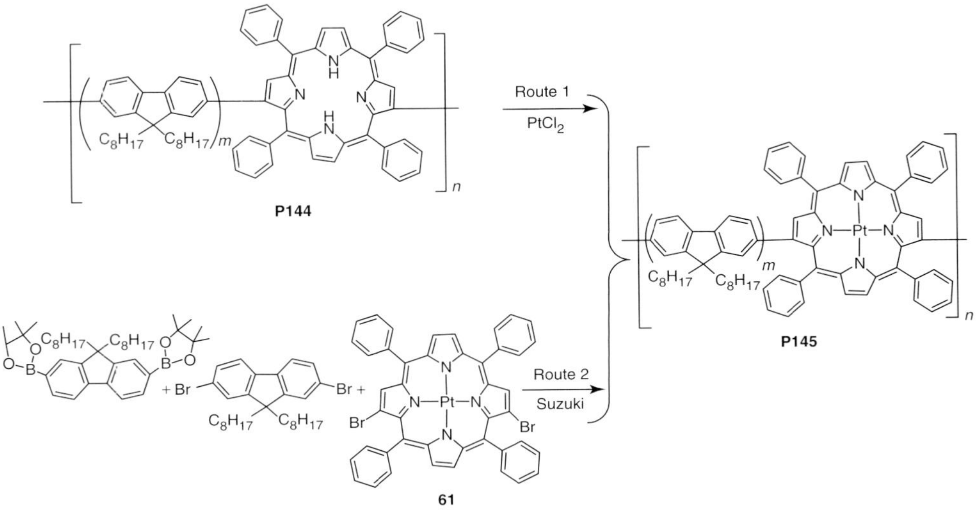

Scheme 9.56 Synthesis routes for **P145**.

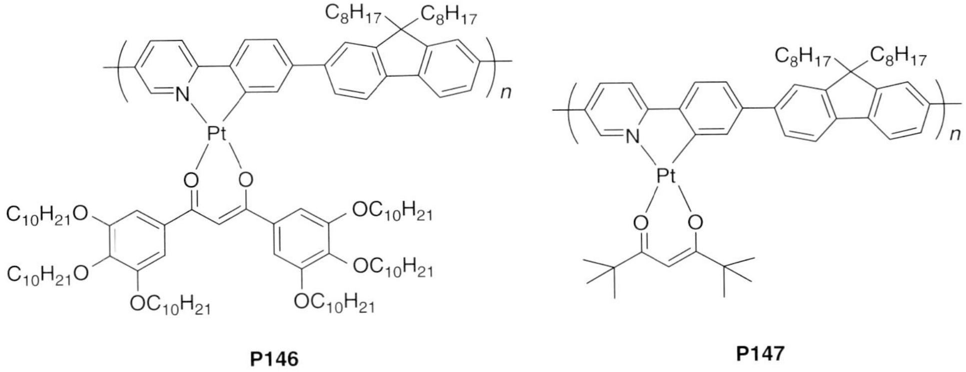

Scheme 9.57 Polymers with Pt(II) complexes for chemosensing.

PFs containing Re(I) or Ru(II) complexes on the chain also have been investigated. Using 2,7-(9,9′-dihexylfluorene) and bpy copolymers as the backbones, Ma *et al.* synthesized a series of π-conjugated copolymers (**P148** and **P149**) incorporating (bpy)Re$(CO)_3$Cl in the backbones of the copolymers through a highly active replacement reaction of bpy elements and CO in Re$(CO)_5$Cl complexes (Scheme 9.58) [167]. Cao *et al.* synthesized bipyridine-based aminoalkyl-PF **P150** containing Re(I) complex by postpolymerization method [168]. The polymer is insoluble in common organic solvents but have a good solubility in polar solvents, such as methanol and DMF. This kind of polymer can be used as the electron-transport layer in PLED.

PF with on-chain Ru(II) complex (polymer **P151**) was synthesized by Suzuki polycondensation (Scheme 9.59) [169]. The PL of the copolymer is slightly blueshifted

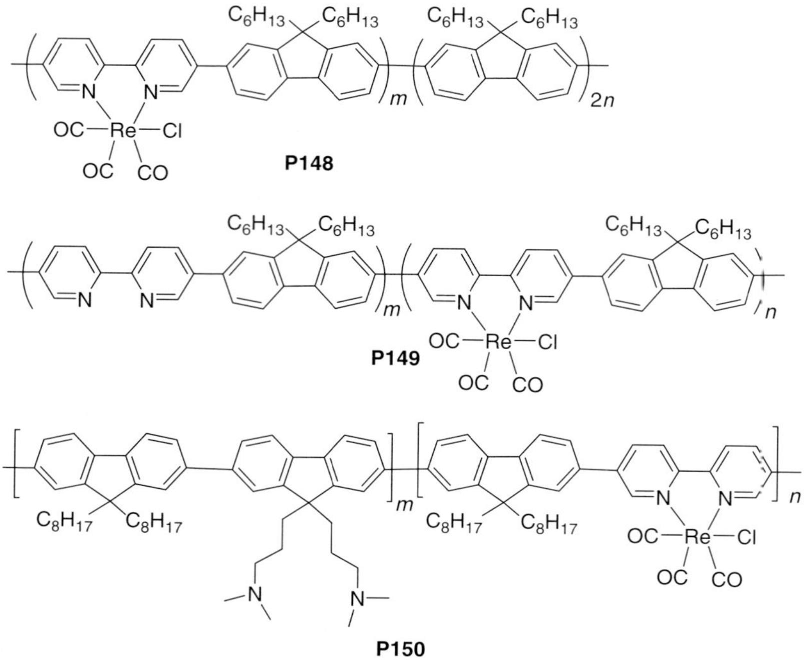

P148

P149

P150

Scheme 9.58 Polymers containing Re(I) complexes.

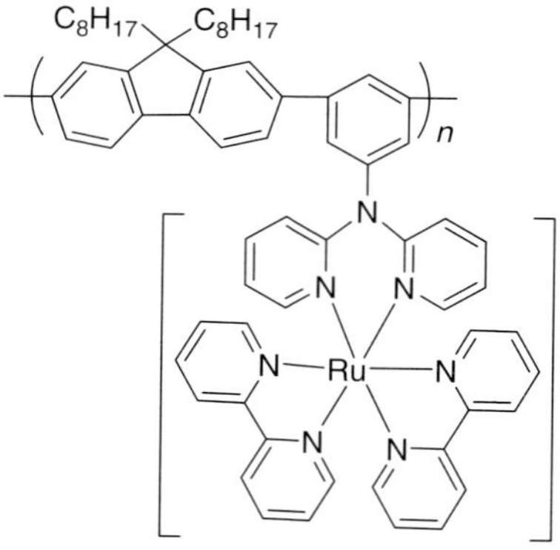

P151

Scheme 9.59 Polyfluorene derivative with Ru(II) complexes.

with increase in the concentration of dipyridylamine. The introduction of dipyridylamine and the Ru(II) complex into the polymer significantly improves the PL efficiency.

Shu, Chi and Chou *et al.* introduced red-emitting Os(II) complex into PF copolymer containing both hole- and electron-transporting pendant groups and prepared

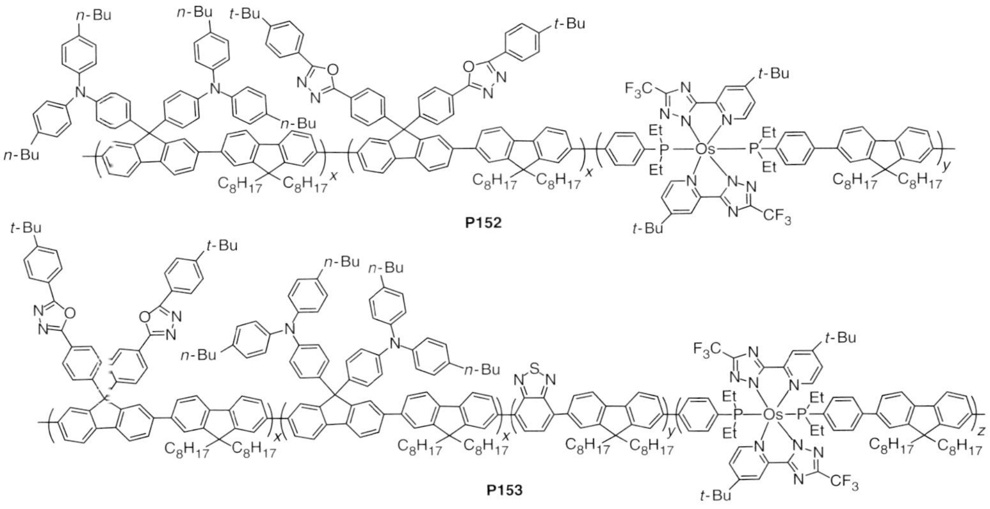

P152

P153

Scheme 9.60 Chemical structures of polymers containing Os(II) complexes.

red-emitting **P152** (Scheme 9.60) [170]. Highly efficient red-emitting PLED based on this polymer that exhibits an external quantum efficiency of 18.0% was realized. Furthermore, they introduced appropriate amounts of green-emitting BT unit into backbone of **P152** and realized efficient white-light EL polymer **P153** that displays simultaneous blue, green, and red emissions.

9.7 Polyfluorenes with Rod–Coil Structure

Rod–coil block copolymers consisting of conjugated segments connected with flexible segments have been the subject of extensive studies because of their unusual morphologies and photophysical behavior in both solution and solid state [171–176]. For this kind of block copolymers, a combination of complementary virtues from both rod and coil blocks can be envisioned. The approach suppresses the unfavorable macrophase separation encountered in many polymer blends, at the same time, presenting unique microphase separation with the formation of highly ordered geometries or patterns at nanoscales, which may lead to additional electronic processes such as exciton confinement and interfacial effects [177, 178]. Therefore, preparation of rod–coil copolymers opens a new channel to manipulate the optoelectronic properties of conjugated materials.

The synthesis of PF copolymers with rod–coil structures includes two steps. First, PF macroinitiators are prepared. Next, atom transfer radical polymerization (ATRP), which has been proven efficient to prepare well-defined PF-based block copolymers, is adopted from the resulting macroinitiators, following which, the fluorene-based rod–coil copolymers can be obtained.

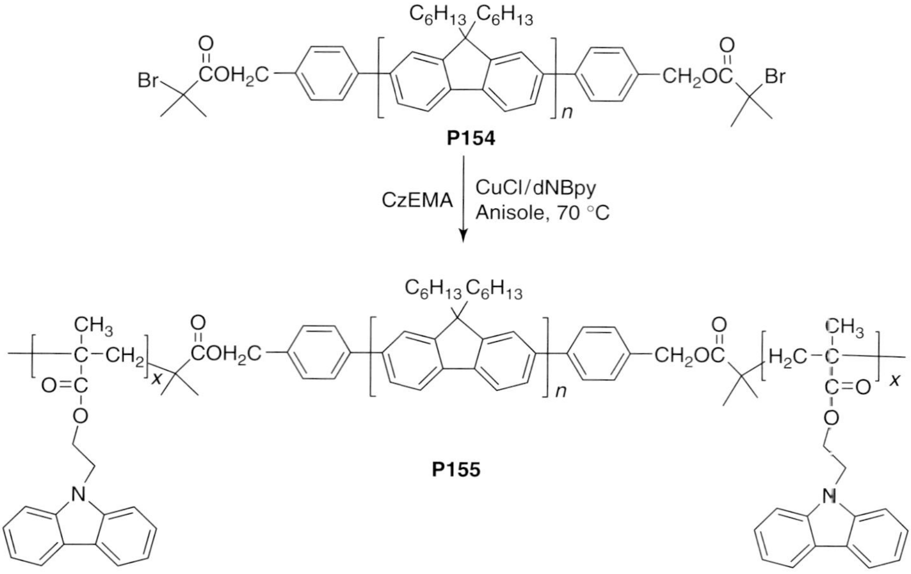

Scheme 9.61 Synthesis of rod–coil polymer **P155**.

Using this method, Huang *et al.* synthesized a series of PF-based triblock copolymers **P155** containing poly(2-(9-carbazolyl)ethyl methacrylate) (PCzEMA) as the flexible block (Scheme 9.61) [179]. The PDI of the block copolymers has narrowed to less than 1.4 as compared with those of the macroinitiators **P154**, indicating the formation of the well-defined block structures. Differential scanning calorimetry (DSC) data show that no crystallization and melting peaks are detected for the samples, indicating that these materials are amorphous. Förster energy transfer from PCzEMA to PF in such architecture is observed and intramolecular energy transfer dominates in solution while intermolecular energy transfer plays a more important role in solid state. The presence of the Cz effectively raises the HOMO level with respect to the PF homopolymer, suggesting better hole-injection properties. Preliminary LED experiments with these polymers indicate enhanced device performance compared to the PF homopolymer. The results imply that rod–coil block PFs with optoelectronic functionality are promising candidates for blue LED applications in terms of the suppression of green emission.

Functional groups can be expediently introduced into the flexible segments of the rod–coil structure. The triblock copolymer **P157** can be synthesized by ATRP with **P156** as the macroinitiator (Scheme 9.62). Then it can be converted to amphiphilic triblock copolymer **P158** by vacuum thermolysis of **P157** [180]. **P158** is soluble in THF, DMSO, and even MeOH with the formation of clear solutions, but cannot be dissolved in $CHCl_3$ and water. The PF segments within the block copolymer are found to form aggregates in water as revealed by electronic spectroscopy and ^1H NMR studies, leading to a blueshift of the absorption maxima, a bathochromic shift of the emission maxima, and a decrease of the quantum efficiency.

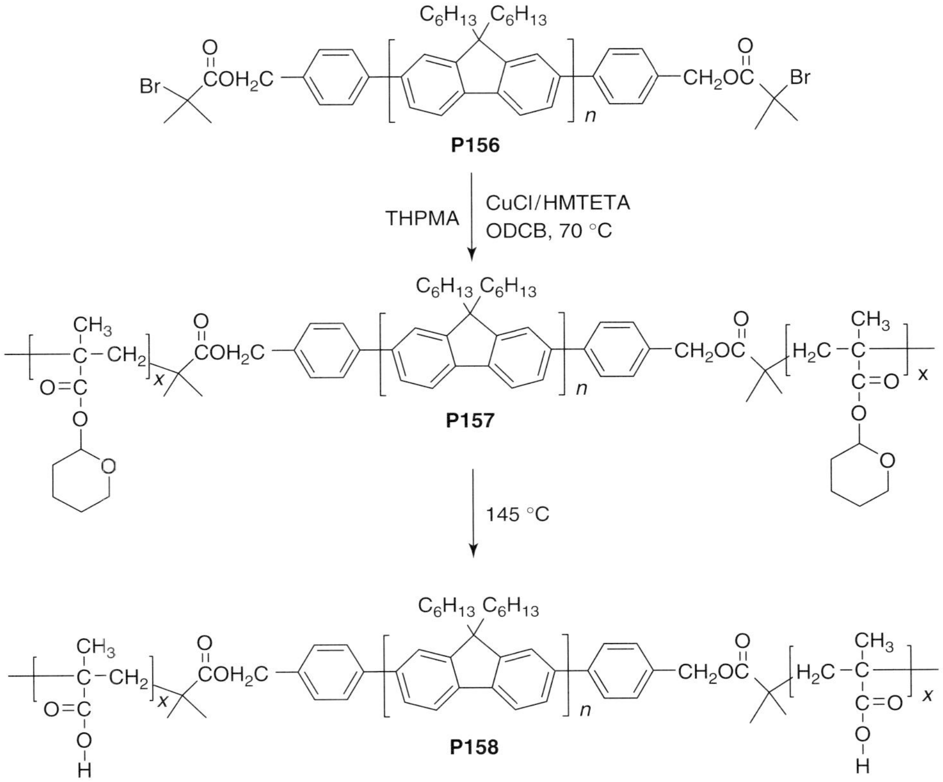

P156

THPMA, CuCl/HMTETA, ODCB, 70 °C

P157

145 °C

P158

Scheme 9.62 Synthesis of amphiphilic triblock copolymer **P158**.

Huang *et al.* chose poly(2-(dimethylamino)ethyl methacrylate) (PDMAEMA) as telechelic coils and synthesized amphiphilic block copolymer **P159** (Scheme 9.63) [181]. The resulting copolymer could be facilely quaternized to produce conjugated-ionic block copolymer **P160** with strongly enhanced hydrophilicity [182]. This kind of rod–coil copolymers show a significant variation in their surface structures and photophysical properties with respect to solvent composition (e.g., water–THF), temperature, and pH, suggesting that they have potential applications as a multifunctional sensory materials toward solvent, temperature, and pH [183]. **P159** can also act as the multidentate ligand of CdSe nanocrystals to form nanocomplexes [184]. Therefore, the interaction between **P159** and CdTe quantum dots (QDs) in aqueous solution was investigated [185]. Efficient fluorescence resonance energy transfer (FRET) occurred from cationic polymer serving as energy donor to anionic charged CdTe as energy acceptor due to the electrostatic attraction between them. A more effective FRET process occurs in the larger-sized QDs and polymer-blended systems. This strategy opens up a broad prospect for developing hybrid materials of semiconductor polymers and inorganic nanocrystals.

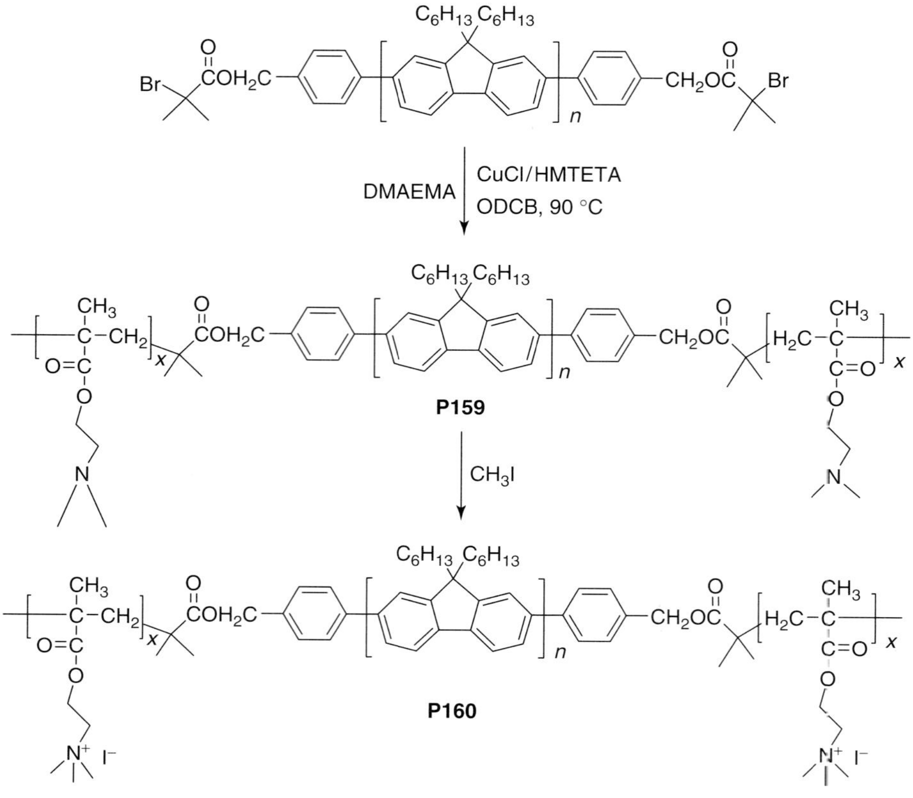

Scheme 9.63 Synthesis of conjugated-ionic block copolymer **P160**.

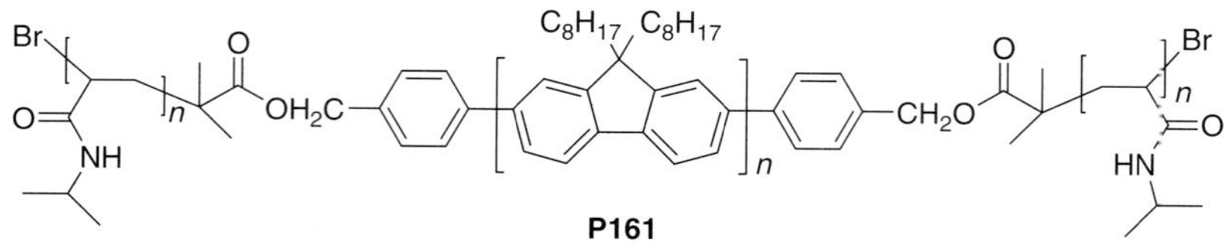

Scheme 9.64 Rod–coil copolymer **P161** based on fluorene and NIPAAm.

Poly(*N*-isopropylacrylamide) (PNIPAAm) is a kind of environmental stimuli-responsive materials, which can be introduced into PFs by ATRP to obtain **P161** (Scheme 9.64) [186]. This polymer combines the optical properties of PF with the stimuli-responsive capabilities of PNIPAAm, resulting in a thermally responsive fluorescent polymer. The light-scattering studies of **P161** in water indicate that the rod–coil copolymer forms a nanoscale core-shell structure in cold water and changes in conformation and solubility at temperatures above its lower critical solution temperature. All the measurements demonstrate that,

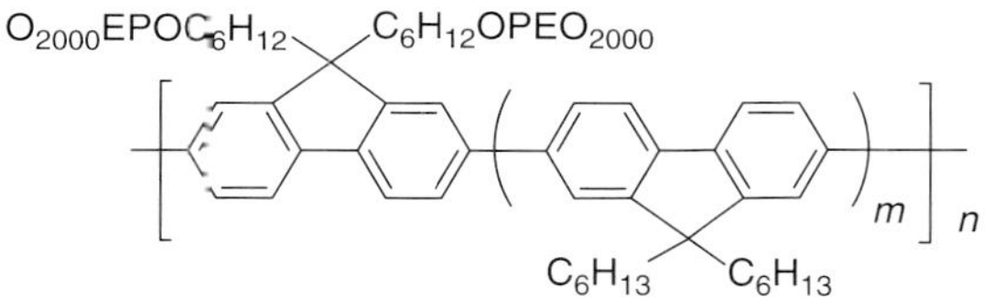

$O_{2000}EPOC_6H_{12}$ $C_6H_{12}OPEO_{2000}$ C_6H_{13} C_6H_{13} m n

P162

Scheme 9.65 Chemical structure of rod-coil polymer **P162** with coil segment in the side chain.

in the region of 25–45 °C, the polymer undergoes a phase transition and the corresponding change in optical properties in water solution. However, **P161** does not show completely reversible behavior upon heating and cooling.

The coil segments can also be introduced into the side chain of PFs. Chen, Li and Loh *et al.* synthesized graft copolymer **P162** (Scheme 9.65) [187]. Poly(ethylene oxide) (PEO) with molecular weight of 2000 Da was selected for use as hydrophilic segments linked to PF side chains. Stable and uniform fluorescent micelles are formed from the fluorescent amphiphiles in aqueous solution with the compact fluorescent hydrophobic core and swollen PEO shell. The average diameter of micelles in aqueous solution is 85 nm. And the stable fluorescent micelle can find applications in biolabeling and drug delivery tracing.

Nomura *et al.* established another facile, efficient synthesis of ABA-type amphiphilic triblock copolymers **P166** in a precise manner by grafting poly(ethylene glycol) (PEG) into both ends of the poly(9,9-di-*n*-octylfluorene-2,7-vinylene) (**P163**) chain (Scheme 9.66) [188]. **P163** is prepared by acrylic diene metathesis polymerization using $Ru(CHPh)(Cl)_2(IMesH_2)(PCy_3)$ [$IMesH_2$ = 1,3-bis (2,4,6-trimethylphenyl)-2-imidazolidinylidene]. Treatment of the vinyl groups at the **P163** chain ends with Mo(1) (to generate Mo-alkylidene moieties) and the subsequent addition of $4\text{-}Me_3SiOC_6H_4CHO$ give **P164** containing Me_3SiO group at both ends of the polymer chain. The $SiMe_3$ group in the resultant **P164** is easily cleaved by treating with aqueous HCl to afford **P165** in high yields (78–99%). The OH groups in the chain ends are then treated with KH in THF, and the subsequent reaction with mesylated poly(ethylene glycol) ($PEGMs_2$) gave ABA-type amphiphilic triblock copolymer **P166** in rather high yields (69–90%). The resultant copolymers possess uniform molecular weight distributions.

9.8
Polyfluorene-Based Conjugated Polyelectrolytes

PF-based conjugated polyelectrolytes are a novel type of conjugated polymers that contain ionic side chains, hence are soluble in water and some other polar organic solvents. These polymers combine the optoelectronic advantages of traditional PFs and the ionic nature of polyelectrolytes. But compared with their neutral

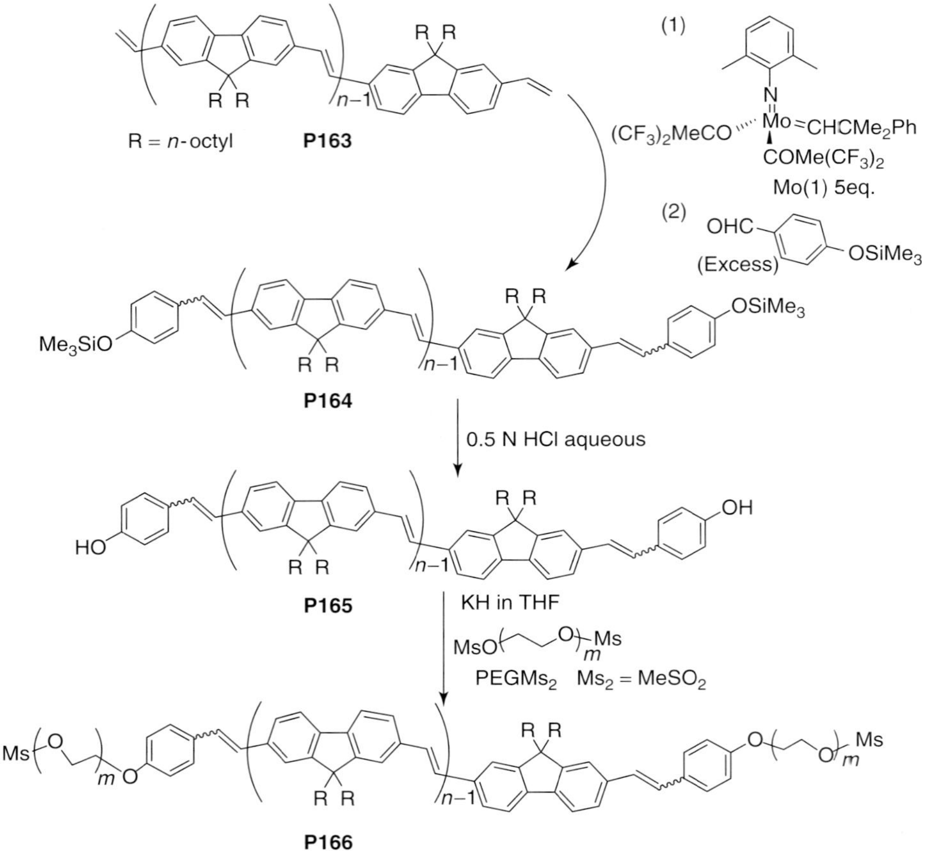

Scheme 9.66 Synthesis of rod-coil polymer **P166**.

precursors in organic solvents, water-soluble conjugated polymers (WSCPs) usually exhibit much lower PL efficiencies in aqueous solutions [189–192]. The water solubility of PF can be achieved by functionalizing the substituted side chains with ammonium (NR_3^+) to form cationic PF or with terminal carboxylate (CO_2^-), sulfonate (SO_3^-), and phosphonate (PO_3^{2-}) groups to form anionic PF. PF-based conjugated polyelectrolytes can be obtained by direct polymerization of quaternized monomer or by quaternization of their neutral precursors. Postquaternization approach is more popular because of the convenient characterization of the polymers and the controllable quaternization degree.

This research field is becoming more and more attractive owing to the unique optoelectronic properties of these polymers that may serve as the basis for a new generation of optoelectronic devices and bio/chemical sensors [193–197].

9.8.1 Cationic Polyfluorene-Based Polymers

Cationic fluorene-based conjugated polyelectrolytes are usually obtained by quaternization of the neutral counterparts. Huang *et al.* synthesized the first cationic fluorene-based WSCP **P168** through a facile postquaternization approach (Scheme 9.67) [198, 199]. The conversion of the neutral polymer **P167** to the final water-soluble polymer was achieved by treatment with bromoethane in DMSO and THF (1 : 4). **P168** emits bright blue fluorescence with Φ_f of 0.25 in aqueous solution. The water solubility of the resultant polymer can be tunable in the postquaternization approach by the controlled quaternization degree, however, better solubility in polar solvents is accompanied by a spectral blueshift for polymers with a higher quaternization degree.

The side chains of NR_3^+ can be easily introduced onto the C-9 positions of fluorene units, thus the water solubility of PFs is realized. Bazan *et al.* synthesized polymer **P170** by quarternization of **P169** with methyliodide in THF/DMF/water (Scheme 9.68) [200]. They investigated the energy transfer from **P170** and its oligomeric counterparts to 1,4-bis(4′ (2′′,4′′- bis(butoxysulfonate)-styryl)styryl)-2-(butoxysulfonate)-5-methoxybenzene and demonstrated that **P170** exhibited maximum sensitivity for the energy transfer compared to oligomeric counterparts.

For the applications in biosensors, the sensitivity of polyelectrolytes can be influenced by the polymer structure. **P172** has the anthracenyl substituent at

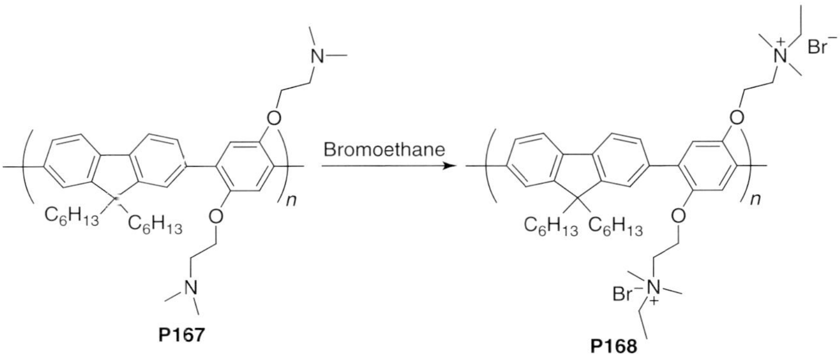

Scheme 9.67 Synthesis of cationic fluorene-based conjugated polymer **P168**.

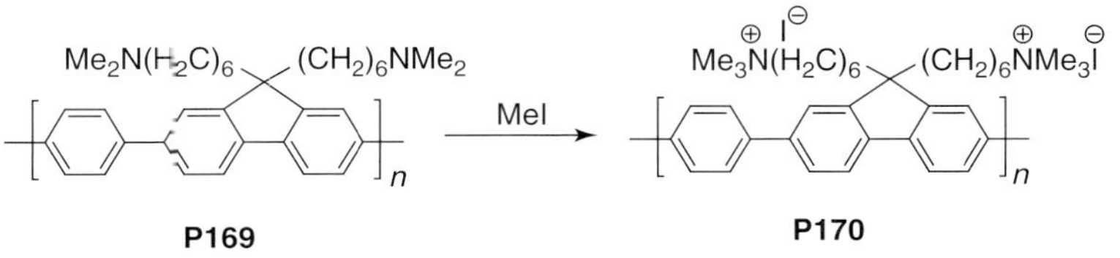

Scheme 9.68 Synthesis of cationic fluorene-based conjugated polymer **P170**.

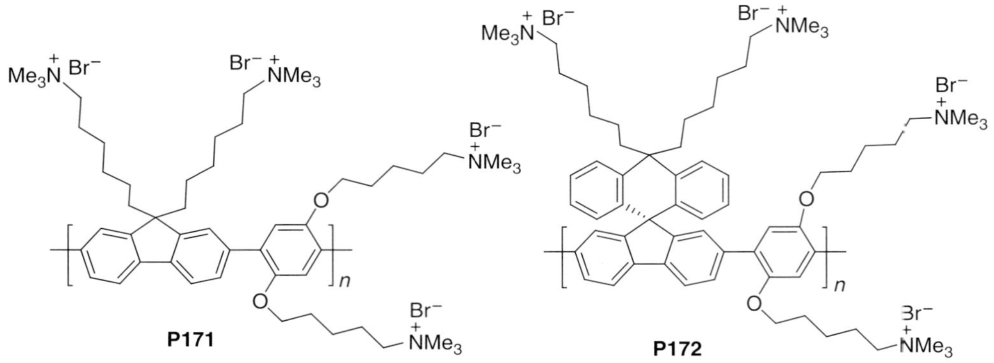

Scheme 9.69 Chemical structures of cationic polymers **P171** and **P172**.

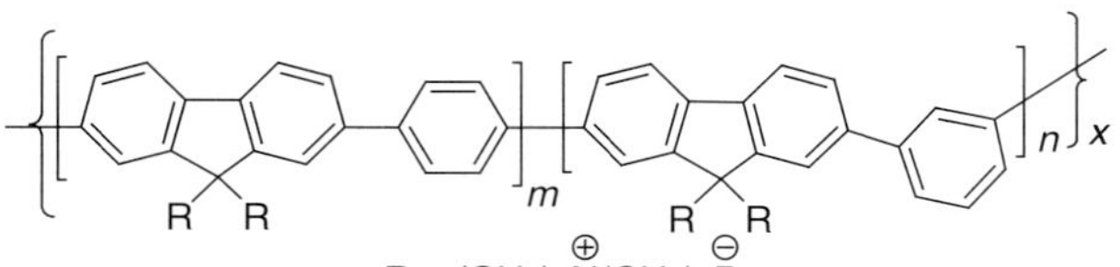

$R = (CH_2)_6\overset{\oplus}{N}(CH_3)_3\overset{\ominus}{Br}$

P173

Scheme 9.70 Chemical structures of cationic polymer with nonlinear "kinks."

the C-9 positions of the fluorene comonomer units, which is orthogonal to the backbone axis and serves to increase separation between chains in aggregated phases (Scheme 9.69). **P172** shows optical bandgaps and orbital energy levels similar to those of **P171** without anthracenyl substituent; however, these two polymers exhibit different phenomenon in DNA sensory method. For **P172**, it is possible to observe emission from ssDNA-fluorescein by FRET with an FRET efficiency of approximately 60% in ssDNA-fluorescein/**P172** complex. However fluorescein is not emissive within the ssDNA-fluorescein/**P171** complex owing to photoinduced charge transfer (PCT) quenching. The authors propose that the presence of the "molecular bumper" increases fluorescein emission by increasing the donor–acceptor distance, which decreases PCT quenching more acutely relative to FRET [201].

To improve conformational freedom and registry with analyte shape, nonlinear "kinks" are introduced along a linear polymer structure (**P173**) (Scheme 9.70) [202]. The ratio of *p*- and *m*-phenyl linkages was tuned at the synthesis stage. There is a progressive blueshift in absorption with increasing meta content for **P173**, consistent with the more effective electronic delocalization across para linkages. Increasing the para content past the 50 : 50 ratio does not perturb the emission maxima. These polymers conform more tightly to the secondary structure of dsDNA, resulting in improved FRET efficiencies.

Because of the high delocalization of single exciton and rapid energy migration along the conjugated backbone, poly(*p*-phenyleneethynylene)s have attracted more attention in chemo/bioanalysis [203, 204]. Although water-soluble PPEs are suitable for studying structure–property relationships, their PL quantum yields (Φ_f) are generally very low in aqueous solution (Φ_f usually hovers about 5%), which is highly disfavored in their applications [205]. Introduction of a strong fluorophore, such as fluorene into polymer backbone tends to increase PL quantum yields of WSCPs [198]. On the basis of these studies, Huang *et al.* [206, 207] synthesized blue-light-emitting fluorene-containing poly(arylene ethynylene)s **P174–P176** with amino-functionalized side groups through the Sonogashira reaction (Scheme 9.71). Through the postquaternization treatment of **P174–P176** with methyl iodide, they obtained cationic water-soluble conjugated polyelectrolytes **P177–P179**. They successfully modulated their water solubility by adjusting the content of hydrophilic side chains and obtained high Φ_f values for some of them by controlling the degrees

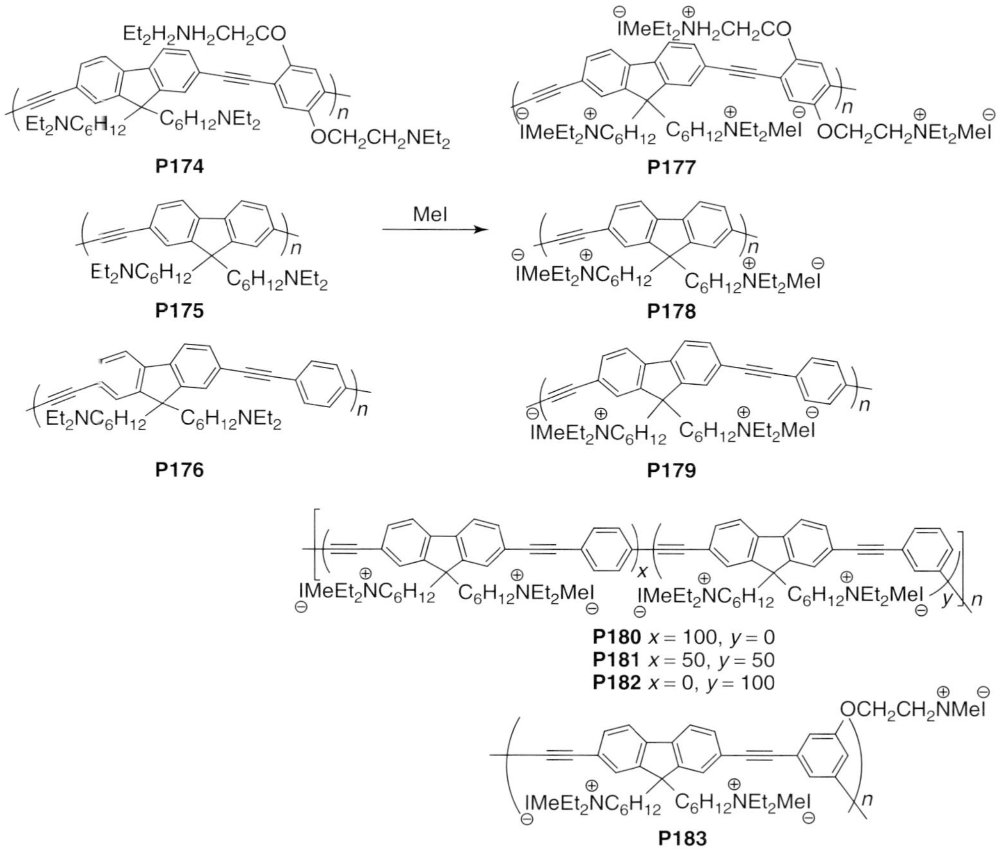

Scheme 9.71 Polyelectrolytes based on fluorene-containing poly(arylene ethynylene)s.

of aggregation in aqueous solutions. The water solubility was gradually improved from **P179** to **P177** with increasing contents of hydrophilic side chains. Accordingly, from **P177** to **P179**, there was a gradually enhanced degree of aggregation in aqueous solutions. The Φ_f values of **P177–P179** are 0.26, 0.22, and 0.08, respectively. Thus, **P179**, with the greatest degree of aggregation in H_2O, exhibited the lowest Φ_f value.

Most WSCPs reported in the study of biological detections are composed of para-linked and rigid backbone structures [208, 209]. It was proposed that some WSCPs with more meta content showed more efficient FRET to dsDNA because they had more conformational freedom and adapted better to the secondary structures of biological macromolecules [202]. Using the synthetic route similar to the one for polymers **P177–P179**, para-linked or meta-linked, cationic, water-soluble polymers **P180–P183** were synthesized (Scheme 9.71) and the effects of conformational changes on biological detection were explored [210]. In pure water, with gradually increasing *meta*-phenylene content (0, 50, and 100%), they undergo a gradual transition process of conformation from disordered aggregate structure to helix structure, which is not compactly folded. Moreover, the polymer with an ammonium-functionalized side chain on the *meta*-phenylene unit appears to adopt a more incompact or extended helix conformation than the corresponding one without this side chain. Furthermore, the conformational changes of these cationic polymers in H_2O are used to study their effects on biological detection. Rubredoxin (Rd), a type of anionic iron–sulfur-based electron transfer protein, was chosen to act as biological analyte in the fluorescence-quenching experiments of these polymers. All polymers exhibit amplified fluorescence quenching, and the polymer with more features of helix conformation tends to be quenched by Rd more efficiently.

Efforts have also been made to tune the conjugated polymer emission through energy transfer between different segments in conjugated polymers. For example, PF-based conjugated polyelectrolytes with BT chromophore (**P184** and **P185** in Scheme 9.72) have been synthesized and efficient energy transfer from the phenylene–fluorene segments (blue-emitting regions) to BT-centered sites is realized at low BT chromophore content [211]. In dilute solution, the emission of polymer **P184** and **P185** containing low BT chromophore content is blue. Upon addition of ssDNA or dsDNA, however, the emission shifts to a green color characteristic of the BT chromophore. Electrostatic complexation with negatively charged DNA brings together polymer segments and results in a reduction of the average intersegment distance and an improved FRET to the green-emitting sites. This kind of polyelectrolytes can efficiently enhance sensibility by the change of emission intensity of blue (phenylene–fluorene segments) and green (BT-centered sites) emissions [212].

Liu *et al.* synthesized a series of cationic porphyrin-containing conjugated polyfluoreneethynylenes (**P186–P189** in Scheme 9.73) for naked-eye detection and quantification of mercury(II) through a new synthetic approach by Sonogashira coupling polymerization between 2,7-dibromo-9,9′-bis(6-bromohexyl)fluorene and Zn(II) coordinated porphyrin, which was followed by demetalation [213]. A small

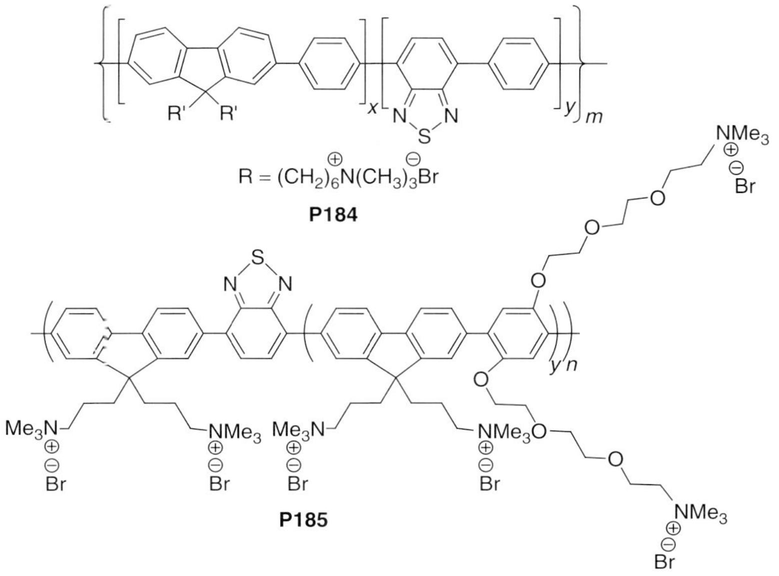

$R = (CH_2)_6N^{\oplus}(CH_3)_3Br^{\ominus}$

P184

P185

Scheme 9.72 Fluorene-based polyelectrolytes containing BT chromophore.

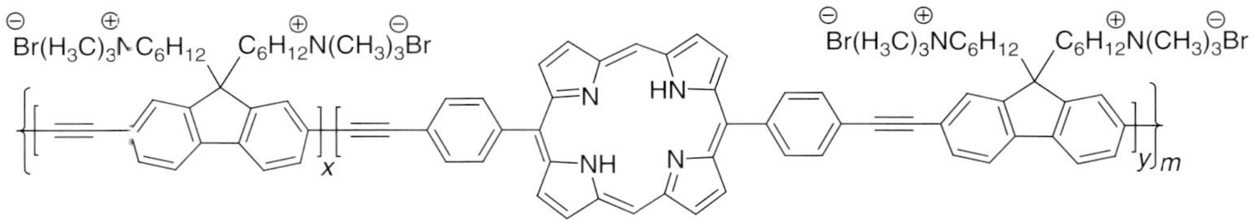

P186 $x = 100\%$, $y = 0\%$
P187 $x = 98\%$, $y = 1\%$
P188 $x = 90\%$, $y = 5\%$
P189 $x = 80\%$, $y = 10\%$

Scheme 9.73 Cationic porphyrin-containing polyfluoreneethynylenes.

fraction of porphyrin units act as both an energy acceptor and a metal-ion recognition element. The polymer offers dual-emissive polyelectrolytes with both blue and red emission bands resulting from incomplete intramolecular energy transfer from the fluoreneethynylene segments to the porphyrin units. The addition of mercury(II) could quench the emission from porphyrin units more efficiently than that from fluoreneethynylene segments, which changes the fluorescent color of the solution.

Cationic PFs can also be used as electron-transport/injection layers (ETLs) in PLEDs [214–218]. Their charged nature allows one to circumvent often quoted guidelines for optimizing charge injection into organic semiconductors that rely on matching the semiconductor LUMO with the electrode work function [219]. Gong and Bazan *et al.* synthesized polymer **P190** comprising alternating fluorene and

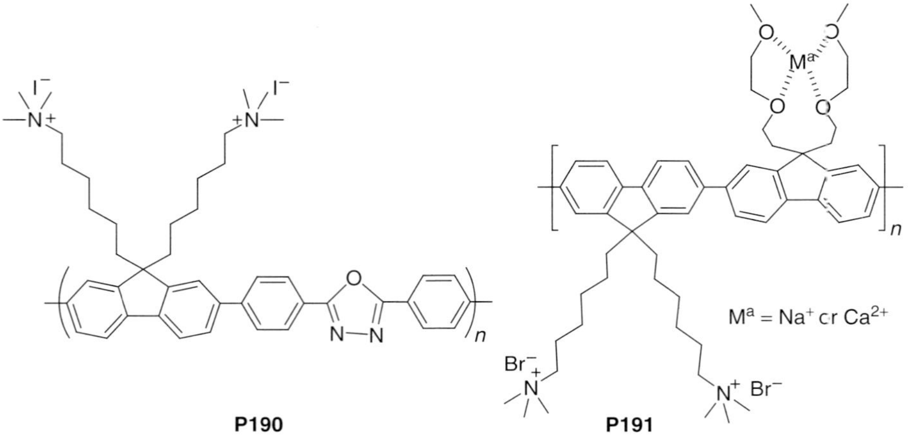

Scheme 9.74 Chemical structures of polyelectrolytes **P190** and **P191**.

phenylene-oxadiazole-phenylene in the main chain (Scheme 9.74) [220]. Multilayer PLEDs are fabricated using a semiconducting polymer (red-, green-, or blue-light emitting) as an emissive layer and **P190** as an electron-transporting layer. The results demonstrated that devices with polymer **P190** as ETL have significantly lower turn-on voltages, higher brightness, and improved luminous efficiency than those without. Kim *et al.* synthesized a novel water-soluble **P191** containing mobile alkali or alkaline earth metal ions bound to ethylene oxide groups (Scheme 9.74) and found it can be used as an efficient electron-injecting layer for high EL efficiency in PLEDs with a high-work-function Al cathode [221]. Facilitated migration of metal ions by ion-transporting ethylene oxide groups dramatically improved the device efficiency by effectively reducing the electron-injection barrier.

Bo *et al.* synthesized a series of water-soluble dendronized poly(fluorene-phenylene)s (**P192–P194**) carrying peripheral charged amino groups (Scheme 9.75) [222]. The introduction of lateral dendrons should not only provide the conjugated polymers with good solubility in water through the charged amino groups but also reduce the aggregation of polymers through the "site-isolation" effect. In solution, all these polymers exhibited bright blue emission. Furthermore, PL quantum efficiencies in water is higher for the polymer with the larger size of the attached dendrons. The second-generation dendronized PFs exhibited a quite high quantum yield ($\Phi_f = 94\%$) in water.

9.8.2
Anionic Polyfluorene-Based Polymers

Burrows and Tapia *et al.* synthesized sulfonato-functionalized polymer **P195** with side chains containing SO_3^- (Scheme 9.76) [223]. Aqueous solutions of **P195** showed a broad absorption around 381 nm and a structured fluorescence (λ_{max} 424, 448, 475 nm (shoulder)). Upon addition of nonionic

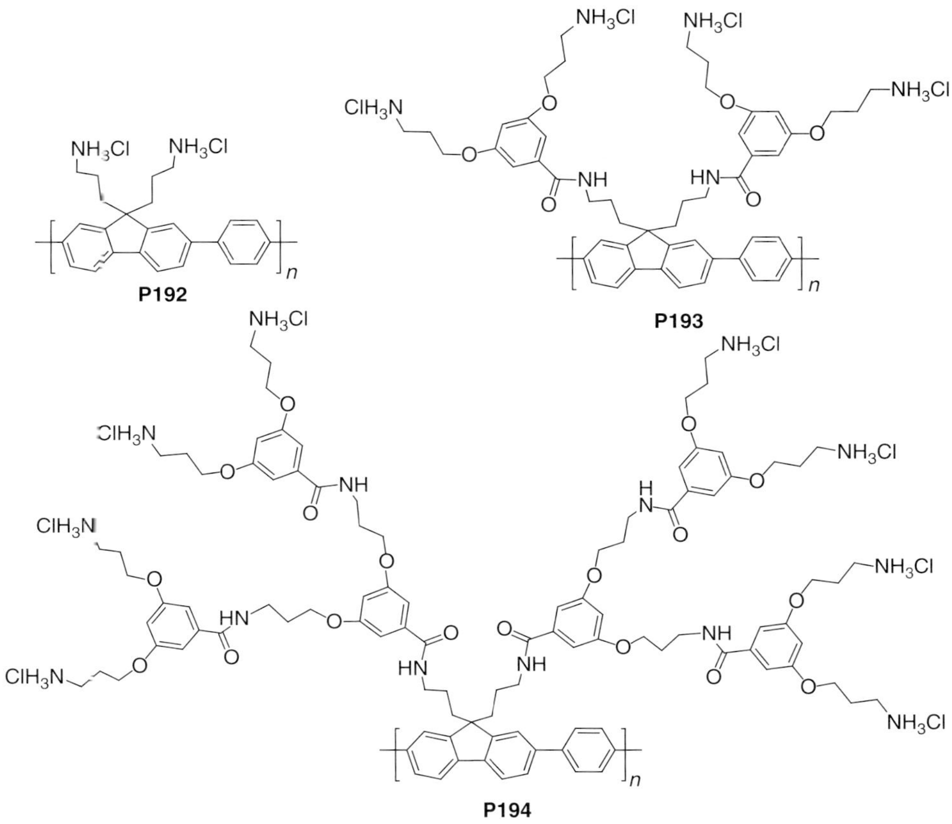

Scheme 9.75 Chemical structures of water-soluble dendronized poly(fluorene-phenylene)s.

surfactant *n*-dodecylpentaoxyethylene glycol ether to aqueous solutions of polymer **P195** results in blueshifts in absorption (11 nm) and emission (13 nm) maxima, and marked increases in absorbance and fluorescence intensity. Wang *et al.* found that **P195** can form a complex with cationic quencher 4-(trimethylammonium)-2,2,6,6-tetramethylpiperidine-1-oxyl iodide (CAT1) through electrostatic interactions [224]. The fluorescence of **P195** is efficiently quenched by CAT1 with a Stern–Volmer constant (K_{SV}) of 2.3×10^7 M^{-1}. Hydrogen abstraction or reduction can transform the paramagnetic nitroxide radical into diamagnetic hydroxylamine, which inhibits the quenching; thereby, the fluorescence of **P195** is recovered, which can be used to probe the processes of hydrogen transfer reaction from antioxidants to radicals and the reduction reaction of radicals by antioxidants.

An anionic PF derivative containing BT units (**P196** in Scheme 9.77) is reported and intramolecular energy transfer from the fluorene units to the BT sites is demonstrated when oppositely charged substrates adenosine triphosphate (ATP)

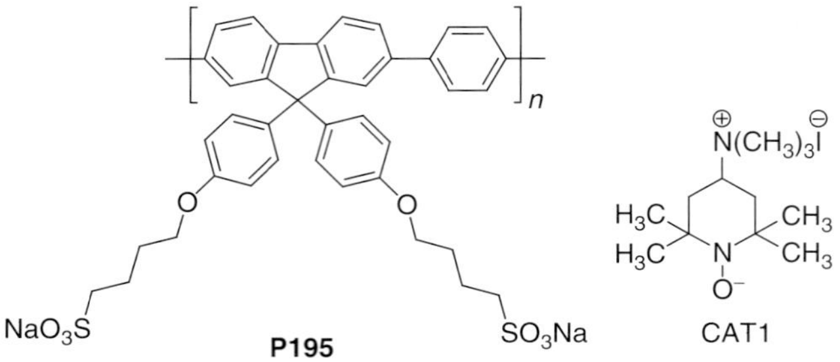

Scheme 9.76 Chemical structures of **P195** and CAT1.

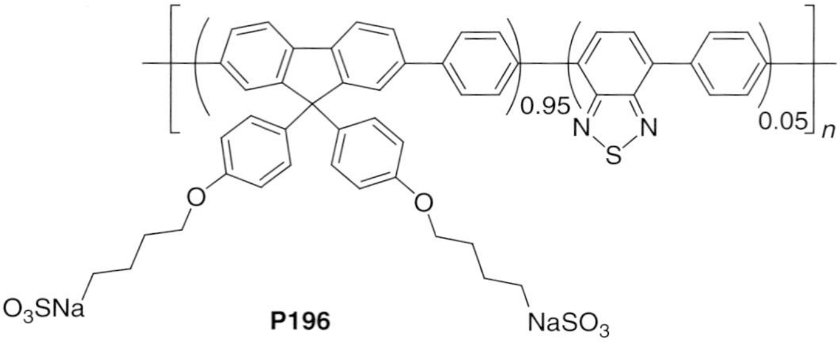

Scheme 9.77 Anionic polyfluorene derivative with BT units.

or polyarginine peptide (Arg6) are added, following by a shift in emission color from blue to green [225]. When the substrates are cleaved into fragments, the relatively weak electrostatic interaction of substrate fragments with conjugated polymers keep their main chains separated, FRET from the fluorene units to the BT sites is inefficient, and the conjugated polymers emit blue fluorescence. By triggering the shift in emission color and the change of emission intensity of conjugated polymers, it is possible to assay the enzyme activity and screen drugs based on the inhibition of enzymes.

By introducing receptors, anionic PFs can realize convenient detection of target molecules or ions in aqueous solution. For example, **P197** containing 2′2-bipyridine units in the main chain as receptors for transition-metal ions are synthesized by Sonogashira reaction with sulfonato-containing fluorene monomer and 2,2-bipyridyl monomer (Scheme 9.78) [226]. Its fluorescence in aqueous solution could be completely quenched upon addition of transition-metal ions. Convenient detection of transition-metal ions in water was thus realized.

Liu *et al.* also synthesized **P201–P203** with terminal carboxylate (CO_2^-) groups through the Suzuki coupling between carboxylate-functionalized fluorene monomers and BT to afford the neutral polymers (**P198–P200**), which was followed by treatment in trifluoroacetic acid to afford the water-soluble polymers of **P201–P203** (Scheme 9.79) [227]. Both the optical spectra and the light-scattering studies show that the polymers are aggregated in water at low pH and the aggregation decreases at high pH. Along with the deprotonation process, the

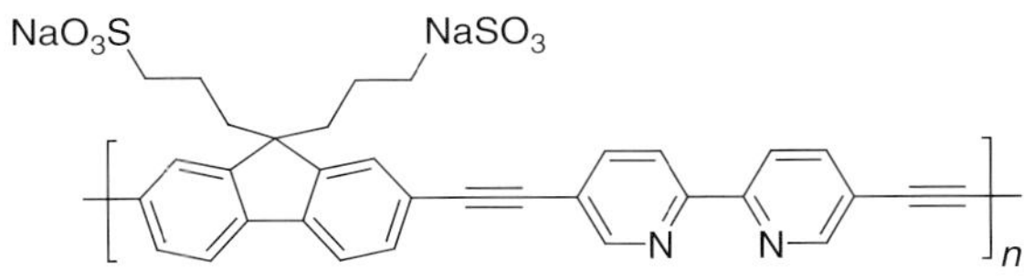

Scheme 9.78 Anionic polyfluorene derivative containing 2′2-bipyridine units.

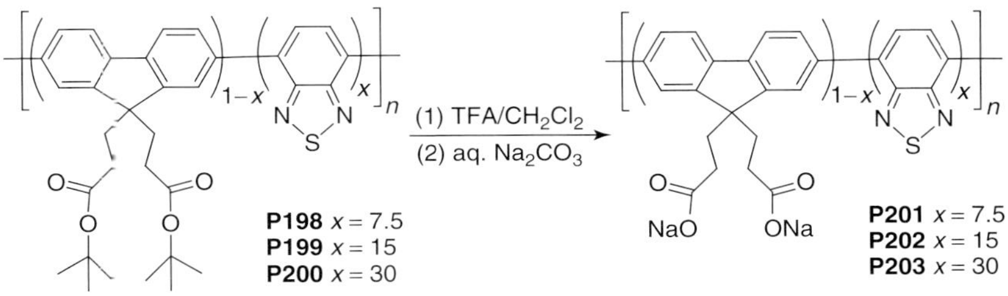

Scheme 9.79 Synthesis of anionic polyfluorene derivatives with terminal carboxylate groups.

polymer emission changed from yellow (pH < 6) to blue (pH > 9). The polymers can be used as multicolor sensors for different proteins, due to the difference in hydrophobic nature, net charge, and the structure among proteins.

9.9 Concluding Remarks

PFs and their derivatives are a family of important optoelectronic materials because of their excellent photonic and electronic properties and have been applied in many fields. This chapter has highlighted recent progress in this area and we hope it will help readers learn more about this fascinating family of materials.

Abbreviations

ATRP	atom transfer radical polymerization
BT	2,1,3-benzothiadiazole
CIE	Commission Internationale de l'Eclairage
Cz	carbazole
DCM	4-dicyanomethylene-6-methyl-4H-pyran
DSC	differential scanning calorimetry
E_g	optical bandgap

EL	electroluminescence
ETLs	electron-transport/injection layers
FETs	field-effect transistors
FRET	fluorescence resonance energy transfer
fwhm	full width at half-maximum
HH	head-to-head
HOMO	highest occupied molecular orbital
ICT	intramolecular charge transfer
K_{sv}	Stern-Volmer constant
LE	local exciton
LUMO	lowest unoccupied molecular orbital
MAOS	microwave-assisted organic synthesis
M_n	number average molecular weights
MW	microwave
NTSC	National Television System Committee
PCs	photovoltaic cells
PCT	photoinduced charge transfer
PDI	polydispersity index
PFs	polyfluorenes
PF8	poly(9,9-di-*n*-octylfluorene)
PL	photoluminescence
PLEDs	polymer light-emitting devices
POSS	polyhedral oligomeric silsesquinoxane
PPEs	poly(phenylene ethynylene)s
PPPs	poly(*para*-phenylene)s
PPV	poly(phenylvinylene)
PT	polythiophene
T_d	thermal decomposition temperature
T_g	glass transition temperature
TGA	thermogravimetric analysis
TT	tail-to-tail
WSCPs	water-soluble conjugated polymers
Φ_f	photoluminescence quantum efficiency

References

1. Burroughes, J.H., Bradley, D.D.C., Brown, A.R., Marks, R.N., Mackay, K., Friend, R.H., Burns, P.L., and Holmes, A.B. (1990) *Nature*, **347**, 539.
2. Akcelrud, L. (2003) *Prog. Polym. Sci.*, **28**, 875.
3. Choi, M.C., Kim, Y., and Ha, C.S. (2008) *Prog. Polym. Sci.*, **33**, 581.
4. Grimsdale, A.C., Chan, K.L., Martin, R.E., Jokisz, P.G., and Holmes, A.B. (2009) *Chem. Rev.*, **109**, 897.
5. Ong, B.S., Wu, Y., Li, Y., Liu, P., and Pan, H. (2008) *Chem. Eur. J.*, **14**, 4766.
6. Ling, M.M. and Bao, Z.N. (2004) *Chem. Mater.*, **16**, 4824.
7. Cicoira, F. and Santato, C. (2007) *Adv. Funct. Mater.*, **17**, 3421.

8. Morana, M., Wegscheider, M., Bonanni, A., Kopidakis, N., Shaheen, S., Scharber, M., Zhu, Z., Waller, D., Gauciana, R., and Brabec, C. (2008) *Adv. Funct. Mater.*, **18**, 1757.
9. Yip, H.L., Hau, S.K., Baek, N.S., Ma, H., and Jen, A.K.Y. (2008) *Adv. Mater.*, **20**, 2376.
10. Leger, J.M., Patel, D.G., Rodovsky, D.B., and Bartholomew, G.P. (2008) *Adv. Funct. Mater.*, **18**, 1212.
11. Soci, C., Hwang, I.W., Moses, D., Zhu, Z., Waller, D., Gaudiana, R., Brabec, C.J., and Heeger, A.J. (2007) *Adv. Funct. Mater.*, **17**, 632.
12. Blouin, N., Michaud, A., and Leclerc, M. 2007) *Adv. Mater.*, **19**, 2295.
13. Ling, Q.D., Liaw, D.J., Zhu, C., Chan, D.S.H., Kang, E.T., and Neoh, K.G. (2008) *Prog. Polym. Sci.*, **33**, 917.
14. Kim, D.Y., Cho, H.N., and Kim, C.Y. (2000) *Prog. Polym. Sci.*, **25**, 1089.
15. Blouin, N. and Leclerc, M. (2008) *Acc. Chem. Res.*, **41**, 1110.
16. Osaka, I. and Mccullough, R.D. (2008) *Acc. Chem. Res.*, **41**, 1202.
17. Liu, C.Y. and Chen, S.A. (2007) *Macromol. Rapid Commun.*, **28**, 1743.
18. Kappaun, S., Slugovc, C., and List, E.J.W. (2008) *Adv. Polym. Sci.*, **212**, 273.
19. Como, E.D., Becker, K., and Lupton, J.M. (2008) *Adv. Polym. Sci.*, **212**, 293.
20. Knaapila, M. and Winokur, M.J. (2008) *Adv. Polym. Sci.*, **212**, 227.
21. Scherf, U. and List, E.J.W. (2002) *Adv. Mater.*, **14**, 477.
22. Zhan, X., Liu, Y., Yu, G., Wu, X., Zhu, D.B., Sun, R., Wang, D., and Epstein, A.J. (2001) *J. Mater. Chem.*, **11**, 1606.
23. Grimsdale, A.C. and Müllen, K. (2006) *Adv. Polym. Sci.*, **199**, 1.
24. Yu, W.L., Pei, J., Huang, W., and Heeger, A.J. (2000) *Adv. Mater.*, **12**, 828.
25. Liu, B., Yu, W.L., Lai, Y.H., and Huang, W. (2001) *Chem. Mater.*, **13**, 1984.
26. Zeng, G., Yu, W.L., Chua, S.J., and Huang, W. (2002) *Macromolecules*, **35**, 6907.
27. Huang, W., Meng, H., Yu, W.L., Gao, J., and Heeger, A.J. (1998) *Adv. Mater.*, **10**, 593.
28. Zhang, F., Mammo, W., Andersson, L.M., Admassie, S., Andersson, M.R., and Inganäs, O. (2006) *Adv. Mater.*, **18**, 2169.
29. Chen, S.A., Lu, H.H., and Huang, C.W. (2008) *Adv. Polym. Sci.*, **212**, 49.
30. Wu, Z.L., Xiong, Y., Zou, J.H., Wang, L., Liu, J.H., Chen, Q.L., Yang, W., Peng, J.B., and Cao, Y. (2008) *Adv. Mater.*, **20**, 2359.
31. Wong, K.T., Chien, Y.Y., Chen, R.T., Wang, C.F., Lin, Y.T., Chiang, H.H., Hsieh, P.Y., Wu, C.C., Chou, C.H., Su, Y.O., Lee, G.H., and Peng, S.M. (2002) *J. Am. Chem. Soc.*, **124**, 11576.
32. Clarkson, R.G. and Gomberg, M. (1930) *J. Am. Chem. Soc.*, **52**, 2881.
33. Wu, F.I., Dodda, R., Reddy, D.S., and Shu, C.F. (2002) *J. Mater. Chem.*, **12**, 2893.
34. Scholl, R., Seer, C., and Weitzenböck, R. (1910) *Ber. Deutsch. Chem. Ges.*, **43**, 2202.
35. Fukuda, M., Sawada, K., and Yoshino, K. (1989) *Jpn. J. Appl. Phys.*, **28**, 1433.
36. Fukuda, M., Sawada, K., and Yoshino, K. (1993) *J. Polym. Sci., Part A: Polym. Chem.*, **31**, 2465.
37. Woo, E.P., Inbasekaran, M., Shiang, W., and Roof, G.R. (1997) PCT. Int. Pat. Appl. Wo. 05184.
38. Bernius, M., Inbasekaran, M., O'Brien, J., and Wu, W. (2000) *Adv. Mater.*, **12**, 1737.
39. Chunwaschirasiri, W., Tanto, B., Huber, D.L., and Winokur, M.J. (2005) *Phys. Rev. Lett.*, **94**, 107402.
40. Bradley, D.D.C., Grell, M., Long, X., Mellor, H., Grice, A., Inbasekaran, M., and Woo, E.P. (1997) *Proc. SPIE*, **3145**, 254.
41. Grell, M., Bradley, D.D.C., Long, X., Chamberlain, T., Inbasekaran, M., Woo, E.P., and Soliman, M. (1998) *Acta Polym.*, **49**, 439.
42. Grell, M., Bradley, D.D.C., Ungar, G., Hill, J., and Whitehead, K.S. (1999) *Macromolecules*, **32**, 5810.
43. Jeffrey, P., Brocker, E., Xu, Y.H., and Guillermo, C.B. (2008) *Adv. Mater.*, **20**, 1882.
44. Ariu, M., Sims, M., Rahn, M.D., Hill, J., Fox, A.M., Lidzey, D.G., Oda, M., Cabanillas-Jonzalez, G.,

and Bradley, D.D.C. (2003) *Phys. Rev. B.*, **67**, 195333.
45. Wohlgenannt, M., Jiang, X.M., Vardeny, Z.V., and Janssen, R.A.J. (2002) *Phys. Rev. Lett.*, **88**, 197401.
46. Hayer, A., Khan, A.L.T., Friend, R.H., and Kohler, A. (2005) *Phys. Rev. B.*, **71**, 241302.
47. Cadby, A.J., Lane, P.A., Mellor, H., Martin, S.J., Grell, M., Griebeler, C., Bradley, D.D.C., Wohlgenannt, M., An, C., and Vardeny, Z.V. (2000) *Phys. Rev. B.*, **62**, 15604.
48. Rothe, C., Galbrecht, F., Scherf, U., and Monkman, A. (2006) *Adv. Mater.*, **18**, 2137.
49. Ryu, G., Xia, R., and Bradley, D.D.C. (2007) *J. Phys. Condens. Matter.*, **19**, 056205.
50. Chen, S.H., Su, A.C., and Chen, S.A. (2005) *J. Phys. Chem. B.*, **109**, 10067.
51. Ariu, M., Lidzey, D.G., Sims, M., Cadby, A.J., Lane, P.A., and Bradley, D.D.C. (2002) *J. Phys. Condens. Matter.*, **14**, 9975.
52. Khan, A.L.T., Sreearunothai, P., Herz, L.M., Banach, M.J., and Köhler, A. (2004) *Phys. Rev. B.*, **69**, 085201.
53. List, E.J.W., Güntner, R., de Freitas, P.S., and Scherf, U. (2002) *Adv. Mater.*, **14**, 374.
54. Daniel, W.B., Fernando, B.D., Frank, G., Ulli, S., and Monkman, A.P. (2009) *Adv. Funct. Mater.*, **19**, 67.
55. Neher, D. (2001) *Macromol. Rapid Commun.*, **22**, 1365.
56. Pei, Q.B. and Yang, Y. (1996) *J. Am. Chem. Soc.*, **118**, 7416.
57. Grice, A.W., Bradley, D.D.C., Bernius, M.T., Inbasekaran, M., Wu, W.W., and Woo, E.P. (1998) *Appl. Phys. Lett.*, **73**, 629.
58. Monkman, A., Rothe, C., King, S., and Dias, F. (2008) *Adv. Polym. Sci.*, **212**, 187.
59. Chen, S.H., Su, A.C., Su, C.H., and Chen, S.A. (2005) *Macromolecules*, **38**, 379.
60. Becker, K. and Lupton, J.M. (2005) *J. Am. Chem. Soc.*, **127**, 7306.
61. List, E.J.W., Kim, C.H., Naik, A.K., Scherf, U., Leising, G., Graupner, W., and Shinar, J. (2001) *Phys. Rev. B.*, **64**, 155204.
62. List, E.J.W., Kim, C.H., Shinar, J., Pogantsch, A., Leising, G., and Graupner, W. (2000) *Appl. Phys. Lett.*, **76**, 2083.
63. List, E.J.W., Pratee, J., Shinar, J., Scherf, U., Müllen, K., Graupner, W., Petritsch, K., Zojer, E., and Leising, G. (2000) *Phys. Rev. B.*, **61**, 10807.
64. Kreyenschmidt, M., Klaerner, G., Fuhrer, T., Ashenhurst, J., Karg, S., Chen, W.D., Lee, V.Y., Scott, J.C., and Miller, R.D. (1998) *Macromolecules*, **31**, 1099.
65. Klärner, G., Lee, J.I., Davey, M.H., and Miller, R.D. (1999) *Adv. Mater.*, **11**, 115.
66. Lee, J.I., Klaerner, G., and Miller R.D. (1999) *Chem. Mater.*, **11**, 1083.
67. Xiao, S., Nguyen, M., Gong, X., Cao, Y., Wu, H., Moses, D., and Heeger, A.J. (2003) *Adv. Funct. Mater.*, **13**, 25.
68. Kulkarni, A.P. and Jenekhe, S.A. (2003) *Macromolecules*, **36**, 5285.
69. Wu, R., Schumm, J.S., Pearson, D.L., and Tour, J.M. (1996) *J. Org. Chem.*, **61**, 6906.
70. Pudzich, R., Fuhrmann-Lieker, T., and Salbeck, J. (2006) *Adv. Polym. Sci.*, **199**, 83.
71. Kreuder, W., Lupo, D., Salbeck, J., Schenk, H., and Stehlin, T. (1997) US Patent 5 621 131.
72. Wu, Y.G., Li, J., Fu, Y.Q., and Bo, Z.S. (2004) *Org. Lett.*, **6**, 3485.
73. Vak, D.J., Chun, C.E., Lee, C.L.Y., Kim, J.J., and Kim, D.Y. (2004) *J. Mater. Chem.*, **14**, 1342.
74. Vak, D.J., Jo, J., Ghim, J., Chun, C., Lim, B., Heeger, A.J., and Kim, D.Y. (2006) *Macromolecules*, **39**, 6433.
75. Luo, J., Zhou, Y., Niu, Z.Q., Zhou, Q.F., Ma, Y.G., and Pei, J. (2007) *J. Am. Chem. Soc.*, **129**, 11314.
76. Lee, J.H. and Hwang, D.H. (2003) *Chem. Commun.*, 2836.
77. McFarlane, S.L., Piercey, D.G., Coumont, L.S., Tucker, R.T., Fleischauer, M.D., Brett, M.J., and Veinot, G.C.J. (2009) *Macromolecules*, **42**, 591.
78. Wu, C.W., Tsai, C.M., and Lin, H.C. (2006) *Macromolecules*, **39**, 4298.

79. Cho, H.J., Jung, B.J., Cho, N.S., Lee, J., and Shim, H.K. (2003) *Macromolecules*, **36**, 6704.
80. Lin, W.J., Chen, W.C., Wu, W.C., Niu, Y.H., and Jen, A.K.Y. (2004) *Macromolecules*, **37**, 2335.
81. Chou, C.H., Hsu, S.L., Dinakaran, K., Chiu, M.Y., and Wei, K.H. (2005) *Macromolecules*, **38**, 745.
82. Teetsov, J. and Fox, M.A.J. (1999) *J. Mater. Chem.*, **9**, 2117.
83. Yu, W.L., Pei, J., Huang, W., Pei, J., Cao, Y., and Heeger, A.J. (1999) *Chem. Commun.*, 1837.
84. Wang, L.Y., Cao, Y., Pei, J., Huang, W., and Heeger, A.J. (1999) *Appl. Phys. Lett.*, **75**, 3270.
85. Liu, B., Yu, W.L., Pei, J., Liu, S.Y., Lai, Y.H., and Huang, W. (2001) *Macromolecules*, **34**, 7932.
86. Gather, M.C., Heeney, M., Zhang, W.M., Whitehead, K.S., Bradley, D.D.C., Culloch, I.M., and Campbell, A.J. (2008) *Chem. Commun.*, 1079.
87. Tang, R.P., Tan, Z.A., Li, Y.F., and Xi, F. (2006) *Chem. Mater.*, **18**, 1053.
88. Huang, F., Zhang, Y., Liu, M.S., Cheng, Y.J., and Jen, A.K.Y. (2007) *Adv. Funct. Mater.*, **17**, 3808.
89. Xiao, X., Fu, Y.Q., Sun, M.H., Li, L., and Bo, Z.S. (2007) *J. Polym. Sci., Part A: Polym. Chem.*, **45**, 2410.
90. Chen, R.F., Fan, Q.L., Zheng, C., and Huang, W. (2006) *Org. Lett.*, **8**, 203.
91. Chen, R.F., Zhu, R., Fan, Q.L., and Huang, W. (2008) *Org. Lett.*, **10**, 2913.
92. Chen, R.F., Fan, Q.L., Liu, S.J., Zhu, R., Pu, K.Y., and Huang, W. (2006) *Synth. Met.*, **156**, 1161.
93. Wang, E., Li, C., Mo, Y.Q., Zhang, Y., Ma, G., Shi, W., Peng, J.B., Yang, W., and Cao, Y. (2006) *J. Mater. Chem.*, **16**, 4133.
94. Liu, J., Zou, J.H., Yang, W., Wu, H.B., Li, C., Zhang, B., Peng, J.B., and Cao, Y. (2008) *Chem. Mater.*, **20**, 4499.
95. King, S.M., Perepichka, I.I., Perepichka, I.F., Dias, F.B., Bryce, M.R., and Monkman, A.P. (2009) *Adv. Funct. Mater.*, **19**, 586.
96. Bonifácio, V.D.B., Morgado, J., Scherf, U. (2008) *J. Polym. Sci., Part A: Polym. Chem.*, **46**, 2878.
97. Gao, C. and Yan, D. (2004) *Prog. Polym. Sci.*, **29**, 183.
98. Xin, Y., Wen, G.A., Zeng, W.J., Zhao, L., Zhu, X.R., Fan, Q.L., Feng, J.C., Wang, L.H., Wei, W., Peng, B., Cao, Y., and Huang, W. (2005) *Macromolecules*, **38**, 6755.
99. Wen, G.A., Xin, Y., Zhu, X.R., Zeng, W.J., Zhu, R., Feng, J.C., Cao, Y., Zhao, L., Wang, L.H., Wei, W., Peng, B., and Huang, W. (2007) *Polymer*, **48**, 1824.
100. Tsai, L.R. and Chen, Y. (2007) *Macromolecules*, **40**, 2984.
101. Liu, X.M., Xu, J.W., Lu, X.H., and He, C.B. (2006) *Macromolecules*, **39**, 1397.
102. Wu, C.W. and Lin, H.C. (2006) *Macromolecules*, **39**, 7232.
103. Tang, D.F., Wen, G.A., Qi, X.Y., Wang, H.Y., Peng, B., Wei, W., and Huang, W. (2007) *Polymer*, **48**, 4412.
104. Tang, D.F., Wen, G.A., Wei, W., and Huang, W. (2008) *Polym. Int.*, **57**, 1235.
105. Tsai, L.R. and Chen, Y. (2008) *Macromolecules*, **41**, 5098.
106. Zhao, L., Li, C., Zhang, Y., Zhu, X.H., Peng, J., and Cao, Y. (2006) *Macromol. Rapid Commun.*, **27**, 914.
107. Geng, Y., Culligan, S.W., Trajkowska, Y., Wallace, J.U., and Chen, S.H. (2003) *Chem. Mater.*, **15**, 542.
108. Lai, W.Y., Chen, Q.Q., He, Q.Y., Fan, Q.L., and Huang, W. (2006) *Chem. Commun.*, 1959.
109. Lai, W.Y., Zhu, R., Fan, Q.L., Hou, L.T., Cao, Y., and Huang, W. (2006) *Macromolecules*, **39**, 3707.
110. Lai, W.Y., He, Q.Y., Zhu, R., Chen, Q.Q., and Huang, W. (2008) *Adv. Funct. Mater.*, **18**, 265.
111. Lai, W.Y., Xia, R.D., He, Q.Y., Levermore, P.A., Huang, W., and Bradley, D.D.C. (2009) *Adv. Mater.*, **21**, 355.
112. McGehee, M.D. and Heeger, A.J. (2000) *Adv. Mater.*, **12**, 1655.
113. Samuel, I.D.W. and Turnbull, G.A. (2007) *Chem. Rev.*, **107**, 1272.
114. Lawrence, J.R., Namdas, E.B., Richards, G.J., Burn, P.L., and Samuel, I.D.W. (2007) *Adv. Mater.*, **19**, 3000.

115. Yap, B., Xia, R., Campoy-Quiles, M., Stavrinou, P.N., and Bradley, D.D.C. (2008) *Nat. Mater.*, **7**, 376.
116. Kanibolotsky, A.L., Berridge, R., Skabara, P.J., Perepichka, I.F., Bradley, D.D.C., and Koeberg, M. (2004) *J. Am. Chem. Soc.*, **126**, 13695.
117. Liu, F., Lai, W.Y., Tang, C., Wu, H.B., Chen, Q.Q., Peng, B., Wei, W., Huang, W., and Cao, Y. (2008) *Macromol. Rapid Commun.*, **29**, 659.
118. Liu, Q.D., Lu, J.P., Ding, J.F., Day, M., Tao, Y., Barrios, P., Stupak, J., Chan, K., Li, J.J., and Chi, Y. (2007) *Adv. Funct. Mater.*, **17**, 1028.
119. Pei, J., Yu, W.L., Huang, W., and Heeger, A.J. (2000) *Chem. Commun.*, 1631.
120. Liu, B., Yu, W.L., Lai, Y.H., and Huang, W. (2000) *Macromolecules*, **33**, 8945.
121. Faïd, K., Cloutier, R., and Leclerc, M. (1993) *Macromolecules*, **26**, 2501.
122. Cloutier, R. and Leclerc, M. (1991) *J. Chem. Soc. Chem. Commun.*, 1194.
123. Wu, S.H., Huang, H.M., Chen, K.C., Hu, C.W., Hsu, C.C., and Tsiang, R.C.C. (2006) *Adv. Funct. Mater.*, **16**, 1959.
124. Liu, J., Bu, L.J., Dong, J.P., Zhou, Q.G., Geng, Y.H., Ma, D.G., Wang, L.X., Jing, X.B., and Wang, F.S. (2007) *J. Mater. Chem.*, **17**, 2832.
125. Ma, Z., Lu, S., Fan, Q.L., Qing, C.Y., Wang, Y.Y., Wang, P., and Huang, W. (2006) *Polymer*, **47**, 7382.
126. Cheon, C.H., Joo, S.H., Kim, K., Jin, J.I., Shin, H.W., and Kim, Y.R. (2005) *Macromolecules*, **38**, 6336.
127. Hsieh, B.Y. and Chen, Y. (2007) *Macromolecules*, **40**, 8913.
128. Cho, N.S., Park, J.H., Lee, S.K., Lee, J., Shim, H.K., Park, M.J., Hwang, D.H., and Jung, B.J. (2006) *Macromolecules*, **39**, 177.
129. Chan, L.H., Lee, Y.D., and Chen, C.T. (2006) *Macromolecules*, **39**, 3262.
130. Li, B.S., Li, J., Fu, Y.Q., and Bo, Z.S. (2004) *J. Am. Chem. Soc.*, **126**, 3430.
131. Chen, Z.H., Bouffard, J., Kooi, S.E., and Swager, T.M. (2008) *Macromolecules*, **41**, 6672.
132. Hsieh, B.Y. and Chen, Y. (2008) *J. Polym. Sci., Part A: Polym. Chem.*, **46**, 3703.
133. Liu, J., Guo, X., Bu, L.J., Xie, Z.Y., Cheng, Y.X., Geng, Y.H., Wang, L.X., Jing, X.B., and Wang, F.S. (2007) *Adv. Funct. Mater.*, **17**, 1917.
134. Liu, J., Cheng, Y.X., Xie, Z.Y., Geng, Y.H., Wang, L.X., Jing, X.B., and Wang, F.S. (2008) *Adv. Mater.*, **20**, 1357.
135. Park, M.J., Lee, J.H., Park, J.H., Lee, S.K., Lee, J.I., Chu, H.Y., Hwang, D.H., and Shim, H.K. (2008) *Macromolecules*, **41**, 3063.
136. Liu, J., Gao, B.X., Cheng, Y.X., Xie, Z.Y., Geng, Y.H., Wang, L.X., Jing, X.B., and Wang, F. (2008) *Macromolecules*, **41**, 1162.
137. Liu, J., Shao, S.Y., Chen, L., Xie, Z.Y., Cheng, Y.X., Geng, Y.H., Wang, L.X., Jing, X.B., and Wang, F. (2007) *Adv. Mater.*, **19**, 1859.
138. Liu, J., Zhou, Q.G., Cheng, Y.X., Geng, Y.H., Wang, L.X., Ma, D.G., Jing, X.B., and Wang, F. (2005) *Adv. Mater.*, **17**, 2974.
139. Luo, J., Li, X.Z., Hou, Q., Peng, J.B., Yang, W., and Cao, Y. (2007) *Adv Mater.*, **19**, 1113.
140. Liu, S.J., Zhao, Q., Mi, B.X., and Huang, W. (2008) *Adv. Polym. Sci.*, **212**, 125.
141. Baldo, M.A., O'Brien, D.F., You, Y., Shoustikov, A., Sibley, S., Thompson, M.E., and Forrest, S.R. (1998) *Nature*, **395**, 151.
142. Lamansky, S., Kwong, R.C., Nugent, M., Djurovich, P.I., and Thompson, M.E. (2001) *Org. Electron.*, **2**, 53.
143. Sandee, A.J., Williams, C.K., Evans, N.R., Davies, J.E., Boothby, C.E., Kohler, A., Friend, R.H., and Holmes, A.B. (2004) *J. Am. Chem. Soc.*, **126**, 7041.
144. Chen, X., Liao, J.L., Liang, Y., Ahmed, M.O., Tseng, H.E., and Chen, S.A. (2003) *J. Am. Chem. Soc.*, **125**, 636.
145. Jiang, J., Jiang, C., Yang, W., Zhen, H., Huang, F., and Cao, Y. (2005) *Macromolecules*, **38**, 4072.
146. Du, B., Wang, L., Wu, H.B., Yang, W., Zhang, Y., Liu, R.S., Sun, M.L.

Peng, J.B., and Cao, Y. (2007) *Chem. Eur. J.*, **13**, 7432.
147. Yang, X.H., Wu, F.I., Neher, D., Chien, C.H. and Shu, C.F. (2008) *Chem. Mater.*, **20**, 1629.
148. Mei, C.Y., Ding, J.Q., Yao, B., Cheng, Y.X., Xie, Z.Y., Geng, Y.H., and Wang, L.X. (2007) *J. Polym. Sci., Part A: Polym. Chem.*, **45**, 1746.
149. Schulz, G.L., Chen, X., Chen, S.A., and Holdcroft, S. (2006) *Macromolecules*, **39**, 9157.
150. Ito, T., Suzuki, S., and Kido, J. (2005) *Polym. Adv. Technol.*, **16**, 480.
151. Zhen, H.Y., Jiang, C.Y., Yang, W., Jiang, J.X., Huang, F., and Cao, Y. (2005) *Chem. Eur. J.*, **11**, 5007.
152. Zhen, H.Y., Luo, C., Yang, W., Song, W., Du, B., Jiang, J., Jiang, C., Zhang, Y., and Cao, Y. (2006) *Macromolecules*, **39**, 1693.
153. Zhang, K., Chen, Z., Zou, Y., Yang, C., Qin, J., and Cao, Y. (2007) *Organometallics*, **26**, 3699.
154. Zhang, K., Chen, Z., Yang, C.L., Gong, S.L., Qin, J.G., and Cao, Y. (2006) *Macromol. Rapid Commun.*, **27**, 1926.
155. Zhang, K., Chen, Z., Yang, C.L., Tao, Y.T., Zou, Y., Qin, J.G., and Cao, Y. (2008) *J. Mater. Chem.*, **18**, 291.
156. Liu, S.J., Zhao, Q., Chen, R.F., Deng, Y., Fan, Q.L., Li, F.Y., Wang, L.H., Huang, C.H., and Huang, W. (2006) *Chem. Eur. J.*, **12**, 4351.
157. Liu, S.J., Zhao, Q., Xia, Y.J., Deng, Y. Lin, J., Fan, Q.L., Wang, L.H., and Huang, W. (2007) *J. Phys. Chem. C*, **111**, 1166.
158. Ling, Q.D., Kang, E.T., Neoh, K.G., and Huang, W. (2003) *Macromolecules*, **36**, 6995.
159. Song, Y., Tan, Y.P., Teo, E.Y.H., Zhu, C.X., Chan, D.S.H., Ling, Q.D., Neoh, K.G., and Kang, E.T. (2006) *J. Appl. Phys.*, **100**, 084508.
160. Ling, Q.D., Song, Y., Teo, E.Y.H., Lim, S.L., Zhu, C.X., Chan, D.S.H., Kwong, D.L., Kang, E.T., and Neoh, K.G. (2006) *Electrochem. Solid-State Lett.*, **9**, G268.
161. Song, Y., Ling, Q.D., Zhu, C., Kang, E.T., Chan, D.S.H., Wang, Y.H., and Kwong, D.L. (2006) *IEEE Electron Device Lett.*, **27**, 154.
162. Ling, Q.D., Song, Y., Ding, S.J., Zhu, C.X., Chan, D.S.H., Kwong, D.L., Kang, E.T., and Neoh, K.G. (2005) *Adv. Mater.*, **17**, 455.
163. Wen, G.A., Zhu, X.R., Wang, L.H., Feng, J.C., Zhu, R., Wei, W., Peng, B., Pei, Q.B., and Huang, W. (2007) *J. Polym. Sci., Part A: Polym. Chem.*, **45**, 388.
164. Hou, Q., Zhang, Y., Li, F.Y., Peng, J.B., and Cao, Y. (2005) *Organometallics*, **24**, 4509.
165. Zhuang, W.L., Zhang, Y., Hou, Q., Wang, L., and Cao, Y. (2006) *J. Polym. Sci., Part A: Polym. Chem.*, **44**, 4174.
166. Thomas, S.W., Yagi, S., and Swager, T.M. (2005) *J. Mater. Chem.*, **15**, 2829.
167. Zhang, M., Lu, P., Wang, X.M., He, L., Xia, H., Zhang, W., Yang, B., Liu, L.L., Yang, L., Yang, M., Ma, Y.G., Feng, J.K., Wang, D.J., and Tamai, N. (2004) *J. Phys. Chem. B*, **108**, 13185.
168. Zhang, Y., Huang, Z., Zeng, W.J., and Cao, Y. (2008) *Polymer*, **49**, 1211.
169. Dinakaran, K., Chou, C.H., Hsu, S.L., and Wei, K.H. (2004) *J. Polym. Sci., Part A: Polym. Chem.*, **42**, 4838.
170. Chien, C.H., Liao, S.F., Wu, C.H., Shu, C.F., Chang, S.Y., Chi, Y., Chou, P.T., and Lai, C.H. (2008) *Adv. Funct. Mater.*, **18**, 1430.
171. Lee, M., Cho, B.K., and Zin, W.C. (2001) *Chem. Rev.*, **101**, 3869.
172. Jenekhe, S.A. and Chen, X.L. (1998) *Science*, **279**, 1903.
173. Kong, X. and Jenekhe, S.A. (2004) *Macromolecules*, **37**, 8180.
174. Surin, M., Marsitzky, D., Grimsdale, A.C., Müllen, K., Lazzaroni, R., and Leclère, P. (2004) *Adv. Funct. Mater.*, **14**, 708.
175. Chochos, C.L., Tsolakis, P.K., Gregoriou, V.G., and Kallitsis, J.K. (2004) *Macromolecules*, **37**, 2502.
176. Liang, Y., Wang, H., Yuan, S., Lee, Y., Gan, L., and Yu, L. (2007) *J. Mater. Chem.*, **17**, 2183.
177. Morteani, A.C., Dhoot, A.S., Kim, J.S., Silva, C., Greenham, N.C., Murphy, C., Moons, E., Cina, S., Burroughes, J.H., and Friend, R.H. (2003) *Adv. Mater.*, **15**, 1708.
178. Chen, X.L. and Jenekhe, S.A. (1996) *Macromolecules*, **29**, 6189.

179. Lu, S., Liu, T.X., Ke, L., Ma, D.G., Chua, S.J., and Huang, W. (2005) *Macromolecules*, **38**, 8494.
180. Lu, S., Fan, Q.L., Liu, S.Y., Chua, S.J., and Huang, W. (2002) *Macromolecules*, **35**, 9875.
181. Lu, S., Fan, Q.L., Chua, S.J., and Huang, W. (2003) *Macromolecules*, **36**, 304.
182. Bütün, V., Armes, S.P., and Billingham, N.C. (2001) *Macromolecules*, **34**, 1148.
183. Lin, S.T., Tung, Y.C., and Chen, W.C. (2008) *J. Mater. Chem.*, **18**, 3985.
184. Fang, C., Qi, X.Y., Fan, Q.L., Wang, L.H., and Huang, W. (2007) *Nanotechnology*, **18**, 035704.
185. Fang, C., Zhao, B.M., Lu, H.T., Sai, L.M., Fan, Q.L., Wang, L.H., and Huang, W. (2008) *J. Phys. Chem. C*, **112**, 7278.
186. Ma, Z., Qiang, L.L., Zheng, Z., Wang, Y.Y., Zhang, Z.J., and Huang, W. (2008) *J. Appl. Polym. Sci.*, **110**, 18.
187. Yao, J.H., Mya, K.Y., Shen, L., He, B.P., Li, L., Li, Z.H., Chen, Z.K., Li, X., and Loh, K.P. (2008) *Macromolecules*, **41**, 1438.
188. Nomura, K., Yamamoto, N., Ito, R., Fujiki, M., and Geerts, Y. (2008) *Macromolecules*, **41**, 4245.
189. Tan, C., Pinto, M.R., and Schanze, K.S. (2002) *Chem. Commun.*, 446.
190. Pinto, M.R., Kristal, B.M., and Schanze, K.S. (2003) *Langmuir*, **19**, 6523.
191. Fan, Q.L., Zhou, Y., Lu, X.M., Hou, X.Y., and Huang, W. (2005) *Macromolecules*, **38**, 2927.
192. Fan, Q.L., Lu, S., Lai, Y.H., Hou, X.Y., and Huang, W. (2003) *Macromolecules*, **36**, 6976.
193. Liu, B. and Bazan, G.C. (2004) *Chem. Mater.*, **16**, 4467.
194. Thomas, S.W.III, Joly, G.D., and Swager, T.M. (2007) *Chem. Rev.*, **107**, 1339.
195. Herland, A. and Inganäs, O. (2007) *Macromol. Rapid Commun.*, **28**, 1703.
196. Feng, F., He, F., An, L., Wang, S., Li, Y., and Zhu, D. (2008) *Adv. Mater.*, **20**, 2959.
197. Peter, K. and Hammarström, R.N.P. (2008) *Adv. Mater.*, **20**, 2639.
198. Liu, B., Yu, W.L., Lai, Y.H., and Huang, W. (2000) *Chem. Commun.*, 551.
199. Liu, B., Yu, W.L., Lai, Y.H., and Huang, W. (2002) *Macromolecules*, **35**, 4975.
200. Stork, M., Gaylord, B.S., Heeger, A.J., and Bazan, G.C. (2002) *Adv. Mater.*, **14**, 361.
201. Woo, H.Y., Vak, D.J., Korystov, D., Mikhailovsky, A., Bazan, G.C., and Kim, D.Y. (2007) *Adv. Funct. Mater.*, **17**, 290.
202. Liu, B., Wang, S., Bazan, G.C., and Mikhailovsky, A. (2003) *J. Am. Chem. Soc.*, **125**, 13306.
203. Zhou, Q. and Swager, T.M. (1995) *J. Am. Chem. Soc.*, **117**, 12593.
204. Bunz, U.H.F. (2000) *Chem. Rev.*, **100**, 1605.
205. Haskins-Glusac, K., Pinto, M.R., Tan, C., and Schanze, K.S. (2004) *J. Am. Chem. Soc.*, **126**, 14964.
206. Huang, Y.Q., Fan, Q.L., Lu, X.M. Fang, C., Liu, S.J., Yu-Wen, L.H., Wang, L.H., and Huang, W. (2006) *J. Polym. Sci., Part A: Polym. Chem.*, **44**, 5778.
207. Huang, Y.Q., Fan, Q.L., Zhang, G.W., Chen, Y., Lu, X.M., and Huang, W. (2006) *Polymer*, **47**, 5233.
208. Gaylord, B.S., Heeger, A.J., and Bazan, G.C. (2002) *Proc. Natl. Acad. Sci. U.S.A.*, **99**, 10954.
209. Xu, H., Wu, H.P., Fan, C.H., Li, W.X., Zhang, Z.Z., and He, L. (2004) *Chin. Sci. Bull.*, **49**, 2227.
210. Huang, Y.Q., Fan, Q.L., Li, S.B., Lu, X.M., Cheng, F., Zhang, G.W., Chen, Y., Wang, L.H., and Huang, W. (2006) *J. Polym. Sci., Part A: Polym. Chem*, **44**, 5424.
211. Liu, B. and Bazan, G.C. (2004) *J. Am. Chem. Soc.*, **126**, 1942.
212. Chi, C., Mikhailovsky, A., and Bazan, G.C. (2007) *J. Am. Chem. Soc.*, **129**, 11134.
213. Fang, Z., Pu, K.Y., and Liu, B. (2008) *Macromolecules*, **41**, 8380.
214. Park, J., Yang, R., Hoven, C.V., Garcia, A., Fischer, D.A., Nguyen, T.Q., Bazan, G.C., and DeLongchamp, D.M. (2008) *Adv. Mater.*, **20**, 2491.

215. Yang, R., Wu, H., Cao, Y., and Bazan, G.C. (2006) *J. Am. Chem. Soc.*, **128**, 14422.
216. Shen, H.L., Huang, F., Hou, L.T., Wu, H.B. Cao, W., Yang, W., and Cao, Y. (2005) *Synth. Met.*, **152**, 257.
217. Hou L.T., Huang, F., Peng, J.B., Wu, H.B., Wen, S.S., Mo, Y.Q., and Cao, Y. (2006) *Thin Solid Films*, **515**, 2632.
218. Hoven, C.V., Garcia, A., Bazan, G.C., and Nguyen, T.Q. (2008) *Adv. Mater.*, **20**, 3793.
219. Parker, I.D.J. (1994) *Appl. Phys.*, **75**, 1656.
220. Ma, W.L., Iyer, P.K., Gong, X., Liu, B., Moses, D., Bazan, G.C., and Heeger, A.J. (2005) *Adv. Mater.*, **17**, 274.
221. Oh, S.H., Vak, D.J., Na, S.I., Lee, T.W., and Kim, D.Y. (2008) *Adv. Mater.*, **20**, 1624.
222. Zhu, B., Han, Y., Sun, M.H., and Bo, Z.S. (2007) *Macromolecules*, **40**, 4494.
223. Burrows, H.D., Lobo, V.M.M., Pina, J., Ramos, M.L., Melo, J.S.D., Valente, A.J.M., Tapia, M.J., Pradhan, S., and Scherf, U. (2004) *Macromolecules*, **37**, 7425.
224. Tang, Y.L., He, F., Yu, M.H., Wang, S., Li, Y.L., and Zhu, D.B. (2006) *Chem. Mater.*, **18**, 3605.
225. An, L., Tang, Y., Feng, F., He, F., and Wang, S. (2007) *J. Mater. Chem.*, **17**, 4147.
226. Chen, Y., Fan, Q.L., Wang, P., Zhang, B., Huang, Y.Q., Zhang, G.W., Lu, X.M., Chan, H.S.O., and Huang, W. (2006) *Polymer*, **47**, 5228.
227. Yu, D.Y., Zhang, Y., and Liu, B. (2008) *Macromolecules*, **41**, 4003.

Index

Design and Synthesis of Conjugated Polymers. Edited by Mario Leclerc and Jean-François Morin

ISBN: 978-3-527-32474-3